Kinship and Behavior
in Primates

Kinship and Behavior in Primates

EDITED BY
Bernard Chapais
Carol M. Berman

UNIVERSITY PRESS

2004

OXFORD
UNIVERSITY PRESS

Oxford   New York
Auckland   Bangkok   Buenos Aires   Cape Town   Chennai
Dar es Salaam   Delhi   Hong Kong   Istanbul   Karachi   Kolkata
Kuala Lumpur   Madrid   Melbourne   Mexico City   Mumbai   Nairobi
São Paulo   Shanghai   Taipei   Tokyo   Toronto

Published by Oxford University Press, Inc.,
198 Madison Avenue, New York, New York 10016

www.oup.com

Oxford is a registered trademark of Oxford University Press

Library of Congress Cataloging-in-Publication Data
Chapais, Bernard.
Kinship and behavior in primates / edited by Bernard Chapais and Carol M. Berman.
p. cm.
Includes bibliographical references
ISBN 0-19-514889-4
1. Primates—Behavior.   2. Kin recognition in animals.   I. Berman,
Carol M.   II. Title.
QL737.P9 C456 2003
599.8′15—dc21   2003002335

9 8 7 6 5 4 3 2 1

Printed in the United States of America
on acid-free paper

To my children, Catherine and Louis-Charles.

B. C.

To my son, Sam, to my father, Horace Berman, who introduced me
to Animal Behavior, and to my mother, Florence Berman, who tolerated my bringing
worms and snails (but not monkeys) into the house.

C. B.

Contents

Contributors

Helen Perich Alvarez
Department of Anthropology
University of Utah
270 South 1400 East, Room 102
Salt Lake City, UT 84112

Patrick Bélisle
Département d'Anthropologie
Université de Montréal
C.P. 6128, Succursale Centre-ville
Montréal, Canada H3C 3J7

Carol M. Berman
Department of Anthropology
State University of New York at Buffalo
380 MFAC, Ellicott Complex
Buffalo, NY 14261

Bernard Chapais
Département d'Anthropologie
Université de Montréal
C.P. 6128, Succursale Centre-ville
Montréal, Canada H3C 3J7

Dorothy L. Cheney
Department of Biology
University of Philadelphia
Philadelphia, PA 19104

Fernando Colmenares
Departamento Psicobiologia
Universidad Complutense de Madrid
28223 Madrid, Spain

James M. Dietz
Department of Zoology
University of Maryland
College Park, MD 20742

Tony L. Goldberg
Department of Veterinary Pathobiology
College of Veterinary Medicine
2001 South Lincoln Avenue
University of Illinois
Urbana, IL 61801

Kristen Hawkes
Department of Anthropology
University of Utah
270 South 1400 East, Room 102
Salt Lake City, UT 84112

David A. Hill
University of Sussex
School of Biological Sciences
Falmer, Brighton BN1 9QG
England

Guy A. Hoelzer
Department of Biology
University of Nevada, Reno
Reno, NV 89557

Lynne A. Isbell
Department of Anthropology
University of California, Davis
One Shields Avenue
Davis, CA 95616

Ellen Kapsalis
University of Miami School of Medicine
Division of Veterinary Resources
12500 SW 152nd Street, Building A
Miami, FL 33177

Jutta Kuester
Abt Neurobiologie
MPI f. Biophysikalische Chemie
Am Fassberg 11
37077 Goettingen, Germany

Don J. Melnick
Center for Environmental Research and
   Conservation
Columbia University
Schermerhorn Extension
1200 Amsterdam Avenue
New York, NY 10027

Juan Carlos Morales
Center for Environmental Research and
   Conservation
Columbia University
Schermerhorn Extension
1200 Amsterdam Avenue
New York, NY 10027

Phillip A. Morin
Max Planck Institute for Evolutionary
   Anthropology
Inselstrasse 22
D-04103 Leipzig, Germany

Leanne T. Nash
Department of Anthropology
Arizona State University
Tempe, AZ 85287–2402

Andreas Paul
Institut für Zoologie und Anthropologie
Universität Göttingen
Berliner Str. 28
D-37073 Göttingen, Germany

Drew Rendall
Department of Psychology and
   Neuroscience
University of Lethbridge
Lethbridge, Canada T1K 3M4

Lars Rodseth
Department of Anthropology
University of Utah
270 South 1400 East, Room 102
Salt Lake City, UT 84112–0060

Robert M. Seyfarth
Department of Psychology
University of Pennsylvania
Philadelphia, PA 19104

Karen B. Strier
Department of Anthropology
University of Wisconsin–Madison
1180 Observatory Drive
Madison, WI 53706

David S. Woodruff
Section of Ecology, Behavior and
   Evolution
Division of Biology
University of California, San Diego
La Jolla, CA 92093–0116

Richard Wrangham
Department of Anthropology
Peabody Museum
Harvard University
11 Divinity Drive
Cambridge, MA 012138

Kinship and Behavior
in Primates

1

Introduction: The Kinship Black Box

Bernard Chapais
Carol M. Berman

The idea that kin relationships play a central role in the organization of non-human primate behavior has become almost a truism over the last 40 years of research. Several excellent reviews have described the ways in which kinship is woven into the fabric of social life of primate species (Gouzoules 1984, Walters 1987, Bernstein 1991, Silk 2001). Indeed, as we gain information about increasing numbers of primate species, the validity of this notion becomes increasingly reinforced. Nevertheless, the status of kinship in the history of nonhuman primate studies is somewhat paradoxical. While a great deal of progress has been made in describing kin-related behavior patterns, our understanding about some of the most basic aspects of the concept (including the mechanisms of kin recognition, the precise categories of kin discriminated, and the role of kin selection) has not progressed at the same pace. In this sense, the concept of kinship is still, to a large extent, a black box. At the same time, this situation has begun to change rapidly over the last 10 years or so, and some light is beginning to penetrate parts of the black box. Given this rapid progress, our goal in this volume is to review current knowledge about kinship in primates with the hope that it will not only inform a broad range of advanced readers, but also inspire and facilitate the continued illumination of the box. We will return to the black box analogy, but let us first examine briefly some of the reasons why kinship has become so central to the study of nonhuman primate behavior.

Although kinship has been profitably studied in a range of taxonomic groups, primates have been particularly suitable subjects for kinship studies. First and most fundamentally, this is because most primates live permanently with several of their kin, a fact that reflects some of the most basic features of the primate order. Group living, combined with sex-biased philopatry (or residency), produces groups in which the related individuals are mostly males, or mostly females, depending on the philopatric sex. Second, the fact that the sex forming enduring social bonds with offspring (i.e., females) is also the philopatric sex in

3

several primate species renders matrilineages recognizable behaviorally, hence whole matrilineal structures visible and amenable to analysis. Where ecological conditions permit, many such primates display life history traits that lead to the co-existence of three to four generations of individuals in the same group. When this occurs, matrilineal structures may be potentially extensive, including several categories of direct and collateral kin. Finally, primates tend to form well-differentiated social relationships based on several of their partners' characteristics such as age and dominance rank. This increases the likelihood that kinship is also a basis for differentiation. The importance of the co-occurrence of these factors in the same mammalian order can hardly be overestimated in trying to understand why kinship has become a central focus in the history of behavioral primatology. Had primate species been mostly male philopatric (and kinship relations much less discernible to observers as a result) or had they been mostly monogamous, the status of kinship studies in nonhuman primates would assuredly be very different today.

No less important in a historical perspective is the fact that the first species whose kinship structures were studied in detail were Japanese macaques (*Macaca fuscata*: Kawai 1958, Kawamura 1958) and Cayo Santiago rhesus macaques (*M. mulatta*: Sade 1965, 1967). These two species happen to belong to the category we now call despotic species, characterized by strongly asymmetrical dominance relations and a profound influence of matrilineal kinship on social relationships. So important is the influence of kinship in these species that it closely predicts the structure of female dominance relations, affiliation, tolerance, and support. The particular populations studied by Kawai, Kawamura, and Sade were provisioned, and hence matrilines were large and extended, rendering the association between behavior and kinship particularly conspicuous. Even so, the importance of kinship in shaping social structure could have been easily missed had these early researchers not had the insight to maintain long-term census records on maternal kin relationships and to focus on social relationships among females. This insight was probably derived from interests in psychoanalytic theory in the case of the Japanese researchers and in population dynamics in the case of Donald Sade. Thus, early in the history of primatology, the impact of kinship on behavior was strikingly evident to those receptive to perceive it. Had these early studies been carried out on more egalitarian species (even other more egalitarian macaques such as *M. tonkeana*) in which kinship affects behavior to a significantly lesser extent (Thierry 2000), the history of kinship studies would have taken a different turn. Thus, the characteristics of primate societies in general, of the particular societies that were studied early in the history of primatology, and of the particular researchers studying them together account for the early recognition of the importance of kinship in the social life of primates.

As birth records and behavioral observations accumulated longitudinally for a few primate groups and species over the 1960s and 1970s, it became increasingly clear that matrilineal kinship was one of the most significant independent variables affecting the behavior of primates even in groups that were not provisioned and did not have extended lineages. Thus, kinship, along with the age, sex, and dominance status of individuals, emerged as what one could call the big four factors structuring social relationships. More specifically, kin were found to interact preferentially with each other—that is, to practice kin bias or nepotism—and to exhibit lower levels of sexual activity during periods of female fertility—that is, to avoid incest. In fact, the co-occurrence of kinship relations with such behavioral biases became the main criteria to infer that kin discrimination was taking place, hence that primates recognized their kin.

In the 1970s, kinship studies in general were given a major impetus with the development by W. D. Hamilton of inclusive fitness or kin selection theory, the first and single evolutionary theory about kin bias or nepotism (Hamilton 1964). Inclusive fitness theory posits that individuals should be evolutionarily selected to behave in ways that promote the reproduction not only of their own offspring, but also that of their other genetic relatives. Since an individual's genetic relatives share in common some of the genes they inherit from a common ancestor, individuals have the potential to promote the reproduction of their own genes by promoting the reproduction of their kin. Thus, an altruistic act (i.e., an act that may benefit the reproduction of a relative, at some reproductive cost to the performer) has the potential to increase the fitness of the performer and hence be selected. Whether this potential is realized depends on the cost to the performer, the benefit to the relative, and their degree of relatedness (the average probability that they share a given gene through common descent). Hamilton expressed this potential in a mathematical formulation, since named Hamilton's rule: $br > c$, where b = the benefit of an act to the recipient, in terms of its fitness gains, c = the fitness cost of the act to the performer, and r = the average degree of relatedness between the performer and the recipient. Thus, for an altruistic act to be selected, the cost to the recipient must be smaller than the benefit to the recipient discounted by their degree of relatedness.

Inclusive fitness theory opened a whole new research avenue on kinship, turning attention from the proximate aspects of nepotism to its functional underpinnings. Using Hamilton's rule (1964), it was potentially possible to test the adaptive function of kin altruism. However, in actuality, the 1970s and 1980s saw only a relatively small number of studies and models concerned specifically with testing kin selection theory in nonhuman primates (e.g., Kurland 1977, Chapais & Schulman 1980, Silk 1982). The paucity of kin selection studies was due primarily to two problems. First, it was difficult to measure adequately the costs and benefits of specific behaviors hypothesized to be altruistic in long-lived species, whether at a proximate or an ultimate level. Second, it was difficult to establish realistic predictions about the distribution of nepotism in species with social organizations as complex as primates (e.g., Altmann 1979, Schulman & Rubenstein 1983, Chapais 2001). Retrospectively, it appears that the impact of kin selection on primate studies was more indirect, the theory essentially providing a new and exciting rationale for probing the proximate aspects of the influence of kinship on behavior. While this increased our understanding of the nature and scope of kin bias tremendously over the next decades, it had the unfortunate consequence of discouraging the consideration of alternative explanations for the origins or maintenance of kin bias, such as mutualism or reciprocity. Rather, an implicit equation between kin bias and kin selection was forged in the minds of many researchers regardless of whether kin-biased interactions were fundamentally altruistic, reciprocal, or cooperative (Chapais 2001).

The Kinship Black Box

Given the importance granted to kinship by primatologists, it may appear somewhat surprising that no single-authored or edited book on kinship and behavior in nonhuman primates has been published until now, although, as we mentioned earlier, a number of excellent review papers have been published (Gouzoules 1984, Walters 1987, Bernstein 1991, Silk

2001). A number of factors may help to explain this. First, until recently the actual number of studies on the impact of kinship (beyond mother-offspring relationships) on primate behavior was fairly limited. A quick search of the PrimateLit database using "kinship" or "relatedness" as keywords reveals that the number of references has grown exponentially over the last 50 years. Between 1952 and 2001, the number of references per 10-year block went from 3 in the 1950s to 31 in the 1960s, 113 in the 1970s, 294 in the 1980s, and 493 in the 1990s. Thus, more studies on kinship were published over the last decade of the twentieth century than over the 40 preceding years. Moreover, until recently the bulk of sufficiently detailed studies on kinship have been carried out on a very small proportion of nonhuman primate species, including chimpanzees, gorillas, and a handful of cercopithecines. Longitudinal data needed to ascertain matrilineal kinship in other species have simply been lacking, Thus, any synthesis admittedly suffered from a strong taxonomic bias. But more important perhaps, even when information on kinship and behavior became relatively abundant, the kinship concept remained, to a large extent, a black box. For decades primatologists have used the concept of kinship despite knowing almost nothing about some of its most fundamental aspects, among them what matrilineal kin are recognized, whether patrilineal kin are also discriminated, and what mechanisms are involved in kin recognition.

Above all, there was, and still is, no consensus about the definition of matrilineal kin—that is, about the specific categories of kin that are treated differentially based on their genetic relatedness. Indeed, very little is known about the nature and limits of kin discrimination. Do primates differentiate a number of distinct kinship categories or do they discriminate only between larger, more inclusive categories? At one extreme, one could envisage a series of categories broken down in a number of ways, for example, by degrees of relatedness and/or other factors such as directness of descent versus collaterality. Alternatively, at another extreme, one could propose a simpler dichotomous system, for example, based only on close kin versus others. To determine which precise categories of kin primates discriminate, researchers need to analyze behavioral biases according to one or more hypothetical categorical schemes. Curiously, only a handful of studies provided such data in the 1970s and 1980s (e.g., Kurland 1977, Massey 1977, Berman 1978, Glick et al. 1986). In the absence of precise information on kin discrimination, researchers most often assessed the impact of kinship on behavior by using a dual classification (e.g., kin and nonkin). Such classifications were arbitrary to a large extent, as illustrated by the variety of definitions of kin that were used, from conservative ones ($r \geq .25$), to progressively more inclusive ones such as $r \geq .125$, or $r \geq .063$, and vague definitions such as "belonging to the same matriline" or "being matrilineally related." Obviously, variation in the definitions of kin has made it extremely difficult to assess the precise influence of kinship on behavior and to compare that influence between groups and species.

A second reason that the kinship concept may be thought of as a black box is that half of the contribution to genetic relatedness, namely the patrilineal component, has been difficult to assess and almost unexplored until recently. As mentioned previously, the enduring focus on matrilineal kinship reflected to a large extent a fundamental sex difference in patterns of parental investment typical of mammals. Due to the absence of preferential, long-term bonds between fathers and offspring in most nonhuman primate species, coupled with the promiscuous nature of mating systems in a large number of species, researchers could not ascertain paternity in most cases. Without information on paternity, other patrilineal kinship relations could not be identified, except by controlling reproduction in captivity and studying, for example, paternal sibships (e.g., Fredrikson & Sackett 1984), or by conducting a posteriori genetic analyses (see below).

Third, the kinship concept can be seen as a black box in the sense that extremely little is known about the mechanisms involved in the recognition of kin in primates, matrilineal or patrilineal. A handful of experimental studies have shown that prior familiarity is a necessary condition for the establishment of behavioral kin biases between maternal or paternal siblings, strongly suggesting that phenotypic matching is either not sufficient or not involved in kin recognition (Fredrikson & Sackett 1984, Sackett & Fredrikson 1987). However, this has been questioned recently by a few observational studies of biases among genetically confirmed patrilineal kin (e.g., Alberts 1999, Widdig et al. 2001, Buchan et al. 2003, Smith et al. 2003). But even assuming that familiarity is a sufficient condition and that phenotypic matching is not involved in kin recognition, the notion of familiarity constitutes a sort of black box in itself, because we know little about the sensory modalities, precise learning mechanisms, and developmental processes involved. Moreover, we know little about the likely variation in these processes among taxa, dispersal regimes, life history traits, and group compositions.

Recent Developments in Nonhuman Primate Kinship Studies

Over the last 10 years significant advances have been made on various fronts of kinship research, shedding some light inside the black box. In the following paragraphs, we briefly outline some of these advances and cite the chapters in which they are discussed in more detail in this volume. First, a true methodological revolution has taken place in our ability to obtain information on genetic relatedness (chapters 2, 3). Molecular genetic methods using various sets of genetic markers allow one to determine paternity, hence to infer some patrilineal kinship relations with much greater efficiency than previously. Similarly, matrilineality can be inferred from molecular methods using nDNA or mtDNA, even in the absence of birth records. Methods for determining all degrees of kinship are developing rapidly. Paralleling these developments, noninvasive genotyping techniques now make it possible to extract the relevant genetic data from extremely small amounts of DNA contained in hair and feces, hence without capturing and manipulating the animals. Together, these methods are rapidly allowing researchers to establish genealogical relations in increasing numbers of wild populations of primates even in the absence of long-term behavioral observations. In addition, these methodological breakthroughs are providing new ways of addressing old questions. For example, dispersal patterns have traditionally been inferred from behavioral data on transfers. Researchers can now enhance their inferential abilities by testing the prediction that the average degree of relatedness of the resident sex is higher than that of the dispersing sex.

Not only has the information on genetic relatedness and kinship structures accumulated extremely rapidly over the last 10 years, but significant progress has also been made regarding the influence of kinship on behavior. A few detailed studies on the impact of patrilineal kinship on behavior are now available, raising new and intriguing questions about the ability to discriminate paternal kin and the mechanisms involved (chapters 8, 13, 14). With respect to matrilineal relatedness, more detailed studies have analyzed behavioral biases according to degree of kinship and specific kin categories (chapters 7, 16). Kin discrimination has been analyzed both by observational studies that controlled for the effect of possible confounding factors such as rank distance, and by experimental studies that controlled for time constraints and the availability of kin in the preferential treatment of specific kin categories. As a result, the domain of matrilineal kindred recognized by primates is better understood. Also, the recognition by individuals of the kin relations of others (chapter 15) has been studied in the wild using playback experiments to test the ability of monkeys to associate callers with their kin, and in captivity by testing the ability of primates to associate portraits of kin.

Data are also accumulating, although at a slower pace, on the mechanisms involved in kin recognition (chapter 13). The acoustic cues used by individuals to recognize kin have received particular attention over recent years. Playback experiments, in which particular calls are presented to an audience in specific social contexts, have proven particularly informative in this context. In a developmental perspective (chapter 14), we are also beginning to outline the ontogenetic processes through which immature animals come to display kin biases that are not an artifact of the mother's control.

In general, data on kinship and behavior are now available for a much larger array of catarrhine and platyrrhine species (chapters 7–11). The sheer accumulation of data on intergroup variation, at both the intraspecific and interspecific levels, has allowed us to refine our hypotheses and models about the ways in which ecology, life history, and demography mediate the impact of kinship on behavior. For example, significant progress has been made in the socioecological modeling of philopatry and dispersal and, hence, on the socioecology of kin group formation (chapter 4). The comparison of demographically different groups of the same species has revealed the profound impact that an individual's genealogical environment has on the development of its social network (chapters 6, 14). As another example, the relative importance of kinship in structuring social relationships has been found to vary considerably between species, even among closely related ones, pointing to the importance of phylogenetic constraints and ecological variation (chapters 6, 14, 16, 20).

More tests of kin selection theory are now available. For example, kin altruism has been analyzed experimentally in the areas of intervention behavior and co-feeding (chapter 16). These studies have provided more precise tests of kin selection because they tested single kin dyads at a time and controlled for various factors that might affect the distribution of altruism according to degree of kinship, including the cost-benefit ratio of behaviors, the potential confounding effect of reciprocal altruism on kin selection, and various temporal, spatial, and demographic constraints limiting the ability of individuals to favor distant kin.

The influence of kinship on mating inhibition (incest avoidance) has been documented for matrilineal kinship (chapter 12). More specifically, more data are available concerning the impact of degree of matrilineal kinship on mating activity, and data on the effect of patrilineal kinship on mating behavior are now available thanks to paternity determination studies. In addition, the relative contributions by male and female kin in mating inhibition are better understood.

Finally, data have accumulated on the distribution of genetic diversity among groups and populations of certain species (chapter 5). We now have a better understanding of the impact of patterns of philopatry and dispersal on intraspecific variation of both nuclear and mitochondrial DNA. These data have significant implications for the management of primate populations for conservation purposes and for understanding the potential for evolutionary change.

Primates and the Origins of Human Kinship

Over the last 10 years, primatologists and anthropologists have also contributed significantly to our understanding of the origins of human kinship. In particular, several researchers have built upon the pioneering work of Robin Fox (1972, 1975, 1979), who first used primate data to characterize human kinship. Benefiting from the increased amount and diversity of primate data available, these researchers have pursued comparisons between nonhuman and human primates in an attempt to identify more precisely the attributes of kinship that we share with other primates and those that are unique to our species. One such influential attempt was that of Rodseth et al. (1991) who embarked on the task of identifying the central tendencies of human kinship systems. Modifying a scheme developed by Foley and Lee (1989), they compared primate societies along two dimensions—whether males and females maintain relationships with their respective kin, and whether they form permanent and sexually exclusive bonds. The comparison revealed that humans are the only known primate species in which groups are composed of conjugal families united by consanguineal (e.g., patrilineal) ties and in which the formation of pair bonds between males and females usually entails the establishment of ties between their respective kin, that is, between in-laws (or affines), and hence between the groups that exchange mates. Indeed, our species is the only one in which the dispersing sex maintains long-term relationships with its own kin despite spatial distance (Rodseth et al. 1991).

From this, Rodseth et al. concluded that the distinctiveness of human kinship systems may not reside in any particular feature, but rather in the co-occurrence of a number of patterns, namely sex-biased philopatry (e.g., patrilocality), the pair bond (marriage), and intergroup affinity. The last section of the book (chapters 17–19) reports additional progress and controversy that has arisen regarding these issues. Chapter 17 argues for phylogenetic continuity between chimpanzee patterns of male alliances and fraternal interest groups in hunter-gatherers, and between consanguineal kinship in nonhuman primates and the extension of kinship well beyond biological consanguinity in our species. Chapter 18 reexamines and questions the widespread notion that patrilocality is the preponderant residence pattern observed among hunter-gatherers, and hence that human patrilocality is homologous with male philopatry/female dispersal in chimpanzees. Finally, chapter 19 critically reviews theory and empirical evidence as they relate to two major competing hypotheses about the evolution of human pair bonds—male parental effort through family provisioning versus male mating effort through family protection.

Organization of the Book

The main goals in organizing the book have been to cover most major aspects of the study of kinship and behavior in primates and to produce a sort of collective advanced textbook on the subject. To do so, we defined a number of themes, and we asked the authors of each chapter to write a critical review of the literature pertaining to that theme.

The resulting book includes five major sections. The first is about methodology. It reviews methods for assessing relatedness from molecular data (chapter 2) and noninvasive genotyping methods (chapter 3). The second section discusses the kin composition of social groups from three different perspectives—ecological determinants of dispersal patterns (chapter 4), population genetics of dispersal (chapter 5), and the influence of demography on kin-related social behavior (chapter 6). The third section summarizes the influence of kinship on behavior for several types of social organization and distinguishes between nepotism and the influence of kinship on mating. Chapters 7 and 8 review the impact of matrilineal and patrilineal kinship respectively, in groups that contain several kin, whether they are unimale or multimale. The following three chapters examine kinship in nongregarious primates (chapter 9), cooperative breeding species (chapter 10), and multilevel societies (chapter 11). The last chapter of the section (chapter 12) reviews the impact of kinship on mating and reproduction.

The fourth section addresses various proximate and functional processes leading to kin biases in behavior, namely kin recognition (chapter 13), developmental processes (chapter 14), recognition of the kinship relationships of others (chapter 15), and constraints acting on the operation of kin selection (chapter 16). The fifth and last section contains three chapters that investigate the evolutionary origins of human kinship and pair bonds (see descriptions above). Finally, in our concluding chapter (chapter 20), we draw together some of the major findings of the book to consider briefly a final area of the kinship black box—the question of interspecies variation in nepotistic regimes and kin recognition.

Acknowledgments   We wish to thank the following researchers who volunteered to review the chapters of various authors: Stuart Altmann, Christopher Boehm, Michael Bruford, Tim Clutton-Brock, Frans de Waal, Barbara DeVinney, Leslie Digby, Robin Dunbar, Melissa Gerald, Sharon Gursky, Warren Holmes, Sarah Hrdy, Jane Lancaster, Joe Manson, Dario Maestripieri, John Mitani, Jim Moore, John Moore, Alexandra Müller, Theresa Pope, Anne Pusey, Joan Silk, Barbara Smuts, Bernard Thierry, Michael Tomasello, and Linda Vigilant. We owe much to their generosity and insightful comments. We also wish to thank authors who have donated photographs from their own collections or arranged to borrow them from others. We are grateful to Leanne Nash for clarifying the common names of prosimians and to Renee Cadzow, Monique Fortunato, and Aviva Kugel for clerical and technical assistance. Finally, we thank Kirk Jensen, our editor at OUP, for support and encouragement throughout the project.

References

Alberts, S. C. 1999. Paternal kin discrimination in wild baboons. *Proc. Royal Soc. Lond., B*, 266, 1501–1506.

Altmann, S. 1979. Altruistic behavior: The fallacy of kin deployment. *Anim. Behav.*, 27, 958–959.

Berman, C. M. 1978. Social relationships among free-ranging infant rhesus monkeys. PhD dissertation, University of Cambridge.

Bernstein, I. S. 1991. The correlation between kinship and behaviour in non-human primates. In: *Kin Recognition* (Ed. by P. G. Hepper), pp. 7–29. Cambridge: Cambridge University Press.

Buchan, J. C., Alberts, S. C., Silk, J. B., & Altmann, J. 2003. True paternal care in a multimale primate society. *Nature*, 425, 179–181.

Chapais, B. 2001. Primate nepotism: what is the explanatory value of kin selection? *Int. J. Primatol.*, 22, 203–229.

Chapais, B. & Shulman, S. 1980. An evolutionary model of female dominance in primates. *J. Theor. Biol.*, 82, 47–89.

Foley, R. A. & Lee, P. C. 1989. Finite social space, evolutionary pathways, and reconstructing hominid behavior, *Science*, 243, 901–906.

Fox, R. 1972. Alliance and constraint: sexual selection in the evolution of human kinship systems. In: *Sexual Selection and the Descent of Man 1871–1971* (Ed. by B. Campbell), pp. 282–331. Chicago: Aldine.

Fox, R. 1975. Primate kin and human kinship. In: *Biosocial Anthropology* (Ed. by R. Fox), pp. 9–35. New York: John Wiley and Sons.

Fox, R. 1979. Kinship categories as natural categories. In: *Evolutionary Biology and Human Social Behavior: An Anthropological Perspective* (Ed. by N. A. Chagnon & W. Irons), pp. 132–144. North Scituate, MA: Duxbury Press.

Frederickson, W. T. & Sackett, G. P. 1984. Kin preferences in primates (*Macaca nemestrina*): relatedness or familiarity? *J. Comp. Psychol.*, 98, 29–34.

Glick, B. B., Eaton, G. G., Johnson, D. F., & Worlein, J. 1986. Development of partner preferences in Japanese macaques (*Macaca fuscata*): effects of gender and kinship during the second year of life. *Int. J. Primatol.*, 7, 467–479.

Gouzoules, S. 1984. Primate mating systems, kin associations and cooperative behavior: evidence for kin recognition? *Ybk. Phys. Anthropol.*, 27, 99–134.

Hamilton, W. D. 1964. The genetical evolution of social behaviour. *J. Theor. Biol.*, 7, 1–51.

Kawai, M. 1958. On the system of social ranks in a natural troop of Japanese monkey (I)—basic rank and dependent rank. *Primates*, 1, 111–130.

Kawamura, S. 1958. Matriarchial social ranks in the Minoo-B troop: a study of the rank system of Japanese monkeys. *Primates*, 1–2, 149–156.

Kurland, J. A. 1977. *Kin Selection in the Japanese Monkey*. Basel: Karger.

Massey, A. 1977. Agonistic aids and kinship in a group of pigtail macaques. *Behav. Ecol. Sociobiol.*, 2, 31–40.

Rodseth, L., Wrangham, R. W., Harrigan, A. M., & Smuts, B. B. 1991. The human community as a primate society. *Curr. Anthropol.*, 32, 221–254.

Sackett, G. P. & Frederickson, W. T. 1987. Social preferences by pigtailed macaques: familiarity versus degree and type of kinship. *Anim. Behav.*, 35, 603–606.

Sade, D. S. 1965. Some aspects of parent-offspring and sibling relations in a group of rhesus monkeys, with a discussion of grooming. *Am. J. Phys. Anthropol.*, 23, 1–17.

Sade, D. S. 1967. Determinants of dominance in a group of free-ranging rhesus monkeys. In: *Social Communication Among Primates* (Ed. by S. A. Altmann), pp. 99–114. Chicago: University of Chicago Press.

Schulman, S. R. & Rubenstein, D. I. 1983. Kinship, need, and the distribution of altruism. *Am. Nat.*, 121, 776–788.

Silk, J. B. 1982. Altruism among adult female bonnet macaques: explanation and analysis of patterns of grooming and coalition formation. *Anim. Behav.*, 79, 162–187.

Silk, J. B. 2001. Ties that bond: the role of kinship in primate societies. In: *New Directions in Anthropological Kinship* (Ed. by L. Stone), pp. 71–92. New York: Rowman & Littlefield.

Smith, K., Alberts, S. C., & Altmann, J. 2003. Wild female baboons bias their social behaviour towards paternal half-sisters. *Proc. Royal Soc. Lond., B*, 270, 503–510.

Thierry, B. 2000. Covariation of conflict management patterns across macaque species. In: *Natural Conflict Resolution* (Ed. by F. Aureli & F. B. M. de Waal), pp. 106–128. Berkeley: University of California Press.

Walters, J. R. 1987. Kin recognition in nonhuman primates. In: *Kin Recognition in Animals* (Ed. by D. F. Fletcher & C. D. Michener), pp. 359–393. New York: John Wiley.

Widdig, A., Nurnberg, P., Krawczak, M., Streich, W. J., & Bercovitch, F. B. 2001. Paternal relatedness and age proximity regulate social relationships among adult female rhesus macaques. *Proc. Natl. Acad. Sci. USA*, 98, 13769–13773.

Part I

Who Are Kin?
Methodological Advances in
Determining Kin Relationships

The role of kinship in social behavior is not well documented in many species, including these Tibetan macaques (*Macaca thibetana*), partly because we lack definitive knowledge about the kin relationships among most study groups. Photo by Carol Berman.

2

Determination of Genealogical Relationships from Genetic Data: A Review of Methods and Applications

Phillip A. Morin
Tony L. Goldberg

The concept of kinship has been central to investigating the remarkably varied social structures of primates. Genealogical relationships between individuals are predicted, from the first principles of evolutionary theory, to be critical influences on the nature of social relationships. Sociobiological/socioecological theory in particular predicts that kinship should have primary importance for the cohesion of groups, dominance, inbreeding avoidance, and coalitional behavior (Hamilton 1964, Wrangham 1980, Trivers 1985, Silk 1987).

Determining kinship is therefore a major focus of many studies of primate sociality. Kinship information has proven indispensable to addressing questions relevant to the evolution of sociality, mate choice, breeding systems, social dominance, and kin selection (Ross 2001). Consequently, investigators have used many direct and indirect methods to try to determine kinship relationships in primate groups. Among these, the most powerful, and currently the most widely used, are the molecular genetic methods.

Molecular genetic methods for determining kinship vary in their accuracy and in the amount of effort, expertise, money, and error involved. This chapter reviews molecular genetic methods that are commonly used or potentially useful in studies of primate kinship. Methods are reviewed with respect to their relative costs and benefits in terms of effort, financial cost, expertise, or specialized equipment, as well as with respect to the limitations of the inferences that can be drawn from them. A short review of the published applications of molecular methods for determining genealogical relationships follows to put the use of these methods into historical as well as methodological perspective.

The following sections describe potential sources of genetic material and the various classes of genetic markers commonly used in studies of primate kinship. Because of the current popularity and accessibility of DNA-based methods, we will not consider genetic methods that make use of other molecules (RNA, proteins). The goal is to provide information that will facilitate the informed choice of an appropriate marker or set of markers. Please refer to table 2.1 for definitions of terms used throughout this chapter.

Genetic Methods

Sources of DNA

Molecular methods that make use of DNA without subsequent amplification by PCR require large amounts of high-quality starting material. Such DNA must be extracted from blood or tissue. These sources yield microgram to milligram quantities of high molecular weight DNA, which is required for direct visualization of the digested or probed fragments. To preserve the quality of the DNA during transport to the laboratory, such samples are typically stored and shipped frozen.

By contrast, PCR allows amplification of specific DNA fragments from as little as one copy of the genomic DNA template. PCR therefore expands the possible sources of DNA significantly, eliminating the need to collect large amounts of blood or tissue. Furthermore, samples collected for PCR need not be frozen immediately. Some degradation of the starting genetic material is acceptable, especially when the PCR amplicon is short (less than approximately 300 bp), as would be the case for most microsatellite loci and SNPs. Sources of DNA that have become accessible to primate geneticists since the invention of PCR now include shed or plucked hair (Vigilant 1999, Morin & Woodruff 1992, Morin et al. 1993, Goossens et al. 1998, Higuchi et al. 1988), feces (Taberlet et al. 1997, Launhardt et al. 1998, Gerloff et al. 1999, Immel et al. 2000, Smith et al. 2000), food wadges (Takasaki & Takenaka 1991), and, in theory, any other source that would contain genetic material from the primate of interest (Morin & Woodruff 1996, Taberlet et al. 1999).

Genomic Components and Modes of Inheritance

Primates have two sets of chromosomes in the nuclear genome, one set from each parent, and a single circular chromosome in each mitochondrion, which are present in thousands of copies per cell.

The mitochondrial genome is passed to offspring in the cytoplasm of the egg and is thus inherited only maternally. Because it is haploid (one copy) and inherited from one sex, the effective population size for mitochondrial genes is one-fourth that of autosomal (nonsex chromosomes) nuclear genes, which are diploid (two copies) and inherited from both parents. Mitochondrial DNA, therefore, reflects only matrilineality. Because of this, and because of its lack of genetic recombination, mitochondrial DNA is also useful for phylogeographic and phylogenetic studies, particularly when the sex bias of dispersal is known.

Nuclear, autosomal loci are found in two copies per cell. Individuals can be either homozygous or heterozygous at any given locus. Examination of nuclear autosomal loci therefore allows for the detection of heterozygotes within individuals, as well as the examination of variation among individuals in populations. Biparental inheritance of the alleles at these loci means that, when polymorphisms are present, the source of the variant alleles can be traced from parent to offspring (figure 2.1).

Table 2.1. Useful Definitions

AFLP	Amplified fragment length polymorphism, a method for generating random, noncodominant variable genotypes based on restriction enzyme digestion and selective amplification of DNA fragments.
Allelic dropout	Artifactual loss of an allele from a genotype because of random PCR amplification of only one of two alleles (usually due to very small amounts of template DNA).
Allele-specific probe hybridization	Single-stranded DNA will hybridize with (bind to) other fragments of single-strand DNA to form double-stranded DNA when the two fragments are complementary. Under stringent conditions, hybridization will only take place when the two fragments are perfectly matched; alleles differing by ≥1 nucleotide can be distinguished by presence or absence of hybridization to DNA probes for each allele sequence.
Allozyme	A variant of a protein that can be differentiated based on protein size, charge, structure, or function, as detected by starch gel electrophoresis and various staining, denaturing, or enzymatic assays.
Amplicon	DNA fragment amplified via polymerase chain reaction (see PCR).
Amplification	Polymerase chain reaction DNA replication (see PCR).
Autosome	Diploid, nonsex chromosomes of the genome.
Codominant	Indicates that both alleles of a heterozygous genotype can be detected.
Control region	The region of the mitochondrial genome known as the origin of replication, and which does not code for a protein; also known as d-loop.
Diploid	Having two copies (maternal and paternal) of a chromosome (e.g., in the case of nuclear autosomal DNA).
DNA fingerprinting	Historically refers to the use of multilocus minisatellite probes to produce a pattern of DNA fragments that is often highly variable among individuals. Also used to describe composite multilocus genotypes.
Electrophoresis	A process of separating molecules by size (as in the case of DNA) or charge (as in the case of proteins) by passing them through a solid gel matrix using an electric current.
Exon	A section of protein-coding DNA. A gene may be made up of one exon, or many exons separated by introns.
Genotype	The combination of alleles present in an individual for a given locus or set of loci.
Haploid	Having one copy of a chromosome (e.g., in the case of mitochondrial DNA).
Hv1	The first hypervariable portion of the mitochondrial control region, most often used for phylogenetic and genealogical studies in primates and many other species.
Intron	A sequence of non–protein-coding DNA that lies between sections of protein-coding DNA (exons) that make up a gene.
Locus	A portion of the genome defined by function, location, or DNA sequence.
Mendelian heritability	Pattern of inheritance of one-half of the genome from each parent. When used to describe codominant alleles of a locus, this refers to the observed inheritance of one allele from each parent.
Microsatellite	Tandemly repeated short DNA sequence motifs. Repeated elements are usually between 1 and 6 nucleotides in length, and are repeated perfectly or imperfectly between 5 and 30 times. Also known as simple sequence repeats (SSRs) or simple tandem repeats (STRs).
Minisatellite	Tandemly repeated DNA sequence motifs, typically between 5 and 50 nucleotides long. Minisatellites are often near the telomeric (end) regions of chromosomes. Also known as variable number tandem repeats (VNTRs).

Table 2.1. Continued

MtDNA	Abbreviation for mitochondrial DNA, the circular, maternally inherited genome of the mitochondria.
Multilocus "fingerprint" or genotype	Compiled alleles from multiple loci that make up the composite genotypes for the individual.
Multiplex genotyping	The ability to combine multiple loci in a single assay and to distinguish individual genotypes from each locus.
PCR	Polymerase chain reaction, an enzymatic process of replicating specific fragments of DNA in vitro using cycles of DNA denaturation, primer annealing, and DNA polymerization (copying) of the template DNA strand.
Polymorphic locus	Any locus for which two or more alleles are present in a population.
Primer	Short synthetic DNA sequence (oligonucleotide) that is complementary to a segment of genomic DNA and is used to initiate replication of the DNA template during PCR.
Probe	DNA segment (oligonucleotide or larger DNA fragment) used to hybridize to a mixed or genomic DNA sample to identify that locus in the sample. The probe is usually labeled with radioactive material or other chemical attachments to facilitate visualization of the hybridized fragment on a solid matrix.
RAPD	Random amplified polymorphic DNA, a process by which short, nonspecific primers are used to amplify unknown regions of genomic DNA and thus to create individual-specific banding patterns.
RFLP	Restriction fragment length polymorphism, fragments of DNA created by the enzymatic hydrolysis of DNA based on the enzymatic recognition of specific DNA sequences.
Single nucleotide polymorphism (SNP)	Variation, within a population, in the nucleotide (A, C, G, or T) at a particular position in a DNA sequence.

FACING PAGE

Figure 2.1. Examples of electrophoretic detection of genetic variation in a pedigree. Black and hatched bands represent maternal and paternal alleles, respectively. Dashed bands can be either maternal or paternal and provide no information in these scenarios for parental exclusion. (A) variation in size (DNA) or charge (protein) of genetic markers. (B) Variation in DNA sequence detected by RFLP analysis. Change in the DNA sequence results in creation of a restriction enzyme recognition site, so the DNA is cleaved into two fragments at the new site. The heterozygous individual exhibits the patterns for the uncut fragment and the two cut fragments. (C) Variation in DNA sequence (SNP) for two loci detected by SBE and electrophoresis. Slight differences in allele size occur sometimes because of the different properties of the various fluorescent dyes attached to the SBE products. L1 and L2 refer to two multiplexed loci.

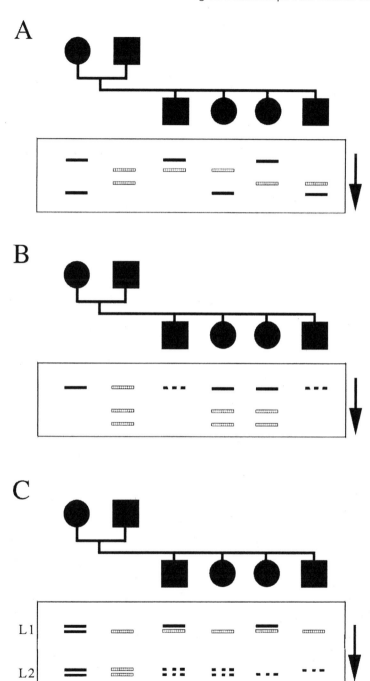

In all primates, two of the nuclear chromosomes are sex chromosomes. The X and Y chromosomes are inherited in a sex-biased fashion, and are thus useful for studies of matrilineality and patrilineality, as well as for sex determination. Y chromosome genetic markers have been developed for humans and some nonhuman primate species (Underhill et al. 1997, Thomas et al. 1999, Stone et al. 2002). However, the process of selecting Y chromosome markers is difficult and time consuming, primarily because of the current paucity of species-specific Y chromosome DNA sequence data for primates other than humans. Nevertheless, Y chromosome markers could provide important information for male dispersal studies and for studies of kinship in males within and between groups (e.g., Thomas et al. 2000). X chromosome markers have not been reported in primate kinship studies to date and provide little information on relatedness that cannot be obtained from the more accessible autosomal markers or the more quickly evolving mitochondrial genome.

Sexing Primate Genetic Samples

The value of genetic information collected during primate kinship studies is often enhanced when the sex of individual samples can be determined. This is particularly true in noninvasive studies of unhabituated primates, when direct observation of individuals is impossible or unreliable.

Sexing of samples can easily be done for most primates using currently available techniques, or using minor adaptations of techniques developed for other mammalian species. Several methods have been described for PCR amplification of the amelogenin gene, for example, which is found on both the X and Y chromosomes, but which often differs in DNA sequence and length on the different chromosomes. A size polymorphism of 6 bp has been reported in humans (Sullivan et al. 1993), and is also present in other apes (Bradley et al. 2001). This assay has the advantage of producing a PCR product from one set of PCR primers for each sex chromosome, so that the assay is internally controlled for amplification. The disadvantage is that these primers have not been tested widely in primates, so it is not yet known whether they will amplify the targeted gene region, or whether the product will vary in size between the X and Y chromosome in other species. A similar assay involving size differences in the ZFX and ZFY genes has been reported for humans and a variety of New and Old World primates (Wilson & Erlandsson 1998). This assay is limited to high-quality DNA (e.g., from blood or tissue) because of the large size of the PCR amplification target.

An alternative method, which has been used on a wide range of mammals, makes use of PCR amplification of a segment of the SRY gene (Griffiths & Tiwari 1993). This assay has the advantage of being widely applicable among species. However, nonamplification of the PCR product is a nonspecific result, in that it can be due either to lack of a Y chromosome (female) or to failure of the PCR for other reasons. In this and other such analyses, one or more internal positive control PCRs should be performed simultaneously to validate the sexing results.

Genetic Marker Systems

Table 2.2 compares the most common or promising marker systems currently used to determine kinship relationships among primates. Genetic markers useful for inferring primate

Table 2.2. Comparison of Molecular Genetic Techniques Available to Study Kinship in Primates[a]

	mtDNA Sequencing[b]	Multilocus Minisatellite	RAPD	AFLP[b]	Protein (Allozyme) Electrophoresis	RFLP (Nuclear Locus)	Microsatellites	SNPs[b] (SBE with Electrophoresis)
Development time[c]	1–4 weeks	1 month	1–2 weeks	2–4 weeks	2–4 weeks	1–4 months	2–6 months	2–6 months
Processing time[d]	2–4 weeks	2–4 weeks	1 week	1 week	2 days	2–4 weeks	2–4 days	2–4 days
Genetic variation	Medium (but variable)	High	Medium	Medium	Low (typically 2–4 alleles)	Low (typically 2–4 alleles)	High (typically 3–15 alleles)	Low (typically 2 alleles)
Risk of anomalous results	Low	Low–moderate	High (dominant alleles, sensitive PCR conditions)	Moderate–high (dominant alleles, sensitive PCR conditions)	Low	Low	Moderate (null alleles, allelic drop-out)	Low
Ease of scoring	Moderate	Moderate–difficult	Difficult	Difficult	Easy–moderate	Easy	Easy–moderate	Easy
Relative cost[e]	High	Moderate	Low	Low–moderate	Low	Low	Moderate	Low–moderate

[a]Modified from Webster and Westneat (1998), with permission from Michael Webster and Brikhäuser Verlag.

[b]Marker types added to table by Webster and Westneat (1998), based on references and authors' experience (mtDNA, SNPs).

[c]Assumes lab is using the technique successfully on another species; if not, add 6–12 months minimum for all of the DNA techniques, 2–6 months for proteins to learn and develop procedures.

[d]For a set of 15–20 samples from collected tissue to bands ready to be scored (includes 1–4 days to isolate DNA for DNA techniques), for one probe, primer, or locus. For loci with low to moderate variability, multiple loci may be required, so time must be added accordingly. If noninvasive samples are used, sufficient replication also needs to be considered.

[e]Relative costs represent per-sample costs and do not include costs of marker development or specialized equipment.

kinship can be divided broadly into two classes, based on the type of laboratory method used. Single-locus methods examine individual loci separately and yield locus-specific genotypes. The major advantage of single-locus methods is analytical; alleles in single-locus systems are inherited in a Mendelian fashion and can be analyzed using powerful statistical methods derived from classical population genetic theory. Single-locus methods often suffer from relatively low individual locus variability, however, so that large numbers of loci are required to establish high confidence levels for unique individual or relationship identification.

Multilocus methods examine several loci simultaneously. Such methods generate patterns of DNA fragments, usually visualized as bands on a gel, that differ among individuals. The major advantage of multilocus methods is that they generate a great deal of information quickly. Their major disadvantage, however, is that the DNA fragments generated cannot be assigned to specific loci, and the data cannot be analyzed in a Mendelian framework. Interpretation of multilocus fingerprint patterns, especially beyond first-order relationships, is therefore difficult or impossible (Jeffreys 1987, Lynch 1988, Lander 1989, Jeffreys et al. 1991, Weatherhead & Montgomerie 1991, Smith et al. 1992; for a more complete review of these methods, see Martin et al. 1992, Smith & Wayne 1996, DeSalle & Schierwater 1998, Hoelzel 1998).

Genetic marker systems can also be divided into those that employ PCR and those that do not. This distinction is useful because the non–PCR-based methods require large amounts of high-quality starting DNA and are therefore of limited utility for studies of most wild primates. The non-PCR methods include, most notably, the classic "DNA fingerprinting" techniques so widely employed in early studies of paternity and reproductive success (see below). Nevertheless, these methods are being almost universally replaced by PCR-based techniques. We therefore concentrate on the PCR-based methods in this chapter.

Like molecular genetic methods in general, PCR-based methods can be divided into those that target specific, known loci, and those that generate random polymorphic DNA profiles from many loci simultaneously. The former include such methods as microsatellite and SNP genotyping, and are described in detail below. The latter include such methods as amplified fragment length polymorphism (AFLP) and random amplified polymorphic DNA (RAPD) analyses. The primary benefit of these multilocus PCR-based systems is that they do not require a priori knowledge of the DNA sequences of the target species. They have been widely used in plant and microbial genetics, but less often in animal studies. The primary problems with these systems are their unreliability with respect to reproducing consistent banding patterns, and the inherent interpretive limitation that the DNA fragments generated are not inherited in a Mendelian fashion. For these reasons, and because they have not been widely used in primate genetics, we will not consider these methods further; the interested reader can obtain more information from these review chapters: Caetano-Anolles (1998), Hoelzel and Green (1998), and Webster and Westneat (1998).

Methods considered further in this chapter are single-locus methods that are amenable to PCR. These methods have the widest potential utility for genetic studies of primate kinship, both in captivity and in the wild. The specific methods that we discuss include mitochondrial DNA sequencing, microsatellite genotyping, and single nucleotide polymorphism (SNP) genotyping. The first two methods are commonly used in primate kinship studies today, and the third method has great promise for the near future.

The three methods have in common their ability to infer primate genotypes at specific loci. A genotype can mean anything from a DNA sequence to a set of allele sizes (microsatellites) or allele nucleotides (SNPs). Obtaining the genotypes represents the bulk of the work in relatedness studies and has been the goal of a correspondingly large variety of methods. For mtDNA sequencing and microsatellite genotyping, the most common methods are well established. For SNPs, the methodologies are evolving rapidly. Each marker type also has its own set of limitations and assumptions for application and analysis (for review, see Sunnucks 2000). The limitations described for each marker type below are especially pertinent to the estimation of genetic relatedness in nonhuman primates.

Mitochondrial DNA Sequencing

Mitochondrial DNA has several characteristics that recommend it strongly for some types of studies. From a practical standpoint, it is relatively easy to obtain from samples, even highly degraded ones, because of the high copy number in each cell. Its maternal inheritance, lack of recombination, and relatively high evolutionary rate also make it suitable for studies of phylogenetics and phylogeography, and for behavioral studies of matrilineality (e.g., Avise et al. 1987, Avise 1989, Morin, Moore, Chakraborty et al. 1994, Hashimoto et al. 1996, Goldberg & Wrangham 1997, Mitani et al. 2000, Pope 2000, Warren et al. 2001).

The mammalian mitochondrial genome contains 35 genes (13 protein-coding genes, 22 tRNAs) and the "control region" (also known as D-loop), or origin of replication, which does not code for a protein or RNA molecule and is highly variable. Different regions of the genome evolve at different rates and can thus be chosen to resolve "shallow" genealogical relationships (e.g., kinship) or "deep" relationships (e.g., systematics), as the situation warrants.

For intraspecific primate studies, the most commonly sequenced segment is the first hypervariable region (Hv1) of the control region (Hv1 is one of two hypervariable regions in the control region). Hv1 is popular as a target locus both because of its convenient size for PCR amplification (typically less than 450 bp), and because there are "universal" primers (Kocher et al. 1989) that work well to amplify it in a variety of species. As an example of the popularity of this segment, the HvrBase database (Burckhardt et al. 1999, www.hvrbase. org) for ape Hv1 sequences contained 9,309 human, 434 chimpanzee (*Pan troglodytes*), 4 bonobo (*P. paniscus*), 28 gorilla (*Gorilla gorilla*), and 3 orangutan (*Pongo pygmaeus*) Hv1 sequences as of September 2001. It should be noted that this database is not updated frequently and only contains unique sequences. Most published sequences are available through GenBank (http://www.ncbi.nlm.nih.gov/Entrez/).

Obtaining or Developing Mitochondrial Markers

The mitochondrial genome has been one of the most thoroughly studied segments of DNA in nonhuman species and has been sequenced in its entirety for more than 30 species, with the number growing on a monthly basis (e.g., Schmitz et al. 2000). Given the amount of information now available, it is relatively easy to find primers for taxa of interest in the literature or to design primers from published and aligned sequences available in public databases (e.g., AMmtDB: http://bighost.area.ba.cnr.it/mitochondriome, Lanave et al. 1999).

Furthermore, regions of mtDNA can be chosen based on the relative amount and patterns of variation desired for a particular study (Pesole et al. 1999).

The control region is the most variable portion of the mitochondrial genome and has been most widely used for investigations of relationships among individuals and groups below the species level. Primers for the entire control region (Kocher et al. 1989) or portions of it (e.g., Hv1 or Hv2) have been designed for various primate species. Nevertheless, sequencing of the control region from a variety of individuals of the target primate species may be required to ensure that species-specific primers are designed that avoid polymorphic sites and amplify the given segment from all (or the majority) of individuals.

Obtaining mitochondrial markers is therefore more straightforward than obtaining any other type of marker. The technical limitations are only those normally associated with DNA extraction, PCR amplification, and DNA sequencing.

Genotyping Methods for Mitochondrial DNA

Early methods of surveying variation in the mitochondrial genome primarily involved restriction fragment length polymorphism (RFLP) analysis. Direct DNA sequencing of PCR products (or cloned PCR products), however, has since become the method of choice (Hoelzel 1998). When little variation is present, some studies have screened for variants quickly using surrogate methods such as allele-specific probe hybridization (Morin et al. 1992) or other nucleotide screening methods (e.g., Amato et al. 1998, Dean & Milligan 1998).

Given the wide use of direct sequencing protocols, and companies or university core labs that perform sequencing efficiently and relatively inexpensively, the primary issues are (1) verifying that the sequence is truly mitochondrial and not a nuclear insert of a mitochondrial sequence (see below), and (2) choosing appropriate methods of sequence alignment and analysis.

Limitations and Assumptions of Mitochondrial DNA

The primary limitation of mtDNA in studies of genealogical relationship is that the entire mitochondrial genome is inherited intact (except for possible mutations) from the mother. As a result, mtDNA offers no information about paternity. Furthermore, the ability to use mtDNA for maternal lineage determination is dependent on the level of sequence variation in the population and its distribution; the fact that individuals in a group share mitochondrial haplotypes may or may not indicate that they are closely related. For example, some species have inherently low levels of variation (and therefore have shared haplotypes regardless of maternal relationship). Similarly, in some social systems, all females in a group may be maternally related. In these cases, mtDNA has little or no discriminatory power within groups. Nevertheless, patterns of mtDNA distribution on larger spatial scales may be very informative for inferring sex-biased dispersal and for making phylogeographic inferences.

Another problem with mtDNA sequence analysis is that portions of the mitochondrial genome have been incorporated into the nuclear genomes of most species. When the nuclear mitochondrial inserts (also called numts; Zischler et al. 1998, Zischler 2000, Bensasson et al. 2001) are amplified instead of the actual mitochondrial DNA, incorrect relationships can be inferred. This is particularly important for phylogenetic analyses, but can also lead to false inferences about individual relationships. Unfortunately, numts may be difficult or impossible to distinguish from actual mitochondrial genes without prior information. Some phylogenetic analysis of the species of interest and its close relatives is usually required to determine whether mtDNA sequences are indeed from the mitochondrial genome. It has also been demonstrated that the amplification of numts may be more likely (in some species) from some DNA sources than from others (e.g., hair vs. blood; Greenwood & Pääbo 1999). Cloning of PCR products, followed by sequencing of multiple clones for each product, can sometimes reveal the presence of both the true mtDNA sequences and numts in a sample and therefore facilitate identification of the numts.

Finally, all studies to date that have used mtDNA to infer matrilineality have relied on DNA sequence data. Animals are assigned to different matrilines if they have different mitochondrial DNA sequences. Because typical mitochondrial sequences used in such studies are between 300 and 400 bp long, even error rates of less than 1% in DNA sequencing could lead to false exclusions of matrilineality. Methods of matrilineality exclusion that take such error into account are needed but have not been developed formally (Goldberg & Wrangham 1997).

Microsatellite Genotyping

The discovery in the 1980s of a class of highly variable nuclear markers, called simple sequence repeats (SSR), simple tandem repeats (STR), or microsatellites (Litt & Luty 1989, Tautz 1989, Weber & May 1989, Weber 1990), that are amenable to amplification by PCR, represented a major step forward in the analysis of individual genetic relationships in a variety of animal species (Queller et al. 1993). Microsatellite loci typically have 5 to 10 alleles varying in size by multiples of the repeat unit (e.g., two base pairs for a dinucleotide repeat, three for a trinucleotide repeat, and four for a tetranucleotide repeat). Microsatellites are abundant in the genomes of humans and many other species (Tautz 1989, Weber & May 1989; reviewed in Zane et al. 2001).

Because microsatellite alleles differ by size, they are amenable to detection and genotyping by polyacrylamide gel electrophoresis of the PCR products (figure 2.1). Initial detection methods involved radioactive tagging of the PCR products, but many methods for nonradioactive detection using fluorescent stains or modified nucleotides are now in use.

After initial studies demonstrated the utility of microsatellites in humans (Edwards et al. 1991, Moore et al. 1991, Schlötterer et al. 1991), their application to nonhuman primates followed quickly. Many loci were conserved between humans and the other primates, and could be amplified from noninvasive samples such as hair (Morin & Woodruff 1992, Morin et al. 1993). This was particularly important because of the desire of field primatologists to determine genetic relationships of their wild populations without disturbing the animals or reversing hard-won habituation.

Some of the first studies of wild primate populations using PCR amplification of microsatellites were in apes (Morin, Moore, Chakraborty et al. 1994, Morin, Moore, Wallis et al. 1994). These and future studies chose microsatellite loci from among those originally discovered in the human genome (Dib et al. 1996). Although this strategy worked well for apes and Old World monkeys (Morin & Woodruff 1992; Morin et al., 1997, 1998; Kayser et al. 1996; Coote & Bruford 1996; Wise et al. 1997), the discovery of markers de novo has been necessary for some species, especially New World monkeys and lemurs (Ellsworth & Hoelzer 1998, Jekielek & Strobeck 1999).

Obtaining or Developing Microsatellite Markers

Obtaining novel microsatellites for a species of interest often involves a laborious and complicated process of creating a genomic library (fragments of genomic DNA inserted into a bacterial or viral DNA vector) and enriching or screening it for the microsatellite repeats of interest (Zane et al. 2001 and references therein). For the majority of primate species, however, screening of previously characterized human markers is a more practical approach. Particularly for apes and Old World monkeys, this has resulted in the discovery of large numbers of highly variable microsatellites in a variety of species (e.g., Morin, Moore, Wallis et al. 1994, Coote & Bruford 1996, Ely et al. 1996, Kayser et al. 1996, Morin et al. 1997, Launhardt et al. 1998, Morin et al. 1998, Clifford et al. 1999, Goossens, Latour et al. 2000, Rogers et al. 2000, Smith et al. 2000, Zhang et al. 2001). The utility of this approach also extends to New World monkey species (Rogers et al. 1995, Ellsworth & Hoelzer 1998), although the success rate of finding variable loci may be lower and the incidence of null alleles (see below) higher. Given the recent increase in the number of companies able to make enriched microsatellite libraries to generate species-specific markers and the decrease in costs to have such libraries made, it may be more practical in some instances to employ a commercial service for this purpose.

Briefly, cross-species amplification involves the selection of a range of PCR conditions (usually covering approximately 10°C temperature range, and sometimes two to three magnesium concentrations from 1 to 3 mM), to try to amplify the target species DNA using human PCR primers. Sequence differences between human and target species DNA at the primer binding sites will reduce or prevent PCR amplification, however. By reducing the stringency of the PCR (lower temperature and/or higher magnesium concentration), reliable amplification of the homologous locus may be achieved despite the presence of some sequence differences. If the priming sites are conserved, one must then evaluate whether the locus itself is polymorphic or not in the target species.

Thousands of microsatellite loci are currently known in humans, and many of them can be purchased as unlabeled primer pairs for screening (e.g., Research Genetics, Huntsville, Alabama). For the majority of studies of nonhuman primates, it will be easier, faster, and more cost effective to screen human markers than to generate species-specific microsatellites using genomic libraries or other genome screening methods (Morin et al. 1998).

Fortunately, many primers have already been validated in nonhuman primate species. Although public databases are still not complete, the Molecular Ecology Notes database is a useful resource for finding published microsatellite markers (http://blackwellpublishing.com/Journals/men).

Genotyping Methods for Microsatellites

Microsatellite loci have the major advantage that alleles vary in size. Electrophoretic separation of DNA fragments is therefore perfectly suited to genotyping individuals at microsatellite loci. For fragments up to about 500 nucleotides, discriminating between fragments that differ in size by even a single nucleotide is possible using denaturing polyacrylamide gels, in conjunction with radioactive or fluorescent labels, or fluorescent or silver staining methods (David & Menotti-Raymond 1998, Schlötterer 1998). For this reason, PCR primers that amplify microsatellite loci are generally designed to yield short amplification products (typically 100–300 bp).

The primary problems with microsatellite genotyping are accurate and reproducible sizing of the alleles, correct interpretation of the allele patterns for genotyping, and generation of accurate genotypes (not incorrectly determined due to contamination, null alleles, allelic dropout, or false alleles; see below). As stated above, genotype inaccuracy can be a major problem, especially when noninvasive samples are used. It is very important that researchers be aware of the possible problems, and that they take appropriate steps to avoid them or to compensate for them (e.g., Morin et al. 2001, Taberlet et al. 1996).

Limitations and Assumptions of Microsatellites

Microsatellites are currently the marker of choice for genetic studies of genealogical relatedness (for reviews, see Jarne & Lagoda 1996, Rosenbaum & Deinard 1998, Schlötterer 1998, Schlötterer & Pemberton 1998). Significant problems must be overcome, however, to ensure data integrity and quality. Among these are discriminatory power (see analysis section), null alleles (nonamplification of some alleles in the population), high mutation rates, and PCR artifacts (allelic stutter, dropout, and false alleles).

Null alleles (Callen et al. 1993) are difficult to detect and can cause incorrect assignment of homozygous genotypes to heterozygous individuals. Because PCR amplification depends on perfect or near perfect match of the PCR primers to the template DNA, nucleotide mismatches in the primer binding sites can cause PCR failure. When these mismatches are polymorphic in the populations, this can cause nonamplification of one allele in a genotype (failure of both alleles would be rarer, and may not be attributed to null alleles unless other reasons for PCR failure can be ruled out). Statistical tests for Hardy-Weinberg equilibrium (HWE) or verification of Mendelian heritability in known families can help to identify the presence of null alleles within a data set (Brookfield 1996).

Allelic dropout (Navidi et al. 1992, Taberlet et al. 1996, Gagneux, Boesch et al. 1997) can similarly produce apparent homozygotes. Allelic dropout results from the failure of PCR to amplify one or more microsatellite alleles at a locus, because of very low amounts of starting template. Studies that use noninvasive samples (e.g., feces, hair, food wadges) may suffer from both allelic dropout and false alleles (figure 2.2). Significant effort must be made to ensure accurate genotyping, including replicate PCRs of samples and/or quantitative evaluation of DNA content and integrity (Taberlet et al. 1996, Goossens et al. 1998, Goossens, Chikhi et al. 2000, Morin et al. 2001).

The high (and variable) mutation rates of microsatellite loci occasionally cause actual mutations (shifts from one allele size to another) between generations. In a study of kinship, this could result in false exclusion of an ancestor (e.g., a parent). Maximum likelihood analysis programs (e.g., Marshall et al. 1998, Goodnight & Queller 1999) employ methods of statistical analysis that correct for both mutation and user errors in the genotype assignment, so that reasonably robust assignments can be made regardless of some level of error in the data. High and variable mutation rates among loci can also be a problem for assessing population-level relationships (i.e., population structure), because models that estimate interpopulational distances from microsatellite data often assume particular mutation rates and mutational patterns (Balloux & Lugon-Moulin 2002).

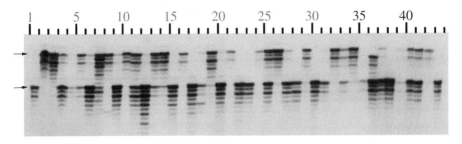

Figure 2.2. Replicate amplifications of the same sample for a single heterozygous locus (true alleles indicated with arrows), reprinted from Taberlet et al. (1996), with permission from Pierre Taberlet and Oxford University Press. This sample had a very low DNA concentration, so allelic dropout is the most common problem (e.g., lanes 1, 2, 3, 5), but false alleles are also evident (lanes 19, 37), and some allelic patterns are very difficult to interpret (e.g., lane 8). All of these problems can vary from locus to locus, and between samples, so choice of loci and quality of DNA are important issues to consider when a study is planned.

Single Nucleotide Polymorphism Genotyping

The type of polymorphism that occurs in the greatest abundance within primate genomes is the single nucleotide polymorphism. An SNP is simply a nucleotide position at which two (or, rarely, more) of the four possible bases (A, C, G, or T) occur within a population. SNPs are estimated to occur on average approximately every 1,000 bp in the human nuclear genome (Kruglyak 1997). SNPs have not been extensively exploited to date because most individual SNPs have only two alleles per site, and because of the limited technologies available for efficiently finding and genotyping many loci. Because of their low numbers of alleles per locus many more independent SNPs than microsatellites would need to be screened to achieve sufficient discrimination for kinship studies. Novel SNP detection is also labor intensive; de novo sequencing of many genome segments in many individuals is necessary. Finally, lack of size differences between alleles means that the SNPs cannot be detected simply by electrophoretic separation of PCR products.

The obvious way to generate SNP genotypes for animals would be to sequence large regions of nuclear DNA from many individuals within a population. The cost of such an endeavor would be high, however, even with today's technology. Fortunately, many methods for generating SNP genotypes have been developed, fueled by the search for genes involved in complex human disease (e.g., Marshall 1997). These methods are reviewed and updated almost monthly in journals.

Obtaining or Developing SNP Markers

SNPs can be found throughout the genome, but there is no way to predict SNP locations or to increase the likelihood of finding them (other than targeting genome segments that are less likely to be functionally constrained). Unlike microsatellites, which often show conserved regions of polymorphism in related species, SNPs may not be shared even between closely related species (Hacia et al. 1999). For this reason, and to obtain markers that are not likely to be subject to strong selection, noncoding DNA is the most logical place to search for SNPs.

In an "unknown" mammalian genome, the method of choice for detecting novel SNPs has been to design PCR primers that bind to conserved regions of protein coding genes but amplify a product that includes an intron. These anchored primer pairs have been called comparative anchor tagged sequences (CATS; Lyons et al. 1997) or exon-primed intron-crossing (EPIC) loci (Palumbi & Baker 1994). Because the primers are designed to anneal to conserved regions, it is likely that they will amplify a homologous product in a wide variety of mammalian species. Recently, a subset of CATS (along with primers just for protein coding sequences) was used to create a nuclear gene phylogeny of the placental mammals, indicating that at least some of these loci will be widely useful (Murphy et al. 2001). At present, there are at least 200 such loci available to test (Venta et al. 1996, Lyons et al. 1997, Bruillette et al. 2000, Shubitowski et al. 2001), so it is likely that a sufficient number will amplify homologous segments in primates to allow rapid screening of sequences for novel SNPs.

Once PCR products are obtained for a given set of CATS, they can be sequenced from individual or pooled DNA samples to detect SNPs. The choice of method for SNP genotyping then dictates the steps for assay development and SNP verification (see below).

Genotyping Methods for SNPs

As mentioned, SNP genotyping is still relatively new and has yet to be applied in primate (or other species) genealogical studies. Critical issues for generating SNP genotypes for a relatively large number of loci and a population of individuals include capital equipment investment, assay component and reagent expenses, accuracy, and efficiency. Most methods are moving toward a "single tube" approach when possible, or at least minimizing the number of steps needed to generate the genotype data (for overviews of possible methods, see Kwok 2000, Gut 2001, Shi 2001, Syvanen 2001).

Since SNPs do not change the length of a DNA sequence, they cannot be detected by traditional electrophoresis. Some nucleotide changes can be detected by methods that combine electrophoretic migration of PCR products with changes in mobility due to denaturation (denaturing gradient gel electrophoresis, or DGGE) or secondary structure of the DNA (single strand conformation polymorphism, or SSCP) (Dean & Milligan 1998). These methods can be reasonably efficient and inexpensive, as they make use of equipment that is also used for sequencing or microsatellite genotyping and can reliably detect most polymorphisms. Not all polymorphisms, however, can be detected, and assays cannot be multiplexed, so each assay requires a separate lane on a gel or a separate capillary electrophoresis run.

A modification of the DNA sequencing chemistry called single base extension (SBE; also called minisequencing) makes use of modified (dideoxy) nucleotides with fluorescent labels to produce small oligonucleotides with fluorescent dyes that can be detected by electrophoresis or on microarrays. The method requires a PCR product large enough to encompass the SNP of interest and also provide a binding site for a single primer immediately "upstream" (5′) of the SNP nucleotide. Highly accurate polymerases that allow incorporation of a single dideoxynucleotide complementary to the SNP produce products labeled with a colored dye, which is different for each possible nucleotide (figure 2.3). These products can vary in size (depending on the oligonucleotide design) and color, so they can be multiplexed

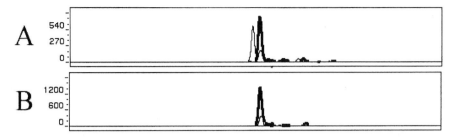

Figure 2.3. Single base extension genotype electropherograms for (A) heterozygous and (B) homozygous individuals for an SNP in the APOB gene. Small secondary peaks are thought to occur because of incomplete removal of primers after the initial PCR or the incomplete removal of dideoxynucleotides.

in electrophoresis for increased genotyping efficiency (e.g., SNaP-Shot SBE kit, Applied BioSystems, Foster City, California). SBE products can also be produced on microarrays (Syvänen & Landegren 1994, Pastinen et al. 1997), or produced in solution and resolved by hybridization to microarrays (Hirschhorn et al. 2000) or "liquid arrays" of microspheres (Chen et al. 2000). These methods require sophisticated equipment for fluorescent signal resolution on microarrays, but provide the opportunity for high levels of multiplexing and miniaturization for more efficient and inexpensive genotyping of many SNPs.

Finally, SNPs can be detected using probe hybridization methods, such as the 5′ exonuclease assay (commonly called Taqman) or "molecular beacons." These methods use highly specific oligonucleotide probes that, under optimal conditions, hybridize only to exactly complementary sequences (Morin et al. 1999, Mhlanga & Malmberg 2001). The high specificity of these probes makes it possible to distinguish between DNA sequences that differ by even a single base pair within the region of probe hybridization (i.e., an SNP). Fluorescent dye systems can be used in conjunction with such probes so that their amplification can be tracked in real time and quantified during PCR amplification, or determined after amplification once systems have been optimized. The growing ease and accessibility of "real-time PCR" makes these approaches especially promising; genotypes can be generated in no more time than it takes to run a standard PCR, and loci can be multiplexed at a low level (two to four loci). These methods are highly reproducible and accurate, but assay reagents are relatively expensive, limiting their use for large numbers of assays, especially with small sample sizes.

Limitations and Assumptions of SNPs

To our knowledge, there have not yet been any demonstrations of the use of SNPs for genealogical relationship inference in primates, and it is not yet clear what the actual limitations will be. It is clear that the low information content of individual SNPs will mean that more SNPs than microsatellites will be needed to resolve kinship relationships. Theoretical evaluations have suggested that approximately 30 to 60 SNPs will be needed to match the power of a panel of 13 to 15 highly polymorphic microsatellite loci used in human forensics (Chakraborty et al. 1999, Krawczak 1999). Most SNP assay methods require amplification of smaller DNA segments than microsatellites, so it is likely that amplification success from degraded DNA samples (e.g., noninvasive samples) will be higher, but this has yet to be shown. Null alleles will be a problem with SNPs as well (as with any PCR-based system); whether null alleles will prove to be less frequent for individual SNP loci than for microsatellites remains to be seen.

Data Analysis

Table 2.3 summarizes the applicability of various markers for studies of kinship. Below we discuss several applications in detail, including paternity exclusion, paternity inclusion, and determination of other kinship relationships.

Paternity Exclusion

The information content of a variable genetic marker is directly proportional to its level of variability (number of alleles) and inversely proportional to the variance in the frequencies of those alleles. For example, a locus that has five alleles of equal frequency is much more informative than a locus with two alleles, one of which is found at a frequency of 90%. This is because the probability that two individuals share an allele by chance alone is higher for the second locus (both individuals are highly likely to share the common allele).

Rare alleles can be very useful in tracing paternity or other relationships, since they are unique characters that are likely to be common to related individuals because of recent shared inheritance. To study many individuals, however, one must develop a suite of polymorphic markers that is likely to produce unique patterns in most or all cases (i.e., the probability of identity, Chakraborty et al. 1999). For paternity, this means assembling a set of markers that, on average, have a very low probability (usually less than 0.1%) of producing any particular pattern of alleles. In other words, two individuals chosen at random will share the haploid set of alleles (that half of the genome inherited from one parent, in the example of parent-offspring pairs) by chance less than one time in 1,000 comparisons (when $P < .1\%$). This is called the exclusion probability (E_H), and is expressed as the probability of excluding a random male who is not the father. Paternity can be considered excluded when E_H is very high (e.g., greater than 0.999).

For a given set of genetic loci, the exclusion probability can be calculated from the allele frequencies in the population (Chakravarti & Li 1983; reviewed in Morin & Woodruff 1992, Morin, Moore, Wallis et al. 1994). Estimates of E_H assume the population is in Hardy-Weinberg equilibrium. E_H is also strongly influenced by allele frequencies, so inaccurate estimates of the allele frequencies in the population will affect the estimates of exclusion probability for a given locus in the population. Nevertheless, if these conditions are satisfied, the individual probability of exclusion of an offspring can be calculated from its genotype and the frequency of alleles in the population (Chakraborty et al. 1988; reviewed in Morin & Woodruff 1992, Morin, Moore, Wallis et al. 1994).

As discussed above, different types of markers have different information content. Microsatellites have proven very useful for paternity studies because they are often highly variable, are present in most species, and can be genotyped without extremely sophisticated equipment. SNPs are much less variable, but are much more common in all genomes, and

Table 2.3. Applicability of Marker Types to Kinship Studies[a]

	mtDNA Sequencing	Multilocus Minisatellite	RAPD	AFLP	Protein (Allozyme) Electrophoresis	RFLP (Nuclear Locus)	Microsatellites	SNPs[b] (SBE with Electrophoresis)
Paternity (known mother)	N/A	Good–moderate	Good–moderate	Good–moderate	Moderate–low (need many loci)	Moderate–low (need many loci)	Good	Good–moderate (need many loci)
Siblings	Moderate	Moderate	Low	Low	Low	Low	Good–moderate	Good–moderate
Higher order relationship	Low	Moderate	Low	Low	Low	Moderate–low	Good–moderate	Low
Population assignment test	Low	Low	Low	Low	Moderate–low	Moderate–low	Good–moderate	Good–moderate
Population structure	Moderate	Low	Moderate	Moderate	Moderate–low	Moderate	Good–moderate	Good
Phylogeography	Good–moderate	Low	Low	Low	Good–moderate	Moderate	Low–moderate	Good–moderate

[a]For loci with few alleles/locus, the number of loci needed is higher. For allozymes and nuclear RFLPs, it is difficult to obtain a large enough number of loci, even though the resolving power is similar to that of SNPs, for which the possible number of loci is very high.

[b]The applicability is estimated by extrapolation from other codominant loci with few alleles, based on availability of large numbers of loci and projected ease of genotyping.

could potentially be genotyped more reliably and inexpensively (though the equipment may be more sophisticated and expensive for some methods).

The relative lack of information content in SNPs can be compensated for by adding more loci. In an analysis of the power of exclusion for a set of 13 standard human microsatellites used for forensics cases, Chakraborty et al. (1999) compared the power of those microsatellites to a hypothetical set of biallelic SNPs to see how many SNPs would be needed to obtain the same exclusion probability. Given a range of allele frequencies from 0.3 to 0.5 (for the less frequent allele), they concluded that 30 to 60 SNP loci would produce the same exclusion probability as these 13 microsatellites ($E_H > 0.99$). Similar conclusions were drawn by Krawczak (1999).

Paternity Inclusion

Exclusion should be used when the E_H is very high and all but one potential father can be excluded (Marshall et al. 1998). In many (but not all, e.g., Vigilant et al. 2001) field studies, this is not the case. When paternity exclusion is not possible in all cases, methods for paternity inclusion can be used.

Maximum likelihood (ML)–based methods for paternity inclusion allow the user to evaluate the data for the most likely father out of several candidates. These methods take into account, in addition to allele frequencies, the estimated genotyping error rates, the presence of null alleles, and the likely portion of potential sires not sampled. Maximum likelihood estimates calculate a statistical probability of paternity and evaluation of the next most likely candidate. This can be particularly important when potential sires are siblings, or when the information content of the markers, or the particular combination of alleles, leads to low resolving power between likely sires.

Because maximum likelihood methods depend on algorithms for fitting data to hypotheses and then comparing the fit, they require computer methods. Several software packages have been developed for paternity analysis, with a variety of options for data quality checking and statistical evaluation (e.g., Marshall et al. 1998, Goodnight & Queller 1999). These programs also typically calculate the average paternity exclusion probability for the loci screened on a given mother-offspring pair. The programs currently available have previously been reviewed in the context of statistical analysis of microsatellite data (Luikart & England 1999).

Analytical methods for estimating parentage are improving steadily. A new program called PAPA (Duchesne et al., 2002), for example, is available for parental allocation by likelihood methods when neither parent is known. This program also allows the user to simulate parental assignment given some knowledge of the allele frequencies in the population, and thus to determine whether the loci will provide sufficient power for parental allocation. This may be helpful for assessment of loci early in a study, thereby allowing researchers to find additional loci if needed.

Determining Other Classes of Relatedness

General patterns of relatedness of individuals in populations have been used to assess social structure, behavior, dispersal, and population structure (e.g., Morin, Moore, Chakraborty et al. 1994, Gagneux et al. 1999, Pope 2000, Constable et al. 2001, Vigilant et al. 2001).

Genetic markers have also been used to infer specific levels of relatedness, when actual genealogies are not known (reviewed in Queller et al. 1993). In practice, this involves determining the level of shared alleles in "unrelated" individuals of the population and in individuals of known relationship (e.g., parent-offspring, full siblings, half siblings), and then inferring the relationships of unknown pairs of individuals. This is a rapidly evolving field, and at least five different methods of inferring relationships from molecular data have been proposed (Queller & Goodnight 1989, Li et al. 1993, Blouin et al. 1996, Ritland 1996, Lynch & Ritland 1999; compared in Van De Casteele et al. 2001).

To date, however, such methods applied to primates have only been able to distinguish close (i.e., first-degree) relatives from other classes of relatives (Bruford & Altmann 1993, Altmann et al. 1996, Pope 1990, Gerloff et al. 1999, Vigilant et al. 2001). No genetic study of primate kinship has yet, in the absence of external data, been able to resolve relatedness on a finer scale (e.g., half versus full sibs, first versus second cousins). The molecular and analytical methods are progressing, however, such that the number of markers required for higher level relatedness inference will be reasonable, and the analytical methods refined for reasonable certainty of determining classes of relationships from samples within a group or population (Van De Casteele et al. 2001).

Applications of Molecular Genetic Methods to Determining Primate Kinship

The "holy grail" for genetic studies of primate kinship would be a complete account of genealogy for all animals in the social unit, extending back to its founding members. Such information would allow hypotheses about the influences of specific degrees of kinship on social behavior to be tested without genealogical error. Extension of such a genealogy across time and space would similarly inform studies of intraspecific differentiation, phylogeography, and population-level social organization.

In theory, a complete and accurate genealogy of living individuals could be reconstructed "blind" from genetic data alone. To date, however, this has never been done for any primate group. The number of variable, independently assorting genetic loci that would be required to achieve such a degree of precision would be impracticably large (Lynch 1988, Queller et al. 1993, Pemberton et al. 1999). Moreover, in all but captive and the most carefully studied wild primate groups, the required complete sampling of individuals would be difficult.

Fortunately, genealogy need not normally be reconstructed in the absence of prior information. Mother-offspring relationships can often be identified on the basis of behavioral observation alone (but see Smith et al. 1999). Behavioral observation can thus generate a preliminary matrilineal genealogy that can be resolved further using genetic data. Father-offspring relationships are less obvious. Especially in species with promiscuous mating systems, paternity is the one genealogical relationship for which behavioral inference should be considered unreliable a priori. Not surprisingly, therefore, the emphasis of DNA studies to date has been on resolving paternity.

Early primate paternity studies tended to focus on macaques (*Macaca* spp.) because of their ubiquity in captive settings. Also, largely as a result of work by Inoue and colleagues (e.g., Inoue et al. 1990, 1992), "DNA fingerprinting" methods (Southern blotting with minisatellite probes; Jeffreys et al. 1985, Jeffreys 1987) were available relatively early for macaques.

Many of the first such studies focused on the relationship between male dominance rank and reproductive success. The results of these studies were variable, sometimes showing strong positive associations between male rank and reproductive success and sometimes showing no association at all (see discussions in Turner et al. 1992, Bauers & Hearn 1994; also see Inoue & Takenaka 1993 and accompanying articles in that issue of *Primates*). Captivity seems insufficient for explaining this variability, since similar results were obtained in semicaptive (von Segesser et al. 1995, Bercovitch & Nurnberg 1997) and wild settings (de Ruiter et al. 1992, 1994; Keane et al. 1997). Positive associations were found most frequently when males maintained high-rank positions for sustained periods of time.

Traditional DNA fingerprinting technologies, as well as extensions of these using PCR and microsatellite loci, have since been applied to the elucidation of paternity in a variety of other species, including other Old World monkeys (Wickings & Dixson 1992, Bruford & Altmann 1993, Dixson et al. 1993, Altmann et al. 1996, Borries et al. 1999, Smith et al. 1999), New World monkeys (Dixson et al. 1992, Ellsworth & Hoelzer 1998, Pope 2000), and lemurs (Turner et al. 1992, Jekielek & Strobeck 1999, Fietz et al. 2000). Techniques for the noninvasive isolation of DNA from such sources as hair, feces, wadges, and chewed fruit have opened the door to studies of paternity in wild apes, which are not amenable to capture (Takasaki & Takenaka 1991, Morin & Woodruff 1992, Morin, Moore, Wallis et al. 1994, Gerloff et al. 1995, Field et al. 1998, Gerloff et al. 1999, Vigilant 1999, Bradley et al. 2000, Immel et al. 2000, Constable et al. 2001, Oka & Takenaka 2001, Vigilant et al. 2001).

Noninvasive paternity studies to date have been somewhat sobering, however. Morin, Moore, Wallis et al. (1994) investigated paternity in chimpanzees (*Pan troglodytes schweinfurthii*) in Gombe, Tanzania. Using PCR-amplified DNA from shed hair and a panel of eight microsatellite loci, they were able to assign paternity in 2 out of 10 cases and, in conjunction with behavioral data, to infer paternity in another 2 cases. Gagneux and colleagues (Gagneux, Woodruff et al. 1997, Gagneux et al. 1999) examined paternity in western chimpanzees (*P. t. verus*) from the Taï forest, Ivory Coast, for 13 infants of known maternity, using DNA from hair and chewed fruit and 11 PCR-amplified microsatellite loci. In seven cases, all community males could be excluded as fathers, indicating a surprising degree of extragroup paternity.

These results have been called into question, however, through analysis of additional and different loci, use of different analytical methods, and collection of new samples. Constable et al. (2001), using a likelihood-ratio approach, assigned paternity to three offspring in Gombe for which these fathers had previously been excluded. Vigilant et al. (2001), using new genetic data for an expanded study of the Taï chimps, were able to assign within-group paternity to four out of seven infants for which extragroup paternity was previously suspected. The phenomenon of allelic dropout, and subsequent false scoring of homozygosity at certain loci, apparently accounts for many of the initial erroneous paternity exclusions. The emerging consensus is that high replication and genotype verification measures to protect against contamination and false genotypes must be taken when primate kinship is inferred from DNA collected noninvasively (Taberlet et al. 1996, Morin et al. 2001).

Noninvasive paternity determinations in wild gorillas (Field et al. 1998), gibbons (*Hylobates muelleri*) (Oka & Takenaka 2001), and hanuman langurs (*Presbytis entellus*) (Borries et al. 1999) have failed thus far to document any significant discordance between social system and mating system. As longer term studies with larger sample sizes are conducted, significant differences between social systems and mating systems may become evident. It will, therefore, be doubly important for these studies to use common genetic markers and genotyping methods that facilitate the comparison and combination of data sets, and to analyze these data using methods that allow for genotyping errors.

The use of genetic technology to elucidate kinship has not been limited to investigating paternity. For example, mitochondrial DNA (because of its unique mode of exclusively maternal inheritance) has been used to exclude and include matrilineality in lion-tailed macaques (*M. silenus*) (Morin & Ryder 1991), common chimpanzees (Goldberg & Wrangham 1997, Mitani et al. 2000), and bonobos (Hashimoto et al. 1996). These studies have shown that social preferences are largely independent of matrilineality in male chimpanzees and female bonobos respectively. Mitochondrial DNA can also be used to reconstruct matrilineality among females in female philopatric primates, when such relationships have not been documented behaviorally (Pope 2000).

To date, few primate studies have used genetic markers to examine kinship relationships other than paternity and matrilineality. Dixson et al. (1992), for example, confirmed the genetic identity of twins in wild common marmosets (*Callithrix jacchus jacchus*) using DNA fingerprinting. Morin, Moore, Chakraborty et al. (1994) and Vigilant et al. (2001) have used allele-sharing methods to examine levels of relatedness among groups of individuals within and between chimpanzee communities. De Ruiter and Geffen (1998) used blood protein markers to infer relatedness among long-tailed macaques (*M. fascicularis*) for which external pedigree data were available. No behavioral studies to date have used Y chromosome data to determine patrilineal relatedness, although this would be feasible.

Genetic methods are well suited to the reconstruction of all classes of kinship, including those that span temporal and spatial scales great enough that their study would more conventionally be considered phylogeography. With the advent of technologies that can examine large numbers of variable loci simultaneously, "blind" genealogy reconstruction may become possible in the future. In the interim, however, the best studies are those that test specific hypotheses about specific degrees of kinship, and that choose genetic markers appropriate to the question at hand.

Conclusion

Behavioral studies tell us a great deal about primate societies, but actual kinship relationships are often difficult or impossible to infer without genetic information. Within the last two decades, technical and analytical advances in genealogy and relatedness detection methods have made it possible for any organism or population to be studied, and for kinship to be resolved at most levels.

Nevertheless, the techniques available today do not have the resolving power to reconstruct kinship reliably in the absence of external information. Genetic markers for study must therefore be chosen wisely, with consideration of their technical limitations and their ability to resolve kinship at the level in question. The best studies are those that define a specific kinship-related hypothesis and choose a marker system suitable for testing it.

Acknowledgments We thank Amy Roeder, Linda Vigilant, Bernard Chapais, Carol Berman, and one anonymous reviewer for helpful critiques of the manuscript, and Steve Smith for generating the SNP genotype figure. Pierre Taberlet and Oxford University Press granted permission to reproduce the allelic dropout gel image, and Michael Webster and Birkhäuser Verlag granted permission to reproduce and modify table 2.2.

References

Altmann, J., Alberts, S. C., Haines, S. A., Dubach, J., Muruthi, P., Coote, T., Geffen, E., Cheesman, D. J., Mututua, R. S., Saiyalel, S. N., Wayne, R. K., Lacy, R. C., & Bruford, M. W. 1996. Behavior predicts genetic structure in a wild primate group. *Proc. Natl. Acad. Sci. USA*, 93, 5797–5801.

Amato, G., Gatesy, J., & Brazaitis, P. 1998. PCR assays of variable nucleotide sites for identification of conservation units: an example from *Caiman*. In: *Molecular Approaches to Ecology and Evolution* (Ed. by R. DeSalle & B. Schierwater), pp. 177–190. Boston: Birkhäuser.

Avise, J. C. 1989. Gene trees and organismal histories: a phylogenetic approach to population biology. *Evolution*, 43, 1192–1208.

Avise, J. C., Arnold, J., Ball, R. M., Bermingham, E., Lamb, T., Neigel, J. E., Reeb, C. A., & Saunders, N. C. 1987. Intraspecific phylogeography: the mitochondrial DNA bridge between population genetics and systematics. *Annu. Rev. Ecol. Syst.*, 18, 489–522.

Balloux, F. & Lugon-Moulin, N. 2002. The estimation of population differentiation with microsatellite markers. *Mol. Ecol.*, 11, 155–165.

Bauers, K. A. & Hearn, J. P. 1994. Patterns of paternity in relation to male social rank in the stumptailed macaque, *Macaca arctoides*. *Behaviour*, 129, 149–176.

Bensasson, D., Zhang, D. X., Hartl, D. L., & Hewitt, G. M. 2001. Mitochondrial pseudogenes: evolution's misplaced witnesses. *Trends Ecol. Evol.*, 16, 314–321.

Bercovitch, F. B. & Nurnberg, P. 1997. Genetic determination of paternity and variation in male reproductive success in two populations of rhesus macaques. *Electrophoresis*, 18, 1701–1705.

Blouin, M. S., Parsons, M., Lacaille, V., & Lotz, S. 1996. Use of microsatellite loci to classify individuals by relatedness. *Mol. Ecol.*, 5, 393–401.

Borries, C., Launhardt, K., Epplen, C., Epplen, J. T., & Winkler, P. 1999. Males as infant protectors in Hanuman langurs (*Presbytis entellus*) living in multimale groups: defense pattern, paternity and sexual behaviour. *Behav. Ecol. Sociobiol.*, 46, 350–356.

Bradley, B. J., Boesch, C., & Vigilant, L. 2000. Identification and redesign of human microsatellite markers for genotyping wild chimpanzee (*Pan troglodytes verus*) and gorilla (*Gorilla gorilla gorilla*) DNA from faeces. *Conserv. Genet.*, 3, 289–292.

Bradley, B. J., Chambers, K. E., & Vigilant, L. 2001. Accurate DNA-based sex identification of apes using noninvasive samples. *Conserv. Genet.*, 2, 179–181.

Brookfield, J. F. Y. 1996. A simple new method for estimating null allele frequency from heterozygote deficiency. *Mol. Ecol.*, 5, 453–455.

Bruford, M. W. & Altmann, J. 1993. DNA fingerprinting and the problems of paternity determination in an inbred captive population of Guinea baboons (*Papio hamadryas papio*). *Primates*, 34, 403–411.

Bruillette, J., Andrew, J., & Venta, P. 2000. Estimate of nucleotide diversity in dogs with a pool-and-sequence method. *Mamm. Genome*, 11, 1079–1086.

Burckhardt, F., von Haeseler, A., & Meyer, S. 1999. HvrBase: compilation of mtDNA control region sequences for primates. *Nucleic Acids Res.*, 27, 138–142.

Caetano-Anolles, G. 1998. Arbitrary oligonucleotides: primers for amplification and direct identification of nucleic acids, genes and organisms. In: *Molecular Approaches to Ecol-*

ogy and Evolution (Ed. by R. DeSalle & B. Schierwater), pp. 107–123. Berlin: Birkhäuser Verlag.

Callen, D. F., Thompson, A. D., Shen, Y., Phillips, H. A., Richards, R. I., Mulley, J. C., & Sutherland, G. R. 1993. Incidence and origin of "null" alleles in the $(AC)_n$ microsatellite markers. *Am. J. Hum. Genet.*, 52, 922–927.

Chakraborty, R., Meagher, T. R., & Smouse, P. E. 1988. Parentage analysis with genetic markers in natural populations. I. The expected proportion of offspring with unambiguous paternity. *Genetics*, 118, 527–536.

Chakraborty, R., Stivers, D. N., Su, B., Zhong, Y., & Budowle, B. 1999. The utility of short tandem repeat loci beyond human identification: implications for development of new DNA typing systems. *Electrophoresis*, 20, 1682–1696.

Chakravarti, A. & Li, C. C. 1983. The effect of linkage on paternity calculations. In: *Inclusion Probabilities in Parentage Testing* (Ed. by H. F. Polesky), pp. 411–420. Arlington, VA: American Association of Blood Banks.

Chen, J., Iannone, M. A., Li, M. S., Taylor, J. D., Rivers, P., Nelsen, A. J., Slentz-Kesler, K. A., Roses, A., & Weiner, M. P. 2000. A microsphere-based assay for multiplexed single nucleotide polymorphism analysis using single base chain extension. *Genome Res.*, 10, 549–557.

Clifford, S. L., Jeffrey, K., Bruford, M. W., & Wickings, E. J. 1999. Identification of polymorphic microsatellite loci in the gorilla (*Gorilla gorilla gorilla*) using human primers: application to noninvasively collected hair samples. *Mol. Ecol.*, 8, 1556–1558.

Constable, J., Ashley, M. V., Goodall, J., & Pusey, A. E. 2001. Noninvasive paternity assignment in the Gombe chimpanzees. *Mol. Ecol.*, 10, 1279–1300.

Coote, T. & Bruford, M. W. 1996. Human microsatellites applicable for analysis of genetic variation in apes and Old World monkeys. *J. Heredity*, 87, 406–410.

David, V. A. & Menotti-Raymond, M. 1998. Automated DNA detection with fluorescence-based technologies. In: *Molecular Genetic Analysis of Populations: A Practical Approach* (Ed. by A. R. Hoelzel), pp. 337–370. New York: Oxford University Press.

Dean, M. & Milligan, B. G. 1998. Detection of genetic variation by DNA conformational and denaturing gradient methods. In: *Molecular Genetic Analysis of Populations: A Practical Approach* (Ed. by A. R. Hoelzel), pp. 263–286. New York: Oxford University Press.

de Ruiter, J. R. & Geffen, E. 1998. Relatedness of matrilines, dispersing males and social groups in long-tailed macaques (*Macaca fascicularis*). *Proc. Royal Soc. Lond., B*, 265, 79–87.

de Ruiter, J. R., Scheffrahn, W., Trommelen, G. J. J. M., Uitterlinden, A. G., Martin, R. D., & van Hooff, J. A. R. A. M. 1992. Male social rank and reproductive success in wild longtailed macaques: paternity exclusions by blood protein analysis and DNA fingerprinting. In: *Paternity in Primates: Genetic Tests and Theories* (Ed. by R. D. Martin, A. F. Dixson, & E. J. Wickings), pp. 175–191. Basel: Karger.

de Ruiter, J. R., Van Hooff, J. A. R. A. M., & Scheffrahn, W. 1994. Social and genetic aspects of paternity in wild long-tailed macaques (*Macaca fascicularis*). *Behaviour*, 129, 203–224.

DeSalle, R. & Schierwater, B. 1998. *Molecular Approaches to Ecology and Evolution.* Boston: Birkhäuser.

Dib, C., Faure, S., Fizames, C., Samson, D., Drouot, N., Vignal, A., Millasseau, P., Marc, S., Hazan, J., Seboun, E., Lathrop, M., Gyapay, G., Morissette, J., & Weissenbach, J. 1996. A comprehensive genetic map of the human genome based on 5264 microsatellites. *Nature*, 380, 152–154.

Dixson, A. F., Anzenberger, G., Monterio Da Cruz, M. A. O., Patel, I., & Jeffreys, A. J.

1992. DNA fingerprinting of free-ranging groups of common marmosets (*Callithrix jacchus jacchus*) in NE Brazil. In: *Paternity in Primates: Genetic Tests and Theories* (Ed. by R. D. Martin, A. F. Dixson, & E. J. Wickings), pp. 192–202. Basel: Karger.

Dixson, A. F., Bossi, T., & Wickings, E. J. 1993. Male dominance and genetically determined reproductive success in the mandrill (*Mandrillus sphinx*). *Primates*, 34, 525–532.

Duchesne, P., Godbout, M. H., & Bernatchez, L. 2002. PAPA (Package for the Analysis of Parental Allocation): a computer program for simulated and real parental allocation. *Mol Ecol Notes*, 2, 191.

Edwards, A., Civitello, A., Hammond, H. A., & Caskey, C. T. 1991. DNA typing and genetic mapping with trimeric and tetrameric tandem repeats. *Am. J. Hum. Genet.*, 49, 746–756.

Ellsworth, J. A. & Hoelzer, G. A. 1998. Characterization of microsatellite loci in a New World primate, the mantled howler monkey (*Alouatta palliata*). *Mol. Ecol.*, 7, 657–658.

Ely, J., Campbell, J. L., Gonzalez, D. L., & Stone, W. H. 1996. Successful amplification of PCR-amplified DNA markers for paternity determination in rhesus monkeys (*Macaca mulatta*) and chimpanzees (*Pan troglodytes*). *Lab. Primate Newsl.*, 35, 1–4.

Field, D., Chemnick, L., Robbins, M., Garner, K., & Ryder, O. 1998. Paternity determination in captive lowland gorillas and orangutans and wild mountain gorillas by microsatellite analysis. *Primates*, 39, 199–209.

Fietz, J., Zischler, H., Schwiegk, C., Tomiuk, J., Dausmann, K. H., & Ganzhorn, J. U. 2000. High rates of extra-pair young in the pair-living fat-tailed dwarf lemur, *Cheirogaleus medius*. *Behav. Ecol. Sociobiol.*, 49, 8–17.

Gagneux, P., Boesch, C., & Woodruff, D. S. 1997. Microsatellite scoring errors associated with noninvasive genotyping based on nuclear DNA amplified from shed hair. *Mol. Ecol.*, 6, 861–868.

Gagneux, P., Boesch, C., & Woodruff, D. S. 1999. Female reproductive strategies, paternity and community structure in wild West African chimpanzees. *Anim. Behav.*, 57, 19–32.

Gagneux, P., Woodruff, D. S., & Boesch, C. 1997. Furtive mating in female chimpanzees. *Nature*, 387, 358–359.

Gerloff, U., Hartung, B., Fruth, B., Hohmann, G., & Tautz, D. 1999. Intracommunity relationships, dispersal pattern and paternity success in a wild living community of Bonobos (*Pan paniscus*) determined from DNA analysis of faecal samples. *Proc. Royal Soc. Lond., B*, 266, 1189–1195.

Gerloff, U., Schlötterer, C., Rassmann, K., Rambold, I., Hohmann, G., Fruth, B., & Tautz, D. 1995. Amplification of hypervariable simple sequence repeats (microsatellites) from excremental DNA of wild living Bonobos (*Pan paniscus*). *Mol. Ecol.*, 4, 515–518.

Goldberg, T. & Wrangham, R. 1997. Genetic correlates of social behaviour in wild chimpanzees: evidence from mitochondrial DNA. *Anim. Behav.*, 54, 559–570.

Goodnight, K. F. & Queller, D. C. 1999. Computer software for performing likelihood tests of pedigree relationship using genetic markers. *Mol. Ecol.*, 8, 1231–1234.

Goossens, B., Latour, S., Vidal, C., Jamart, A., Ancrenaz, M., & Bruford, M. W. 2000. Twenty new microsatellite loci for use with hair and faecal samples in the chimpanzee (*Pan troglodytes troglodytes*). *Folia Primatol.*, 71, 177–180.

Goossens, B., Waits, L. P., & Taberlet, P. 1998. Plucked hair samples as a source of DNA: reliability of dinucleotide microsatellite genotyping. *Mol. Ecol.*, 7, 1237–1241.

Goossens, B. G., Chikhi, L., Utami, S. S., de Ruiter, J. R., & Bruford, M. W. 2000. Multiple-samples and multiple-extracts approach for microsatellite analysis of faecal samples in an arboreal ape. *Cons. Genet.*, 1, 157–162.

Greenwood, A. & Pääbo, S. 1999. Nuclear insertion sequences of mitochondrial DNA predominate in hair but not in blood of elephants. *Mol. Ecol.*, 8, 133–137.

Griffiths, R. & Tiwari, B. 1993. Primers for the differential amplification of the sex-determining region Y gene in a range of mammal species. *Mol. Ecol.*, 2, 405–406.

Gut, I. G. 2001. Automation in genotyping of single nucleotide polymorphisms. *Hum. Mutat.*, 17, 475–492.

Hacia, J. G., Fan, J.-B., Ryder, O. A., Jin, L., Edgemon, K., Ghandour, G., Mayer, R. A., Sun, B., Hsie, L., Robbins, C. M., Brody, L. C., Wang, D., Lander, E. S., Lipshutz, R., Fodor, S. P. A., & Collins, F. S. 1999. Determination of ancestral alleles for human single-nucleotide polymorphisms using high-density oligonucleotide arrays. *Nat. Genet.*, 22, 164–167.

Hamilton, D. W. 1964. The genetical evolution of social behaviour 1 & 2. *J. Theor. Biol.*, 7, 1–52.

Hashimoto, C., Furuichi, T., & Takenaka, O. 1996. Matrilineal kin relationship and social behavior of wild bonobos: sequencing the D-loop region of mitochondrial DNA. *Primates*, 37, 305–318.

Higuchi, R., von Beroldingen, C. H., Sensabaugh, G. F., & Erlich, H. A. 1988. DNA typing from single hairs. *Nature*, 332, 543–546.

Hirschhorn, J. N., Sklar, P., Lindblad-Toh, K., Lim, Y.-M., Ruiz-Gutierrez, M., Bolk, S., Langhorst, B., Schaffner, S., Winchester, E., & Lander, E. S. 2000. SBE-TAGS: an array-based method for efficient single-nucleotide polymorphism genotyping. *Proc. Natl. Acad. Sci USA*, 97, 12164–12169.

Hoelzel, A. R. 1998. *Molecular Genetic Analysis of Populations: A Practical Approach.* Oxford: Oxford University Press.

Hoelzel, A. R. & Green, A. 1998. PCR protocols and population analysis by direct DNA sequencing and PCR-based DNA fingerprinting. In: *Molecular Genetic Analysis of Populations* (Ed. by A. R. Hoelzel), pp. 201–236. Oxford: Oxford University Press.

Immel, U. D., Hummel, S., & Herrmann, B. 2000. Reconstruction of kinship by fecal DNA analysis of orangutans. *Anthropol. Anz.*, 58, 63–67.

Inoue, M., Mitsunaga, F., Ohsawa, H., Takenaka, A., Sugiyama, Y., Soumah, A. G., & Takenaka, O. 1992. Paternity testing in captive Japanese macaques (*Macaca fuscata*) using DNA fingerprinting. In: *Paternity in Primates: Genetic Tests and Theories* (Ed. by R. D. Martin, A. F. Dixson, & E. J. Wickings), pp. 131–140. Basel: Karger.

Inoue, M., Takenaka, A., Tanaka, S., Kominami, R., & Takenaka, O. 1990. Paternity discrimination in a Japanese macaque troop by DNA fingerprinting. *Primates*, 31, 563–570.

Inoue, M. & Takenaka, O. 1993. Japanese macaque microsatellite PCR primers for paternity testing. *Primates*, 34, 37–45.

Jarne, P. & Lagoda, P. J. L. 1996. Microsatellites, from molecules to populations and back. *Trends Ecol. Evol.*, 11, 424–429.

Jeffreys, A. J. 1987. Highly variable minisatellites and DNA fingerprints. *Biochem. Soc. Trans.*, 15, 309–317.

Jeffreys, A. J., Turner, M., & Debenham, P. 1991. The efficiency of multilocus DNA fingerprint probes for individualization and establishment of family relationships determined from extensive casework. *Am. J. Hum. Genet.*, 48, 824–840.

Jeffreys, A. J., Wilson, V., & Thein, S. L. 1985. Hypervariable "minisatellite" regions in human DNA. *Nature*, 314, 67–73.

Jekielek, J. & Strobeck, C. 1999. Characterization of polymorphic brown lemur (*Eulemur fulvus*) microsatellite loci and their amplification in the family Lemuridae. *Mol. Ecol.*, 8, 901–903.

Kayser, M., Ritter, H., Bercovitch, F., Mrug, M., Roewer, L., & Nurnberg, P. 1996. Identification of highly polymorphic microsatellites in the rhesus macaque *Macaca mulatta* by cross-species amplification. *Mol. Ecol.*, 5, 157–159.

Keane, B., Dittus, W. P., & Melnick, D. J. 1997. Paternity assessment in wild groups of toque macaques *Macaca sinica* at Polonnaruwa, Sri Lanka using molecular markers. *Mol. Ecol.*, 6, 267–282.

Kocher, T. D., Thomas, W. K., Meyer, A., Edwards, S. V., Pääbo, S., Villablanca, F. X., & Wilson, A. C. 1989. Dynamics of mitochondrial DNA evolution in animals: amplification and sequencing with conserved primers. *Proc. Natl. Acad. Sci. USA*, 86, 6196–6200.

Krawczak, M. 1999. Informativity assessment for biallelic single nucleotide polymorphisms. *Electrophoresis*, 20, 1676–1681.

Kruglyak, L. 1997. The use of a genetic map of biallelic markers in linkage studies. *Nat. Genet.*, 17, 21–24.

Kwok, P.-Y. 2000. High-throughput genotyping assay approaches. *Pharmacogenomics*, 1, 95–100.

Lanave, C., Attimonelli, M., De Robertis, M., Licciulli, F., Liuni, S., Sbisa, E., & Saccone, C. 1999. Update of AMmtDB: a database of multi-aligned metazoa mitochondrial DNA sequences. *Nucleic Acids Res.*, 27, 134–137.

Lander, E. S. 1989. DNA fingerprinting on trial. *Nature*, 339, 501–505.

Launhardt, K., Epplen, C., Epplen, J. T., & Winkler, P. 1998. Amplification of microsatellites adapted from human systems in faecal DNA of wild hanuman langurs (*Presbytis entellus*). *Electrophoresis*, 19, 1356–1361.

Li, C. C., Weeks, D. E., & Chakravarti, A. 1993. Similarity of DNA fingerprints due to chance and relatedness. *Hum. Hered.*, 43, 45–52.

Litt, M. & Luty, J. A. 1989. A hypervariable microsatellite revealed by in vitro amplification of a dinucleotide repeat within the cardiac muscle actin gene. *Am. J. Hum. Genet.*, 44, 397–401.

Luikart, G. & England, P. R. 1999. Statistical analysis of microsatellite DNA data. *Trends Ecol. Evol.*, 14, 253–256.

Lynch, M. 1988. Estimation of relatedness by DNA fingerprinting. *Mol. Biol. Evol.*, 5, 584–599.

Lynch, M. & Ritland, K. 1999. Estimation of pairwise relatedness with molecular markers. *Genetics*, 152, 1753–1766.

Lyons, L. A., Laughlin, T. F., Copeland, N. G., Jenkins, N. A., Womack, J. E., & O'Brien, S. J. 1997. Comparative anchor tagged sequences (CATS) for integrative mapping of mammalian genomes. *Nat. Genet.*, 15, 47–56.

Marshall, E. 1997. The hunting of the SNP. *Science*, 278, 2047.

Marshall, T. C., Slate, J., Kruuk, L. E. B., & Pemberton, J. 1998. Statistical confidence for likelihood-based paternity inference in natural populations. *Mol. Ecol.*, 7, 639–655.

Martin, R. D., Dixson, A. F., & Wickings, E. J. 1992. *Paternity in Primates: Genetic Tests and Theories*. Basel: Karger.

Mhlanga, M. M. & Malmberg, L. 2001. Using molecular beacons to detect single-nucleotide polymorphisms with real-time PCR. *Methods*, 25, 463–471.

Mitani, J., Merriwether, D. A., & Zhang, C. 2000. Male affiliation, cooperation, and kinship in wild chimpanzees. *Anim. Behav.*, 59, 885–893.

Moore, S. S., Sargeant, L. L., King, T. J., Mattick, J. S., Georges, M., & Hetzel, D. J. S. 1991. The conservation of dinucleotide microsatellites among mammalian genomes allows the use of heterologous PCR primer pairs in closely related species. *Genomics*, 10, 654–660.

Morin, P. A., Chambers, K. E., Boesch, C., & Vigilant, L. 2001. Quantitative PCR analysis of DNA from noninvasive samples for accurate microsatellite genotyping of wild chimpanzees (*Pan troglodytes*). *Mol. Ecol.*, 10, 1835–1844.

Morin, P. A., Kanthaswamy, S., & Smith, D. G. 1997. Simple sequence repeat (SSR) polymorphisms for colony management and population genetics in Rhesus macaques (*Macaca mulatta*). *Am. J. Primatol.*, 42, 199–213.

Morin, P. A., Mahboubi, P., Wedel, S., & Rogers, J. 1998. Rapid screening and comparison of human microsatellite markers in baboons: allele size is conserved, but allele number is not. *Genomics*, 53, 12–20.

Morin, P. A., Moore, J. J., Chakraborty, R., Jin, L., Goodall, J., & Woodruff, D. S. 1994. Kin selection, social structure, gene flow, and the evolution of chimpanzees. *Science*, 265, 1193–1201.

Morin, P. A., Moore, J. J., Wallis, J., & Woodruff, D. S. 1994. Paternity exclusion in a community of wild chimpanzees using hypervariable simple sequence repeats. *Mol. Ecol.*, 3, 469–478.

Morin, P. A., Moore, J. J., & Woodruff, D. S. 1992. Identification of chimpanzee subspecies with DNA from hair and allele specific probes. *Proc. Royal Soc. Lond., B*, 249, 293–297.

Morin, P. A. & Ryder, O. A. 1991. Founder contribution and pedigree inference in a captive breeding colony of lion-tailed macaques, using mitochondrial DNA and DNA fingerprint analyses. *Zoo Biol.*, 10, 341–352.

Morin, P. A., Saiz, R., & Monjazeb, A. 1999. High-throughput single nucleotide polymorphism genotyping by fluorescent 5′ exonuclease assay. *BioTechniques*, 27, 538–540, 542, 544.

Morin, P. A., Wallis, J., Moore, J. J., Chakraborty, R., & Woodruff, D. S. 1993. Noninvasive sampling and DNA amplification for paternity exclusion, community structure, and phylogeography in wild chimpanzees. *Primates*, 34, 347–356.

Morin, P. A. & Woodruff, D. S. 1992. Paternity exclusion using multiple hypervariable microsatellite loci amplified from nuclear DNA of hair cells. In: *Paternity in Primates: Genetic Tests and Theories* (Ed. by R. D. Martin, A. F. Dixson, & E. J. Wickings), pp. 63–81. Basel: Karger.

Morin, P. A. & Woodruff, D. S. 1996. Noninvasive sampling for vertebrate conservation. In: *Molecular Approaches in Conservation* (Ed. by R. Wayne & T. Smith), pp. 298–313. Oxford: Oxford University Press.

Murphy, W. J., Eizirik, E., Johnson, W. E., Zhang, Y. P., Ryder, O. A., & O'Brien, S. J. 2001. Molecular phylogenetics and the origins of placental mammals. *Nature*, 409, 614–618.

Navidi, W., Arnheim, N., & Waterman, M. S. 1992. A multiple-tubes approach for accurate genotyping of very small DNA samples by using PCR: statistical considerations. *Am. J. Hum. Genet.*, 50, 347–359.

Oka, T. & Takenaka, O. 2001. Wild gibbon's parentage tested by noninvasive DNA sampling and PCR-amplified polymorphic microsatellites. *Primates*, 42, 67–73.

Palumbi, S. R. & Baker, C. S. 1994. Contrasting population structure from nuclear intron sequences and mtDNA of humpback whales. *Mol. Biol. Evol.*, 11, 426–435.

Pastinen, T., Kurg, A., Metspalu, A., Peltonen, L., & Syvänen, A.-C. 1997. Minisequencing: a specific tool for DNA analysis and diagnostics on oligonucleotide arrays. *Genome Res.*, 7, 606–614.

Pemberton, J. M., Coltman, D. W., Smith, J. A., & Pilkington, J. G. 1999. Molecular analysis of a promiscuous, fluctuating mating system. *Biol. J. Linnean Soc.*, 68, 289–301.

Pesole, G., Gissi, C., De Chirico, A., & Saccone, C. 1999. Nucleotide substitution rate of mammalian mitochondrial genomes. *J. Mol. Evol.*, 48, 427–434.

Pope, T. R. 1990. The reproductive consequences of male cooperation in the red howler monkey: paternity exclusion in multi-male troops using genetic markers. *Behav. Ecol. Sociobiol.*, 27, 439–446.

Pope, T. R. 2000. Reproductive success increases with degree of kinship in cooperative coalitions of female red howler monkeys (*Alouatta seniculus*). *Behav. Ecol. Sociobiol.*, 48, 253–267.

Queller, D. C. & Goodnight, K. F. 1989. Estimating relatedness using genetic markers. *Evolution*, 43, 258–275.

Queller, D. C., Strassmann, J. E., & Hughes, C. R. 1993. Microsatellites and kinship. *Trends Ecol. Evol.*, 8, 285–288.

Ritland, K. 1996. Estimators for pairwise relatedness and individual inbreeding coefficients. *Genet. Res.*, 67, 175–185.

Rogers, J., Mahaney, M. C., Witte, S. M., Nair, S., Newman, D., Wedel, S., Rodriguez, L. A., Rice, K. S., Slifer, S. H., Perelygin, A., Slifer, M., Palladino-Negro, P., Newman, T., Chambers, K., Joslyn, G., Parry, P., & Morin, P. A. 2000. A genetic linkage map of the baboon (*Papio hamadryas*) genome based on human microsatellite polymorphisms. *Genomics*, 67, 237–247.

Rogers, J., Witte, S. M., & Slifer, M. A. 1995. Five new microsatellite DNA polymorphisms in squirrel monkey (*Saimiri boliviensis*). *Am. J. Primatol.*, 36, 151.

Rosenbaum, H. C. & Deinard, A. S. 1998. Caution before claim: an overview of microsatellite analysis in ecology and evolutionary biology. In: *Molecular Approaches to Ecology and Evolution* (Ed. by R. DeSalle & B. Schierwater), pp. 87–106. Boston: Birkhäuser.

Ross, K. G. 2001. Molecular ecology of social behaviour: analyses of breeding systems and genetic structure. *Mol. Ecol.*, 10, 265–284.

Schlötterer, C. 1998. Microsatellites. In: *Molecular Genetic Analysis of Populations* (Ed. by A. R. Hoelzel), pp. 237–262. New York: Oxford University Press.

Schlötterer, C., Amos, B., & Tautz, D. 1991. Conservation of polymorphic simple sequences in cetacean species. *Nature*, 354, 63–65.

Schlötterer, C. & Pemberton, J. 1998. The use of microsatellites for genetic analysis of natural populations—a critical review. In: *Molecular Approaches to Ecology and Evolution* (Ed. by R. DeSalle & B. Schierwater), pp. 71–86. Berlin: Birkhäuser Verlag.

Schmitz, J., Ohme, M., & Zischler, H. 2000. The complete mitochondrial genome of *Tupaia belangeri* and the phylogenetic affiliation of Scandentia to other eutherian orders. *Mol. Biol. Evol.*, 17, 1334–1343.

Shi, M. M. 2001. Enabling large-scale pharmacogenetic studies by high-throughput mutation detection and genotyping technologies. *Clin. Chem.*, 47, 164–172.

Shubitowski, D., Venta, P., Douglass, C., Zhou, R.-X., & Ewart, S. 2001. Polymorphism identification within 50 equine gene-specific sequence tagged sites. *Anim. Genet.*, 32, 78–88.

Silk, J. B. 1987. Social behavior in evolutionary perspective. In: *Primate Societies* (Ed. by B. B. Smuts, D. L. Cheney, R. M. Seyfarth, R. W. Wrangham, & T. T. Struhsaker), pp. 318–329. Chicago: University of Chicago Press.

Smith, D.-G., Kanthaswamy, S., Disbrow, M., & Wagner, J.-L. 1999. Reconstruction of parentage in a band of captive hamadryas baboons. *Int. J. Primatol.*, 20, 415–429.

Smith, D. G., Rolfs, B., & Lorenz, J. 1992. A comparison of the success of electrophoretic methods and DNA fingerprinting for paternity testing in captive groups of rhesus macaques. In: *Paternity in Primates: Genetic Tests and Theories* (Ed. by R. D. Martin, A. F. Dixson, & E. J. Wickings), pp. 32–52. Basel: Karger.

Smith, K. L., Alberts, S. C., Bayes, M. K., Bruford, M. W., Altmann, J., & Ober, C. 2000. Cross-species amplification, noninvasive genotyping, and non-Mendelian inheritance of human STRPs in Savannah baboons. *Am. J. Primatol.*, 51, 219–227.

Smith, T. B. & Wayne, R. 1996. *Molecular Genetic Approaches in Conservation.* New York: Oxford University Press.

Stone, A. C., Griffiths, R. C., Zegura, S. L., & Hammer, M. F. 2002. High levels of Y-chromosome nucleotide diversity in the genus Pan. *Proc. Natl. Acad. Sci. USA*, 99, 43–48.

Sullivan, K. M., Mannucci, A., Kimpton, C. P., & Gill, P. 1993. A rapid and quantitative DNA sex test: fluorescence-based PCR analysis of X-Y homologous gene amelogenin. *BioTechniques*, 15, 637–641.

Sunnucks, P. 2000. Efficient genetic markers for population biology. *Trends Ecol. Evol.*, 15, 199–203.

Syvanen, A. C. 2001. Accessing genetic variation: genotyping single nucleotide polymorphisms. *Nat. Rev. Genet.*, 2, 930–942.

Syvänen, A.-C. & Landegren, U. 1994. Detection of point mutations by solid-phase methods. *Hum. Mutat.*, 3, 172–179.

Taberlet, P., Camarra, J.-J., Griffin, S., Uhrés, E., Hanotte, O., Waits, L. P., Dubois-Paganon, C., Burke, T., & Bouvet, J. 1997. Noninvasive genetic tracking of the endangered Pyrenean brown bear population. *Mol. Ecol.*, 6, 869–876.

Taberlet, P., Griffin, S., Goossens, B., Questiau, S., Manceau, V., Escaravage, N., Waits, L. P., & Bouvet, J. 1996. Reliable genotyping of samples with very low DNA quantities using PCR. *Nucleic Acids Res.*, 24, 3189–3194.

Taberlet, P., Waits, L. P., & Luikart, G. 1999. Noninvasive genetic sampling: look before you leap. *Trends Ecol. Evol.*, 14, 323–327.

Takasaki, H. & Takenaka, O. 1991. Paternity testing in chimpanzees with DNA amplification from hairs and buccal cells in wadges: a preliminary note. In: *Primatology Today: Proceedings of the XIIIth Congress of the IPS, Nagoya and Kyoto* (Ed. by M. Iwamoto), pp. 613–616. Amsterdam: Elsevier.

Tautz, D. 1989. Hypervariability of simple sequences as a general source for polymorphic DNA markers. *Nucleic Acids Res.*, 17, 6463–6471.

Thomas, M. G., Bradman, N., & Flinn, H. M. 1999. High throughput analysis of 10 microsatellite and 11 diallelic polymorphisms on the human Y-chromosome. *Hum. Genet.*, 105, 577–581.

Thomas, M. G., Parfitt, T., Weiss, D. A., Skorecki, K., Wilson, J. F., Le Roux, M., Bradman, N., & Goldstein, D. B. 2000. Y-chromosome traveling south: the Cohen modal haplotype and the origins of the Lemba—the Black Jews of Southern Africa. *Am. J. Hum. Genet.*, 66, 674–686.

Trivers, R. L. 1985. *Social Evolution.* Menlo Park: Benjamin/Cummings.

Turner, T. R., Weiss, M. L., & Pereira, M. E. 1992. DNA fingerprinting and paternity assessment in Old World monkeys and ringtailed lemurs. In: *Paternity in Primates: Genetic Tests and Theories* (Ed. by R. D. Martin, A. F. Dixson, & E. J. Wickings), pp. 96–112. Basel: Karger.

Underhill, P. A., Jin, L., Lin, A. A., Mehdi, S. Q., Jenkins, T., Vollrath, D., Davis, R. W., Cavalli-Sforza, L. L., & Oefner, P. J. 1997. Detection of numerous Y-chromosome biallelic polymorphisms by denaturing high-performance liquid chromatography. *Genome Res.*, 7, 996–1005.

Van De Casteele, T., Galbusera, P., & Matthysen, E. 2001. A comparison of microsatellite-based pairwise relatedness estimators. *Mol. Ecol.*, 10, 1539–1549.

Venta, P., Brouillette, J., Yuzbasiyan-Gurkan, V., & Brewer, G. 1996. Gene-specific universal mammalian sequence tagged sites: application to the canine genome. *Biochem. Genet.*, 34, 321–341.

Vigilant, L. 1999. An evaluation of techniques for the extraction and amplification of DNA from naturally shed hairs. *Biol. Chem.*, 380, 1329–1331.

Vigilant, L., Hofreiter, M., Siedel, H., & Boesch, C. 2001. Paternity and relatedness in wild chimpanzee communities. *Proc. Natl. Acad. Sci. USA*, 98, 12890–12895.

von Segesser, F., Scheffrahn, W., & Martin, R. D. 1995. Parentage analysis within a semi-free-ranging group of Barbary macaques *Macaca sylvanus*. *Mol. Ecol.*, 4, 115–120.

Warren, K. S., Verschoor, E. J., Langenhuijzen, S., Heriyanto, Swan, R. A., Vigilant, L., & Heeney, J. L. 2001. Speciation and intrasubspecific variation of Bornean orangutans, *Pongo pygmaeus pygmaeus*. *Mol. Biol. Evol.*, 18, 472–480.

Weatherhead, P. J. & Montgomerie, R. D. 1991. Good news and bad news about DNA fingerprinting. *Trends Ecol. Evol.*, 6, 173–174.

Weber, J. L. 1990. Informativeness of human (dC-dA)n.(dT-dG)n polymorphisms. *Genomics*, 7, 524–530.

Weber, J. L. & May, P. E. 1989. Abundant class of human DNA polymorphisms which can be typed using the polymerase chain reaction. *Am. J. Hum. Genet.*, 44, 388–396.

Webster, M. S. & Westneat, D. F. 1998. The use of molecular markers to study kinship in birds: techniques and questions. In: *Molecular Approaches to Ecology and Evolution* (Ed. by R. DeSalle & B. Schierwater), pp. 7–35. Berlin: Birkhäuser Verlag.

Wickings, E. J. & Dixson, A. F. 1992. Application of DNA fingerprinting to familial studies of Gobonese primates. In: *Paternity in Primates: Genetic Tests and Theories* (Ed. by R. D. Martin, A. F. Dixson, & E. J. Wickings), pp. 113–130. Basel: Karger.

Wilson, J. F. & Erlandsson, R. 1998. Sexing of human and other primate DNA. *Biol. Chem.*, 379, 1287–1288.

Wise, C. A., Michaela, S., Rubinsztein, D. C., & Easteal, S. 1997. Comparative nuclear and mitochondrial genome diversity in humans and chimpanzees. *Mol. Biol. Evol.*, 14, 707–716.

Wrangham, R. W. 1980. An ecological model of female-bonded primate groups. *Behaviour*, 75, 262–300.

Zane, L., Bargelloni, L., & Patarnello, T. 2001. Strategies for microsatellite isolation: A review. *Mol. Ecol.*, 11, 1–16.

Zhang, Y. W., Morin, P. A., Ryder, O. A., & Zhang, Y. P. 2001. Microsatellite loci screening in gorillas (*Gorilla gorilla gorilla*) and orangutans (*Pongo pygmaeus*) using human primers prescreened on baboons. *Cons. Genet.*, 2, 391–395.

Zischler, H. 2000. Nuclear integrations of mitochondrial DNA in primates: inference of associated mutational events. *Electrophoresis*, 21, 531–536.

Zischler, H., Geisert, H., & Castresana, J. 1998. A hominoid-specific nuclear insertion of the mitochondrial D-loop: implications for reconstructing ancestral mitochondrial sequences. *Mol. Biol. Evol.*, 15, 463–469.

3

Noninvasive Genotyping and Field Studies of Free-Ranging Nonhuman Primates

David S. Woodruff

The evolution of complex societies in which individuals rely on assistance from others involves kin selection, mutualism, and competition (Wilson 2000, Adcock 2001, Clutton-Brock 2002, Kappeler & van Schaik 2002). Establishing relationships among free-ranging primates to test sociobiological hypotheses has traditionally taken years of painstaking fieldwork, and the relationship between social and genetic relationships was rarely verified. This changed in the 1990s with the introduction of noninvasive genotyping. Hypotheses based on decades of field observations became testable, and long-standing debates about kinship and behavior were, at least in theory, resolvable within months. In this chapter, I briefly review the first applications of noninvasive genotyping, the lessons learned in the past 10 years, and the prospects for the method's second decade. The challenges for the next decade are to refine the methods so that reliable genotyping becomes an economic and required adjunct of behavioral studies whose interpretation depends on a knowledge of individual pairwise relatedness. As explained below, noninvasive genotyping represented a revolutionary solution to the dilemma that arose when the effects on behavioral studies of shooting, trapping and bleeding, or biopsy darting animals became unacceptable.

Credit for the development of noninvasive genotyping goes to forensic scientist Cecilia von Beroldingen, who first demonstrated that single human hair roots contained enough DNA for analysis (von Beroldingen et al. 1987, Higuchi et al. 1988). Her discovery took advantage of the recent introduction of the method of DNA amplification by the polymerase chain reaction (PCR) and of electrophoretic methods of direct sequencing and visualizing DNA. The PCR is a process that enables the quick production of a billion copies of a short sequence of interesting DNA once the sequences on either side of the target sequence are known. These flanking sequences are used to construct oligonucleotide primer pairs that, if properly designed, permit the amplification of only the target sequence from a mixture containing all the nuclear and mitochondrial DNA in an extract. The PCR process involves

46

the use of an enzyme, DNA polymerase, from a thermophilic bacterium, *Thermus aquaticus* (hence the trade name *Taq* polymerase), to repeatedly double the number of copies of the target sequence in a test tube. The process can be so sequence specific that genotyping is possible from trace or degraded samples in which the target sequence is present in single or very low copy numbers. For example, the two copies of a specific short nuclear sequence can be found in the extract of a single human cell that contains the entire 3 billion bp genome coding for more than 30,000 genes.

In 1988, based on von Beroldingen's discovery, Phillip Morin and I used human sequence primers to amplify homologous genes from hair samples of captive chimpanzees (*Pan troglodytes*), gorillas (*Gorilla gorilla*), orangutans (*Pongo pygmaeus*), gibbons (*Hylobates* spp.), and macaques (*Macaca* spp.). We also published a first demonstration that microsatellite loci could be amplified from gibbon hair (Woodruff 1990). Microsatellites, which have subsequently proven extraordinarily useful in resolving questions of paternity, kinship, and population structure, are discussed in detail below. Suffice it to say that by 1990 others studying nonhuman primates were also experimenting with the new methods; Takasaki and Takenaka (1991) reported amplifying a nuclear locus from hair shed by wild chimpanzees at Mahale, Tanzania. By 1992 there were reports of successful hair-based genotyping of chimpanzees (Morin & Woodruff 1992), gorillas (Garner & Ryder 1992), gibbons (*Hylobates nomascus*) (Garza & Woodruff 1992), and macaques (Mubumbila et al. 1992). Although early noninvasive studies were based almost exclusively on shed and plucked hair samples, alternative DNA sources were quickly introduced; buccal cells by Takasaki and Takenaka (1991) and feces by Höss et al. (1992), Sugiyama (1993), and Takenaka et al. (1992). These early reports were reviewed by Woodruff (1993) and Morin and Woodruff (1996).

Without diminishing the significance of the foregoing reports, the first demonstration of the full power of noninvasive genotyping was Morin's (1992) study of wild chimpanzees. He used shed hair collected from night nests to genotype an entire community of free-ranging chimpanzees at Gombe, Tanzania. Hypervariable microsatellite loci were used to provide individual—specific multilocus genotypes of all members of the Kasakela community. Data on relatedness, paternity, and population structure were provided for comparison with Goodall's (1986) 30 years of behavioral observations. Morin also provided the first data on female gene flow and on intersubspecific genetic differences using two mitochondrial (mtDNA) sequences amplified from hair samples collected opportunistically across Africa. He found the West African subspecies, *Pan troglodytes verus*, was very different from the other subspecies, and some subsequent workers have treated *P. t. verus* as a separate species. The results of this pioneering study are reported by Morin et al. (1992, 1993, 1995; Morin, Moore et al. 1994; Morin, Wallis et al. 1994). His approach was quickly applied to other habituated communities of chimpanzees and to numerous other wild primates.

Sampling and Genotyping Methods

Before reviewing the later noninvasive studies of free-ranging primates, it is worth digressing briefly into the terminology associated with sampling and genotyping methods. There are two approaches to DNA acquisition that are sometimes confused: noninvasive and nondestructive. Noninvasive genotyping has been based on shed hair, buccal cells, and cells extracted from feces and urine. Nondestructive methods, on the other hand, involve small tissue samples, typically blood, plucked hair, or tail, ear, and toe clips. Blood and nonlethal tissue biopsy samples are obtained by live trapping and by darting. The difference between the two approaches hinges on whether or not the animals are physically handled. The advantage of nondestructive sampling is that it provides a much richer source of high-quality DNA of the quantity required for protein or allozyme electrophoresis, DNA fingerprinting based on minisatellite or variable number tandem repeats (VNTR), and random amplified polymorphic DNA (RAPD) analyses. Noninvasive sampling, in contrast, provides trace or forensic quantities of frequently degraded DNA. Nondestructive sampling, despite its very significant DNA quality advantage, has two disadvantages. First, such tissue samples may require special treatment in the field and, if preserved in alcohol, are hazardous and may be more difficult to move back to the laboratory by air. Second, the collection of these samples exposes the subjects to stress and the collectors to various risks. Nondestructive sampling may affect both individual and social group behaviors and can quickly undo years of painstaking habituation. Noninvasive samples, in contrast, may be collected without disturbance of the subjects and, in fact, can be collected from animals that the collectors may never see.

The fact that the first applications of noninvasive genotyping involved chimpanzees rather than other primates is understandable in terms of the then available genetic information base. My research group simply took advantage of the human genome projects and applied the associated discoveries to our closest living relatives. Primer sequences designed to amplify specific human genes or microsatellite loci had a good chance of amplifying homologous loci in chimpanzees but a diminishing likelihood of amplifying DNA in more distantly related primates (Coote & Bruford 1996, Garza & Freimer 1996, Clisson et al. 2000). This was not a problem with the more conserved mtDNA sequences but a significant impediment to the use of heterologous primers for the highly informative microsatellite loci required for population and behavioral studies. But, as most primates have never been characterized genetically, researchers working with new species for the first time typically had to undertake months of preparative cloning and sequencing to develop species-specific primer pairs (Ellsworth & Hoelzer 1998, Jekeilek & Strobeck 1999). Today, after a decade of very significant technical progress these problems are diminished, but genetic research on little-studied genera may still require months of demanding and expensive preparatory work (see chapter 2).

Most behavioral analyses of interest to primatologists can be addressed using within-gene sequence variation and microsatellite repeat length polymorphisms. Originally, mtDNA sequences were favored, as they were easier to amplify from hair samples, being 100 to 1,000 times more abundant than the rarer nuclear nDNA sequences. Mitochondria are transmitted maternally and so mtDNA sequences are especially useful in confirming mother-offspring relatedness, estimating historical patterns of female gene flow, and phylogeographic analyses. But nucleotide sequences of, for example, cytochrome *b* and cytochrome oxidase II are relatively conserved across related species and therefore of little use in population-level studies. In contrast, the mitochondrial control region is typically more variable and can be informative at the levels of within-species variation, as some parts evolve faster than other regions of the mitochondrial genome. Some earlier reports casually equated the approximately 1,000 bp control region with the central and more conserved D-loop. The control region is now known to have more variable right and left domains on either side of the D-loop, which comparative studies have shown to be the only nonclocklike evolving part of the mitochondrial genome. In summary, the early work on noninvasive genotyping favored mtDNA over nDNA as the mitochondrial sequences were thought to evolve up to 10 times faster, on average, than nuclear genes and because low copy number made it harder to amplify nDNA.

Nuclear DNA is, however, pivotal for behavioral analyses. Unlike mtDNA, which is essentially haploid and typically behaves as a single linkage group or haplotype, the nuclear genome is diploid and recombination creates enormously informative variation. Within the nuclear genome, investigators have a number of different types of DNA to choose from. Y chromosome–linked markers are clearly useful in sexing samples of unknown origin and in tracing patrilineal inheritance, historical patterns of male gene flow, and reconstructing a species phylogeography and phylogeny (Ke et al. 2001, Gusmão et al. 2002, Stone et al. 2002). Autosomal functional genes are typically not variable enough to use in population and behavioral studies, although some MHC loci are exceptionally variable. The critically conserved and informative sequences of DNA base pairs that code for particular proteins, termed exons, are, however, typically broken up into many shorter sequences along the chromosome, separated by bits of apparently extraneous DNA termed introns. In contrast to exon sequences, introns and interspersed "junk" DNA have proven extremely informative in studies of population structure and sociobiology. Interest focused initially on long repetitive sequences called minisatellites or variable number tandem repeats, but these are too large (10–50 kbp) to amplify from degraded samples obtained noninvasively. The resulting DNA fingerprints are also difficult to interpret genetically. Shorter microsatellites (also called simple sequence repeats [SSR] and simple tandem repeats [STR]) have, in contrast, attributes that make them almost perfect for pedigree and population-level analyses. Microsatellites are short 70 to 200 bp sequences of tandem repeats of a core motif of 2 to 6 bp (e.g., di-, trinucleotide, etc. repeats). There are tens of millions of these SSRs scattered along all the chromosomes of all eukaryotes, and each locus has a unique flanking sequence that provides the basis for locus-specific amplification by the PCR. Each locus typically exhibits sequence length polymorphism involving variation in the number of times the basic motif is repeated. Furthermore, as sequence length polymorphism alleles are inherited in a codominant manner, a homozygous individual will have two alleles of equal length and a heterozygote will have two alleles of different lengths. Following electrophoretic separation of alleles by size on an agarose or polyacrylamide gel, an individual's genotype can quickly be established. For example, if the father's genotype at a dinucleotide locus was $(CA)_4$ $(CA)_4$ and the mother's genotype was $(CA)_{10}$ $(CA)_{10}$, their offspring would be heterozygous $(CA)_4$ $(CA)_{10}$. If one were to screen all members of a population for variation at 6 to 100 unlinked loci, one would quickly be able to ascertain relationships based on individually unique combinations of alleles at these loci with a high degree of probability. The real power of microsatellite genotyping arises, however, from the fact that at most loci there may be 5 to 10 or more alleles segregating in a single primate population. This extraordinary variability makes microsatellites ideal Mendelian markers for sociobiological analyses. Methods of identifying microsatellites, designing locus-specific primer pairs to amplify them, and scoring their variability are described elsewhere (Smith & Wayne 1996, Hoelzel 1998, Rosenbaum & Deinard 1998, Schlötterer 1998, Schlötterer & Pemberton 1998, Beaumont & Bruford 1999, Goldstein & Schlötterer 1999, Taberlet & Luikart 1999, Sunnucks 2000).

In addition to direct sequencing of mtDNA and nDNA, and microsatellite (nDNA) sizing, three other molecular genetic methods have been used to genotype individuals whose DNA has been sampled noninvasively. These methods involve either whole genomic DNA extracts or mtDNA or nDNA only. First, AFLP (amplified fragment length polymorphism) analysis involves the PCR and so can be used on trace samples (Mueller & Wolfenbarger 1999). It is a relatively cheap, easy, fast, reliable, and replicable method to generate hundreds of dominant multilocus nDNA markers. It allows the detection of polymorphisms of genomic restriction fragments by PCR amplification. Advantages include its ability to resolve extremely small genetic differences and its utility in linkage mapping of quantitative trait loci (QTLs). Variation in most phenotypic and behavioral characters is continuous rather than discrete and controlled by multiple QTLs (see Roff 2002 for an account of the relevance of QTLs to, for example, kinship, reproductive rates, and behavior). Although AFLP markers segregate in Mendelian fashion, their typically dominant nature (resolved as presence or absence of bands) does not permit some of the more powerful genetic analyses conducted with codominant chromosomally mapped microsatellite markers. A method of focusing on the approximately 10% of the AFLP markers that are codominant, termed microsatellite AFLP or SAMPL (selective amplification of microsatellite polymorphic loci), may provide an important extension of the approach (Mueller & Wolfenbarger 1999). The second method, SSCP (single strand conformational polymorphisms) analysis, can also be performed with trace samples. Genetic diversity among PCR products from large numbers of individuals at, for example, variable MHC loci or the mitochondrial control region can be detected without sequencing. Vigilant et al. (1991) provided an early but powerful example of the application of SSCP analysis to human populations. As with AFLP, the method's apparent efficiencies in providing a one-time picture of variation within a population have to be balanced against the difficulties of extending a study in time and space. Methods that produce archival sequence or microsatellite genotype data are easily extended as new individuals or populations may be added to the database by the original investigators or by others. In contrast, methods that provide snapshots of variation and relatedness (e.g., genetic fingerprints based on multilocus fragment patterns and overall band sharing) are more difficult to replicate and extend. The third method, SNP (single nucleotide polymorphism) analysis (Wang et al. 1998), may further revolutionize the way some species are genotyped. There are millions of SNP loci in every genome and although variation at any one locus is not enough to establish population genealogies (there being typically only two alleles per locus or SNP site), this can be countered by the simultaneous assay of variation at very large numbers of SNP loci (see chapter 2). Such multilocus nDNA genotyping can be performed automatically using fluorescent reactions on customized DNA microarrays (Morin et al. 1999, Gut 2001, Jain 2001, Syvanen 2001). The development of such arrays for studies of gene expression in humans, chimpanzees, and macaques (see, for example, Enard et al. 2002) may enable field researchers to take advantage of this otherwise expensive technology, but it remains to be seen if these advances can be used with noninvasive samples.

Genotyping by these various methods is still too expensive for casual investigations. Sequences and fragment sizes are detected using radiometric or fluorescent-tagged DNA migrating through slab or capillary gels or past detectors in mass spectrometers. In my earlier reviews, I offered estimates of the costs of genotyping associated with behavioral and phylogeographic studies. Rapid technological advances have rendered such estimates wrong by orders of magnitude, so I shall not offer more. Nevertheless, primatologists should consult widely and recognize the costs in time and money before embarking on a significant noninvasive genotyping study.

Noninvasive Genotyping: Results of the First Decade

Following Morin's analysis of relationships and social structure of the Gombe chimpanzees, there have been a growing number of comparable studies of other primate communities and phylogeography. The following list is not exhaustive but illustrates the range of taxa examined. In some studies not all individuals were free-ranging and in others a mix of noninvasive and nondestructive methods were employed. Some studies are very narrow (e.g., descriptions of primers for new marker loci), but a few involve attempts to genotype entire free-ranging social communities. The latter include Gagneux, Woodruff et al. (1999), Gerloff et al. (1999), Morin, Moore et al. (1994), Nievergelt et al. (2000), Nievergelt, Pastorini et al. (2002), and Nievergelt, Mutschler et al. (2002).

Noninvasive studies of *Pan* (chimpanzees and bonobos) include Constable et al. (1995, 2001), Constable (2000), Gerloff et al. (1995, 1999), Hashimoto et al. (1996), Goldberg (1997), Goldberg and Ruvolo (1997a, 1997b), Goldberg and Wrangham (1997), Gonder et al. (1997), Gonder (2000), Houlden et al. (1997), Gagneux (1997, 2002), Gagneux et al. (1997a, 2001), Gagneux, Woodruff et al. (1999), Gagneux, Wills et al. 1999), Hohmann et al. (1999), Kaessmann et al. (1999), Goossens, Latour et al. (2000), Mitani et al. (2000), and the above-mentioned Morin et al. (1993, 1995), Morin, Moore et al. (1994), and Morin, Wallis et al. (1994). Noninvasive studies of gorillas include Garner and Ryder (1992, 1996), Field et al. (1998), Saltonstall et al. (1998), Clifford et al. (1999), and Bradley et al. (2001). Muir et al. (1994, 1998, 2000), Goossens, Chikhi et al. (2000), Immel et al. (1999, 2000), and Warren et al. (2001) studied orangutans. Garza and Woodruff (1992, 1994), Monda (1996), Andayani et al. (2001), Oka and Takenaka (2001), Roos and Geissmann (2001), and Reichard et al. (in preparation: *Hylobates lar* at Khao Yai, Thailand) studied gibbons. Frantzen et al. (1998) and Smith et al. (2000) used fecal genotyping to study baboons (*Papio cynocephalus*), and comparable studies based on nondestructive genotyping include Altmann et al. (1996), Morin et al. (1998), and Smith et al. (2000). Noninvasive studies of other primates include von Segesser et al. (1995), Chu et al. (1999), and Lathuilliere et al. (2001) on macaques, Launhardt (1998), Borries et al. (1999), and Little et al. (2002) on langurs (*Presbytis entellus*), Nievergelt et al. (1998, 2000) on marmosets (*Callithrix* spp.), Surridge et al. (2002) on tamarins (*Saguinus* spp.), Lawler et al. (2001) on sifakas (*Propithecus verreauxi*), Nievergelt, Pastorini et al. (2002), and Nievergelt, Mutschler et al. (2002) on gentle lemurs (*Hapalemur griseus*).

Collectively, these references demonstrate the revolutionary impact of noninvasive and nondestructive genotyping on studies of free-ranging primates. Each detailed study has shed light on some questions that have frustrated observers for decades by providing definitive answers based on genetic relationships rather than conjecture. Hypotheses about cooperation between kin, about the relative reproductive fitness of alpha males and females, about inbreeding avoidance, about sex-biased dispersal and gene flow, and about infanticide can now be tested with statistically acceptable rigor. I will not present a review of all the fascinating behavioral findings reported in the above works here but, by way of an example, let me draw attention to Nievergelt's work on the little-known and endangered Alaotran gentle lemur (*Hapalemur griseus alaotrensis*). She has elucidated the social system of this primate, which lives only in the marshes of Lake Alaotra, by genetically monitoring all 99 individuals in one population comprising 22 neighboring social groups. These territorial lemurs live in small groups of one or two breeding females, their offspring, and one reproducing male. In the 40% of the groups that contained two breeding females, Nievergelt found that the females were mother and daughter or full sisters. Females form the core of each social group, and the intergroup transfer of adult males is relatively frequent. The mating system is variable and ranges from serial monogamy to polygyny. Extragroup males were involved in 8% of the paternities. Genetic data enabled her to reconstruct demographic events that could not have been observed directly and to characterize a previously unknown mating system. Although males had been thought to pursue a resource defense strategy to control access to females, Nievergelt found no evidence for male philopatry. Males apparently compete for control of two-female groups but have no incentive to stay if the younger female is a daughter. These discoveries were all based on the analysis of individual multilocus genotypes derived from plucked hair samples of animals that are impossible to observe directly during much of the year, when water levels in Lake Alaotra preclude human approach.

One other important lesson remains to be learned from this long and growing list of publications representing the successes of noninvasive genotyping, but it is rarely mentioned in the literature. Although authors always describe their sample collection methods, they do so in language that belies the fact that sample acquisition is a nontrivial activity. For example, anyone interested in collecting shed hair samples from chimpanzee night nests in the canopy 50 meters above the forest floor will need to know how to use jumars and how to search a nest designed to support an animal half the weight of the investigator. I suspect that much vitally important information has gone unreported in the tersely written published reports. Anyone contemplating noninvasive or nondestructive genotyping should contact the pioneers directly to learn what was really involved in acquiring the hair, buccal cell, and fecal samples.

Problems Encountered with Noninvasive Genotyping During the First Decade

As soon as noninvasive microsatellite genotyping was introduced, it became clear that the interpretation of results based on degraded, low-concentration DNA templates was going to be difficult. Various scoring errors were encountered and described by Gerloff et al. (1995), Taberlet et al. (1996, 1997), Gagneux et al. (1997a, b), and Goossens et al. (1998). Navidi et al. (1992) were among the first to detail the types of genotyping errors to be expected and the precautions needed to avoid them. These errors are more invidious and generic than the periodic failure of PCR due to bad *Taq* polymerase or the occasional malfunctioning of PCR thermal cyclers. With hindsight we can group these scoring errors into three categories: allelic dropout, stutter bands, and contamination. Before describing the steps that must be taken to minimize each source of error, I shall illustrate their significance by describing an experience in my own laboratory.

A good example of the challenges of noninvasive genotyping based on shed hair samples is provided by a study conducted in my own laboratory on the mating systems in wild chimpanzees (Gagneux et al. 1997a, Gagneux, Woodruff et al. 1999, 2001). In a genetic analysis of one community of chimpanzees in the Taï forest, West Africa, we concluded that 7 out of 13 offspring were sired by males not found in the mother's social group (Gagneux et al. 1997a). A newer study of paternity in three social communities, including the original one, shows that the incidence of extragroup paternity is much lower (1 out of 14 offspring) (Vigilant et al. 2001). This second study was based mostly on DNA extracted from fecal samples prescreened for total DNA quantity, by a method described below. Genotypes were based on PCR amplification of 14 highly polymorphic tri- and tetranucleotide microsatellite loci, using fluorescent-tagged primers, separation of products by capillary electrophoresis, and computer-assisted genotype scoring. The study was conducted in a different laboratory than the original work, which involved amplification with radionucleotide-tagged primers for mostly longer dinucleotide repeat loci, manual electrophoresis using sequencing gels, and visual gel interpretation. A direct comparison at the only microsatellite locus reexamined (of 11 in the original study) revealed that 10 of 66 alleles (15%) and 9 of 33 individuals (27%) were incorrectly genotyped. Allelic dropout in the amplification of nuclear DNA from field-collected shed hair samples and miscalled genotypes due to PCR artifacts (stutter bands) were the principle causes of scoring error. This is ironic, as we had presented one of the first analyses of the possible significance of this problem in genotyping from degraded template (Gagneux et al. 1997b). It is now clear that we did not repeat our determinations a sufficient number of times (we averaged about five times for homozygotes) and did not then know that the quantity of DNA needed to avoid allelic dropout was at least 50 pg/µL instead of 25 pg/µL. Clearly, the genotyping methods we used in 1994 were much less reliable than we believed. The new analysis confirms that extragroup paternity can occur in nature, but shows that it is rare and that the social community probably corresponds to the reproductive unit in chimpanzees, as was long assumed. It is now clear that some results of our early microsatellite genotyping from shed hair projects (and possibly those published by some others) may not be replicable.

Navidi et al. (1992) developed a multiple tubes approach to avoid scoring errors due to allelic dropout resulting from very low concentrations of DNA in the initial template. Allelic dropout occurs when one of the two alleles at a heterozygous locus fails to amplify, by chance due to its minute concentration, and gives the appearance of being a homozygote. Navidi et al. found that PCR amplifications should be replicated 10 or more times to ensure that heterozygotes were not incorrectly scored. Taberlet et al. (1996) further refined this guideline by recommending two separate replications for each locus initially determined to be heterozygous and seven separate replications for each locus initially determined to be homozygous. As a result of such considerations, many laboratories adopted noninvasive genotyping protocols requiring two extractions and four PCRs for all loci, followed by three more PCRs to confirm apparent homozygotes. Despite general appreciation of the problem, the multitubes solution was only slowly adopted as it greatly increased the time and cost of an investigation. Furthermore, the quantity of tissue available for extraction, especially with shed hair, was often less than that required for the necessary replications.

Morin et al. (2001) provided an extract sorting solution to the difficulties associated with the multitubes approach by quantifying the amount of amplifiable DNA in extracts obtained noninvasively. Prescreening DNA extracts of shed chimpanzee hair and feces allowed them to formulate a more efficient approach to microsatellite genotyping. Although repetitions are still necessary for genotyping with minute quantities of initial DNA, they become less necessary as DNA template quantity increases. Their experimental results are worth considering in detail and, for perspective, it should be noted that they found single freshly plucked human hairs contain on average 326 pg/μL nDNA (range: 24–1202), and that 7% of such hairs contained no amplifiable nDNA. In contrast, they found single shed chimpanzee hairs averaged only 21 pg/μL DNA and 79% of the extracts contained no amplifiable nDNA. For chimpanzees, a 200 μL fecal extraction contained, on average, 192 pg/μL nDNA (range: 0–2550), but a similar shed hair extract contained only 4.4 pg/μL nDNA (range: 0–228). Furthermore, 49% of the fecal extracts but only 2% of the hair extracts contained more than 50 pg DNA. They found that the remaining 21% of the shed chimpanzee hair extracts contained, on average, about 600 diploid cells and that a typical 1 g fecal sample contained about 55,000 cells. In their 1,300 PCR reactions involving fecal samples, they found the dropout rate at heterozygous loci to average 24% (12–35% per locus). Based on their experiments, they concluded that when template DNA is less than 25 pg per reaction, the dropout rate averages 68% and that more than 12 repetitions per locus are necessary to obtain reliable results. With template concentrations per reaction in the ranges of 26–100 pg, 101–200 pg, and more than 200 pg, they found seven, four, and two repetitions, respectively, were necessary to obtain genotypes with 99% confidence.

Morin et al.'s (2001) extract sorting method is based on DNA concentrations determined fluorescently with a 5′ nuclease assay targeting an 81 bp portion of the *c-myc* proto-oncogene. This method permits an assay of total species-specific DNA concentration in each specific extract. This is an improvement over the total DNA fluorescent method used by Gagneux et al. (1997b), which was always known to be inappropriate for fecal extracts, because of the presence of extraneous plant, fungal, and bacterial DNA. The species-specific quantification method is also preferable when samples contain PCR inhibitors like melanin in some hair. Fecal samples, in particular, may contain PCR-inhibiting polysaccharides, but their impact can be minimized by the use of bovine serum albumin. In conclusion, Morin et al. (2001) found that fecal samples were about 10 times more likely to yield enough nDNA for reliable multilocus microsatellite genotyping. Finally, they also pointed out that their results apply to mtDNA sequence analyses as well as nuclear microsatellites despite the fact that mitochondrial genomes are often 1,000 times more abundant than nuclear genomes in many cell extracts. The need for more care with mtDNA sequencing when very few sets of template molecule are present arises because nucleotide misincorporation in the first amplification cycles can be a real problem.

Thus, after a decade of effort it is clear that the serious allelic dropout problem associated with noninvasive microsatellite genotyping has been recognized and largely resolved. For chimpanzees, Morin et al.'s (2001) analysis shows that a 200 μL volume of fecal extract yields enough nDNA for 100 PCR reactions. In contrast, a 5 μL shed hair extract would permit only 40 reactions and would be insufficient for genotyping 10 loci using the traditional (duplicate extracts and triplicate amplifications) multitubes approach. Morin et al. now recommend preparing single extractions from each hair or fecal sample and sorting these extracts after quantification. Only extracts with detectable DNA (31% of hair and 93% of fecal) are used to produce duplicate PCRs for each locus. Thus extract sorting used in conjunction with a modified multitubes approach to PCR can provide reliable noninvasive multilocus microsatellite genotypes with improved efficiency.

In addition to allelic dropout, two other sources of scoring error are associated with noninvasive genotyping: stutter bands and contaminants. Stutter or shadow bands occur as PCR artifacts (*Taq* polymerase-generated slippage products) and give false heterozygotes and three-banded (triallelic) individuals. This is a serious problem with dinucleotide loci scored by eye from radiometrically labeled autoradiographs. Most researchers now avoid this problem by using tri- and tetranucleotide microsatellite markers and automated sequencers with genotyping software to alleviate the problem of interpreting stutter bands. Scoring errors of this type amounted to less than 5% of the errors detected by Taberlet et al. (1996) and less than 1% of those found by Morin et al. (2001). The third class of errors complicating noninvasive microsatellite genotyping involves genuine contamination of the template sample with extraneous DNA. The contaminant DNA can come from other extracts of the same individual, from other samples of the same species, and from exotic DNA in the laboratory environment. Its significance varies greatly among laboratories and investigators, and it can be a sporadic or pervasive problem. Fortunately, stringent protocols have been developed for handling ancient DNA (Pääbo 1990, Herrmann & Hummel 1994) that can alleviate the problems. These include the physical isolation of extraction and amplification procedures, the protection of extracts from sources of contamination, the use of dedicated micropipettes with aerosol-resistant disposable tips, the use of multiple negative controls, and the care with which laboratory personnel handle the materials and themselves. Contamination by researchers is especially problematic in studies of close human relatives, like chimpanzees, but genotyping all personnel permits the determination of the source and corrective action.

Finally, three other problems were recognized during the 1990s that affect the interpretation of results of all molecular genotyping projects, whether noninvasive or not. Two involve unexpected behavior of mtDNA and the third arises when microsatellite primers are not specific enough. The first complicating factor involves the possibility of occasional recombination between parental mtDNA chromosomes. It is widely assumed that in primates and most animals, mtDNA haplotypes are exclusively maternally inherited. However, the occurrence of very low levels of recombination has been recognized in humans (Eyre-Walker 2000), so this possibility must be borne in mind when interpreting patterns in nonhuman primates. The second complicating factor is far more pervasive and involves the detection of multiple mtDNA haplotypes in a single individual. In probably the worst case reported to date, 120 different 12s RNAs were found in one individual macaque (Vartanian & Wain-Hobson 2002). This arises from the occasional and repeated movements of mtDNA sequences into the cell's nucleus and their incorporation into and transmission with nDNA (Bensasson et al. 2001). Such sequences are termed numts (nuclear sequences of mitochondrial origin; pronounced "new mites") and are often amplified by the same primer pairs as the original mtDNA sequence. They are pseudogenes in the sense that they have lost their function and are consequently freer to evolve than the ancestral gene. They have been described in more than 26 mammal species and probably occur in all primates. Over 350 different numts have already been found in humans. Mundy et al. (2000) discovered four nuclear paralogs of a 380 bp segment of cytochrome *b* in marmosets and tamarins. They are actually more likely to be encountered in samples obtained noninvasively than they are in blood, because nuclei are more common in hair and fecal samples than they are in mammal blood, where most cells are enucleate (Greenwood & Pääbo 1999). Various tests for detecting, and procedures for avoiding, numts have been developed (Bensasson et al. 2001), but clearly ambiguous results are usually the first clue to their presence. Although numts are very useful markers of phylogenetically significant changes and molecular evolution (Zischler 2000), they are more typically nuisances in population-level studies. The final problem to be noted here is associated with microsatellites and the observation that the same primer pairs may amplify nonhomologous sequences. This phenomenon is typically detected when bimodal clusters of fragment lengths are found in a population, or more than two alleles are found in repeated amplification from the same individual. Given the ubiquity of microsatellites, this problem can be avoided by choosing alternative loci or prevented by more stringent primer design. Again, these three problems are not peculiar to noninvasive genotyping but have contributed to the errors encountered during the method's first decade of use.

Noninvasive Microsatellite Genotyping: Challenges of the Second Decade

As noninvasive genotyping's first decade drew to a close, Taberlet and Waits (1998) argued that controversy had arisen between researchers who suggested that the method would permit definitive genetic analyses and researchers who demonstrated that the frequency of method-associated genotyping errors was unacceptable. This was not really the case, as the method's advocates and critics were one and the same people. Gagneux, Gerloff, Morin, Taberlet, Constable, Vigilant, and others contributed both to the demonstration of the method's power and to the identification and resolution of the problems. In a more detailed assessment, Taberlet et al. (1999) admit that the early difficulties associated with noninvasive genotyping will probably be overcome with improved methodology. This has, in fact, already occurred and although the potential for errors remains high, the stringent adoption of the lessons learned in the first decade bodes well for the future.

The speed with which technical advances have been incorporated in sociobiological studies is almost without precedent. Given the enormous resources devoted to the various model species genome projects and to forensic genetics, even greater advances can be expected in the automation of genotyping of primates. This will affect the cost of noninvasive genotyping projects, which are still 10 to 100 times more expensive than they need to be. Parallel improvements can also be expected in the areas of sample preservation and DNA extraction (see, for example, Kohn & Wayne 1997, Frantzen et al. 1998, Flagstad et al. 1999, Vigilant 1999, Whittier et al. 1999, Valiere & Taberlet 2000). Whether noninvasive genotyping will still be based primarily on microsatellites or whether SNPs or other markers become more important remains to be seen. Finally, it is possible that national and international regulations controlling the movement of DNA samples may change in ways that will facilitate more noninvasive genotyping. Currently, treated fecal samples are relatively easy to move between countries but hair samples are not, as they fall under the purview of CITES (Convention on International Trade in Endangered Species), the U.S. Endangered Species Act, and other nations' laws.

Given the high cost of the early noninvasive genotyping studies, the method was criticized by some for diverting scarce funds from traditional observational fieldwork. Such tensions are to be expected but will diminish as costs decline and as the method's promises are realized. Nevertheless, the need for fieldworkers to seek genetic collaborators forced a cultural change on the behavioral ecologists. The idea of sharing one's animals with laboratory scientists of a different background, who had little appreciation for the animals themselves or the frustrations of fieldwork, took a few years to catch on. These tensions were exacerbated by the technical problems and delays and cost overruns experienced by the laboratory geneticists. Despite these frustrations, it is now clear that noninvasive genotyping can provide answers to questions that simply could not be asked 15 years ago. Although it is undeniable that nondestructive genotyping provides better quality DNA, acquisition of blood or tissue biopsy is simply out of the question with most free-ranging primates. So the higher laboratory costs associated with noninvasive genotyping have to be balanced against the field costs of nondestructive tissue collection. As costs of the former come down and costs of the latter go up, noninvasive genotyping will become more and more important. But it should never be assumed that genetics alone will solve behavioral questions—ignoring social structure (natural patterns of mating and dispersal) will confound even the best genetic analyses (Sugg et al. 1996). Genetics and behavior have to be considered simultaneously.

As the molecular methods of noninvasive genotyping were refined, a parallel revolution occurred in the analytical and statistical methods used to interpret the data. The reader is referred to Luikart and England (1999) and Taberlet et al. (2001) for summaries of the methods and computer programs currently used to analyze allele frequency data. Statistical methods based on maximum likelihood, coalescent, and Markov chain Monte Carlo algorithms have been used with some success, and Bayesian methods promise to become more significant in the next decade. Bayesian inference involves both real data and subjective information about the past probability distribution of various parameters of interest. Combining best-guess information about prior distribution of such parameters as population size, generation time, and inbreeding coefficient permits the calculation of likelihoods of current demographic parameters from population genetic snapshots. Among the many valuable parameters that can be estimated from noninvasive population-wide surveys are: effective population size, detection of changes in the effective population size and the dating of such changes, sex ratio, mating system and population structure, inbreeding and outbreeding coefficients, home ranges of individuals, dispersal and gene flow including sex-biased dispersal, and population assignment. Various maximum likelihood computer programs are available for assigning paternity (including CERVUS, Marshall et al. 1998) and calculating coefficients of relatedness (including RELATEDNESS, Queller & Goodnight 1989, and KINSHIP, Goodnight & Queller 1999). Such analytical methods and computer programs are still under near-continuous refinement as the underpinning assumptions about mutations and linkage of genetic markers and about population structure and dispersion of individuals are accounted for more realistically. However, it will always be the case, given the small populations studied by primatologists, that tests involving hypervariable microsatellite loci (whose variability is a boon to some analyses) will often have limited statistical power. Similarly, empty cells in a matrix of genotype data from a small population can sometimes produce results that are statistically significant but of no real biological significance (Hedrick 1999). Nevertheless, within the next decade debates regarding current analytical methods should be resolved and the interpretation of population genotype data should become more routine.

Similar advances occurred in the 1990s in the analysis of DNA sequence data, although some of these are of less relevance to the behavioral issues of concern in this volume. However, one area in which further advances in DNA sequence analysis can be expected involves the comparison of gene trees using coalescence models of neutral evolution (Avise 2000). Gene trees show the genealogy of sequence variants (alleles) discovered in current populations and, under coalescence models of mutational change, one can trace all today's sequences back through time to a single ancestral sequence. Analysis of current sequence variants and the shape of a gene tree provide evidence for the history of today's populations. Observable characteristics of a gene tree and departure from expectations under the neutral evolution model permit estimation of the genetic effective size of a population, changes in population size and structure (including periods of demographic collapse or rapid growth), mutation rates, episodes of strong natural selection including selective sweeps, rates of gene flow, incidents of natural hybridization, and divergence times. Although most published analyses have involved the interpretation of mtDNA patterns, greater emphasis on nuclear gene trees can be anticipated.

Both phylogeographic and phylogenetic inferences can be based on allele frequency data and sequence data, especially mtDNA sequence data, and are important components of most noninvasive genotyping projects. Algorithms based on maximum likelihood, neighbor joining, and Bayesian inference are used to establish the evolutionary history of related populations (Nei & Kumar 2000, Huelsenbeck et al. 2001, Felsenstein 2002). Such history is critical in recognizing species and other evolutionarily significant units used in comparative studies. The genetic characterization of such taxa is especially important in primates, for which numerous recent studies have shown that traditional taxonomy may be a poor indicator of true species-level diversity. Examples for which genetic insights have necessitated taxonomic revision include chimpanzees (Morin, Moore et al. 1994, Gagneux, Wills et al. 1999), gorillas (Garner & Ryder 1996), orangutans (Muir et al. 2000), and lemurs (Rasoloarison et al. 2000, Yoder et al. 2000).

There are three areas in which greater application of noninvasive genotyping to free-ranging primate populations can be expected. First, under certain circumstances it is now possible to monitor evolutionary changes in natural populations as they occur. Pemberton et al. (1999) provide two examples of the use of microsatellite variation to study the fitness consequences of inbreeding and outbreeding in natural populations. Assuming a stepwise model for microsatellite mutation, they showed how allele lengths can be used to estimate an intrapopulation genetic distance measure, mean d^2, which represents the average time to coalescence of the maternal and paternal microsatellite alleles carried by an individual. In red deer and in harbour seals, neonate fitness traits (birth weight and survival) were better explained by mean d^2 than by individual heterozygosity. Mean d^2 permits analysis of each individual's fitness status without a formal pedigree. Coupled with noninvasive genotyping, such methods may permit primatologists to monitor various fitness-related traits in free-ranging populations. Second, in some primate populations where feces can be sampled, it will be possible to monitor changes in population size and sex ratio without actually seeing the animals. Genetic censusing based on hair or fecal multilocus microsatellite genotyping has now been demonstrated in several secretive animals including fossorial wombats (*Lasiorhinus krefftii*) (Taylor et al. 1994, 1997; Sloan et al. 2000), largely nocturnal coyotes (*Canis latrans*) (Kohn et al. 1999), and unapproachable grizzly bears (*Ursus arctos*) (Poole et al. 2001) and forest elephants (*Loxodonta africana*) (Eggert et al. 2002, 2003). The species identification of fecal samples can be confirmed by mtDNA analysis prior to developing the individual-specific microsatellite profile (see, for example, Foran, Crooks et al. 1997, Foran, Minta et al. 1997, Paxinos et al. 1997, Reed et al. 1997). Although the current analytical methods of estimating census size from genetic data involve assumptions about the population's dispersion and vagility, and formalization of the sampling grid, we can anticipate that by 2010 these methods will become standardized. A third area of increasing importance to primatologists involves the recent demonstration that it is possible to monitor genetic erosion in isolated populations. Genetic erosion is the process by which small isolated populations lose variability due to genetic drift and inbreeding. The loss of genetic variability affects the future evolvability of a population, so genetic erosion is both a symptom and a cause of genetic endangerment. As once-continuous primate populations become fragmented, smaller and more isolated, their innate genetic variability will be of interest to anyone studying their evolution and sociobiology. As very small populations can lose half their genetic variability in the first 20 generations following isolation, future studies of "natural" behavior will increasingly have to be calibrated against each population's innate variability. Interpopulation comparisons may be misleading unless the genetic underpinnings of the traits of interest are also taken into account. As the genetics of most traits of interest to behaviorists are unknown, we currently take a population's overall genetic variability (measured across as many loci as have been surveyed) as an indicator of its genetic health. Srikwan and Woodruff (2000) showed how microsatellites are variable enough to monitor genetic erosion in isolated populations of rats and mice within a few years of range fragmentation. Although that study and a parallel one involving tree shrews (Srikwan et al. 2002) were based on nondestructive tissue sampling, they could easily have been performed noninvasively. The combination of nondestructive tissue sampling of museum specimens and noninvasive sampling of living animals, albeit birds rather than primates, shows how genetic erosion can be monitored over periods of tens to hundreds of years (Mundy et al. 1997, Bouzat et al. 1998). The impact of genetic erosion on social behavior is still to be investigated but, if population genetic variability affects behavior and changes in behavior, comparative studies will have to allow for differences in innate variability.

The next decade will therefore see the further refinement of the methods of noninvasive genotyping and their widespread application to behavioral studies of free-ranging primates. The data generated by these studies will permit the characterization of the social behavior of the more than 300 species of primates surviving today. In addition, these data can be used to address more general behavioral questions of interest to ecologists and sociobiologists. My own personal favorite puzzle today concerns kinship and the evolution of cooperation in chimpanzees: why is it that even the most trustworthy results (e.g., Vigilant et al. 2001) have not found males to be more related to one another than are females at either Taï or Gombe? Female dispersal and male philopatry would lead one to predict a difference between the sexes. We can expect such issues to be solved in the next few years using noninvasive genotyping methods. More significantly, perhaps, noninvasive genotyping will also permit the monitoring of a primate population's responses to hunting, habitat fragmentation, and climate change (Cowlishaw & Dunbar 2000). With half the species of primates under threat of extinction in the present century (at least nine now have populations of fewer than 400 individuals), there is an urgent need to monitor the remaining populations' viability (Woodruff 2001a). Noninvasive genotyping thus promises to be far more useful than in simply describing the behavior of different primate species and resolving various sociobiological questions. Multilocus genotype data will also permit population viability analyses and alert us to the risk of extirpation due to genetic problems (Frankham et al. 2002). The collection and archiving of genetic data on today's populations will ultimately facilitate our ability to conserve these populations in the future. Action taken over the next few decades will determine how impoverished the biosphere will be in 2100 when most primate species and populations will suffer reduced evolvability and require genetic and ecological management. The survival of small populations (of often closely related individuals) of primates in fragmented habitats will increasingly require interventive management to ensure their long-term viability. Bioneering, the interventive genetic and ecological management of species, communities, and ecosystems in a postnatural world, is poised to become a growth industry (Woodruff 2001b). Noninvasive genotyping, associated with the behavioral studies we still have the luxury of conducting, will provide us with baseline data for the future management of selected populations. No other method is currently available to provide the data required to ensure the sound conservation of our surviving relatives.

Acknowledgments   I thank Pascal Gagneux, Phillip Morin, and Caroline Nievergelt for their helpful comments and Romel Hokanson for assistance with manuscript preparation. I am indebted to many primatologists including Jane Goodall, Jim Moore, and Janette Wallis at Gombe, Christophe Boesch at Taï, Warren Brockelman in Thailand, and Leslie Digby in Brazil. My own collaborative studies would not have been possible without their assistance and the support of the U.S. National Science Foundation.

References

Adcock, J. 2001. *The Triumph of Sociobiology*. Oxford: Oxford University Press.

Altmann, J. Alberts, S. C., Haines, S. A., Dubach, J., Muruthi, P., Coote, T., Geffen, E., Cheesman, D. J., Mututua, R. S., Saiyalel, S. N., Wayne, R. K., Lacy, R. C., & Bruford, M. W. 1996. Behavior predicts genetic structure in a wild primate group. *Proc. Natl. Acad. Sci. USA*, 93, 5797–5801.

Andayani, N., Morales, J. C., Forstner, M. R. J., Supriantna, J., & Melnick, D. J. 2001. Genetic variability in mtDNA of the silvery gibbon: implications for the conservation of a critically endangered species. *Conserv. Biol.*, 15, 770–775.

Avise, J. C. 2000. *Phylogeography: The History and Formation of Species*. Cambridge, MA: Harvard University Press.

Beaumont, M. A. & Bruford, M. W. 1999. Microsatellites in conservation genetics. In: *Microsatellites: Evolution and Applications*. (Ed. by D. B. Goldstein & C. Schlötterer), pp. 165–182. Oxford: Oxford University Press.

Bensasson, D., Zhang, D. X., Hartl, D. L., & Hewitt, G. M. 2001. Mitochondrial pseudo-genes: evolution's misplaced witnesses. *Trends Ecol. Evol.*, 16, 314–321.

Borries, C., Launhardt, K., Epplen, C., Epplen, J. T., & Winkler, P. 1999. Males as infant protectors in hanuman langurs (*Presbytis entellus*) living in multimale groups: defense pattern, paternity and sexual behaviour. *Behav. Ecol. Sociobiol.*, 46, 350–356.

Bouzat, J. L., Lewin, H. A., & Paige, K. N. 1998. The ghost of genetic diversity past: historical DNA analysis of the greater prairie chicken. *Am. Nat.*, 152, 1–6.

Bradley, B. J., Boesch, C., & Vigilant, L. 2001. Identification and redesign of human micro-satellite markers for genotyping wild chimpanzee (*Pan troglodytes verus*) and gorilla (*Gorilla gorilla gorilla*) DNA from faeces. *Conserv. Genet.*, 1, 289–292.

Chu, J. H., Wu, H. Y., Yang, Y. J., Takenaka, O., & Lin, Y. S. 1999. Polymorphic micro-satellite loci and low-invasive DNA sampling in *Macaca cyclopis*. *Primates*, 40, 573–580.

Clifford, S. L., Jeffrey, K., Bruford, M. W., & Wickings, E. J. 1999. Identification of poly-morphic microsatellite loci in the gorilla (*Gorilla gorilla gorilla*) using human primers: application to noninvasively collected hair samples. *Mol. Ecol.*, 8, 1556–1558.

Clisson, M., Lathuilliere, B., & Crouau-Roy, B. 2000. Conservation and evolution of microsatellite loci in primate taxa. *Am. J. Primatol.*, 50, 205–214.

Clutton-Brock, T. 2002. Breeding together: kin selection and mutualism in cooperative vertebrates. *Science*, 296, 69–72.

Constable, J. 2000. Reproductive strategies among Gombe chimpanzees evaluated by paternity assessment from fecal and hair DNA. PhD dissertation, University of Minnesota, Minneapolis-St. Paul.

Constable, J. J., Packer, C., Collins, D. A., & Pusey, A. E. 1995. Nuclear DNA from primate dung. *Nature*, 373, 393.

Constable, J. L., Ashley, M. V., Goodall, J., & Pusey, A. E. 2001. Noninvasive paternity assignment in Gombe chimpanzees. *Mol. Ecol.*, 10, 1279–1300.

Coote, T. & Bruford, M. W. 1996. A set of human microsatellites amplify polymorphic markers in Old World apes and monkeys. *J. Hered.*, 87, 406–410.

Cowlishaw, G. & Dunbar, R. 2000. *Primate Conservation Biology.* Chicago: University of Chicago Press.

Eggert, L. S., Eggert, J. A., & Woodruff, D. S. 2003. Estimating population sizes for elusive animals: the forest elephants of Kakum National Park, Ghana. *Mol. Ecol.* 12, 1389–1402.

Eggert, L. S., Rasner, C. A., & Woodruff, D. S. 2002. The evolution and phylogeography of the African elephant (*Loxodonta africana*) inferred from mitochondrial DNA sequence and nuclear microsatellite markers. *Proc. Royal Soc. Lond., B*, 269, 1993–2006.

Ellsworth, J. A. & Hoelzer, G. A. 1998. Characterization of microsatellite loci in a New World primate, the mantled howler monkey (*Alouatta palliata*). *Mol. Ecol.*, 7, 657–658.

Enard, W., Khaitovich, P., Klose, J., Zollner, S., Heissig, F., Giavalisco, P., Nieselt-Struwe, K., Muchmore, E., Varki, A., Ravid, R., Doxiadis, G. M., Bontrop, R. E., & Pääbo, S. 2002. Intra- and interspecific variation in primate gene expression patterns. *Science*, 296, 340–343.

Eyre-Walker, A. 2000. Do mitochondria recombine in humans? *Phil. Trans. Royal Soc. Lond., B*, 355, 1573–1580.

Felsenstein, J. 2002. *Inferring Phylogenies.* Sunderland, MA: Sinauer.

Field, D., Chemnick, L., Robbins, M., Garner, K., & Ryder, O. A. 1998. Paternity determination in captive lowland gorillas and orangutans and wild mountain gorillas by microsatellite analysis. *Primates*, 39, 199–209.

Flagstad, O., Roed, K., Stacy, J. E., & Jakobsen, K. S. 1999. Reliable noninvasive genotyping based on excremental PCR of nuclear DNA purified with a magnetic bead protocol. *Mol. Ecol.*, 8, 879–883.

Foran, D. R., Crooks, K. R., & Minta, S. C. 1997. Species identification from scat: an unambiguous genetic method. *Wildl. Soc. Bull.*, 25, 835–839.

Foran, D. R., Minta, S. C., & Heinemeyer, K. S. 1997. DNA-based analysis of hair to identify species and individuals for population research and monitoring. *Wildl. Soc. Bull.*, 25, 840–847.

Frankham, R., Ballou, J. D., & Briscoe, D. A. 2002. *Introduction to Conservation Genetics.* Cambridge: Cambridge University Press.

Frantzen, M. A., Silk, J. B., Ferguson, J. W., Wayne, R. K., & Kohn, M. H. 1998. Empirical evaluation of preservation methods for faecal DNA. *Mol. Ecol.*, 7, 1423–1428.

Gagneux, P. 1997. Sampling rapidly dwindling chimpanzee populations. *Pan Africa News*, 4, 12–15.

Gagneux, P. 2002. The genus *Pan*: population genetics of an endangered outgroup. *Trends Genet.*, 18, 327–330.

Gagneux, P., Gonder, M. K., Goldberg, T. A., & Morin, P. A. 2001. Gene flow in wild chimpanzees. What genetic data tell us about chimpanzee movements over space and time. *Phil. Trans. Royal Soc. Lond., B,* 356, 889–897.

Gagneux, P., Wills, C., Gerloff, U., Tautz, D., Morin, P. A., Boesch, C., Fruth, B., Hohmann, G., Ryder, O. A., & Woodruff, D. S. 1999. Mitochondrial sequences show diverse evolutionary histories of African hominoids. *Proc. Natl. Acad. Sci. USA,* 96, 5077–5082.

Gagneux, P., Woodruff, D. S., & Boesch, C. 1997a. Furtive female chimpanzees. *Nature,* 387, 358–359.

Gagneux, P., Woodruff, D. S., & Boesch, C. 1997b. Microsatellite scoring errors associated with noninvasive genotyping based on nuclear DNA amplified from shed hair. *Mol. Ecol.,* 6, 861–868.

Gagneux, P., Woodruff, D. S., & Boesch, C. 1999. Female reproductive strategies, paternity and community structure in a community of wild West African chimpanzees. *Anim. Behav.,* 57, 19–32.

Gagneux, P., Woodruff, D., & Boesch, C. 2001. Furtive mating in female chimpanzees. *Nature,* 414, 508.

Garner, K. J. & Ryder, O. A. 1992. Some applications of PCR to studies of wildlife genetics. *Symp. Zool. Soc. Lond.,* 64, 167–181.

Garner, K. J. & Ryder, O. A. 1996. Mitochondrial DNA diversity in gorillas. *Mol. Phylogenet. Evol.,* 6, 39–48.

Garza, J. C. & Freimer, N. B. 1996. Homoplasy for size at microsatellite loci in humans and chimpanzees. *Genome Res.,* 6, 211–217.

Garza, J. C. & Woodruff, D. S. 1992. A phylogenetic study of the gibbons (*Hylobates*) using DNA obtained non-invasively from hair. *Mol. Phylogenet. Evol.,* 1, 202–210.

Garza, J. C. & Woodruff, D. S. 1994. Crested gibbon (*Hylobates* [*Nomascus*]) identification using noninvasively obtained DNA. *Zoo Biol.,* 13, 383–387.

Gerloff, U., Hartung, B., Fruth, B., Hohmann, G., & Tautz, D. 1999. Intracommunity relationships, dispersal pattern and paternity success in a wild living community of bonobos (*Pan paniscus*) determined from DNA analysis of faecal samples. *Proc. Royal Soc. Lond. B,* 266, 1189–1195.

Gerloff, U., Schlötterer, C., Rassmann, K., Rambold, I., Hohmann, G., Fruth, B., & Tautz, D. 1995. Amplification of hypervariable simple sequence repeats (microsatellites) from excremental DNA of wild living bonobos (*Pan paniscus*). *Mol. Ecol.,* 4, 515–518.

Goldberg, T. L. 1997. Inferring the geographic origins of "refugee" chimpanzees in Uganda from mitochondrial DNA sequences. *Conserv. Biol.,* 11, 1441–1446.

Goldberg, T. L. & Ruvolo, M. 1997a. The geographic apportionment of mitochondrial genetic diversity in east African chimpanzees, *Pan troglodytes schweinfurthii. Mol. Biol. Evol.,* 14, 976–984.

Goldberg, T. L. & Ruvolo, M. 1997b. Molecular phylogenetics and historical biogeography of east African chimpanzees. *Biol. J. Linnean Soc.,* 61, 301–324.

Goldberg, T. L. & Wrangham, R. W. 1997. Genetic correlates of social behaviour in wild chimpanzees: evidence from mitochondrial DNA. *Anim. Behav.,* 54, 559–570.

Goldstein, D. B. & Schlötterer, C. (eds.) 1999. *Microsatellites: Evolution and Applications.* Oxford: Oxford University Press.

Gonder, M. K. 2000. Evolutionary genetics of chimpanzees (*Pan troglodytes*) in Nigeria and Cameroon. PhD dissertation, City University of New York, New York.

Gonder, M. K., Oates, J. F., Disotell, T. R., Forstner, M. R. J., Morales, J. C., & Melnick, D. J. 1997. A new west African chimpanzee subspecies? *Nature,* 388, 337.

Goodall, J. 1986. *The Chimpanzees of Gombe: Patterns of Behaviour.* Cambridge: Harvard University Press.

Goodnight, K. F. & Queller, D. C. 1999. Computer software for performing likelihood tests of pedigree relationship using genetic markers. *Mol. Ecol.*, 8, 1231–1234.

Goossens, B., Chikhi, L., Utami, S. S., de Ruiter, J. R., & Bruford, M. W. 2000. Multiple-samples and multiple-extracts approach for microsatellite analysis of faecal samples in an arboreal ape. *Conserv. Genet.*, 1, 157–162.

Goossens, B., Latour, S., Vidal, C., Jamart, A., Ancrenaz, M., & Bruford, M. W. 2000. Twenty new microsatellite loci for use with hair and fecal samples in the chimpanzee (*Pan troglodytes troglodytes*). *Folia Primatol.*, 71, 177–180.

Goossens, B., Waits, L. P., & Taberlet, P. 1998. Plucked hair samples as a source of DNA: reliability of dinucleotide microsatellite genotyping. *Mol. Ecol.*, 7, 1237–1241.

Greenwood, A. & Pääbo, S. 1999. Nuclear insertion sequences of mitochondrial DNA predominate in hair but not in blood of elephants. *Mol. Ecol.*, 8, 133–137.

Gusmão L., González-Neira, A., Alves, C., Sánchez-Diz, P., Dauber, E. M., Amorim, A., & Carracedo, A. 2002. Genetic diversity of Y-specific STRs in chimpanzees (*Pan troglodytes*). *Am. J. Primatol.*, 57, 21–29.

Gut, I. G. 2001. Automation in genotyping of single nucleotide polymorphisms. *Hum. Mutat.*, 17, 475–492.

Hashimoto, C., Furuichi, T., & Takenaka, O. 1996. Matrilineal kin relationship and social behavior of wild bonobos (*Pan paniscus*): sequencing the D-loop region of mitochondrial DNA. *Primates*, 37, 305–318.

Hedrick, P. W. 1999. Perspective: highly variable loci and their interpretation in evolution and conservation. *Evolution*, 53, 313–318.

Higuchi, R., von Berholdingen, C. H., Sensabaugh, G. F., & Erlich, H. A. 1988. DNA typing from single hairs. *Nature*, 332, 543–546.

Herrmann, B. & Hummel, S. (eds.) 1994. *Ancient DNA*. New York: Springer.

Hoelzel, A. R. (ed.) 1998. *Molecular Genetic Analysis: A Practical Approach*. 2nd ed. Oxford: Oxford University Press.

Hohmann, G., Gerloff, U., Tautz, D., & Fruth, B. 1999. Social bonds and genetic ties: kinship association and affiliation in a community of bonobos (*Pan paniscus*). *Behaviour*, 136, 1219–1235.

Höss, M., Kohn, M., Pääbo, S., Knauer, F., & Schroder, W. 1992. Excrement analysis by PCR. *Nature*, 359, 199.

Houlden, B. A., Woodworth, L., & Humphreys, K. 1997. Captive breeding, paternity determination and genetic variation in chimpanzees (*Pan troglodytes*) in the Australasian region. *Primates*, 38, 341–347.

Huelsenbeck, J. P., Ronquist, F., Nielsen, R., & Bollback, J. P. 2001. Bayesian inference of phylogeny and its impact on evolutionary biology. *Science*, 294, 2310–2314.

Immel, U. D., Hummel, S., & Herrmann, B. 1999. DNA profiling of orangutan (*Pongo pygmaeus*) feces to prove descent and identity in wildlife animals. *Electrophoresis*, 20, 1768–1770.

Immel, U. D., Hummel, S., & Herrmann, B. 2000. Reconstruction of kinship by fecal DNA analysis of orangutans. *Anthropologischer Anzeiger*, 58, 63–67.

Jain, K. K. 2001. Biochips for gene spotting. *Science*, 294, 621–623.

Jekeilek, J. & Strobeck, C. 1999. Characterization of polymorphic brown lemur (*Eulemur fulvus*) microsatellite loci and their amplification in the family Lemuridae. *Mol. Ecol.*, 8, 901–903.

Kaessmann, H., Wiebe, V., & Pääbo, S. 1999. Extensive nuclear DNA sequence diversity among chimpanzees. *Science*, 286, 1159–1162.

Kappeler, P. M. & van Schaik, C. P. 2002. Evolution of primate social systems. *Int. J. Primatol.*, 23, 707–740.

Ke, Y. H., Su, B., Song, X. F., Lu, D. R., Chen, L. F., Li, H. Y., Qi, C. J., Marzuki, S., Deka, R., Underhill, P., Xiao, C. J., Shriver, M., Lell, J., Wallace, D., Wells, R. S., Seielstad, M., Oefner, P., Zhu, D. L., Jin, J. Z., Huang, W., Chakraborty, R., Chen, Z., & Jin, L. 2001. African origin of modern humans in East Asia: a tale of 12,000 Y chromosomes. *Science*, 292, 1151–1153.

Kohn, M. H. & Wayne, R. K. 1997. Facts from feces revisited. *Trends Ecol. Evol.*, 12, 223–227.

Kohn, M. H., York, E. C., Kamradt, D. A., Haught, G., Sauvajot, R. M., & Wayne, R. K. 1999. Estimating population size by genotyping faeces. *Proc. Royal Soc. Lond., B*, 226, 657–663.

Lathuillière, M., Ménard, N., Gautier-Hion, A., & Crouau-Roy, B. 2001. Testing the reliability of noninvasive genetic sampling by comparing analyses of blood and fecal samples in Barbary macaques (*Macaca sylvanus*). *Am. J. Primatol.*, 55, 151–158.

Launhardt, K., Epplen, C., Epplen, J. T., & Winkler, P. 1998. Amplification of microsatellites adapted from human systems in faecal DNA of wild hanuman langurs (*Presbytis entellus*). *Electrophoresis*, 19, 1356–1361.

Lawler, R. R., Richard, A. F., & Riley, M. A. 2001. Characterization and screening of microsatellite loci in a wild lemur population (*Propithecus verreauxi verreauxi*). *Am. J. Primatol.*, 55, 253–259.

Little, K., Sommer, V., & Bruford, M. W. 2002. Genetics and relatedness: a test of hypotheses using wild hanuman langurs (*Presbytis entellus*). Caring for Primates. *Abstracts, XIXth Congress. The International Primatological Society*, p. 107. Beijing: Mammalogical Society of China.

Luikart, G. & England, P. R. 1999. Statistical analysis of microsatellite DNA data. *Trends Ecol. Evol.*, 14, 253–256.

Marshall, T. C., Slate, J., Kruuk, L. E. B., & Pemberton, J. 1998. Statistical confidence for likelihood-based paternity inference in natural populations. *Mol. Ecol.*, 7, 639–655.

Mitani, J. C., Merriwether, D. A., & Zhang, C. B. 2000. Male affiliation, cooperation and kinship in wild chimpanzees. *Anim. Behav.*, 59, 885–893.

Monda, K. L. 1996. A phylogenetic study of the gibbons (*Hylobates*) using the control region of the mitochondrial genome. Master's thesis, University of California, San Diego.

Morin, P. A. 1992. Population genetics of chimpanzees. PhD dissertation, University of California, San Diego.

Morin, P. A., Chambers, K. E., Boesch, C., & Vigilant, L. 2001. Quantitative polymerase chain reaction analysis of DNA from noninvasive samples for accurate microsatellite genotyping of wild chimpanzees (*Pan troglodytes verus*). *Mol. Ecol.*, 10, 1835–1844.

Morin, P. A., Mahboubi, P., Wedel, S., & Rogers, J. 1998. Rapid screening and comparison of human microsatellite markers in baboons—allele size is conserved, but allele number is not. *Genomics*, 53, 12–20.

Morin, P. A., Moore, J. J., Chakraborty, R., Jin, L., Goodall, J., & Woodruff, D. S. 1994. Kin selection, social structure, gene flow, and the evolution of chimpanzees. *Science*, 265, 1193–1201.

Morin, P. A., Moore, J. J., & Woodruff, D. S. 1992. Identification of chimpanzee subspecies with DNA from hair and allele specific probes. *Proc. Royal Soc. Lond., B*, 249, 293–297.

Morin, P. A., Moore, J. J., & Woodruff, D. S. 1995. Chimpanzee kinship—reply. *Science*, 268, 186–188.

Morin, P. A., Saiz, R., & Monjazeb, A. 1999. High throughput single nucleotide polymorphism genotyping by fluorescent 5′ exonuclease assay. *BioTechniques*, 27, 538–552.

Morin, P. A., Wallis, J., Moore, J. J., Chakraborty, R., & Woodruff, D. S. 1993. Noninvasive sampling and DNA amplification for paternity exclusion, community structure, and phylogeography in wild chimpanzees. *Primates*, 34, 347–356.

Morin, P. A., Wallis, J., Moore, J. J., & Woodruff, D. S. 1994. Paternity exclusion in a community of wild chimpanzees using hypervariable simple sequence repeats. *Mol. Ecol.*, 3, 469–477.

Morin, P. A. & Woodruff, D. S. 1992. Paternity exclusion using multiple hypervariable microsatellite loci amplified from nuclear DNA of hair cells. In: *Paternity in Primates: Genetic Tests and Theories* (Ed. by R. D. Martin, A. F. Dixson, & E. J. Wickings), pp. 63–91. Basel: Karger.

Morin, P. A. & Woodruff, D. S. 1996. Noninvasive genotyping for vertebrate conservation. In: *Molecular Genetic Approaches in Conservation* (Ed. by T. B. Smith & R. K. Wayne), pp. 298–313. Oxford: Oxford University Press.

Mubumbila, M. V., Duclaud, S., Toussaint, J. L., & Kempf, J. 1992. PCR amplification and identification of D-loop region in primate mtDNA. *Abstracts of XIVth Congress International Primatological Society*, p. 370. Strasbourg: IPS.

Mueller, U. G. & Wolfenbarger, L. L. R. 1999. AFLP genotyping and fingerprinting. *Trends Ecol. Evol.*, 14, 389–394.

Muir, C. C., Galdikas, B. M. F., & Beckenback, A. T. 1994. Genetic variability in orangutans. In: *The Neglected Ape* (Ed. by R. E. A. Nadler), pp. 267–272. New York: Plenum.

Muir, C. C., Galdikas, B. M. F., & Beckenback, A. T. 1998. Is there sufficient evidence to elevate the orangutan of Borneo and Sumatra to separate species? *J. Mol. Evol.*, 46, 378–381.

Muir, C. C., Galdikas, B. M. F., & Beckenback, A. T. 2000. mtDNA sequence diversity of orangutans from the islands of Borneo and Sumatra. *J. Mol. Evol.*, 51, 471–480.

Mundy, N. I., Pissinatti, A., & Woodruff, D. S. 2000. Multiple nuclear insertions of mitochondrial cytochrome *b* sequences in callitrichine primates. *Mol. Biol. Evol.*, 17, 1075–1080.

Mundy, N. I., Winchell, C. S., Burr, T., & Woodruff, D. S. 1997. Microsatellite variation and microevolution in the critically endangered San Clemente Island loggerhead shrike (*Lanius ludovicianus mearnsi*). *Proc. Royal Soc. Lond. B*, 264, 869–875.

Navidi, W., Arnheim, N., & Waterman, M. S. 1992. A multiple-tubes approach for accurate genotyping of very small DNA samples by using PCR: statistical considerations. *Am. J. Hum. Genet.*, 50, 347–359.

Nei, M. & Kumar, S. 2000. *Molecular Evolution and Phylogenetics*. New York: Columbia University Press.

Nievergelt, C. M., Digby, L. J., Ramakrishnan, U., & Woodruff, D. S. 2000. Genetic analysis of group composition and breeding system in a wild common marmoset (*Callithrix jacchus*) population. *Int. J. Primatol.*, 21, 1–20.

Nievergelt, C. N., Mundy, N. I., & Woodruff, D. S. 1998. Microsatellite primers for genotyping common marmosets (*Callithrix jacchus*) and other callitrichids. *Mol. Ecol.*, 7, 1432–1434.

Nievergelt, C. M., Mutschler, T., Feistner, A. T. C., & Woodruff, D. S. 2002. The social system of the Alaotran gentle lemur (*Hapalemur griseus alaotrensis*): genetic characterization of group compositions and mating systems. *Am. J. Primatol.*, 57, 157–176.

Nievergelt, C. M., Pastorini, J. & Woodruff, D. S. 2002. Genetic variability and phylogeography in the wild Alaotran gentle lemur population. In: *Primatology and Anthropology: Into the Third Millennium* (ed. by C. Soligo, G. Anzenberger, & R. D. Martin), pp. 175–179. Evol. Anthropol. Suppl. New York: Wiley.

Oka, T. & Takenaka, O. 2001. Wild gibbon's parentage tested by noninvasive DNA sampling and PCR-amplified polymorphic microsatellites. *Primates*, 42, 67–73.

Pääbo, S. 1990. Amplifying ancient DNA. In: *PCR Protocols* (ed. by M. A. Innis, D. H. Gelfand, J. J. Sninsky, & T. J. White), pp. 159–166. San Diego: Academic Press.

Paxinos, E., McIntosh, C., Ralls, K., & Fleischer, R. 1997. A noninvasive method for distinguishing among canid species: amplification and enzyme restriction of DNA from dung. *Mol. Ecol.*, 6, 483–486.

Pemberton, J. M., Coltman, D. W., Couson, T. N., & Slate, J. 1999. Using microsatellites to measure the fitness consequences of inbreeding and outbreeding. In: *Microsatellites: Evolution and Applications* (ed. by D. B. Goldstein & C. Schlötterer), pp. 150–164. Oxford: Oxford University Press.

Poole, K.G., Mowat, G., & Fear, D.A. 2001. DNA-based population estimate for grizzly bears *Ursus arctos* in northeastern British Columbia, Canada. *Wildl. Biol.*, 7, 105–115.

Queller, D. C. & Goodnight, K. F. 1989. Estimating relatedness using genetic markers. *Evolution*, 43, 258–275.

Rasoloarison, R. M., Goodman, S. M. & Ganzhorn, J. U. 2000. Taxonomic revision of mouse lemurs (*Microcebus*) in the western portions Madagascar. *Int. J. Primatol.*, 21, 963–1019.

Reed, J. Z., Tollit, D. J., Thompson, P. M., & Amos, W. 1997. Molecular scatology: the use of molecular genetic analysis to assign species, sex and individual identity to seal faeces. *Mol. Ecol.*, 6, 225–234.

Roff, D. A. 2002. *Life History Evolution*. Sunderland, MA: Sinauer.

Roos, C. & Geissmann, T. 2001. Molecular phylogeny of the major hylobatid divisions. *Mol. Phylogenet. Evol.*, 19, 486–494.

Rosenbaum, H. C. & Dienard, A. S. 1998. Caution before claim: an overview of microsatellite analysis in ecology and evolutionary biology. In: *Molecular Approaches to Ecology and Evolution* (ed. by R. DeSalle & B. Schierwater), pp. 87–106. Berlin: Birkhauser.

Saltonstall, K., Amato, G., & Powell, J. 1998. Mitochondrial DNA variability in Grauer's gorillas of Kahuzi-Biega National Park. *J. Hered.*, 89, 129–135.

Schlötterer, C. 1998. Microsatellites. In: *Molecular Genetic Analysis: A Practical Approach*, 2nd ed. (ed. by A. R. Hoelzel), pp. 237–262. Oxford: Oxford University Press.

Schlötterer, C. & Pemberton, J. 1998. The use of microsatellites for genetic analysis of natural populations—a critical review. In: *Molecular Approaches to Ecology and Evolution* (ed. by R. DeSalle & B. Schierwater), pp. 71–86. Berlin: Birkhauser.

Sloane, M. A., Sunnucks, P., Alpers, D., Beheregaray, L. B., & Taylor, A. C. 2000. Highly reliable genetic identification of individual northern hairy-nosed wombats from single remotely collected hairs: a feasible censusing method. *Mol. Ecol.*, 9, 1233–1240.

Smith, K. L., Alberts, S. C., Bayes, M. K., Bruford, M. W., Altmann, J., & Ober, C. 2000. Cross-species amplification, noninvasive genotyping, and non-Mendelian inheritance of human STRPs in savannah baboons. *Am. J. Primatol.*, 51, 219–227.

Smith, T. B. & Wayne, R. K. (eds.) 1996. *Molecular Genetic Approaches in Conservation*. New York: Oxford University Press.

Srikwan, S., Hufford, K., Eggert, L., & Woodruff, D. S. 2002. Variable microsatellite markers for genotyping tree shrews, *Tupaia*, and their potential use in genetic studies of fragmented populations. *ScienceAsia*, 28, 93–97.

Srikwan, S. & Woodruff, D. S. 2000. Monitoring genetic erosion in mammal populations following tropical forest fragmentation. In: *Genetics, Demography and Viability of Fragmented Populations* (ed. by A. G. Young & G. M. Clarke), pp. 149–172. Cambridge: Cambridge University Press.

Stone, A. C., Griffiths, R. C., Zegura, S. L., & Hammer, M. F. 2002. High levels of Y-chromosome nucleotide diversity in the genus *Pan*. *Proc. Natl. Acad. Sci. USA*, 99, 43–48.

Sugg, D. W., Chesser, R. K., Dobson, F. S., & Hoogland, J. L. 1996. Population genetics meets behavioral ecology. *Trends Ecol. Evol.*, 11, 338–342.

Sugiyama, Y., Kawamoto, S., Takenaka, O., Kumazaki, K., & Miwa, N. 1993. Paternity discrimination and inter-group relationships of chimpanzees at Bossou. *Primates*, 34, 545–552.

Sunnucks, P. 2000. Efficient genetic markers for population biology. *Trends Ecol. Evol.*, 15, 199–203.

Surridge, A. K., Smith A. C., Buchanan-Smith, H. M., & Mundy, N. I. 2002. Single-copy nuclear DNA sequences obtained from noninvasively collected primate feces. *Am. J. Primatol.*, 56, 185–190.

Syvanen, A. C. 2001. Accessing genetic variation: genotyping single nucleotide polymorphisms. *Nat. Rev. Genet.*, 2, 930–942.

Taberlet, P., Camarra, J. J., Griffin, S., Uhres, E., Hanotte, O., Waits, L. P., Dubois-Paganon, C., Burke, T., & Bouvert, J. 1997. Noninvasive genetic tracking of the endangered Pyrenean brown bear population. *Mol. Ecol.*, 6, 869–876.

Taberlet, P., Griffin, S., Goossens, B., Questiau, S., Manceau, V., Escaravage, N., Waits, L. P., & Bouvet, J. 1996. Reliable genotyping of samples with very low DNA quantities using PCR. *Nucleic Acids Res.*, 24, 3189–3194.

Taberlet, P. & Luikart, G. 1999. Noninvasive genetic sampling and individual identification. *Biol. J. Linnean Soc.*, 68, 41–55.

Taberlet, P., Luikart, G., & Gefen, E. 2001. New methods for obtaining and analyzing genetic data from free-ranging carnivores. In: *Carnivore Conservation* (Ed. by J. L. Gittleman, S. M. Funk, D. Macdonald, & R. K. Wayne), pp. 313–334. Cambridge: Cambridge University Press.

Taberlet, P. & Waits, L. P. 1998. Noninvasive genetic sampling. *Trends Ecol. Evol.*, 13, 26–27.

Taberlet, P., Waits, L. P., & Luikart, G. 1999. Noninvasive genetic sampling: look before you leap. *Trends Ecol. Evol.*, 14, 323–327.

Takasaki, H. & Takenaka, O. 1991. Paternity testing in chimpanzees with DNA amplification from hairs and buccal cells in wadges: a preliminary note. In: *Primatology Today: Proceedings of the XIIIth Congress International Primatological Society* (Ed by A. Ehara, T. Kimura, O. Takenaka, & M. Iwamoto), pp. 613–616. Amsterdam: Elsevier.

Takenaka, O., Takasaki, S., Kawamoto, S., & Takenaka, A. 1992. Polymorphic microsatellite DNA amplification customized for chimpanzee paternity testing. *Abstracts of XIVth Congress International Primatological Society*, p. 137. Strasbourg: IPS.

Taylor, A. C., Horsup, A., Johnson, C. N., Sunnucks, P., & Sherwin, W. B. 1997. Relatedness structure detected by microsatellite analysis and attempted pedigree reconstruction in an endangered marsupial, the northern hairy-nosed wombat *Lasiorhinus krefftii*. *Mol. Ecol.*, 6, 9–19.

Taylor, A. C., Sherwin, W. B., & Wayne, R. K. 1994. Genetic variation of microsatellite loci in a bottlenecked species: the northern hairy-nosed wombat *Lasiorhinus krefftii*. *Mol. Ecol.*, 3, 277–290.

Valiere, N. & Taberlet, P. 2000. Urine collected in the field as a source of DNA for species and individual identification. *Mol. Ecol.*, 9, 2150–2152.

Vartanian, J. P. & Wain-Hobson, S. 2002. Analysis of a library of macaque nuclear mitochondrial sequences confirms macaque origin of divergent sequences from old oral polio vaccine samples. *Proc. Natl. Acad. Sci. USA*, 99, 7566–7569.

Vigilant, L. 1999. An evaluation of techniques for the extraction and amplification of DNA from naturally shed hairs. *Biol. Chem.*, 380, 1329–1331.

Vigilant, L., Hofreiter, M., Siedel, H., & Boesch, C. 2001. Paternity and relatedness in wild chimpanzee communities. *Proc. Natl. Acad. Sci. USA*, 98, 12890–12895.

Vigilant, L., Stoneking, M., Harpending, H. Hawkes, K., & Wilson, A. C. 1991. African populations and the evolution of human mitochondrial DNA. *Science*, 253, 1503–1507.

von Berholdingen, C. H., Higuchi, R. G., Sensabaugh, G. F., & Erlich, H. A. 1987. Analysis of enzymatically amplified HLA-DQalpha DNA from single human hairs. *Am. J. Hum. Genet.*, 41, 725.

von Segesser, F., Scheffrahn, W., & Martin, R. D. 1995. Parentage analysis within a semi-free-ranging group of Barbary macaques *Macaca sylvanus*. *Mol. Ecol.*, 4, 115–120.

Wang, D. G., Fan, J. B., Siao, C. J., Berno, A., Young, P., Sapolsky, R., Ghandour, G., Perkins, N., Winchester, E., Spencer, J., Kruglyak, L., Stein, L., Hsie, L., Topaloglou, T., Hubbell, E., Robinson, E., Mittmann, M., Morris, M. S., Shen, N. P., Kilburn, D., Rioux, J., Nusbaum, C., Rozen, S., Hudson, T. J., Lipshutz, R., Chee, M., & Lander, E. S. 1998. Large-scale identification, mapping, and genotyping of single-nucleotide polymorphisms in the human genome. *Science*, 280, 1077–1082.

Warren, K. S., Verschoor, E. J., Langenhuijzen, S., Heriyanto, Swan, R. A., Vigilant, L., & Heeney, J. L. 2001. Speciation and intrasubspecific variation of Bornean orangutans, *Pongo pygmaeus pygmaeus*. *Mol. Biol. Evol.*, 18, 472–480.

Whittier, C. A., Dhar, A. K., Stem, C., Goodall, J., & Alcivar-Warren, J. 1999. Comparison of DNA extraction methods for PCR amplification of mitochondria cytochrome c oxidase subunit II (COII) DNA from primate fecal samples. *Biotech. Tech.*, 13, 771–779.

Wilson, E. O. 2000. *Sociobiology. Twenty-fifth Anniversary Edition.* Cambridge, MA: Harvard University Press.

Woodruff, D. S. 1990. Genetics and demography in the conservation of biodiversity. *J. Sci. Soc. Thailand*, 16, 117–132.

Woodruff, D. S. 1993. Non-invasive genotyping of primates. *Primates*, 34, 333–346.

Woodruff, D. S. 2001a. The archetypal flagship adrift. *Conserv. Biol.*, 15, 1189–1192.

Woodruff, D. S. 2001b. Declines of biomes and biotas and the future of evolution. *Proc. Natl. Acad. Sci. USA*, 98, 5471–5476.

Yoder, A. D., Rasoloarison, R. M., Goodman, S. M., Irwin, J. A., Atsalis, S., Ravosa, M. J., & Ganzhorn, J. U. 2000. Remarkable species diversity in Malagasy mouse lemurs (primates, *Microcebus*). *Proc. Natl. Acad. Sci. USA*, 97, 11325–11330.

Zischler, H. 2000. Nuclear integrations of mitochondrial DNA in primates: inference of associated mutational events. *Electrophoresis*, 21, 531–536.

Part II

Kin Compositions: Ecological Determinants, Population Genetics, and Demography

A rhesus (*Macaca mulatta*) matriarch (facing the camera) draws her adult offspring into a grooming cluster on Cayo Santiago, Puerto Rico. Photo by Carol Berman.

4

Is There No Place Like Home? Ecological Bases of Female Dispersal and Philopatry and Their Consequences for the Formation of Kin Groups

Lynne A. Isbell

In L. Frank Baum's century-old tale *The Wizard of Oz*, Dorothy, a female primate, disperses with an ally from her natal home range and group to an unfamiliar area. While in this new area, she encounters unfamiliar plants, potential predators, and aggressive strangers, dangers she never faced at home. Dorothy succeeds by establishing relationships with new allies. Despite her successes in the new area, Dorothy has an overwhelming desire to return home. Baum did send Dorothy home eventually, and back to her kin group. Had she stayed in Oz, she would have left her home range as well as her kin group. Decisions about dispersal from the natal home range are important for understanding the evolution of kin groups, but they are not addressed in two widely recognized models that have been developed to explain the evolution of kin groups.

These two models focus on the benefits of grouping with kin. They are the intergroup competition model, in which inclusive fitness benefits are gained by forming groups of relatives in competition against nonrelatives for food (Wrangham 1980), and the predation/intragroup competition model, in which groups evolve in response to predation and females remain in their natal groups for the inclusive fitness benefits that are gained by forming coalitions of relatives within groups for competition for food (van Schaik 1989). Both of these models assume that group living is inherently costly and that cooperation with relatives outweighs this cost.

A third model contrasts with these models in its focus on the costs of dispersal and their effects on kin grouping. In the dispersal model, high costs of dispersal cause reproductive daughters to stay in the home range. Kin groups then form by default when there is an advantage to living in groups (Isbell 1994, Isbell & Van Vuren 1996). In this model, inclusive fitness benefits of helping kin in intergroup competition are a secondary advantage of living in groups with kin.

These three models are built largely on assumptions, some more so than others, because data are still in short supply. Some of these assumptions may eventually prove to have real-life support, but it is important to be cognizant of each assumption and to be cautious in accepting any without critical examination. To compare the relative merits of the models, it is useful to identify their assumptions, some of which were carried over from previous influential works. Thus, I begin with a chronological overview of the salient points of the models as they pertain to the evolution of kin groups. I then extend the dispersal model by focusing on the ecological bases of dispersal and philopatry from the mother's perspective. The result is the dispersal/foraging efficiency model, which describes the evolution of kin groups as a series of small, incremental steps that happen as individuals attempt to maintain or improve their foraging efficiency.

I focus on females because when kin groups form, they usually form around females and because female reproductive success is affected more than male reproductive success by ecological influences. Nonetheless, because there are times when male dispersal decisions affect female decisions to remain in or leave the natal group, I discuss males when appropriate, mainly but not entirely in the context of sex-biased dispersal. Phylogeny may play a role in dispersal patterns (Di Fiore & Rendall 1994, Isbell & Young 2002), but I emphasize the ecological influences on dispersal. I limit discussion to dispersal from the natal group or home range, because natal dispersal breaks up kin groups whereas subsequent movements simply maintain nonkin groups.

A Brief History of Group Living, Female Kin Groups, and Dispersal

The evolution of kin groups has spawned a lively debate since Alexander (1974) challenged the prevailing view that group living is inherently beneficial to all group members. Alexander uncompromisingly stated that "there is no automatic or universal benefit from group living. Indeed, the opposite is true: there are automatic and universal detriments, namely, increased intensity of competition for resources, including mates, and increased likelihood of disease and parasite transmission" (p. 328). Recognizing that there must be benefits that offset the disadvantages of group living, Alexander maintained that three advantages could be gained: (1) reduced susceptibility to predation through cooperative defense, dilution, or selfish herd effects (Hamilton 1971); (2) improved ability to get food, through either group hunting or better detection of scattered foods; and (3) shared use of a large and highly restricted resource such as a location for sleeping sites. These three advantages would not necessarily promote similar kinds of social groups, however. Individuals attempting to reduce predation or improve food finding might be attracted to others, whereas individuals needing a sleeping cliff might aggregate and yet still not be attracted to others. Alexander ruled out all but predation as the selective force favoring grouping in primates because (1) primates do not hunt in groups (with the exceptions of modern humans [*Homo sapiens*] and chimpanzees, [*Pan troglodytes*]); (2) they do not seem to use each other to locate food (although there is now evidence that in some primates, for example rhesus macaques [*Macaca mulatta*] and chimpanzees, individuals alert others to the locations of rich food sources [Hauser & Wrangham 1987, Hauser & Marler 1993, Hauser et al. 1993]); and (3) they do not have sufficiently restricted and critical resources (even including the sleeping cliffs of hamadryas baboons [*Papio cynocephalus hamadryas*]) to force them to aggregate.

Alexander contributed the assumption that group living is inherently costly and that the only benefit that outweighs the cost of living in groups is predation. The assumption that group living is inherently costly has been repeated often and by now permeates our views of primate sociality (e.g., Wrangham 1980, Terborgh & Janson 1986, Janson & van Schaik 1988, van Schaik 1989, Isbell 1991, Sterck et al. 1997). It has been questioned only recently (Isbell & Young 2002). More intensively debated has been the importance of predation as the selective factor that favors group living.

The evolution of group living became closely linked with the evolution of kin groups when Wrangham (1980) proposed that groups evolved in response to food competition, not predation. He argued that when animals are faced with large, clumped, and defendable foods, individuals that cooperate with others can outcompete those that do not cooperate. All else being equal, the best ones with whom to cooperate are kin because helping kin can increase one's inclusive fitness whereas helping nonkin cannot. According to Wrangham, kin groups evolved in the context of intergroup competition. For females living in nonkin groups, Wrangham suggested that male harassment of females favors aggregations of females around protective males (figure 4.1).

Partly because it could explain both the evolution of grouping and the evolution of kin groups, Wrangham's scenario was attractive. It was appealing also because it used the

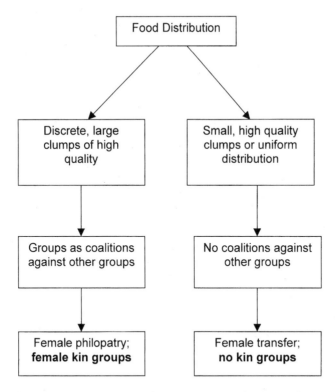

Figure 4.1. A schematic representation of the factors leading to female kin group according to the intergroup competition model (Wrangham 1980).

same theoretical reasoning previously applied to bats (Bradbury & Vehrencamp 1977) and birds (Emlen & Oring 1977) that since female reproductive success is most limited by food (Trivers 1972), females should act in ways that maximize their access to food. The intergroup competition model was not based on information from all primates, however; 23 of 29 species (79%) in the model were catarrhine primates, and 16 of the 23 catarrhines (70%) were cercopithecines, a reflection of the emphasis at that time on studies of Old World primates. It also suffered from a lack of information on female movements. At the time of the model's publication, females were known to breed in their natal groups in only four species, all of them cercopithecines. To overcome the lack of direct evidence, Wrangham used male dispersal as indirect evidence of female residence in the natal group because the two appeared to be closely associated, and indeed they are in cercopithecines. In a few additional species, no information existed on dispersal patterns of males and females. Interactions between females within groups that involved grooming, huddling together, coalition formation, and agonistic interactions were then considered to be representative of female kin groups. A near absence of data thus tied male dispersal to female philopatry and female kin groups to clearly delineated female relationships within groups, that is, easily determined, or strong, dominance hierarchies. Studies since then have shown, however, that females and males both disperse in many primate species (Moore 1984; Strier 1994, 1999), and that in some species, females have dominance hierarchies that are difficult to detect even though they live in kin groups (Isbell & Pruetz 1998, Cords 2000).

Wrangham's model was challenged by van Schaik (1983), who argued against intergroup competition and in favor of predation as the primary selective factor favoring group living. Following Alexander's assumption that group living is always costly, and extending it by assuming that living in larger groups is always costlier than living in smaller groups, van Schaik examined infant/adult female ratios across 14 species (27 data points). He specifically restricted his analysis to those species that were thought at the time to live in female kin groups so that he could test the intergroup competition model. He found that in most cases the number of infants decreased as group size increased, a pattern that would not be expected if the benefits of intergroup competition outweigh the cost of intragroup competition. He interpreted this as evidence against the intergroup competition hypothesis for the evolution of group living, and as support for the predation hypothesis.

As was the case for Wrangham (1980), van Schaik's (1983) approach was undermined by lack of information about female movements. Most important, females in three genera that comprised nearly half of the data points (12 of 27) in van Schaik's analysis are now known to disperse regularly (*Alouatta*) or at least occasionally (*Presbytis* and *Trachypithecus*), substantially weakening the purpose of the analysis as a test of the intergroup competition model. In addition, while fewer infants per female in larger groups may be a real phenomenon, other factors, such as infanticide, can also plausibly explain van Schaik's results (Isbell 1991, Crockett & Janson 2000, Steenbeek 2000). Interestingly, infanticide is now being considered one of the main factors influencing female dispersal decisions, particularly in *Alouatta*, *Presbytis*, and *Trachypithecus* (Isbell 1991, Isbell & Van Vuren 1996, Sterck et al. 1997, Crockett & Janson 2000, Sterck & Korstjens 2000).

The renewed emphasis on predation as the ultimate selective pressure favoring group living provided the opportunity for alternative models for the evolution of female kin groups. If, as van Schaik (1983) argued, intergroup competition was not an important positive force in the evolution of group living, then female kin groups must have evolved for other reasons.

In van Schaik's (1989) predation/intragroup competition model, female kin groups ultimately evolved because predation forces females to live in groups, and the inevitable intragroup competition that occurs with group living favors coalition formation by kin when foods are clumped and therefore monopolizable (figure 4.2). Here van Schaik (1989) agreed with Wrangham (1980) that females form coalitions with kin because they gain inclusive fitness benefits by doing so. Van Schaik (1989) differed from Wrangham (1980), however, by proposing that coalition formation with kin depends not only on the distribution of foods, but also on the intensity of predation. Moreover, coalitions were argued to occur largely within groups rather than between groups in response to monopolizable foods (figure 4.2). Van Schaik (1989) allowed intergroup competition to favor coalitions and therefore kin groups only in rare situations where low predation risk allows females to live in less spa-

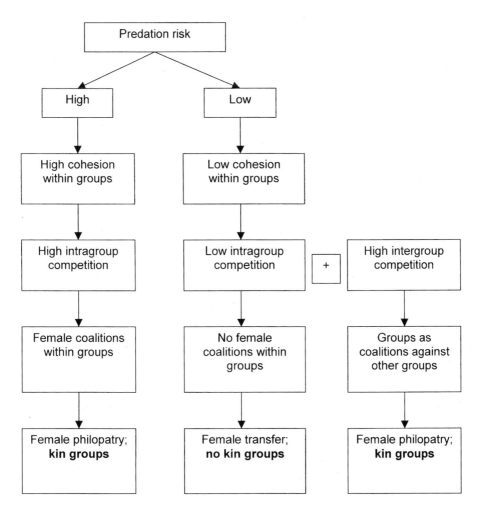

Figure 4.2. A schematic representation of the factors leading to female kin groups according to the predation/intragroup competition model (van Schaik 1989, Sterck et al. 1997).

tially cohesive groups. Increased interindividual distances would decrease intragroup competition to such an extent that it would become less important than intergroup competition (figure 4.2).

Eight years later, Sterck et al. (1997) modified the predation/intragroup competition model to incorporate the growing evidence of female dispersal in species with male dispersal. Though earlier van Schaik (1989) had criticized the intergroup competition model for being unable to explain why females that disperse from their natal groups nonetheless live in groups, Sterck et al. (1997) now converged with it by suggesting that such groups evolve in response to a selective pressure other than the one that favors female kin groups. Whereas the intergroup competition model held that nonkin groups evolve because females benefit from aggregating around males that can protect them from harassment, the modified predation/intragroup competition model proposed that nonkin groups evolve because females benefit from aggregating around males that can protect them from infanticide. Thus, both models agree not only that female kin groups form because of the benefits of coalitions (of one kind or another), they also agree that nonkin groups form because of the benefits of aggregating around males.

Isbell (1991) and colleagues (Isbell & Van Vuren 1996, Isbell & Pruetz 1998, Isbell, Pruetz, & Young 1998, Pruetz & Isbell 2000, Isbell & Enstam 2002, Isbell & Young 2002, Mathy & Isbell 2002) have been working on an alternative model that differs from both the intergroup competition model and the predation/intragroup competition model. Analyzing quantitative data, Isbell (1991) found that species in which groups of females are aggressive toward other groups also expand home ranges with increasing group size (and likewise decrease them with decreasing group size). In nearly all catarrhine species surveyed, females that display intergroup aggression typically remain in their natal groups, a finding consistent with expectations of the intergroup competition model. She suggested that females are aggressive toward other groups because aggression helps to minimize losses of food resources to groups that would otherwise expand into their home ranges. Though her conclusion was consistent with Wrangham's emphasis on intergroup competition for the evolution of kin groups, it differed in the ecological basis for intergroup competition. The combination of home range expansion with increasing group size and invariable female aggression between groups suggested that food abundance, as opposed to food distribution, affects female reproductive success in most species. This inference has subsequently been supported by data from hanuman langurs (*Semnopithecus entellus*) at Ramnagar, Nepal (Koenig 2000) and squirrel monkeys (*Saimiri* spp.: Boinski et al. 2002).

Among populations of females apparently constrained by food abundance, Isbell (1991) found that daily travel distance increases with increasing group size in primates that feed extensively on fruits but not in those that feed extensively on leaves or arthropods (see also Janson & Goldsmith 1995). Nearly all species surveyed that increase daily travel distance with group size also have strong female dominance hierarchies within groups, whereas nearly all species that do not adjust daily travel distance to group size have much less obvious dominance hierarchies. Because fruits are thought to be clumped, and leaves and arthropods ubiquitous or dispersed, she inferred, as did Wrangham (1980) and van Schaik (1989), that the spatial distribution of food resources determines competitive relationships within groups. In fact, more recent investigations of the ecological determinants of variation in competitive relationships among females within groups suggest that the critical characteristic is temporal rather than strictly spatial. Larger food size or, more important, longer feeding site depletion time makes foods more usurpable than do shorter distances between foods (i.e., more clumped distributions; Shopland 1987, Janson 1990, Isbell & Pruetz 1998, Isbell, Pruetz, & Young 1998, Mathy & Isbell 2002).

These results disconnected female philopatry from strong dominance hierarchies for the first time. Isbell's model described a type of female that the intergroup competition model did not recognize and that the predation/intragroup competition model recognized only after changing predation pressure, intragroup competition, and intergroup competition. Isbell's model was more parsimonious than the predation/intragroup competition model in that the existence of this type of female required only a change in the usurpability of food among females limited by food abundance. These species are represented by patas monkeys (*Erythrocebus patas*) and at least two other species of African guenon (blue monkeys [*Cercopithecus mitis*] and redtailed monkeys [*C. ascanius*]). Patas monkeys have female philopatry, intergroup aggression, large interindividual distances within groups, dominance hierarchies that are difficult to detect, and spatially dispersed foods, but they also have heavy predation (Chism et al. 1984, Chism & Rowell 1988, Isbell & Pruetz 1998, Isbell, Pruetz, & Young 1998, Pruetz & Isbell 2000, Isbell & Enstam 2002). Their heavy predation is contrary to what the predation/intragroup competition model predicts, making them more accurately described by Isbell's (1991) model than the predation/intragroup competition model.

In a few species, for example red colobus (*Procolobus badius*), Isbell (1991) found no behavioral indicators of competition, that is, no significant changes in home range size and daily travel distance with group size and little aggression between females either within or between groups. The lack of behavioral indicators of competition led Isbell to begin to question the assumption that increased competition is inherent in group living.

Like Sterck et al. (1997), Isbell and Van Vuren (1996) investigated the growing reports of greater variability in female dispersal patterns but again from a different perspective, that of differential costs of locational and social dispersal to individuals. Locational dispersal involves movement away from a familiar place, whereas social dispersal involves movement away from familiar conspecifics. The main potential costs of locational dispersal are increased risk of predation and poorer access to foods through lack of knowledge about the new environment. The main potential cost of social dispersal is increased aggression, coming from strangers. For group-living animals, there are three potential combinations of dispersal: locational dispersal without social dispersal, locational dispersal coupled with social dispersal, and social dispersal without locational dispersal, each of which has a different set of costs. Only the last two types have an impact on the evolution of kin groups.

Isbell and Van Vuren (1996) found that for catarrhine primates, regular female dispersal from the natal group to another group (transfer) is most likely when the costs of dispersal are minimal, that is, when there is little aggression between females of different groups and females are able to remain in much, if not all, of their natal home ranges when they transfer (e.g., Kibale red colobus and mountain gorillas [*Gorilla gorilla berengei*]). In some catarrhine species (e.g., banded leaf monkeys [*Presbytis melalophos*] and capped langurs [*Trachypithecus pileata*]), female transfer is "occasional," that is, not regular but occurring more often than can be called exceptional (see Isbell & Van Vuren 1996 for quantitative cutoffs for regular, occasional, and exceptional transfer). In these species, females also face little aggression from females of other groups and so face minimal costs of social dispersal. They often also have extensively overlapping home ranges with neighboring groups and so face minimal costs of locational dispersal (Isbell & Van Vuren 1996).

Dispersal by individual females in species with regular and occasional female transfer was suggested to be heavily dependent on the chances of successful reproduction in their natal groups. These females are the same as those catarrhines suggested by Isbell (1991) to not be limited in their reproductive success by food abundance and by Wrangham (1980) and Sterck et al. (1997) as aggregating around males for protection from harassment or infanticide. Isbell and Van Vuren (1996) suggested that female reproductive success in these species depends less on food abundance than on attributes of individual males. Such females would be expected to leave their natal groups when their reproductive success is threatened by infanticidal males or incestuous matings with fathers and brothers. Females would also be expected to leave if they fail to reproduce for reasons unrelated to male behavior, however, because there are additional causes of reproductive failure in females, for example hormonal insufficiency or fetal damage (e.g., Albrecht et al. 2000).

Isbell and Van Vuren (1996) suggested that in those catarrhine primates with only exceptional female dispersal at most, for example vervets (*Cercopithecus aethiops*) and macaques (*Macaca* spp.), females are philopatric because aggression from strangers and/or movement into unfamiliar areas make dispersal too costly. These exceptional cases of female dispersal occur only in unusual situations when females fail to reproduce or are very unlikely to reproduce compared to others in their current groups.

In contrast to catarrhines, platyrrhine females are actually more likely to disperse from their natal groups despite aggression from strangers (costs of locational dispersal could not be examined in New World primates because all the species meeting the criterion for inclusion, that is, cohesive multifemale groups, had no variation in home range overlap, the measure used to estimate costs of locational dispersal). In the absence of more inclusive data, Isbell and Van Vuren (1996) speculated that although both catarrhine primates with female philopatry and platyrrhine primates have costs of social dispersal, platyrrhines might have lower costs of locational dispersal than catarrhines with female philopatry, making dispersal costs lower overall and thus making dispersal more likely in platyrrhines. They suggested that the potential for lower costs of locational dispersal among platyrrhines might exist because extensive home range overlap minimizes unfamiliarity with new areas and there were no reports of platyrrhines experiencing a New World equivalent of the leopard (*Panthera pardus*), a mammalian predator that can decimate primate groups in a short period of time (e.g., Isbell 1990, Isbell & Enstam 2002).

The focus on costs of dispersal to individuals questions the scenario that inclusive fitness benefits from helping kin defend food resources from nonkin (within or between groups) was the selective advantage behind kin group formation. Given that the costs of dispersal would have been sufficient to keep solitarily foraging females in the natal home range, Isbell and Van Vuren (1996) suggested that kin groups would have formed by default once there was an overall advantage to living in groups. Inclusive fitness benefits would be gained as a secondary advantage of living in groups.

This model, like the others, was hindered by insufficient information. Gaps existed particularly in the natural history of New World primates, Malagasy prosimians, and colobines. The long debate over the selective pressures favoring kin groups has, fortunately, helped to generate studies of some of the less well known taxa, providing an opportunity for further modification and refinement of models. In the next section, I summarize the dispersal model from the usual perspective of the disperser and then extend it by taking the mother's perspective and adding data from hitherto underrepresented taxa.

Dispersal from the Offspring's Perspective

Because primates are mammals, male reproductive success is ultimately dependent on females: if females fail to reproduce, males also fail. Males are constrained to react to, rather than determine, female decisions to disperse. When costs of social and locational dispersal exist, individual females should attempt to remain philopatric, but only as long as those individuals can reproduce successfully at home. If a female cannot reproduce where she is, she should take her chances with the potential costs of dispersal in her current social and ecological milieu and leave. Sometimes no costs of dispersal exist that are strong enough to affect reproductive success. Nonetheless, the same bottom line applies: a female should disperse when her chances of reproducing are better elsewhere. This bottom line also applies to males, but because the causes of reproductive failure are more numerous in females, female dispersal is more complex than that of males.

Solitary foraging is generally viewed as the ancestral mammalian foraging/social system (Charles-Dominique 1978, Eisenberg 1981). To understand the evolution of kin groups, it might be profitable to examine the ecological differences between having exclusive access to one's home range (which requires dispersal of offspring) and sharing it with other reproductive females (which allows philopatry of female offspring). Surviving in the home range without reproducing can be as evolutionarily insignificant as dying while dispersing. Thus, when females face costs of dispersal and still disperse, it is likely to be because they would not have reproduced had they remained. If, however, daughters are presented with the opportunity to reproduce in the natal home range or group, they should stay. This opportunity may arise if mothers are able to share their home ranges with their daughters.

Dispersal from the Mother's Perspective

Since the reproductive success of mothers depends not only on the survival and reproduction of their offspring but also on the mother's own ability to obtain sufficient food for future reproduction, mothers are expected to share their home ranges with their reproductive daughters only when both the costs of dispersal make it unlikely that their daughters will reproduce in a new area and they can maintain their own reproductive output. If mothers can expand their home ranges to accommodate their reproductive daughters, their own reproductive success is expected not to be diminished. Primate mothers appear to differ, however, in the extent to which they are able to expand their home ranges to accommodate their reproductive daughters. Five different types of mothers can be distinguished on the basis of home range overlap and expansion: stingy mothers, generous mothers, incomplete suppressors, facilitators, and indifferent mothers.

"Stingy Mothers": Female Dispersal Required

From the perspective of dispersal, there is little difference between females that are traditionally considered solitary, females that live in monogamous groups, and females that form cohesive groups with other adult, but nonbreeding, females. In all cases, only one female in a given "group" typically reproduces, and in all cases, females that do not reproduce typically disperse socially when they have the opportunity to reproduce elsewhere. In many of these cases, reproduction is also limited to one female within a local area. Such females do not share the resources within their home ranges with other reproductive females. These females are referred to as stingy mothers here.

The most obvious examples of stingy mothers are socially monogamous species and solitarily foraging species with minimally overlapping home ranges and aggression between reproductive females. Female pottos (*Perodicticus potto*) are solitary foragers with home ranges that overlap only minimally with those of other females (Charles-Dominique 1977, Bearder 1987). Female western tarsiers (*Tarsius bancanus*) and aye-ayes (*Daubentonia madagascariensis*) also forage alone in nonoverlapping home ranges (Bearder 1987, Sterling 1993). The socially monogamous gibbons (*Hylobates* spp.), titi monkeys (*Callicebus* spp.), owl monkeys (*Aotus* spp.), indri (*Indri indri*), woolly lemurs (*Avahi*), and bamboo lemurs (*Hapalemur* spp.) are territorial, with minimally overlapping home ranges (Wright 1986, Leighton 1987, Palombit 1994, Nievergelt et al. 1998, Fuentes 2000, Kappeler 2000, Thalmann 2001, Bossuyt 2002).

Also considered stingy are females that breed to the exclusion of other females even when multiple adult females share the home range. They are stingy mothers because shared food resources are not typically converted to offspring for any but the one reproductive female. Thus, although more than one adult female may be present in family groups of tamarins (*Saguinus*), lion tamarins (*Leontopithecus*), marmosets (*Callithrix*), and pygmy marmosets (*Cebuella*), they are considered to have stingy mothers because reproduction is limited to one female, with informative exceptions (Goldizen 1987, Ferrari & Lopes Ferrari 1989, Savage 1990, Garber 1993, Rylands 1993, Soini 1993, Digby & Ferrari 1994, Goldizen et al. 1996, Savage et al. 1996). Suppression of reproduction is well documented in female marmosets and tamarins and continues until either the reproductively active female disappears from the group or the suppressed females leave (Abbott et al. 1993). It has not been considered for other stingy females, but in gibbons, if the mother disappears, the daughter will not disperse and will reproduce (Leighton 1987), suggesting a release from some sort of reproductive suppression.

Complete reproductive suppression in philopatric adult female offspring may represent a balance for mothers between the costs of sending their daughters out into the world (high risk of mortality) and the cost of sharing resources with their daughters' offspring (reduction of mothers' future reproduction). Adult daughters may "agree" to suppress their own reproduction if it means they can remain in the natal home range until an opportunity to reproduce arises either in the natal home range or in another home range as a result of the disappearance or displacement of the resident female reproducer on that home range.

Callitrichid groups can at times have more than one reproductive female (Goldizen 1987, Rothe & Koenig 1991, Dietz & Baker 1993, Digby 1995, Goldizen et al. 1996). In many of these cases, the dominant female attempts to kill the offspring of the other females (Digby 2000), a behavior that would be expected of stingy mothers. In lion tamarins (*Leontopithecus rosalia*) without female infanticide, the reproductive females were known to be mothers and daughters, and although the daughters had poorer reproductive success than their mothers, they had greater reproductive success than females that dispersed. Polygyny in lion tamarins was positively correlated with quality of home ranges and home range size (Dietz & Baker 1993), suggesting that kin groups can form at least temporarily if mothers are willing and able to expand their home ranges to accommodate their grandoffspring. This does not seem to be the norm with stingy mothers, however.

Stingy mothers do not often appear to take opportunities to expand their home ranges or they seem unable to expand their home range boundaries because of aggression by neighbors. One female potto left her home range to her daughter rather than expand it (Charles-Dominique 1977). Other female pottos did not expand their home ranges when openings arose but instead left their home ranges and moved to an entirely different home range (Charles-Dominique 1977). Similarly, groups of pygmy marmosets (*Cebuella pygmaea*) moved to new home ranges instead of expanding their old home ranges when their food resources declined (Soini 1993). Saddle-back tamarins (*Saguinus fuscicollis*) at Manu, Peru, did not pass beyond their home range boundaries into another group's home range even when the other group was away (Terborgh 1983). Home range boundaries persisted over many years in owl monkeys (*Aotus trivirgatus*; Terborgh 1983, Wright 1986, Peres 2000). For at least four years, female siamangs (*Hylobates syndactylus*) did not expand into home ranges left open when an illness went through the population (Palombit 1994). Upper limits to home range size are not restricted to primates but are common among terrestrial mammals (Kelt & Van Vuren 2001). Reluctance to expand into available areas is surprising given that expansion presumably increases access to food resources and should therefore be beneficial to mothers' reproductive success. I will suggest ecological reasons for this reluctance after I describe the four other types of mothers.

"Generous Mothers": More Options for Reproduction

Contrasted with stingy mothers are populations in which solitarily foraging reproductive females have overlapping home ranges and thus share food resources in the common area. These are referred to here as generous mothers. To share their home ranges with their adult daughters, mothers must be willing and able to expand their home ranges beyond what they need for their own reproduction. The prevalence of shared home ranges among solitarily foraging mammals (Waser & Jones 1983) suggests that this is often achieved, but in solitarily foraging primates, it appears to be less common than exclusively used home ranges (table 4.1). Female galagos and bush babies (*Galago*, *Galagoides*, and *Otolemur*) and mouse lemurs (*Microcebus murinus*) have extensively overlapping home ranges (Charles-Dominique 1977, Bearder 1987, Radespiel 2000, Eberle & Kappeler 2002), which indicates that they share their resources to some degree. Co-inhabitants of the shared home ranges are expected to be daughters in most cases. Reproductive female galagos share the mothers' home ranges and therefore, their food (Charles-Dominique 1977, Bearder 1987). Close genetic relatedness among female mouse lemurs that have extensive home range overlap has also been confirmed (Wimmer et al. 2002).

"Incomplete Suppressors": Limited Tolerance of Reproduction in Multifemale Groups

Incomplete suppressors are defined here as females that live in home ranges sufficiently large to enable other females to reproduce, but only up to a point. They are similar to stingy mothers in having home ranges with minimal overlap, but they differ in that multiple females defend the same home range. Incomplete suppressors are different from generous mothers in that multiple females often travel together in their shared home range. Though females living in multireproductive female groups have overcome complete reproductive suppression, they will, like stingy females with complete suppression, disperse if their chances of reproducing in their natal groups are poor.

Among incomplete suppressors, female red howlers (*Alouatta seniculus*) that succeed in reproducing in their natal group do not disperse; females that disperse have not yet reproduced. Female red howlers that disperse are frequently targets of aggression by unrelated female group mates before they disperse (Crockett 1984; Pope 2000a, b), and the process of targeted aggression eventually results in groups consisting of single matrilines (Pope 2000b). There are reports of targeted aggression in some group-living Malagasy prosimians (e.g., *Lemur* and *Propithecus*) in some populations, and females targeted with aggression also disperse from their groups (Vick & Pereira 1989, Pereira 1993, Wright 1999). Targeted aggression and dispersal of targeted females may be a mechanism for reducing food competition caused by increasing group size in species that have fixed home ranges. Unlike many other species in which the size of the home range changes with changes in group size (see below), home ranges of ring-tailed lemurs (*Lemur catta*) and Milne-Edwards's sifakas (*Propithecus diadema*) have been extraordinarily stable in size for up to three decades and counting, regardless of changes in group size or population density (Wright 1995, Jolly & Pride 1999).

Targeted aggression with eviction may be a characteristic of incomplete suppressors. Unfortunately, incomplete reproductive suppression makes it impossible to determine what a dispersing female's reproductive success would have been had she remained. In captivity, where home ranges cannot possibly expand and the option to disperse is nonexistent without human intervention, the poorer reproducers are often those that are recipients of aggression (Silk et al. 1981, Wasser & Barash 1983, Silk 1988, Vick & Pereira 1989, Pereira 1993). Without human intervention, targeted aggression can even become fatal in captivity (McGrew 1997). In the wild, females that stay in their groups despite being targeted with aggression could suffer the same fate. Given such dire odds, targeted females may be better off taking their chances with dispersal. At worst, dispersers that die would break even and at best, dispersers that survive would eventually reproduce. Even if their reproductive success is lower than that of females that stay, it is likely to be greater than if they themselves had stayed.

"Facilitators": Greater Tolerance of Reproduction in Multifemale Groups

In some species, mothers do not normally target females in their groups with aggression but allow them to stay and may even facilitate their reproduction through preferential treatment (Fairbanks 2000). Such mothers are called facilitators here. Female yellow baboons (*Papio cynocephalus cynocephalus*) and macaques live in groups of related matrilines; females rarely disperse in these species (Pusey & Packer 1987a, Isbell & Van Vuren 1996). These are equivalent to the species called female-bonded by Wrangham (1980). These are also the species in Isbell's (1991) model for which home range size increases with larger group size. Expansive home ranges are difficult to keep exclusive. Thus, facilitators differ from stingy mothers and incomplete suppressors by having overlapping and indeterminately growing home ranges. Costs of daily travel do not limit home range expansion in these species because daily travel distance and home range size are independent (Isbell 1991). Facilitators differ from generous mothers by invariably traveling within sight of other adult females in the shared home range.

"Indifferent Mothers"

Because it has become so ingrained in the literature that females are limited in their reproductive success by food (Trivers 1972), it may be difficult by now to imagine females that are not limited by food. The fifth type of mother appears, however, to be less responsive than the other four types to differences in food resources.

In these species, females appear more indifferent to than concerned about the presence of other females. They neither force dispersal nor facilitate philopatry. Such females are referred to as indifferent mothers here. In many colobines, such as red colobus, capped langurs, banded leaf monkeys, Thomas's langurs (*Presbytis thomasi*), and Nilgiri langurs (*Trachypithecus johnii*), females commonly or at least occasionally disperse, and they emigrate without aggression from other group members. They can either transfer directly to an existing group or create a new group by joining a male (Struhsaker 1975, Marsh 1979, Stanford 1991, Starin 1991, Bennett & Davies 1994, Oates 1994, Steenbeek et al. 2000). Females also experience little aggression from neighboring groups when they immigrate. Colobines often, though not always, have extensively overlapping home ranges, and females often disperse to groups whose home ranges overlap extensively with those of their natal groups (Isbell & Van Vuren 1996). Aggression from strangers and unfamiliarity with new areas, the main costs of dispersal, thus appear to be minimal for indifferent females. Low costs of dispersal may also reduce pressure on mothers to provide a place in their own home ranges for their daughters.

Although most female red colobus at Abuko, Gambia, leave their natal groups, they often stay within their former group's home range (Starin 1981, 1991). During a five-year study of olive colobus (*Procolobus concolor*) at Taï National Park, Ivory Coast, at least 8 of 16 females changed groups (Korstjens 2001, Korstjens & Schippers 2003). One female whose dispersal history was well documented left her group when the group followed the resident male as he moved back into the home range (but not the group) from which he had come. She returned to her former home range as a solitary female but joined the group again when the male returned with the rest of the group. The only other adult female in the group also left during the shift in range use, but her fate was unknown (Korstjens 2001, Korstjens & Schippers 2003). Site fidelity may be more important than group fidelity in colobines. Similarly, mountain gorillas (*Gorilla gorilla*) have little female aggression within or between groups, extensive home range overlap, and female transfer (Harcourt 1978, Watts 1990, Yamagiwa & Kahekwa 2001).

Table 4.1 provides a summary list of primate genera (sometimes species when they appear to differ) for which there are data to classify them into the five types discussed above using information first and foremost on the extent of home range overlap with other reproductive females, followed by (1) female social dispersal, (2) targeted aggression, and (3) presence or absence of contest competition among females between groups.

With few exceptions, it is fairly easy to classify genera as long as sufficient data are available. The genus *Pan* is a difficult one to classify largely because great individual variation exists in female ranging behavior. For example, in the same population, some female

Table 4.1. Genera Categorized as Stingy Mothers, Generous Mothers, Incomplete Suppressors, Facilitators, and Indifferent Mothers[a]

Stingy Mothers	Generous Mothers	Incomplete Suppressors	Facilitators	Indifferent Mothers
Avahi[b]	*Microcebus*	*Eulemur*	*Cebus*	*Brachyteles*
Cheirogaleus	*Galago*	*Hapalemur*	*Saimiri (boliviensis)*	*Saimiri (oerstedii)*
Daubentonia	*Galagoides*	*Lemur*	*Cercopithecus*	*Nasalis*
Eulemur	*Otolemur*	*Propithecus*	*Erythrocebus*	*Presbytis*
Hapalemur	*Pongo?*[c]	*Varecia*	*Macaca*	*Pygathrix*
Indri	*Pan?*[c]	*Alouatta*	*Mandrillus*	*Trachypithecus*
Lepilemur			*Papio*	*Simias*
Perodicticus			*Theropithecus*	*Colobus*
Phaner			*Semnopithecus* (Ramnagar)	*Procolobus*
Varecia			*Pongo?*[c]	*Gorilla*
Tarsius			*Pan?*[c]	
Alouatta				
Aotus				
Callicebus				
Callithrix				
Cebuella				
Leontopithecus				
Saguinus				
Hylobates				
Pan?[c]				

[a]See text for the criteria used in classifying each category. Varying contributions of goal-directed travel and wandering are predicted to determine the extent to which mothers can expand their home ranges to accommodate reproduction by their daughters. Although phylogenetic niche conservatism probably plays a major role in establishing the relative contributions of goal-directed travel and foraging behavior in most cases, the potential exists for females in different populations within a species to express different types because travel and foraging behavior may also be affected by local conditions. Similarities in energetic constraints make it more likely that when a genus is listed under two types, the combination will be stingy mothers and incomplete suppressors (e.g., *Varecia variegata* and *Alouatta pigra*) or generous mothers and facilitators (e.g., *Pongo?*). Additional references not in text: *Avahi*: Thalmann 2001; *Cheirogaleus*: Fietz 1999; *Eulemur*: Overdorff 1996; *Hapalemur*: Nievergelt et al. 1998, Mutschler et al. 2000; *Lepilemur*: Thalmann 2001; *Tarsius*: Gursky 2000; *Microcebus*: Fietz 1999; *Mandrillus, Theropithecus*: Stammbach 1987, Isbell & Van Vuren 1996; *Brachyteles*: Printes & Strier 1999; *Colobus*: Fashing 2001; *Nasalis, Pygathrix, Simias*: Yeager & Kool 2000.

[b]Genera are listed in most cases, unless variation has been observed at lower taxonomic levels.

[c]*Pan* and *Pongo* have question marks because there is evidence that different types may be represented at the same time by different individuals in the same population.

chimpanzees might be considered generous mothers since they most often forage alone in overlapping home ranges (Williams et al. 2002). Some individuals might, however, be considered stingy mothers because about half of all natal females disperse socially and locationally (Williams et al. 2002). Yet others might be considered facilitators because daughters that do not permanently leave their natal communities return to settle in their mothers' home ranges if their mothers are still alive. In such cases, mothers and daughters also travel together in their shared home ranges (Williams et al. 2002). They can probably all be safely ruled out as incomplete suppressors because they do not travel in cohesive multifemale groups. They can also all be ruled out as indifferent mothers because female reproductive success is positively correlated with high rank (Pusey et al. 1997), and indifferent mothers cannot be ranked. The variability of types of mothers within a single population of chimpanzees may exist partly because female chimpanzees are, unusually for primates, influenced not only by food competition but also by male aggression and the ranging behavior of sons (Williams et al. 2002). Male chimpanzees sometimes employ violent coercion of females that affects their ranging behavior (Wrangham 1979, Smuts & Smuts 1993), and mothers sometimes follow their growing sons as the sons become more involved with adult males (Williams et al. 2002).

Pongo is another genus that may be difficult to classify. Female orangutans might be considered generous mothers because females typically forage alone in overlapping home ranges (Rodman & Mitani 1987, Rodman 1988a, Singleton & van Schaik 2001). The observation of adult females occasionally traveling together (Rodman & Mitani 1987) suggests, however, that individual variation in ranging behavior may also exist in orangutans. Male harassment also occurs in orangutans (Rodman & Mitani 1987, Smuts & Smuts 1993) and affects the ranging behavior of females (Fox 2002). Perhaps male harassment contributes more to patterns of association among female orangutans than is currently recognized.

Some genera have been placed in two different types not because individual females in the same population range differently but because different types can be expressed in (1) different species within the same genus, (2) different populations within the same species, and even (3) the same group during different seasons. For example, female black howlers (*Alouatta pigra*) that live at high densities appear to be incomplete suppressors. At low densities, and when whole groups have been translocated to suitable but unpopulated habitats, females become stingy by sorting themselves into one per home range (Ostro et al. 1999, 2001). Female ruffed lemurs (*Varecia variegata*) change from stingy mothers to incomplete suppressors when they change their ranging patterns in different seasons (Morland 1991, Rigamonti 1993).

Ecological Underpinnings of the Five Types of Mothers

Are Stingy Mothers Highly Goal-Directed Travelers?

Why stingy mothers do not take advantage of openings to expand their home ranges is an unexplored research question. One possibility is that they face energetic constraints that the other types of mothers do not face. If this is the case, the constraint does not appear to come from reliance on any particular diet. Stingy mothers include primates typically classified as frugivores (e.g., gibbons), insectivores (western tarsiers), gummivores (marmosets and pygmy marmosets), and folivores (indris and bamboo lemurs). The constraint may rather come from having a great need to minimize either time or energy in travel. In some cases, this need could arise from having a small body that requires small and frequent feedings, which appears to be the case for very small mammals, such as rodents and shrews (Zynel & Wunder 2002). Though the limited sample on basal metabolic rates for primates calls for caution, stingy mothers appear to have among the lowest basal metabolic rates for their body sizes (35–95% of the expected value; Genoud et al. 1997, Power et al. 2003). Basal metabolic rates measure the energy needed for minimal bodily maintenance, and the low basal metabolic rates that are found in stingy mothers are generally considered to be adaptations for conserving energy under conditions of severe environmental stress (Jolly 1984, Müller 1985, Richard 1987, Wright 1999). Some stingy primates have specialized adaptations for locomotion compared to their closest nonstingy relatives (e.g., extreme brachiation in gibbons but not orangutans). I suggest that along with their severe energetic constraint comes a distinctive traveling style that affects the ability of mothers to expand their home ranges (figure 4.3).

Stingy mothers are often described as highly goal directed in their movements, that is, traveling directly from one food site to the next without foraging between sites (e.g., *Phaner*, *Indri*, *Callicebus*, *Saguinus*, and *Hylobates*; Rodman 1988b, Garber 1989, Wright 1994, Kappeler 2000, Garber & Bicca-Marques 2002, Schülke in press). Some of these (e.g., *Callicebus*, *Phaner*, and *Avahi*) can even be considered trapliners (Wright 1994, Thalmann 2001, Schülke in press), which involves making repeated visits over several hours or days to sequential food sites (Garber 2000, Milton 2000). When an individual is highly goal directed, it acts as if it knows where it is going and that the food will be there. Highly goal-directed animals give the impression that they are minimizing the time or the energetic expense of travel between food sites. Movements between distant food sites are direct and efficient.

Their normal efficiency can be seen by observing individuals that do move to new surroundings or that are faced with a familiar area that has been experimentally altered. A female indri that had apparently recently established herself in a new area traveled two to three times farther to go the same horizontal distance as a female that was familiar with her home range (Pollock 1979). Experimental removal of a food tree resulted in apparent confusion in fork-marked lemurs (*Phaner furcifer*) that had traveled quickly to it (Petter et al. 1975). Golden lion tamarins (*Leontopithecus rosalia*) that had been experimentally introduced to a new area became more efficient at getting to food sites over time (Menzel & Beck 2000). Pottos took 5 to 10 days to find a new food site in their home ranges, while galagos (*Galago alleni* and *Galagoides demidovii*), which share their home ranges with other females, took only 3 to 5 days (Charles-Dominique 1977). The difference between pottos and galagos is not likely to be because pottos are slower and thus cannot cover their home ranges as quickly as galagos; indeed, unlike galagos, they seem to have no difficulty excluding others from their home ranges, a feat that appears to require a certain degree of mobility and the ability to monitor daily (or nightly in the case of pottos) their home range boundaries (Mitani & Rodman 1979). The gibbon's specialization for brachiation has been suggested as a more energetically efficient way to move in an arboreal environment than walking because it allows direct travel (Parsons & Taylor 1977).

Stingy mothers may not expand their home ranges even if given the opportunity because going into new areas requires exploring for food with uncertain success. The "downtime" that is an integral aspect of exploration may decrease foraging efficiency beyond that which highly goal-directed primates can tolerate. Stingy mothers may also be unable share their home ranges with their daughters because additional females could reduce the predictability of food locations (if they do not feed in groups, see below). Consider the energetic cost for females relying on predictably located foods if they were to travel directly and repeatedly to those food sites after others have already reached and depleted them. Stingy mothers may simply be unable to allow their adult daughters to remain and reproduce in a shared home range without suffering a cost to their own reproductive success (figure 4.3). Finally, even if mothers could expand their home ranges to enable their daughters to reproduce in the

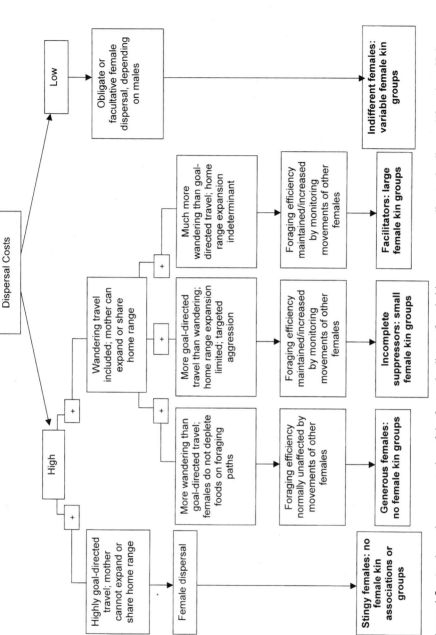

Figure 4.3. A schematic representation of the factors leading to female kin groups according to the dispersal/foraging efficiency model presented in this chapter.

natal home range, such expansion might increase the risk of infiltration by additional, unrelated females, further reducing the predictability of food locations.

It may be informative that stingy mothers also occur among nonprimate mammals that live close to the energetic edge of death. Female voles (*Microtus*), elephant shrews (*Elephantulus*), and tree shrews (*Tupaia*) are stingy: mothers often maintain nonoverlapping home ranges or, if they share their home ranges with other females, only one female typically breeds (Jannett 1978, Rathbun 1979, Getz et al. 1993, Emmons 2000). When nest sharing does occur between reproductive female voles, it usually occurs under high densities coupled with high resource availability (e.g., prairie voles [*Microtus ochrogaster*]; Getz et al. 1993, Cochran & Solomon 2000] when it may be difficult for offspring to find their own home ranges and less costly for mothers to allow others to reproduce in the home range. When densities are low, mothers often abandon their home ranges rather than share them with maturing daughters (e.g., *M. montanus*; Jannett 1978). There is also substantial evidence that offspring of female voles are reproductively suppressed (Wasser & Barash 1983, Solomon & French 1997, but see Wolff et al. 2001). Though use of the same nests repeatedly in some of these species suggests that individuals are goal directed in their travel, the extent to which females are goal directed between food sites remains to be determined. Tree shrews, which are perhaps more closely related to primates than to other taxa (Fleagle 1999) typically eat fruit and slow-moving, hidden arthropods. Emmons (2000) provides evidence that the fruit component of their diet is critical to their survival and reproduction. When they feed on fruit, they do travel in a highly goal-directed manner.

Do Generous Mothers Wander More Than Stingy Mothers?

How might generous mothers differ from stingy mothers to allow them to expand their home ranges so that their daughters can reproduce while remaining philopatric? I suggest that generous mothers are less energetically constrained than stingy mothers. Again, although the data are sparse, generous mothers appear to have somewhat higher basal metabolic rates (84–114% of expected) than stingy mothers (Genoud et al. 1997). The relaxation of energetic constraints may afford them the opportunity to be less goal directed and more exploratory in their ranging behavior (figure 4.3). Exploratory movements involve some degree of wandering as animals forage for food. Compared to goal-directed travel, wandering involves slower travel speeds, more frequent short-term changes in direction, more stops to search for food, and less success at finding food at each stop. In the jargon of feeding ecology, animals that "forage" (as opposed to "feed") engage in wandering as a mode of travel. Unlike goal-directed travel, which takes individuals quickly and directly to a productive food site, wandering can take individuals to places that are unproductive. Home range expansion may be possible for individuals that wander more than they engage in goal-directed travel because both wandering and home range expansion involve going into areas where productivity is uncertain.

The diet of most generous mothers appears to assist their wandering. Galagos move quickly and pursue fast-moving arthropods (Bearder 1987). Unlike plants and slow-moving arthropods, highly mobile arthropods have the ability to move quickly out of the reach of their predators. Predators of mobile prey must be able to follow their prey. By pursuing mobile arthropods, generous mothers are likely to find themselves in areas of unknown productivity.

An interesting consequence of classifying mothers by whether they share home ranges with other reproductive females is that differences in male ranging behavior become more obvious. It has always been puzzling why males of some primate species range almost entirely within a given female's home range (regardless of whether the male and female travel separately or together), when they would theoretically do better reproductively if they were to range over partial home ranges of multiple females. The latter strategy need not include an increase in their own home ranges. It is worth noting that male ranging behavior seems to be fairly well predicted by whether mothers are stingy or generous. This holds for mammals as diverse as rodents, shrews, and primates. In species with stingy mothers, the home range of a given male is almost always shared with only one reproductive female (e.g., elephant shrews, Rathbun 1979; prairie voles, Carter & Getz 1993; gibbons, Leighton 1987). By contrast, in species with generous mothers, the home range of a given male usually overlaps those of multiple females (e.g., meadow voles [*Microtus pennsylvanicus*], Madison 1980, Boonstra et al. 1993, Bowers et al. 1996; galagos, Bearder 1987). It is possible that the energetic constraints that are suggested to operate on stingy mothers also operate on male conspecifics, with similar results. Although this possibility deserves attention, further discussion is beyond the scope of this chapter.

Are Incomplete Suppressors Goal-Directed Wanderers?

How might incomplete suppressors differ from stingy and generous mothers in their movements? The interchangeability of females between stingy mothers and incomplete suppressors at a level as small as that of the group suggests that the major difference between incomplete suppressors and stingy mothers is more ecological than physiological. If this is the case, basal metabolic rates of incomplete suppressors should be similar to those of stingy mothers and different from those of generous mothers. Basal metabolic rates are available for only two incomplete suppressors (35% and 55% of expected), and it is dangerous to draw conclusions on such a small sample size. Nonetheless, since most incomplete suppressors are Malagasy prosimians, and all Malagasy prosimians that have been studied have lower basal metabolic rates than expected (Jolly 1984, Richard 1987, Genoud et al. 1997, Wright 1999), it is possible that more data will support the prediction that incomplete suppressors are more similar to stingy mothers than to generous mothers.

If incomplete suppressors are more energetically limited than generous mothers, they may be more constrained to minimize energy spent in travel. Incomplete suppressors may be able to wander more than stingy mothers but may travel in a goal-directed manner more than generous mothers (figure 4.3). More extensive wandering may enable mothers to share home ranges with their reproductive daughters, while their goal-directed travel may make it difficult for mothers to expand their home ranges indeterminately as additional daughters are born. A high degree of goal-directed travel should also result in little home range overlap with other groups as females attempt to maintain the predictability of their food locations.

Goal-directed travel has been reported for species considered here to be incomplete suppressors. Mantled howlers (*Alouatta palliata*), for example, often move directly to preferred food sites within their home ranges (Milton 1980, 2000). Ring-tailed lemurs (*Lemur catta*) have also been reported to move directly to specific food sites, at times going outside their normal home range to do so (Jolly & Pride 1999). The wandering component may be under-reported in these species because goal-directed travel is more impressive and is often interpreted as an indication of advanced cognitive ability. Nonetheless, ring-tailed lemurs can forage in a broad front as they move (Klopfer & Jolly 1970), which suggests a kind of wandering. Groups of ruffed lemurs move more extensively over their home range during the seasons when they live together as cohesive groups and are incomplete suppressors, but restrict their movements to subranges within their group's home range when they split up into single-female units and become stingy (Morland 1991, Rigamonti 1993). The switch from incomplete suppressor to stingy is predicted to be accompanied by an increase in the percentage of goal-directed travel. The relative contributions of wandering and goal-directed travel should become clearer in the future when more studies of the micromovements of individuals are conducted.

Are Facilitators Predominantly Wanderers?

Facilitators may differ from the other mothers in being the least energetically restricted of the primates; they have the highest basal metabolic rates among the primates (114–142% of expected; Genoud et al. 1997). A lesser need to conserve energy may mean that facilitators are better able than stingy females and incomplete suppressors to spend time actively searching for food when it becomes scarce. Whereas stingy mothers (e.g., fat-tailed dwarf lemurs [*Cheirogaleus medius*]) and incomplete suppressors (e.g., Verreaux's sifakas [*Propithecus verreauxi*]) may increase their resting time when food becomes less abundant (Richard 1978, 1987), facilitators (e.g., brown capuchins [*Cebus apella*]), tend to increase their time spent foraging and feeding (Terborgh 1983). Facilitators may thus be able to wander much more than the other types of mothers (figure 4.3). They may also travel in a goal-directed manner (Janson 1998, Pochron 2001), albeit to a lesser extent than the other types of mothers. A large wandering component should allow mothers to expand their home ranges so that their daughters can reproduce in the natal home range without reducing mothers' foraging efficiency and future reproductive success. Indeed, there seems to be no limit to the size of the home range; home range size typically expands with increasing group size in yellow baboons, macaques, and guenons (Isbell 1991), all of which are considered to be facilitators.

Patas monkeys provide an extreme example of the ranging behavior of facilitators. Patas have extraordinarily large home ranges, which increase as group sizes increase (Chism & Rowell 1988). They have also been described as "feeding at a steady walk" (Hall 1965), and their long stride has been interpreted as an adaptation for efficient foraging over long distances (Chism & Rowell 1988, Isbell, Pruetz, Lewis et al. 1998). Patas typically travel circuitously except when they travel to water during the dry season (Hall 1965, Isbell et al. 1999). Their locations during the day are difficult to anticipate. Indeed, of all the species that I have followed, I have found it most difficult to predict the general location of patas groups later in the day. With one possible exception (vervets), these other species (i.e., red colobus, sifakas, brown lemurs [*Eulemur fulvus*], and ring-tailed lemurs) are not facilitators.

Data from brown capuchins provide another example of facilitators' greater emphasis on wandering. At Iguazu National Park, Argentina, they spent 47% of their time in "slow foraging" (equivalent to wandering) and 6% in "fast travel" (equivalent to goal-directed travel). Although both modes of travel occurred while the animals moved between feeding platforms (Janson & Di Bitetti 1997, C. H. Janson personal communication), wandering involved many short-term changes in direction, presumably as a result of searching for and foraging on insects, whereas goal-directed travel involved direct, nonstop movement to feeding platforms (Janson & Di Bitetti 1997, Janson 1998).

Intriguingly, if basal metabolic rates are a measure of the degree to which primates are energetically constrained, humans and chimpanzees are well within the range for facilitators (basal metabolic rates of 128% and 141% of expected, respectively; Genoud et al. 1997).

Are Indifferent Mothers Not Limited by Food Abundance?

Colobines and mountain gorillas are fairly folivorous, with slow gut passage times that require the animals to rest while digesting their food (Bauchop 1978, Parra 1978, Kay & Davies 1994, Kirkpatrick et al. 2001). Slow digestive rates raise the possibility that most folivores may be more constrained by digestion time than by food abundance (see also Zynel & Wunder 2002 for herbivorous voles). In other words, they may run out of time in a day before they run out of food to eat. If this is the case, the presence of other females should make little difference in their ability to obtain food. When female reproductive success is limited more by digestion than by food abundance, females should either be indifferent to the presence of other females or avoid competing with them over food, particularly if it interferes with digestion.

Although a positive correlation exists between colobine biomass and leaf protein-fiber ratios (Davies 1994, Chapman et al. 2002), the correlation does not necessarily mean that food limits female reproductive success in colobine populations. If the correlation exists as a result of recruitment of infants, then food may indeed limit both female reproductive success and the size of populations. On the other hand, if the correlation exists as a result of movement of individuals to areas with high densities of food, then different factors might limit female reproductive success and populations. Consider animals that are highly mobile, such as many of the larger ungulates. Aggregation of individuals in areas of high food density could produce a positive correlation between food density and numbers of animals without any increase in infants. The same might be said for female colobines because they are less constrained to remain in a particular group than many other female primates. Without knowing details about individuals within populations, that is, their movements and their reproductive success, a correlation between food density and population size cannot be used to infer that food abundance limits reproductive success of females in those populations.

Though all females need sufficient food to give birth, having enough food is not always sufficient for keeping infants alive. This may be most obvious when female reproductive success is limited more by digestion time (or food quality) than food abundance. Under such conditions, other factors, such as infanticide, could replace food abundance as ultimately limiting female reproductive success (see Wolff 1993 for a similar argument for rodents), but these factors would remain hidden if they mirror the density dependence of food limitation. Indeed, in Thomas's langurs, infanticide is more frequent in larger groups than in smaller groups (Steenbeek 2000). More explicitly, in mountain gorillas, infanticide accounts for at least 37% of deaths of animals to age 3 (Watts 1989, see also Fossey 1984). If the frequency of predation were that high, we might have little hesitation in suggesting that predation limits the Virunga gorilla population. Unlike facilitator females, the relative fitness of many indifferent mothers may be determined less by food than by differential ability to anticipate and respond appropriately to the risk of infanticide (figure 4.3). If so, infanticide should be a strong factor in female dispersal decisions in these species. Infanticide has, in fact, been suggested as the cause of female dispersal and small female group sizes in Thomas's langurs (Steenbeek & van Schaik 2001).

Among species in which gut passage times are not slow, female reproductive success is more likely to be limited by food abundance, and infanticide is probably less important than the mother's nutritional condition in contributing to infant mortality. For example, female patas monkeys are facilitators whose ranging behavior appears to be more finely tuned to food abundance than to other factors (Isbell & Enstam 2002). Infanticide was implicated in the death of only 1 of 85 (1.2%) infant patas monkeys over a 10-year period (Enstam et al. 2002, Isbell unpublished data).

Species in which female reproductive success is limited by non–food-related factors other than infanticide are also expected to have indifferent mothers. Squirrel monkeys (*Saimiri oerstedii*) in Corcovado, Costa Rica, and red colobus in Kibale, Uganda, and Gombe, Tanzania, may be examples of indifferent mothers whose reproductive success is limited more by predation than by food. In these species, home ranges overlap extensively, females commonly disperse, and there appears to be little competition for food either within or between groups (Struhsaker 1975, Boinski 1999), all of which are characteristics of indifferent mothers. In contrast, the risk of predation appears to be very high (Stanford 1995, Boinski 1999). Many of the behaviors of female squirrel monkeys, for example, highly synchronized births within groups, spatial associations of multiple mothers and infants, and coordinated group movements, have been interpreted as adaptations for reducing predation (Boinski 1987, 1999; Boinski et al. 2000).

The classification of Corcovado squirrel monkeys as indifferent mothers is, at first glance, questionable because they are more frugivorous than folivorous and have a congener (*S. boliviensis* at Manu, Peru) with facilitator females (Boinski 1999). Nonetheless, interbirth intervals half as long at Corcovado (one year) as at Manu (two years), despite a richer food supply at Manu (Boinski 1999, Boinski et al. 2002), suggest that different factors may indeed limit female reproductive success in these congeners. Unlike many other species with indifferent mothers, the Corcovado females appear to be limited more by predation than infanticide. Although infanticide can shorten interbirth intervals (Hrdy 1974), there have been no cases of infanticide at Corcovado (Boinski 1999) to explain their short interbirth intervals. In contrast, their high infant mortality (50% within the first six months of age) has been attributed mainly to predation (Boinski 2000).

The ecological conditions faced by females whose reproductive success is limited by something other than food abundance may make female social dispersal no more costly than philopatry (see Watts 2000 for gorillas) and perhaps more beneficial than costly. Since for indifferent females food is not as crucial as, for example, avoiding infanticide or predation, their dispersal decisions are not expected to be based on maintaining foraging efficiency (figure 4.3). Their dispersal decisions are instead expected to be determined by their ability to keep their offspring alive, staying if they succeed and leaving if they fail. Thus, although juvenile female dispersal is the norm in Corcovado squirrel monkeys, adult females also transfer between groups if their infants die (Boinski et al. 2002).

Consequences of Foraging Efficiency and Costs of Dispersal on the Formation of Female Kin Groups

The costs of dispersal may have favored the willingness of mothers to allow daughters to remain in the natal home range, the probable first step in the evolution of kin groups (Pusey & Packer 1987b, Isbell & Van Vuren 1996). If mothers could expand their home ranges without sacrificing foraging efficiency and their future reproduction, daughters would be able to stay and reproduce in the natal home range. This problem might have been solved initially by mothers acquiring larger home ranges than were required for their own maintenance and reproduction. If they could not do this, female kin groups would likely not have evolved.

Once females began to share home ranges, there would have been three alternatives available to females whose reproductive success was limited by food abundance. The alternative taken would have depended on the relative contributions of wandering and goal-directed travel, and on the mobility of the food. Two alternatives involve group living, and they differ from each other in the size that the group can become. The third alternative does not involve group living. With this alternative, females can share a common area while foraging alone. Thus, philopatry is not equivalent to feeding and traveling together in a cohesive group.

I suggest that kin groups become beneficial only when it is important to avoid feeding in places recently visited by others that share the same home range. If females must minimize foraging in areas already covered by others in order to maintain their foraging efficiency, females that feed on relatively immobile food may need to monitor the movements of their relatives in some way. Multiple senses are available for monitoring the whereabouts of others. Since primates generally are visually oriented animals, monitoring may be best done visually. Visual monitoring requires fairly close proximity. Primates that appear to be less visually oriented (e.g., the nocturnal and some of the cathemeral species), might also monitor the movements of others in the shared home range through vocal or olfactory cues. Only visual monitoring would require females to remain near each other while moving, however. Thus, the second step in the evolution of kin groups might have involved visually coordinated traveling and feeding together to enable females to avoid places already harvested by others (Cody 1971, Altmann 1974, Rodman 1988b). At this stage, they would be recognizable as groups (figure 4.3). Species differences in the modal size of groups would be determined by the extent to which home ranges could be expanded (and for incomplete suppressors and facilitators, also daily travel costs; Wrangham et al. 1993, Janson & Goldsmith 1995), which would depend to a large degree on the extent of wandering in the travel/feeding repertoire. Aggressive interactions with larger neighboring groups might also contribute to restricting home range expansion.

In some species, for example galagos, females would not need to monitor the movements of their relatives to maintain their foraging efficiency because their food is highly mobile. Flying and hopping arthropods, which can move in and out of the paths of foraging females fairly easily, may render monitoring the movements of others in a shared home range unnecessary and perhaps even detrimental, unless the movements of others help to flush up arthropods. But unless monitoring the movements of others at least maintains foraging efficiency for mothers, mothers are not likely to sacrifice attention to their surroundings to live with others. Such an explanation would be consistent with the absence of group foraging in female galagos even though they share their home ranges with other females (figure 4.3). Galagos often double back on their path during the night's foraging (Charles-Dominique 1977). It can be argued that if a female is willing to forage again along her previous foraging path, she does not perceive a decline in her foraging efficiency. The mobility of their food may enable female galagos to double back without reducing their foraging efficiency. By the same reasoning, foraging efficiency may not be reduced if a female were to forage in areas that have already been visited by another female. Such females would not need to live in groups although they benefit from remaining in a familiar area.

Contrary to the assumption that group living always involves a cost, the dispersal/foraging efficiency model suggests that when there is a reproductive cost to females of having others around, even kin, they do not share the home range with reproductive daughters. Home range sharing only occurs when females do not actually incur reproductive costs. When mothers allow reproductive daughters to share their home ranges, neither mothers nor daughters need incur automatic costs, because the home range will be large enough for them and their offspring. Group living becomes merely an efficient way for visually oriented primates that feed on immobile foods to share their home ranges with other individuals.

In this model, groups evolve through a series of small and incremental steps in which the predominant selection pressure is the maintenance of foraging efficiency. Kin selection is involved only to the extent that a mother tolerates or facilitates reproduction by her offspring in her home range. Intergroup competition, one of the two alternatives that have been invoked in the past as selection pressures favoring the evolution of kin groups (Wrangham 1980), is not a necessary component in the evolution of kin groups according to the dispersal/foraging efficiency model, although it might have been a relatively small step for groups that already travel and feed together to begin cooperating in keeping other groups from exploiting the foods in their home range. Success in intergroup competition is viewed by the dispersal/foraging efficiency model as an additional benefit of living in kin groups, not the primary benefit (Isbell & Van Vuren 1996). Interestingly, within the Cercopithecoidea and Ceboidea, the species with female philopatry have larger home ranges per individual than species with frequent female dispersal (see Milton & May 1976). Perhaps once facilitator females formed kin groups, they became even more acquisitive, an act that reinforces intergroup competition.

Intragroup competition and predation, the other proposed selective pressures (van Schaik 1989, Sterck et al. 1997), also have little influence on the evolution of female kin groups in this model. Intragroup competition is, instead, largely a function of the depletion time of foods. If foods are depleted slowly, they can be usurped and females will interact agonistically, even to the extent of forming coalitions, if doing so helps females usurp the foods (Isbell & Pruetz 1998, Isbell, Pruetz, & Young 1998, Mathy & Isbell 2002). Of course, when coalitions form to help individuals usurp food from others, they will most often form with kin to reap the benefits of inclusive fitness.

The evidence that locational dispersal increases the risk of predation relative to philopatry suggests that predation was a strong selective pressure on mothers to allow daughters to remain in the natal home range. However, although predation may have helped to set the stage for the evolution of kin groups, according to the dispersal/foraging efficiency model, no amount of predation would favor home range sharing with reproductive daughters if mothers' foraging efficiency were compromised. In this model, if foraging efficiency cannot

be maintained, kin grouping does not occur. Kin grouping occurs when there are no reproductive costs, that is, when foraging efficiency can be maintained to enable females to
reproduce. Reproductive costs become apparent only after kin groups become very large
and energy intake cannot keep up with the energetic cost of increasing daily travel distances.
The maintenance of foraging efficiency is sufficient to explain the evolution of kin groups,
and predation need not be invoked. Indeed, evidence that polyspecific groups reduce the
risk of predation for group members, which are clearly not related (e.g., Noë & Bshary
1997), provides perhaps the most convincing evidence that kin groups are not required for
animals to reduce their risk of predation.

Benefits of Dispersal, or Why Dispersal Is Often Sex Biased

I have focused on the costs of dispersal to explain the tendency of females to remain philopatric. Now I discuss the benefits of dispersal, because in some species females commonly
or occasionally disperse, and in the long run there must be an advantage to female dispersal
that outweighs its costs for such species. Some of these benefits appear to be created by
males. To fully understand female decisions to disperse or remain philopatric, it is necessary
to discuss male dispersal, particularly for those species in which female reproductive success
does not appear to be limited by food abundance.

The benefits of dispersal are intensely debated with no real consensus yet, but the disagreement centers on inbreeding avoidance (Pusey & Packer 1987a, Clutton-Brock 1989, Pusey &
Wolf 1996) and increased opportunities for breeding (Moore & Ali 1984; Moore 1988, 1992).
These hypotheses are largely designed to explain sex-biased dispersal, but particularly male
dispersal, because male dispersal with female philopatry has usually been considered the norm
among mammals in general and primates in particular (Clutton-Brock & Harvey 1976, Greenwood 1980, Wrangham 1980, Pusey & Packer 1987a, Clutton-Brock 1989). I argue here that
both advantages accrue, one for females and the other for males.

The adaptationist approach assumes that dispersal had to have a net benefit in order to
evolve, but this does not mean that all individuals will always gain. Dispersers take their
chances, and some succeed whereas others do not. For example, sightings of leopards or their
signs are punctuated by long periods of no sightings and no disappearances of vervets and patas
monkeys in Laikipia, Kenya (Isbell unpublished data), which suggests that leopards are not
always a danger to them (see also Isbell 1990). If leopards had not been present and actively
hunting monkeys when vervets in Amboseli dispersed to new areas, the dispersers might not
have suffered higher mortality despite their ignorance of the new home ranges (Isbell et al.
1990). Anderson (1987) suggested that female dispersal in chacma baboons (*P. ursinus*) at
Suikerbosrand was common because leopards had not been present in the area for over 50
years. The outcome of individuals' decisions to disperse clearly depends on the local social and
ecological milieu (Emlen 1984, Van Vuren & Armitage 1994, Isbell & Van Vuren 1996).

If mothers cannot increase the size of their home ranges to accommodate their daughters,
their daughters will leave their natal home ranges despite the costs of dispersal if their
chances of reproducing elsewhere are greater. Because inbreeding is more costly to females
than to males (Clutton-Brock & Harvey 1976), selection should favor females that minimize
incestuous matings (e.g., Packer 1979). Males disperse because limited mating opportunities
in their natal groups or home ranges create greater mating opportunities in other groups or

home ranges, all else being equal. Male dispersal may thus be driven not by their own avoidance of inbreeding but by reduced mating opportunities in the natal group or home range. In rare cases, males may remain when females disperse and the costs of social dispersal are so high that dispersing males have no chance of increasing their matings elsewhere. Chimpanzees may exemplify this situation.

If females are able to reproduce while remaining in their natal groups or home ranges, males then typically leave. In several species with male-biased dispersal, for example, olive baboons (*P. anubis*), yellow baboons, chacma baboons, gray-cheeked mangabeys (*Lophocebus albigena*), and Tibetan macaques (*Macaca thibetana*), males disperse to groups having more estrous females or more females than their current groups (Packer 1979, Zhao 1994, Alberts & Altmann 1995, Henzi et al. 1998, Olupot & Waser 2001). Though this does not necessarily mean more actual matings, evidence from baboons suggests that socially dispersing males do have greater mating success than philopatric males (Packer 1979, Alberts & Altmann 1995). In other species with male-biased social dispersal, for example, vervets, long-tailed macaques (*Macaca fascicularis*), and hanuman langurs, males do not disperse to groups with more females or more females in estrus (Henzi & Lucas 1980, Cheney & Seyfarth 1983, van Noordwijk & van Schaik 1985, Borries 2000). Whether this still holds when female relatives are excluded has not been considered, however. Since female relatives are largely unavailable for mating, discounting those females may well reveal that males in these species actually do disperse to groups with greater numbers of available females.

When females suffer no costs of dispersal that are sufficient to affect reproductive success (as appears to be the case for indifferent mothers), dispersal patterns are expected to be variable, and dependent upon the conditions facing individual females. Males can then influence female dispersal decisions. Males may respond by remaining philopatric unless they recognize better mating opportunities elsewhere, in which case they may also disperse. It is worth considering that the tendency for males to remain philopatric in large groups (e.g., Kibale red colobus and Costa Rican squirrel monkeys) occurs partly because they become less able to count or compare their relative breeding opportunities as numbers of females in groups increase (Hauser et al. 1996, Brannon & Terrace 1998, Wilson et al. 2001). If males remain, females will be forced to disperse socially because inbreeding depression is more costly to females than to males (Clutton-Brock & Harvey 1976).

In small to midsized multifemale groups in which females experience no significant costs of dispersal, only one male typically mates even if more than one remains. When most males disperse, females have more options. Some females leave their groups while others remain; hence the lower frequency of female compared to male social dispersal in single male, multifemale species with indifferent mothers. Such females should leave their groups if doing so enables them to avoid incestuous matings or infanticidal males, or to return to familiar areas (e.g., gorillas, Tana River red colobus, Thomas's langurs, and olive colobus), or when reproduction has failed for other reasons that are also unrelated to competition for food (Isbell & Van Vuren 1996, Steenbeek 2000; reviewed in Sterck & Korstjens 2000).

Testing the Dispersal/Foraging Model

The opportunity to reproduce while remaining philopatric could arise when mothers are able to accommodate their daughters by expanding their home ranges while still maintaining their foraging efficiency. Whether mothers could do this would depend on the relative importance of goal-directed travel and wandering, with a larger proportion of wandering enabling mothers to enlarge their home ranges without sacrificing foraging efficiency. Kin groups could occur if maintenance of mothers' foraging efficiency also requires visual monitoring of daughters' movements.

The data presented here support the dispersal/foraging efficiency model, but they were not collected specifically to test the model. More direct tests of the model could be developed by collecting comparative data on the percentage of time spent wandering and in goal-directed travel. With its distinction between slow foraging and fast travel, the work of Janson and Di Bitetti (1997) on brown capuchins shows nicely that the data can be collected. Other measures of the micromovements of individuals might also be incorporated into tests of the model. For instance, a higher percentage of unsuccessful stops for food indicates that wandering is more prevalent than goal-directed travel, whereas a higher percentage of successful stops indicates that goal-directed travel is more prevalent than wandering. Stingy mothers are predicted, therefore, to have the highest percentage of successful stops for food, followed by incomplete suppressors, generous mothers, and finally facilitators (figure 4.3). Obviously, we will not gain a full sense of the biological meaning of the data until comparative data from each of these types of mothers are available. Except for stingy mothers and incomplete suppressors (whose basal metabolic rates may be similar), basal metabolic rates are expected to mirror this order, once the effects of body size, and perhaps phylogeny, are removed. Though indifferent mothers cannot be directly compared with the other four types (since their reproductive success is not as dependent on food abundance), they can still contribute to tests of the model if they can be studied sufficiently long and intensively to determine what does limit their reproductive success. As the data trickle in, the weaknesses of this model will undoubtedly become more apparent and a more accurate model will replace it. I look forward to that day.

A Storybook Ending Either Way (as Long as She Reproduces)

Had L. Frank Baum known what we know now about female primates, he might have developed another, equally happy ending for Dorothy. Consider the alternative: if Dorothy had found a mate in Oz, she might have been content to remain. Sometimes for female primates, "somewhere over the rainbow" holds the promise of reproductive success, but when reproductive success is more likely in their natal home ranges, females will respond as if "there's no place like home" whether they live in Madagascar, the neotropics, Africa/Asia, or Kansas. The opportunity and ability to reproduce in the natal home range may have been prerequisites for the formation of kin groups.

Acknowledgments   In memory of Francis Bossuyt, who conducted his study of dispersal in titi monkeys in the best possible way: with patience, endurance, and a love for the animals and their homes.

I thank C. Berman and B. Chapais for inviting me to contribute to their volume. C. Berman, B. Chapais, T. Young, and two anonymous reviewers thoughtfully suggested many modifications to the first attempt, which resulted a very different and (I hope) much improved revision. C. Berman, C. Borries, B. Chapais, S. Harcourt, C. Janson, C. Jones, A.

Koenig, M. Korstjens, D. Van Vuren, and R. Wrangham kindly added comments that helped to fine tune the final product.

References

Abbott, D. H., Barrett, J., & George, L. M. 1993. Comparative aspects of the social suppression of reproduction in female marmosets and tamarins. In: *Marmosets and Tamarins: Systematics, Behaviour, and Ecology* (Ed. by A. B. Rylands), pp. 152–163. New York: Oxford University Press.

Alberts, S. C. & Altmann, J. 1995. Balancing costs and opportunities: dispersal in male baboons. *Am. Nat.*, 145, 279–306.

Albrecht, E. D., Aberdeen, G. W., & Pope, G. J. 2000. The role of estrogen in the maintenance of primate pregnancy. *Am. J. Obstet. Gynecol.*, 182, 432–438.

Alexander, R. D. 1974. The evolution of social behavior. *Ann. Rev. Ecol. Sys.*, 5, 324–382.

Altmann, S. A. 1974. Baboons, space, time and energy. *Am. Zool.*, 14, 221–248.

Anderson, C. M. 1987. Female transfer in baboons. *Am. J. Phys. Anthropol.*, 73, 241–250.

Bauchop, T. 1978. Digestion of leaves in vertebrate arboreal folivores. In: *The Ecology of Arboreal Folivores* (Ed. by G. G. Montgomery), pp. 193–204. Washington, DC: Smithsonian Institution Press.

Bearder, S. K. 1987. Lorises, bushbabies, and tarsiers: diverse societies in solitary foragers. In *Primate Societies* (Ed. by B. B. Smuts, D. L. Cheney, R. M. Seyfarth, R. W. Wrangham, & T. T. Struhsaker), pp. 11–24. Chicago: University of Chicago Press.

Bennett, E. L. & Davies, A. G. 1994. The ecology of Asian colobines. In: *Colobine Monkeys: Their Ecology, Behaviour and Evolution* (Ed. by A. G. Davies & J. F. Oates), pp. 129–171. New York: Cambridge University Press.

Boinski, S. 1987. Birth synchrony in squirrel monkeys (*Saimiri oerstedi*): a strategy to reduce neonatal predation. *Behav. Ecol. Sociobiol.*, 21, 393–400.

Boinski, S. 1999. The social organizations of squirrel monkeys: implications for ecological models of social evolution. *Evol. Anthropol.*, 8, 101–112.

Boinski, S. 2000. Social manipulation within and between troops mediates primate group movement: In: *On the Move: How and Why Animals Travel in Groups* (Ed. by S. Boinski & P. A. Garber), pp. 421–469, Chicago: University of Chicago Press.

Boinski, S., Sughre, K., Selvaggi, L., Quatrone, R., Henry, M., & Cropp, S. 2002. An expanded test of the ecological model of primate social evolution: competitive regimes and female bonding in three species of squirrel monkeys (*Saimiri oerstedii, S. boliviensis*, and *S. sciureus*). *Behaviour*, 139, 227–261.

Boinski, S., Treves, A., & Chapman, C. A. 2000. A critical evaluation of the influence of predators on primates: effects on group travel. In: *On the Move: How and Why Animals Travel in Groups* (Ed. by S. Boinski & P. A. Garber), pp. 43–72. Chicago: University of Chicago Press.

Boonstra, R., Xia, X., & Pavone, L. 1993. Mating system of the meadow vole, *Microtus pennsylvanicus*. *Behav. Ecol.*, 4, 83–89.

Borries, C. 2000. Male dispersal and mating season influxes in Hanuman langurs living in multimale groups. In: *Primate Males: Causes and Consequences of Variation in Group Composition* (Ed. by P. M. Kappeler), pp. 146–158. Cambridge: Cambridge University Press.

Bossuyt, F. 2002. Natal dispersal of titi monkeys (*Callicebus moloch*) at Cocha Cashu, Manu National Park, Peru. *Am. J. Phys. Anthropol. Suppl.*, 34, 47.

Bowers, M. A., Gregario, K., Brame, C. J., Matter, S. F., & Dooley, J. L. Jr. 1996. Use of space and habitats by meadow voles at the home range, patch and landscape scales. *Oecologia*, 105, 107–115.

Bradbury, J. W. & Verhencamp, S. L. 1977. Social organisation and foraging in emballonurid bats. *Behav. Ecol. Sociobiol.*, 2, 1–17.

Brannon, E. M. & Terrace, H. S. 1998. Ordering of the numerosities 1 to 9 by monkeys. *Science*, 282, 746–749.

Carter, C. S. & Getz, L. L. 1993. Monogamy and the prairie vole. *Sci. Am.*, 268, 100–106.

Chapman, C. A., Chapman, L. J., Bjorndal, K. A., & Onderdonk, D. A. 2002. Application of protein-to-fiber ratios to predict colobine abundance on different spatial scales. *Int. J. Primatol.*, 23, 283–310.

Charles-Dominique, P. 1977. *Ecology and Behaviour of Nocturnal Primates*. New York: Columbia University Press.

Charles-Dominique, P. 1978. Solitary and gregarious prosimians: evolution of social structures in primates. In: *Recent Advances in Primatology, Vol. 3* (Ed. by D. J. Chivers & K. A. Joysey), pp. 139–149. London: Academic Press.

Cheney, D. L. & Seyfarth, R. M. 1983. Nonrandom dispersal in free-ranging vervet monkeys: social and genetic consequences. *Am. Nat.*, 122, 392–412.

Chism, J. & Rowell, T.E. 1988. The natural history of patas monkeys. In: *A Primate Radiation: Evolutionary Biology of the African Guenons* (Ed. by A. Gautier-Hion, F. Bourlière, J.-P. Gautier, & J. Kingdon), pp. 412–438. Cambridge: Cambridge University Press.

Chism, J., Rowell, T. E., & Olson, D. K. 1984. Life history patterns of female patas monkeys. In: *Female Primates: Studies by Women Primatologists* (Ed. by M. D. Small), pp. 175–190. New York: Alan R. Liss.

Clutton-Brock, T. H. 1989. Female transfer and inbreeding avoidance in social mammals. *Nature*, 337, 70–71.

Clutton-Brock, T. H. & Harvey, P. H. 1976. Evolutionary rules and primate societies. In: *Growing Points in Ethology* (Ed. by P. G. G. Bateson & R. A. Hinde), pp. 195–237. Cambridge: Cambridge University Press.

Cochran, G. R. & Solomon, N. G. 2000. Effects of food supplementation on the social organization of prairie voles (*Microtus ochrogaster*). *J. Mammal.*, 81, 746–757.

Cody, M. L. 1971. Finch flocks in the Mohave desert. *Theor. Popul. Biol.*, 2, 142–158.

Cords, M. 2000. The agonistic and affliative relationships of adult females in a blue monkey group. In: *Old World Monkeys* (Ed. by C. Jolly & P. Whitehead), pp. 453–479. Cambridge: Cambridge University Press.

Crockett, C. M. 1984. Emigration by female red howler monkeys and the case for female competition. In: *Female Primates: Studies by Women Primatologists* (Ed. by M. F. Small), pp. 159–173. New York: Alan R. Liss.

Crockett, C. M. & Janson, C. H. 2000. Infanticide in red howlers: female group size, male membership, and a possible link to folivory. In: *Infanticide by Males and Its Implications* (Ed. by C. P. van Schaik & C. H. Janson), pp. 75–98. Cambridge: Cambridge University Press.

Davies, A. G. 1994. Colobine populations. In: *Colobine Monkeys: Their Ecology, Behaviour, and Evolution* (Ed. by A. G. Davies & J. F. Oates), pp. 285–310. Cambridge: Cambridge University Press.

Dietz, J. M. & Baker, A. J. 1993. Polygyny and female reproductive success in golden lion tamarins. *Anim. Behav.*, 46, 1067–1078.

Di Fiore, A. F. & Rendall, D. 1994. Evolution of social organization: a reappraisal for primates by using phylogenetic methods. *Proc. Natl. Acad. Sci. USA*, 91, 9941–9945.

Digby, L. J. 1995. Infant care, infanticide, and female reproductive strategies in polygynous groups of common marmosets (*Callithrix jacchus*). *Behav. Ecol. Sociobiol.*, 37, 51–61.

Digby, L. J. 2000. Infanticide by female mammals: implications for the evolution of social systems. In: *Infanticide by Males and Its Implications* (Ed. by C. P. van Schaik & C. H. Janson), pp. 423–446. Cambridge: Cambridge University Press.

Digby, L. J. & Ferrari, S. F. 1994. Multiple breeding females in free-ranging groups of *Callithrix jacchus. Int. J. Primatol.*, 15, 389–397.

Eberle, M. & Kappeler, P. M. 2002. Mouse lemurs in space and time: a test of the socioecological model. *Behav. Ecol. Sociobiol.*, 51, 131–139.

Eisenberg, J. F. 1981. *The Mammalian Radiations: An Analysis of Trends in Evolution, Adaptation, and Behavior.* Chicago: University of Chicago Press.

Emlen, S. T. 1984. Cooperative breeding in birds and mammals. In: *Behavioural Ecology: An Evolutionary Approach* (Ed. by J. R. Krebs & N. B. Davies), pp. 305–339. Sunderland, MA: Sinauer Associates.

Emlen, S. T. & Oring, L. 1977. Ecology, sexual selection, and the evolution of mating systems. *Science*, 197, 215–223.

Emmons, L. H. 2000. *Tupai: A Field Study of Bornean Treeshrews.* Berkeley, CA: University of California Press.

Enstam, K. L., Isbell, L. A., & de Maar, T. W. 2002. Male demography, female mating behavior, and infanticide in wild patas monkeys (*Erythrocebus patas*). *Int. J. Primatol.*, 23, 85–104.

Fairbanks, L. A. 2000. Maternal investment throughout the life span in Old World monkeys. In: *Old World Monkeys* (Ed. by P. F. Whitehead & C. J. Jolly), pp. 341–367. Cambridge: Cambridge University Press.

Fashing, P. J. 2001. Activity and ranging patterns of guerezas in the Kakamega Forest: intergroup variation and implications for intragroup feeding competition. *Int. J. Primatol.*, 22, 549–577.

Ferrari, S. F. & Lopes Ferrari, M. A. 1989. A re-evaluation of the social organisation of the Callitrichidae, with special reference to the ecological differences between genera. *Folia Primatol.*, 52, 132–147.

Fietz, J. 1999. Monogamy as a rule rather than exception in nocturnal lemurs: the case of the fat-tailed dwarf lemur, *Cheirogaleus medius. Ethology*, 105, 259–272.

Fleagle, J. G. 1999. *Primate Adaptation and Evolution,* 2nd ed. New York: Academic Press.

Fossey, D. 1984. Infanticide in mountain gorillas (*Gorilla gorilla berengei*) with comparative notes on chimpanzees. In: *Infanticide: Comparative and Evolutionary Perspectives* (Ed. by G. Hausfater & S. B. Hrdy), pp. 217–235. Hawthorne, NY: Aldine.

Fox, E. A. 2002. Female tactics to reduce sexual harassment in the Sumatran orangutan (*Pongo pygmaeus abelii*). *Behav. Ecol. Sociobiol.*, 52, 93–101.

Fuentes, A. 2000. Hylobatid communities: changing views on pair bonding and social organization in hominoids. *Ybk. Phys. Anthropol.*, 43, 33–60.

Garber, P. A. 1989. Role of spatial memory in primate foraging patterns: *Saguinus mystax* and *Saguinus fuscicollis. Am. J. Primatol.*, 19, 203–216.

Garber, P. A. 1993. Feeding ecology and behaviour of the genus *Saguinus*. In: *Marmosets and Tamarins: Systematics, Behaviour, and Ecology* (Ed. by A. B. Rylands), pp. 273–295. New York: Oxford University Press.

Garber, P. A. 2000. Evidence for the use of spatial, temporal, and social information by primate foragers. In: *On the Move: How and Why Animals Travel in Groups* (Ed by S. Boinski & P. A. Garber), pp. 261–298. Chicago: University of Chicago Press.

Garber, P. A. & Bicca-Marques, J. C. 2002. Evidence of predator sensitive foraging and traveling in single and mixed species tamarin groups. In: *Eat or Be Eaten: Predator*

Sensitive Foraging in Primates (Ed. by L. E. Miller), pp. 138–153. Cambridge: Cambridge University Press.

Genoud, M., Martin, R. D., & Glaser, D. 1997. Rate of metabolism in the smallest simian primate, the pygmy marmoset (*Cebuella pygmaea*). *Am. J. Primatol.*, 41, 229–245.

Getz, L. L., McGuire, B., Pizzuto, T., Hofmann, J. E., & Frase, B. 1993. Social organization of the prairie vole (*Microtus ochrogaster*). *J. Mammal.*, 74, 44–58.

Goldizen, A. W. 1987. Tamarins and marmosets: communal care of offspring. In: *Primate Societies* (Ed. by B. B. Smuts, D. L. Cheney, R. M. Seyfarth, R. W. Wrangham, & T. T. Struhsaker), pp. 34–43. Chicago: University of Chicago Press.

Goldizen, A. W., Mendelson, J., van Vlaardingen, M., & Terborgh, J. 1996. Saddle-back tamarin (*Saguinus fuscicollis*) reproductive strategies: evidence from a thirteen-year study of a marked population. *Am. J. Primatol.*, 38, 57–83.

Greenwood, P. J. 1980. Mating systems, philopatry, and dispersal in birds and mammals. *Anim. Behav.*, 28, 1140–1162.

Gursky, S. 2000. Sociality in the spectral tarsier, *Tarsius spectrum*. *Am. J. Primatol.*, 51, 89–101.

Hall, K. R. L. 1965. Behaviour and ecology of the wild patas monkey, *Erythrocebus patas*, in Uganda. *J. Zool. Lond.*, 148, 15–87.

Hamilton, W. D. 1971. Geometry for the selfish herd. *J. Theor. Biol.*, 31, 295–311.

Harcourt, A. H. 1978. Strategies of emigration and transfer by primates, with particular reference to gorillas. *Z. Tierpsychol.*, 48, 401–420.

Hauser, M. D., MacNeilage, P., & Ware, M. 1996. Numerical representations in primates. *Proc. Natl. Acad. Sci. USA*, 93, 1514–1517.

Hauser, M. D. & Marler, P. 1993. Food-associated calls in rhesus macaques (*Macaca mulatta*). I. Socioecological factors. *Behav. Ecol.*, 4, 191–205.

Hauser, M. D., Teixidar, P., Field, P., & Flaherty, R. 1993. Food-elicited calls in chimpanzees: effects of food quantity and divisibility. *Anim. Behav.*, 45, 817–819.

Hauser, M. D. & Wrangham, R. W. 1987. Manipulation of food calls in captive chimpanzees: a preliminary report. *Folia Primatol.*, 48, 207–210.

Henzi, S. P. & Lucas, J. W. 1980. Observations on the inter-troop movement of adult vervet monkeys (*Cercopithecus aethiops*). *Folia Primatol.*, 33, 220–235.

Henzi, S. P., Lycett, J. E., & Weingrill, T. 1998. Mate guarding and risk assessment by male mountain baboons during inter-troop encounters. *Anim. Behav.*, 55, 1421–1428.

Hrdy, S. B. 1974. Male-male competition and infanticide among the langurs (*Presbytis entellus*) of Abu, Rajasthan. *Folia Primatol.*, 22, 19–58.

Isbell, L. A. 1990. Sudden short-term increase in mortality of vervet monkeys (*Cercopithecus aethiops*) due to leopard predation in Amboseli National Park, Kenya. *Am. J. Primatol.*, 21, 41–52.

Isbell, L. A. 1991. Contest and scramble competition: patterns of female aggression and ranging behavior in primates. *Behav. Ecol.*, 2, 143–155.

Isbell, L. A. 1994. Predation on primates: ecological patterns and evolutionary consequences. *Evol. Anthropol.*, 3, 61–71.

Isbell, L. A., Cheney, D. L., & Seyfarth, R. M. 1990. Costs and benefits of home range shifts among vervet monkeys (*Cercopithecus aethiops*) in Amboseli National Park, Kenya. *Behav. Ecol. Sociobiol.*, 27, 351–358.

Isbell, L. A. & Enstam, K. L. 2002. Predator (in)sensitive foraging in sympatric vervets (*Cercopithecus aethiops*) and patas monkeys (*Erythrocebus patas*). In: *Eat or Be Eaten: Predator Sensitive Foraging in Primates* (Ed. by L. E. Miller), pp. 154–168. Cambridge: Cambridge University Press.

Isbell, L. A. & Pruetz, J. D. 1998. Differences between vervets (*Cercopithecus aethiops*) and patas monkeys (*Erythrocebus patas*) in agonistic interactions between adult females. *Int. J. Primatol.*, 19, 837–855.

Isbell, L. A., Pruetz, J. D., Lewis, M., & Young, T. P. 1998. Locomotor activity differences between sympatric patas monkeys (*Erythrocebus patas*) and vervet monkeys (*Cercopithecus aethiops*): implications for the evolution of long hindlimb length in *Homo*. *Am. J. Phys. Anthropol.*, 105, 199–207.

Isbell, L. A., Pruetz, J. D., Nzuma, B. M., & Young, T. P. 1999. Comparing measures of travel distances in primates: methodological considerations and socioecological implications. *Am. J. Primatol.*, 48, 87–98.

Isbell, L. A., Pruetz, J. D., & Young, T. P. 1998. Movements of adult female vervets (*Cercopithecus aethiops*) and patas monkeys (*Erythrocebus patas*) as estimators of food resource size, density, and distribution. *Behav. Ecol. Sociobiol.*, 42, 123–133.

Isbell, L. A. & Van Vuren, D. 1996. Differential costs of locational and social dispersal and their consequences for female group-living primates. *Behaviour*, 133, 1–36.

Isbell, L. A. & Young, T. P. 2002. Ecological models of female social relationships in primates: similarities, disparities, and some directions for future clarity. *Behaviour*, 139, 177–202.

Jannett, F. J. Jr. 1978. The density-dependent formation of extended maternal families of the montane vole, *Microtus montanus nanus*. *Behav. Ecol. Sociobiol.*, 3, 245–263.

Janson, C. H. 1990. Ecological consequences of individual spatial choice in foraging groups of brown capuchin monkeys, *Cebus apella*. *Anim. Behav.*, 40, 922–934.

Janson, C. H. 1998. Experimental evidence for spatial memory in foraging wild capuchin monkeys, *Cebus apella*. *Anim. Behav.*, 55, 1229–1243.

Janson, C. H. & Di Bitetti, M. S. 1997. Experimental analysis of food detection in capuchin monkeys: effects of distance, travel speed, and resource size. *Behav. Ecol. Sociobiol.*, 41, 17–24.

Janson, C. H. & van Schaik, C. P. 1988. Recognizing the many faces of competition: methods. *Behaviour*, 105, 165–186.

Janson, C. H. & Goldsmith, M. 1995. Predicting group size in primates: foraging costs and predation risks. *Behav. Ecol.*, 6, 326–336.

Jolly, A. 1984. The puzzle of female feeding priority. In: *Female Primates: Studies by Women Primatologists* (Ed. by M. D. Small), pp. 197–215. New York: Alan R. Liss.

Jolly, A. & Pride, E. 1999. Troop histories and range inertia of *Lemur catta* at Berenty, Madagascar: a 33-year perspective. *Int. J. Primatol.*, 20, 359–373.

Kappeler, P. M. 2000. Causes and consequences of unusual sex ratios among lemurs. In: *Primate Males: Causes and Consequences of Variation in Group Composition* (Ed. by P. M. Kappeler), pp. 55–63. Cambridge: Cambridge University Press.

Kay, R. & Davies, A. G. 1994. Digestive physiology. In: *Colobine Monkeys: Their Ecology, Behaviour, and Evolution* (Ed. by A. G. Davies & J. F. Oates), pp. 229–249. Cambridge: Cambridge University Press.

Kelt, D. A. & Van Vuren, D. H. 2001. The ecology and macroecology of mammalian home range area. *Am. Nat.*, 157, 637–645.

Kirkpatrick, R. C., Zou, R. J., Dierenfeld, E. S., & Zhou, H. W. 2001. Digestion of selected foods by Yunnan snub-nosed monkey *Rhinopithecus bieti* (Colobinae). *Am. J. Phys. Anthropol.*, 114, 156–162.

Klopfer, P. H. & Jolly, A. 1970. The stability of territorial boundaries in a lemur troop. *Folia Primatol.*, 12, 199–208.

Koenig, A. 2000. Competitive regimes in forest-dwelling Hanuman langur females (*Semnopithecus entellus*). *Behav. Ecol. Sociobiol.*, 48, 93–109.

Korstjens, A. H. 2001. The mob, the secret sorority, and the phantoms: an analysis of the socio-ecological strategies of three colobines of Taï. PhD dissertation, University of Utrecht, Utrecht, Netherlands.

Korstjens, A. H. & Schippers, E. P. 2003. Dispersal patterns among olive colobus in Taï National Park. *Int. J. Primatol.*, 2, 515–539.

Leighton, D. R. 1987. Gibbons: territoriality and monogamy. In: *Primate Societies* (Ed. by B. B. Smuts, D. L. Cheney, R. M. Seyfarth, R. W. Wrangham, & T. T. Struhsaker), pp. 135–145. Chicago: University of Chicago Press.

Madison, D. M. 1980. Space use and social structure in meadow voles, *Microtus pennsylvanicus*. *Behav. Ecol. Sociobiol.*, 7, 65–71.

Marsh, C. W. 1979. Female transference and mate choice among Tana River red colobus. *Nature*, 281, 568–569.

Mathy, J. W. & Isbell, L. A. 2002. The relative importance of size of food and interfood distance in eliciting aggression in captive rhesus macaques (*Macaca mulatta*). *Folia Primatol.*, 72, 268–277.

McGrew, W. C. 1997. Sex differences in the family life of cotton-top tamarins: socioecological validity in the laboratory. In: *New World Primates: Ecology, Evolution, and Behavior* (Ed. by W. G. Kinzey), pp. 95–107. Hawthorne, NY: Aldine.

Menzel, C. R. & Beck, B. B. 2000. Homing and detour behavior in golden lion tamarin social groups. In: *On the Move: How and Why Animals Travel in Groups* (Ed. by S. Boinski & P. A. Garber), pp. 299–326. Chicago: University of Chicago Press.

Milton, K. 1980. *The Foraging Strategy of Howler Monkeys: A Study in Primate Economics.* New York: Columbia University Press.

Milton, K. 2000. Quo vadis? Tactics of food search and group movement in primates and other animals. In: *On the Move: How and Why Animals Travel in Groups* (Ed by S. Boinski & P. A. Garber), pp. 375–417. Chicago: University of Chicago Press.

Milton, K. & May, M. L. 1976. Body weight, diet and home range area in primates. *Nature*, 259, 459–462.

Mitani, J. C. & Rodman, P. S. 1979. Territoriality: the relation of ranging patterns and home range size to defendability, with an analysis of territoriality among primate species. *Behav. Ecol. Sociobiol.*, 5, 241–251.

Moore, J. 1984. Female transfer in primates. *Int. J. Primatol.*, 5, 537–589.

Moore, J. 1988. Primate dispersal. *Trends Ecol. Evol.*, 3, 144–145.

Moore, J. 1992. Dispersal, nepotism, and primate social behavior. *Int. J. Primatol.*, 13, 361–378.

Moore, J. & Ali, R. 1984. Are dispersal and inbreeding avoidance related? *Anim. Behav.*, 32, 94–112.

Morland, H. S. 1991. Preliminary report on the social organization of ruffed lemurs (*Varecia variegata variegata*) in a northeast Madagascar rain forest. *Folia Primatol.*, 56, 157–161.

Müller, E. F. 1985. Basal metabolic rates in primates—the possible role of phylogenetic and ecological factors. *Comp. Biochem. Physiol.*, 81A, 707–711.

Mutschler, T., Nievergelt, C. M., & Feistner, A. T. C. 2000. Social organization of the Alaotran gentle lemur (*Hapalemur griseus alaotrensis*). *Am. J. Primatol.*, 50, 9–24.

Nievergelt, C. M., Mutschler, T., & Feistner, A. T. C. 1998. Group encounters and territoriality in wild Alaotran gentle lemurs (*Hapalemur griseus alaotrensis*). *Am. J. Primatol,*, 46, 251–258.

Noë, R. & Bshary, R. 1997. The formation of red colobus-diana monkey associations under predation pressure from chimpanzees. *Proc. Royal Soc. Lond., B*, 264, 253–259.

Oates, J. F. 1994. The natural history of African colobines. In: *Colobine Monkeys: Their Ecology, Behaviour, and Evolution* (Ed. by A. G. Davies & J. F. Oates), pp. 75–128. Cambridge: Cambridge University Press.

Olupot, W. & Waser, P. M. 2001. Activity patterns, habitat use and mortality risks of mangabey males living outside social groups. *Anim. Behav.*, 61, 1227–1235.

Ostro, L. E. T., Silver, S. C., Koontz, F. W., Horwich, R. H., & Brockett, R. 2001. Shifts in social structure of black howler (*Alouatta pigra*) groups associated with natural and experimental variation in population density. *Int. J. Primatol.*, 22, 733–748.

Ostro, L. E. T., Silver, S. C., Koontz, F. W., Young, T. P., & Horwich, R. H. 1999. Ranging behavior of translocated and established groups of black howler monkeys *Alouatta pigra* in Belize, Central America. *Biol. Conserv.*, 87, 181–190.

Overdorff, D. J. 1996. Ecological correlates to social structure in two prosimian primates in Madagascar. *Am. J. Phys. Anthropol.*, 100, 487–506.

Packer, C. 1979. Inter-troop transfer and inbreeding avoidance in *Papio anubis*. *Anim. Behav.*, 27, 1–36.

Palombit, R. 1994. Dynamic pair bonds in hylobatids: implications regarding monogamous social systems. *Behaviour*, 128, 65–101.

Parra, R. 1978. Comparison of foregut and hindgut fermentation in herbivores. In: *The Ecology of Arboreal Folivores* (Ed. by G. G. Montgomery), pp. 205–229. Washington, DC: Smithsonian Institution Press.

Parsons, P. E. & Taylor, C. R. 1977. Energetics of brachiation versus walking: a comparison of a suspended and inverted pendulum mechanism. *Physiol. Zool.*, 50, 182–188.

Pereira, M. E. 1993. Agonistic interaction, dominance relations, and ontogenetic trajectories in ringtailed lemurs. In: *Juvenile Primates: Life History, Development, and Behavior* (Ed. by M. E. Pereira & L. A. Fairbanks), pp. 285–305. New York: Oxford University Press.

Peres, C. 2000. Territorial defense and the ecology of group movements in small-bodied neotropical primates. In: *On the Move: How and Why Animals Travel in Groups* (Ed. by S. Boinski & P. A. Garber), pp. 100–123. Chicago: University of Chicago Press.

Petter, J. J., Schilling, A., & Pariete, G. 1975. Observations on behavior and ecology of *Phaner furcifer*. In: *Lemur Biology* (Ed. by I. Tattersall & R. W. Sussman), pp. 209–218. New York: Plenum Press.

Pochron, S. 2001. Can concurrent speed and directness of travel indicate purposeful encounters in the yellow baboons (*Papio hamadryas cynocephalus*) of Ruaha National Park, Tanzania? *Int. J. Primatol.*, 22, 773–785.

Pollock, J. I. 1979. Spatial distribution and ranging behavior in lemurs. In: *The Study of Prosimian Behavior* (Ed. by G. A. Doyle & R. D. Martin), pp. 359–409. New York: Academic Press.

Pope, T. R. 2000a. The evolution of male philopatry in Neotropical monkeys. In: *Primate Males: Causes and Consequences of Variation in Group Composition* (Ed. by P. M. Kappeler), pp. 219–235. Cambridge: Cambridge University Press.

Pope, T. R. 2000b. Reproductive success increases with degree of kinship in cooperative coalitions of female red howler monkeys (*Alouatta seniculus*). *Behav. Ecol. Sociobiol.*, 48, 253–267.

Power, M. L., Tardiff, S. D., Power, R. A., & Layne, D. G. 2003. Resting energy metabolism of Goeldi's monkey (*Callimico goeldii*) is similar to that of other callitrichids. *Am. J. Primatol.*, 60, 57–67.

Printes, R. C. & Strier, K. B. 1999. Behavioral correlates of dispersal in female muriquis (*Brachyteles arachnoides*). *Int. J. Primatol.*, 20, 941–960.

Pruetz, J. D. & Isbell, L. A. 2000. Correlations of food distribution and patch size with agonistic interactions in female vervets (*Chlorocebus aethiops*) and patas monkeys (*Erythrocebus patas*) living in simple habitats. *Behav. Ecol. Sociobiol.*, 49, 38–47.

Pusey, A. E. & Packer, C. 1987a. Dispersal and philopatry. In: *Primate Societies* (Ed. by

B. B. Smuts, D. L. Cheney, R. M. Seyfarth, R. W. Wrangham, & T. T. Struhsaker), pp. 250–266. Chicago: University of Chicago Press.

Pusey, A. E. & Packer, C. 1987b. The evolution of sex-biased dispersal in lions. *Behaviour*, 101, 275–310.

Pusey, A. E., Williams, J. M., & Goodall, J. 1997. The influence of dominance rank on the reproductive success of female chimpanzees. *Science*, 277, 828–831.

Pusey, A. E. & Wolf, M. 1996. Inbreeding avoidance in animals. *Trends Ecol. Evol.*, 11, 201–206.

Radespiel, U. 2000. Sociality in the gray mouse lemur (*Microcebus murinus*) in northwestern Madagascar. *Am. J. Primatol.*, 51, 21–40.

Rathbun, G. B. 1979. The social structure and ecology of elephant-shrews. *Adv. Ethol.*, 20, 1–77.

Richard, A. F. 1978. *Behavioral Variation: Case Study of a Malagasy Lemur*. Lewisburg, PA: Bucknell University Press.

Richard, A. F. 1987. Malagasy prosimians: female dominance. In: *Primate Societies* (Ed. by B. B. Smuts, D. L. Cheney, R. M. Seyfarth, R. W. Wrangham, & T. T. Struhsaker), pp. 25–33. Chicago: University of Chicago Press.

Rigamonti, M. M. 1993. Home range and diet in the red ruffed lemurs (*Varecia variegata rubra*) on the Masoala Peninsula, Madagascar. In: *Lemur Social Systems and Their Ecological Bases* (Ed. by P. M. Kappeler & J. U. Ganzhorn), pp. 25–39. New York: Plenum Press.

Rodman, P. S. 1988a. Diversity and consistency in ecology and behavior. In: *Orangutan Biology* (Ed. by J. H. Schwartz), pp. 31–51. New York: Oxford University Press.

Rodman, P. S. 1988b. Resources and group sizes of primates. In: *The Ecology of Social Behavior* (Ed. by C. N. Slobodchikoff), pp. 83–108. San Diego: Academic Press.

Rodman, P. S. & Mitani, J. C. 1987. Orangutans: sexual selection in a solitary species. In: *Primate Societies* (Ed. by B. B. Smuts, D. L. Cheney, R. M. Seyfarth, R. W. Wrangham, & T. T. Struhsaker), pp. 146–154. Chicago: University of Chicago Press.

Rothe, H. & Koenig, A. 1991. Variability of social organization in captive common marmosets (*Callithrix jacchus*). *Folia Primatol.*, 57, 28–33.

Rylands, A. B. 1993. The ecology of the lion tamarins, *Leontopithecus*: some intrageneric differences and comparisons with other callitrichids. In: *Marmosets and Tamarins: Systematics, Behaviour, and Ecology* (Ed. by A. B. Rylands), pp. 296–313. New York: Oxford University Press.

Savage, A. 1990. The reproductive biology of the cotton-top tamarin (*Saguinus oedipus*) in Colombia. PhD diss., University of Wisconsin, Madison, WI.

Savage, A., Giraldo, L. H., Soto, L. H., & Snowdon, C. T. 1996. Demography, group composition, and dispersal in wild cotton-top tamarin (*Saguinus oedipus*) groups. *Am. J. Primatol.*, 38, 85–100.

Schülke, O. In press. To breed or not to breed—food competition and other factors involved in female breeding decisions in the pair-living nocturnal fork-marked lemur (*Phaner furcifer*). *Behav. Ecol. Sociobiol.*

Shopland, J. M. 1987. Food quality, spatial deployment and the intensity of feeding interferences in yellow baboons (*Papio cynocephalus*). *Behav. Ecol. Sociobiol.*, 21, 149–156.

Silk, J. B. 1988. Social mechanisms of population regulation in a captive group of bonnet macaques (*Macaca radiata*). *Am. J. Primatol.*, 14, 111–124.

Silk, J. B., Clark-Wheatley, C. B., Rodman, P. S., & Samuels, A. 1981. Differential reproductive success and facultative adjustment of sex ratios among captive female bonnet macaques (*Macaca radiata*). *Anim. Behav.*, 29, 1106–1120.

Singleton, I. & van Schaik, C. P. 2001. Orangutan home range size and its determinants in a Sumatran swamp forest. *Int. J. Primatol.*, 22, 877–911.

Smuts, B. B. & Smuts, R. W. 1993. Male aggression and sexual coercion of females in nonhuman primates and other mammals: evidence and theoretical implications. *Adv. Study Behav.*, 22, 1–63.

Soini, P. 1993. The ecology of the pygmy marmoset, *Cebuella pygmaea*: some comparisons with two sympatric tarmarins. In: *Marmosets and Tamarins: Systematics, Behaviour, and Ecology* (Ed. by A. B. Rylands), pp. 257–261. New York: Oxford University Press.

Solomon, N. G. & French, J. A. 1997. *Cooperative Breeding in Mammals.* Cambridge: Cambridge University Press.

Stammbach, E. 1987. Desert, forest and montane baboons: multilevel-societies. In: *Primate Societies* (Ed. by B. B. Smuts, D. L. Cheney, R. M. Seyfarth, R. W. Wrangham, & T. T. Struhsaker), pp. 112–120. Chicago: University of Chicago Press.

Stanford, C. B. 1991. The capped langur in Bangladesh: behavioral ecology and reproductive tactics. *Contrib. Primatol.*, 26, i-xvii, 1–179.

Stanford, C. B. 1995. The influence of chimpanzee predation on group size and anti-predator behaviour in red colobus monkeys. *Anim. Behav.*, 49, 577–587.

Starin, E. D. 1981. Monkey moves. *Nat. Hist.*, 9, 36–43.

Starin, E. D. 1991. Socioecology of the red colobus monkey in the Gambia with particular reference to female-male differences and transfer patterns. PhD thesis, City University of New York.

Steenbeek, R. 2000. Infanticide by males and female choice in wild Thomas's langurs. In: *Infanticide by Males and Its Implications* (Ed. by C. P. van Schaik & C. H. Janson), pp. 153–177. Cambridge: Cambridge University Press.

Steenbeek, R., Sterck, E. H. M., de Vries, H., & van Hooff, J. A. R. A. M. (2000). Costs and benefits of the one-male, age-graded, and all-male phases in wild Thomas's langur groups. In: *Primate Males: Causes and Consequences of Variation in Group Composition* (Ed. by P. M. Kappeler), pp. 130–145. Cambridge: Cambridge University Press.

Steenbeek, R. & van Schaik, C. P. 2001. Competition and group size in Thomas's langurs (*Presbytis thomasi*): the folivore paradox revisited. *Behav. Ecol. Sociobiol.*, 49, 100–110.

Sterck, E. H. M. & Korstjens, A. H. 2000. Female dispersal and infanticide avoidance in primates. In: *Infanticide by Males and Its Implications* (Ed. by C. P. van Schaik & C. H. Janson), pp. 293–321. Cambridge: Cambridge University Press.

Sterck, E. H. M., Watts, D. P., & van Schaik, C. P. 1997. The evolution of female social relationships in nonhuman primates. *Behav. Ecol. Sociobiol.*, 41, 291–309.

Sterling, E. J. 1993. Patterns of range use and social organization in aye-ayes (*Daubentonia madagascariensis*) on Nosy Mangabe. In: *Lemur Social Systems and Their Ecological Basis* (Ed. by P. M. Kappeler & J. U. Ganzhorn), pp. 1–10. New York: Plenum Press.

Strier, K. B. 1994. Myth of the typical primate. *Ybk. Phys. Anthropol.*, 37, 233–271.

Strier, K. B. 1999. Why is female kin bonding so rare? Comparative sociality of neotropical primates. In: *Comparative Primate Socioecology* (Ed. by P. C. Lee), pp. 300–319. Cambridge: Cambridge University Press.

Struhsaker, T. T. 1975. *The Red Colobus Monkey.* Chicago: University of Chicago Press.

Terborgh, J. 1983. *Five New World Primates: A Study in Comparative Ecology.* Princeton, NJ: Princeton University Press.

Terborgh, J. & Janson, C. H. 1986. The socioecology of primate groups. *Ann. Rev. Ecol. Syst.*, 17, 111–135.

Thalmann, U. 2001. Food resource characteristics in two nocturnal lemurs with different social behavior: *Avahi occidentalis* and *Lepilemur edwardsi*. *Int. J. Primatol.*, 22, 287–324.

Trivers, R. L. 1972. Parental investment and sexual selection. In: *Sexual Selection and the Descent of Man, 1871–1972* (Ed. by B. Campbell), pp. 136–179. Chicago: Aldine.

van Noordwijk, M. A. & van Schaik, C. P. 1985. Male migration and rank acquisition in wild long-tailed macaques (*Macaca fascicularis*). *Anim. Behav.*, 33, 849–861.

van Schaik, C. P. 1983. Why are diurnal primates living in groups? *Behaviour*, 87, 120–144.

van Schaik, C. P. 1989. The ecology of social relationships amongst female primates. In: *Comparative Socioecology: The Behavioural Ecology of Humans and Other Mammals* (Ed. by V. Standen & R. A. Foley), pp. 195–218. Oxford: Blackwell Scientific Publications.

Van Vuren, D. & Armitage, K. B. 1994. Survival of dispersing and philopatric yellow-bellied marmots: what is the cost of dispersal? *Oikos*, 69, 179–181.

Vick, L. G. & Pereira, M. E. 1989. Episodic targeting aggression and the histories of lemur social groups. *Behav. Ecol. Sociobiol.*, 25, 3–12.

Waser, P. M. & Jones, W. T. 1983. Natal philopatry among solitary mammals. *Q. Rev. Biol.*, 58, 355–390.

Wasser, S. K. & Barash, D. P. 1983. Reproductive suppression among female mammals: implications for biomedicine and sexual selection theory. *Q. Rev. Biol.*, 58, 513–538.

Watts, D. P. 1989. Infanticide in mountain gorillas: new cases and a reconsideration of the evidence. *Ethology*, 81, 1–18.

Watts, D. P. 1990. Ecology of gorillas and its relation to female transfer in mountain gorillas. *Int. J. Primatol.*, 11, 21–45.

Watts, D. P. 2000. Mountain gorilla habitat use strategies and group movements. In: *On the Move: How and Why Animals Travel in Groups* (Ed by S. Boinski & P. A. Garber), pp. 351–374. Chicago: University of Chicago Press.

Williams, J. M., Pusey, A. E., Carlis, J. V., Farm, B. P., & Goodall, J. 2002. Female competition and male territorial behaviour influence female chimpanzees' ranging patterns. *Anim. Behav.*, 63, 347–360.

Wilson, M., Hauser, M. D., & Wrangham, R. W. 2001. Does participation in intergroup conflict depend on numerical assessment, range location, or rank for wild chimpanzees? *Anim. Behav.*, 61, 1203–1216.

Wimmer, B., Tautz, D., & Kappeler, P. M. 2002. The genetic population structure of the gray mouse lemur (*Microcebus murinus*), a basal primate from Madagascar. *Behav. Ecol. Sociobiol.*, 52, 166–175.

Wolff, J. O. 1993. Why are female small mammals territorial? *Oikos*, 68, 364–370.

Wolff, J. O., Dunlap, A. S., & Ritchhart, E. 2001. Adult female prairie voles and meadow voles do not suppress reproduction in their daughters. *Behav. Proc.*, 55, 157–162.

Wrangham, R. W. 1979. On the evolution of ape social systems. *Soc. Sci. Info.*, 18, 335–368.

Wrangham, R. W. 1980. An ecological model of female-bonded primate groups. *Behaviour*, 75, 262–300.

Wrangham, R. W., Gittleman, J. L., & Chapman, C. A. 1993. Constraints on group size in primates and carnivores: population density and day range as assays of exploitation competition. *Behav. Ecol. Sociobiol.*, 32, 199–209.

Wright, P. C. 1986. Ecological correlates of monogamy in *Aotus* and *Callicebus moloch*. In: *Primate Ecology and Conservation* (Ed. by J. Else & P. C. Lee), pp. 159–167. New York: Cambridge University Press.

Wright, P. C. 1994. The behavior and ecology of the owl monkey. In: *Aotus: The Owl Monkey* (Ed. by J. F. Baer, R. E. Weller, & I. Kakoma), pp. 97–112. New York: Academic Press.

Wright, P. C. 1995. Demography and life history of free-ranging *Propithecus diadema edwardsi* in Ranomafana National Park, Madagascar. *Int. J. Primatol.*, 16, 835–854.

Wright, P. C. 1999. Lemur traits and Madagascar ecology: coping with an island environment. *Ybk. Phys. Anthropol.*, 42, 31–72.

Yamagiwa, J. & Kahekwa, J. 2001. Dispersal patterns, group structure, and reproductive parameters of eastern lowland gorillas at Kahuzi in the absence of infanticide. In: *Mountain Gorillas: Three Decades of Research at Karisoke* (Ed. by M. M. Robbins, P. Sicotte, & K. J. Stewart), pp. 89–122. Cambridge: Cambridge University Press.

Yeager, C. O. & Kool, K. 2000. The behavioral ecology of Asian colobines. In: *Old World Monkeys* (Ed. by P. F. Whitehead & C. J. Jolly), pp. 496–521. Cambridge: Cambridge University Press.

Zhao, Q.-K. 1994. Mating competition and intergroup transfer of males in Tibetan macaques (*Macaca thibetana*) at Mt. Emei, China. *Primates*, 35, 57–61.

Zynel, C. A. & Wunder, B. A. 2002. Limits to food intake by the prairie vole: effects of time for digestion. *Func. Ecol.*, 16, 58–66.

5

Dispersal and the Population Genetics of Primate Species

Guy A. Hoelzer
Juan Carlos Morales
Don J. Melnick

Patterns of genetic diversity within species mark both familial relationships and subpopulation membership. They can also influence ongoing evolutionary processes. For these reasons, data on intraspecific genetic diversity are important to the study of primate social structure and historical biogeography, as well as to the design of management plans for species conservation (see many chapters in Clobert et al. 2001).

Dispersal of individuals from their birthplaces to where they reproduce can have a profound effect on both the amount and kind of genetic variation at particular localities. From a population genetics perspective, dispersal and subsequent reproduction results in what is called gene flow. Dispersal determines the spatial limitations on interactions among relatives and the degree to which relatives are spatially concentrated. The associated flow of genes determines the degree of genetic divergence at different spatial scales, and limitations to gene flow may ultimately lead to significant population subdivision, loss of heterozygosity, and speciation. The impacts of dispersal on the evolution of wild primate populations, and its implications for primate conservation, are the focus of this chapter.

We begin with a review of the theoretical principles and predictions with which population genetic data are interpreted. First, we provide a brief discussion of historical biogeography, an approach that places the diversity of spatial population structures studied in population genetics into a dynamic, geographical context. Most spatial models in population genetics assume static environmental conditions and focus on equilibrium spatial structures for simplicity, but the reader should keep in mind that these structures are typically fluid in nature.

Second, we describe in more detail the most commonly used models in population genetics, including the panmictic, island, steppingstone, and metapopulation models. We highlight two of the most important continuous variables affecting spatial patterns of genetic variation: partial barriers to dispersal and mutation rate. Throughout this section, we relate each concept to the methods of analysis commonly employed in the spatial analysis of population genetic data. This introduction to population genetics theory is followed by a third section that compares and contrasts the information contained in nuclear versus mitochondrial DNA (mtDNA) data.

Fourth, we present a section with examples of published data that relate dispersal patterns to spatial genetic structures in New and Old World primate populations. We end with a section on the potential impacts of dispersal and genetic diversity on social behavior, evolutionary potential, and conservation biology. The extent of local diversity is distinguished from the extent of population substructure in each of these categories.

Theory

A large body of theoretical work relates dispersal to population genetics. However, most of it has involved construction and exploration of models in which the continuum of space has been simplified into discrete compartments. Indeed, the most influential model used in the genetic analysis of dispersal has been the so-called island model of migration (Wright 1931), in which it is assumed that every compartmentalized subpopulation or deme samples the same pool of immigrants regardless of location. This is an example of a mean field approximation, because the universal immigrant pool assumed under this model is taken to be the average (or sum) of all the genetic types represented by emigrants leaving demes.

Mean field approximations can be useful ways of simplifying nature for the purpose of modeling, but they can sometimes result in significant distortions of the way populations evolve in nature. Thus, assuming a mean field model in the analysis of empirical data can sometimes result in biased estimates of critical parameter values, such as dispersal rates or distances. It is currently a major challenge in the field of population genetics to ascertain the conditions under which the island model of migration provides a reasonable approximation of the evolutionary process, and to explore the evolutionary effects of spatial continuity when the mean field approximation breaks down.

The island model, and some spatially explicit models, are discussed in more detail below. But before we concern ourselves with genetic variation, we delve into purely demographic patterns and processes of spatial population structure.

Dispersal and Historical Biogeography

How did the individuals that constitute a species come to be geographically distributed as they are? Under most definitions of biological species (see de Quiroz 1998), each must originate as a group of interacting (i.e., interbreeding) individuals without regional subdivision in its gene pool (but see Schluter et al. 2000). If species origins are generally confined to spatially coherent, freely interacting groups of individuals, it is interesting that we so commonly observe species that have escaped these restrictions. For example, simple range expansion can lead distance itself to provide an effective limitation to gene flow. In addition, many species, including the vast majority of widespread species, are spread across landscapes characterized by localized barriers to dispersal (Brown & Lomolino 1998) that interfere with the free interactions among individuals and divide the population into a number of semi-isolated subpopulations. It is important to keep in mind that these "barriers" often reduce dispersal by degree, rather than completely. For example, a raging river might prevent nearly any successful dispersal across it by a primate, but reducing the rate of water flow in the river by some fraction might reduce the degree of obstruction to dispersal by the same species by, say, 50%. Nevertheless, it can be useful to focus on absolute barriers as a starting point for thinking about the ways in which species ranges, and the distribution of genetic variation therein, can evolve.

Species ranges can go from spatially coherent to spatially subdivided (i.e., disjunct) in two ways. They are most commonly described using the terms *vicariance* (or *fragmentation*) and *colonization* (figure 5.1). Vicariance usually implies the imposition of a barrier upon a previously contiguous range, such as occurs during the isolation of islands due to rising sea level. The term *fragmentation* is usually used in reference to the isolation of refugial subpopulations during periods of habitat destruction (natural or anthropogenic) and decreasing population size (note that population density can remain high within subpopulations). Either way, the spatial cohesion of interactions is lost within the range that was previously connected, and the overall extent of the range is either reduced or remains roughly the same in the process.

In contrast, colonization involves successful dispersal into an area that had not been previously occupied by members of a species. Therefore, it results in range expansion. The area colonized might be at the front edge of an advancing wave of colonization, in which case interaction between the colonists (or their descendents) and the rest of the species is never really lost, and it is fully reestablished as the advancing wave sweeps over this area. Alternatively, colonization might represent a rare occurrence of successful long-distance dispersal (e.g., island colonization from the mainland), in which case the new colony can be as isolated from the rest of the species as a subpopulation cut off by vicariance.

Dispersal and Genetic Diversity

These two paths to the formation of disjunct distributions, vicariance and colonization, can partition the genetic variation of a species in very different ways, despite the fact that the resulting spatial distributions of individuals may be indistinguishable (figure 5.1). If, for example, the genetic variation present in a species was spatially well mixed (i.e., the panmictic model; see below) prior to a vicariant event, then at the inception of their separation each subpopulation would be genetically almost indistinguishable from the other. The only factor that might result in differences between the gene pools would be sampling error,

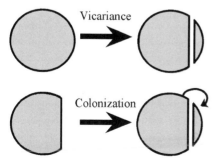

Figure 5.1. A schematic view of the distinction between vicariance and colonization as two modes of creating a disjunct distribution from a coherent distribution.

which would be negligible unless one of the newly formed subpopulations happened to be very small. The importance of sampling error is mentioned in several contexts in this chapter. It simply refers to the statistical fact that small, randomly chosen subsets of genes from a larger population usually differ in frequency from the population itself, and does not imply some sort of mistake. The expected degree of difference between the sample and the parent population depends almost entirely on the size of the sample, rather than the fraction of the total population sampled (Thompson 1992).

Although it is unlikely that sampling error would have much effect during vicariance, it is almost certain to have a large "founder" effect during a rare colonization event, which would generally involve a small number of individuals. This means that the allele and genotype frequencies of the population of colonists would differ substantially from those of the source population, although the most common alleles would be present in both populations unless the number of colonists was extremely small.

Both vicariance and colonization reflect nonequilibrium conditions in factors influencing the distribution of genetic variation. Critical parameters, such as population size and connectedness, change while these processes occur. Indeed, it is reasonable to believe that biological populations are generally not in equilibrium with regard to these parameters. Nevertheless, equilibrium is typically assumed in both of these parameters for the purposes of population genetic modeling, thus making the models more tractable and easier to interpret. Working with equilibrium models provides real insight into the way natural systems work, even though they are idealized abstractions. However, one should always be aware that the accuracy of empirical estimates of dispersal based on these models might be compromised by the degree to which a particular natural system deviates from equilibrium.

The Panmictic Model

The simplest possible model relating dispersal to population genetics is the panmictic model (i.e., the deme model), in which typical dispersal distances are so large compared with the geographical range of the population that the union of gametes approximates that which would occur if mating were random among individuals. There is no compartmentalization of the population due to spatial constraints, so it is treated as a single indivisible object regarding its evolutionary dynamics. Mutation adds genetic variation, genetic drift (sampling error in the transmission of alleles across generations) reduces genetic variation by the stochastic extinction of alleles, and natural selection can do either, although it usually reduces variation in a discriminatory fashion.

Selection can complicate temporal and spatial patterns of genetic variation in so many different ways that it is usually left out of models relating dispersal to population genetics in order to avoid obscuring this fundamental relationship. We adopt this strategy for most of this chapter while remembering that the specter of selection could invalidate our models in particular instances. For this reason, the sources of genetic data gathered to estimate dispersal parameters are generally chosen in part because they are thought to be selectively neutral. For example, noncoding DNA sequences should generally be preferred over coding DNA sequences because they are believed to be selectively neutral or near neutral. When using coding DNA sequences (e.g., mtDNA genes) or their products (e.g., proteins or tRNAs), the allelic variants are assumed to have so little effect on fitness that any spatial variation in selection would have a negligible effect on the spatial distributions of alleles.

Distance is rendered ineffectual by dispersal in the panmictic model, so, ignoring selection, the evolutionary dynamics of the population are simply a result of the tension between mutation and drift (Kimura 1968). Mutation pumps genetic variation into the population at the same rate as it is removed by drift when the population is at equilibrium, and subpopulations do not become differentiated no matter how large the panmictic population. As a conceptual framework, this simplest of models defines an extreme relationship between dispersal and the potential for population subdivision. Thus, it provides a general set of expectations from which one can compare the effects of deviations from its underlying assumptions.

Isolation by Distance

In contrast to the panmictic model, dispersal distances are almost always small compared with the extent of a species' range, so that interactions happen on local scales among spatially restricted subsets of individuals. This was described in its most general form, without recourse to localized dispersal barriers or other particular forms of environmental heterogeneity, as isolation by distance by Sewall Wright (1943). Wright's main conclusion was that such a system would evolve so as to render widely separated samples of "the gene pool" significantly different from one another. Despite the ubiquity of limitations to dispersal distances, evolutionary dynamics under this model are poorly understood. Perhaps the most important result that has been obtained is that isolation by distance can result in the emergence of spatial boundaries between discretely differentiated subpopulations (Sawyer 1977, Slatkin 1985a, Rousset 1997, Hardy & Vekemans 1999). Although this result is not widely appreciated, it can be deduced from Wright's original description of the model in which he argues that drift acts to reduce genetic variation locally, while at the same time causing divergence between gene pools in different localities. These two trends cannot happen simultaneously unless a spatial gradient of transition sharpens between relatively homogeneous gene pools, which is the same thing as saying that a boundary has formed. Therefore, isolation by distance alone can result in the evolutionary origin of population subdivision, and it is clearly a realistic way to relate dispersal to the spatial distribution of genetic variation. We are not aware of any analytical methods for inferring dispersal parameters under the isolation by distance model at this time, but this is an area that is ripe for theoretical development.

Localized, Partial Barriers to Dispersal

As indicated above, models relating dispersal to population genetics to date most often have assumed that subpopulation structure is imposed on the population by the environment, rather than emerging as an outcome of the model. While this seems rather arbitrary, it has been elegantly formulated as the island model of migration (Wright 1931, figure 5.2). The

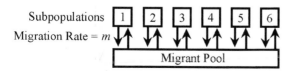

Figure 5.2. A schematic view of the island model of migration (Wright 1931).

island model assumes that a number of demes, all of size N, contribute to and receive a constant number of migrants (M) from the same general migrant pool at rate m (=M/N).

Despite the absence of spatial representation in the island model, it underlies most methods of empirical analysis currently employed in spatial population genetics, such as those that infer migration rates and the degree of population subdivision (Slatkin 1985a, Hartl & Clarke 1997), which is usually characterized using Wright's F_{ST} (Wright 1931) or some analog. F_{ST} is a statistical measure of the degree to which the genetic variation contained in all geographical locations represented in the data is partitioned among those locations. In other words, F_{ST} is a relatively large number when local subpopulations are distinctive, and it is smaller when the variation in the whole population is represented similarly at each site.

The original F_{ST} statistic varied from 0 to 1 (Wright 1931), but some of the newer analogs of this statistic are not bounded in the same way. The general (approximate) equation relating F_{ST} to deme sizes and migration under the island model is

$$F_{ST} \approx 1/(1 + 4Nm),$$

which can be written more simply as

$$F_{ST} \approx 1/(1 + 4M).$$

Inspection of these approximations reveals how the dynamics of the island model create a degree of subpopulation structure. Local genetic drift increases, reducing genetic variation locally, as both deme sizes and effective dispersal between demes decreases. However, the evolutionary dynamics of different demes becomes uncoupled as migration rate decreases, so demes can lose different aspects of the global gene pool. The localization of drift is what drives genetic divergence among demes under the island model.

Because this model does not include mutation as a source of genetic variation, the whole system (all demes combined) is destined to eventually drift to fixation of one allele or another when M > 0 and the number of demes is finite. So the above equations merely describe the way in which existing genetic diversity is expected to be distributed among demes until it erodes away. The island model is highly idealized, making simplifying assumptions that differ markedly from the conditions of real biological populations. However, only recently have population geneticists begun to question the accuracy of estimates based on the island model, when the dynamics of a real system of populations deviate from the model's highly simplified assumptions (Whitlock & McCauley 1999).

Perhaps the most useful way to begin making the island model more realistic is to allow spatial constraints on dispersal (a form of isolation by distance). Steppingstone models of migration were developed to represent spatial effects, while retaining links to the body of analytical theory that had been created to understand genetic evolution within demes (Malecot 1948, Kimura 1953, Kimura & Weiss 1964). In these models, the rate of migration between any pair of demes can vary among pairs, presumably as a function of the distance and potential obstacles to dispersal between them. Some insights into general evolutionary patterns and processes under steppingstone models have been derived from purely analytical approaches (Sawyer 1976, 1977), but the study of these models is most often done through computer simulation.

One interesting observation from a computational steppingstone model is that the isolation by distance represented among demes can lead to the evolution of spatial gene pool boundaries that contain neighboring sets of demes just as they can emerge under isolation by distance without deme structure (Hoelzer et al. unpublished data). Unfortunately, the slow development of a general analytical framework for evolution under steppingstone models has hindered development of analytical methods for empirical estimation of parameters that assume this model. This has not impeded development of computational methods of analysis that permit the use of idiosyncratically imposed population architecture (Beerli & Felsenstein 1999, 2001), although the computational time required can be prohibitive and use of such methods has not yet become common.

A body of theory closely related to steppingstone models is metapopulation theory. This paradigm started as a purely demographic one in which the sizes of local, partially isolated subpopulations fluctuated in part due to migration among them, including local subpopulation extinction and recolonization (Hanski & Gilpin 1997). However, the similarity between this paradigm and steppingstone models in population genetics was quickly recognized, and the effects of fluctuating subpopulation sizes on the amount and geographical distribution of genetic variation have been modeled (Barton & Whitlock 1997, Hedrick & Gilpin 1997, Whitlock 2001). While controversy remains regarding the general population genetic effects of fluctuating subpopulation sizes, including extinction and recolonization, current research suggests that the effects are contingent upon details of the metapopulation architecture. This conclusion illustrates how complex the relationship between dispersal and population genetics can be and indicates that genetic approaches to estimating dispersal parameters within metapopulations will be sensitive to estimates of many other demographic parameters (e.g., number of subpopulations, extinction/colonization rates, mean and variance in number of colonists per event, and mean and variance in sizes of subpopulations).

On the other hand, significant progress has been made recently regarding the role of natural selection in metapopulation evolution. While it has long been known that immigration can offset the process of local adaptation (Haldane 1931), it has recently become clear that it can also counteract the decline in local mean population fitness caused by inbreeding depression (Ingvarsson & Whitlock 2000, Richards 2000, Whitlock et al. 2000, Ebert et al. 2001).

The Effect of Mutation Rate on Estimates of Population Subdivision and Dispersal Rate

Mutation also plays a role in the evolution of population subdivision, although it is often argued that its effect is usually overwhelmed by the influence of migration. Aside from the fact that mutation provides the genetic variation that is the focus of all population genetics, it can contribute directly to divergence among subpopulations when migration is rare and mutation rates are relatively high. The requirement of high mutation rate is not as restrictive as it might appear, because mutation can be defined at any scale. Mutation rate is always higher at the gene level than at the single nucleotide level because a gene encompasses many nucleotides, higher still at the chromosomal level, and so on. Indeed, detection of recent mutations is sufficiently common that a method has been developed for inferring migration rates from the frequencies of alleles confined to single subpopulations, so-called private alleles (Slatkin 1985b, Slatkin & Takahata 1985). The idea behind this method is simple. When new mutations increase in frequency in viscous or deme-structured populations, they spread locally before they "leak" into distant areas. The greater the migration rate, the sooner they spread away from the site of origin and cease being private alleles, thus creating a negative correlation between the frequencies of private alleles and the migration rate. This is a powerful method when the data reveal a sufficient number of private alleles for analysis (Hedrick 1999).

While the private allele approach is hampered by the generally low mutation rates in structural genes, mutation rates in another class of nuclear DNA sequences hold great promise. The discovery of loci characterized by short, tandemly repeated nucleotide sequences or microsatellites (Goldstein & Schlotterer 1999) that are hypervariable for number of repeats has revolutionized empirical population genetics because these sequences offer the first source of nuclear markers that are so rich in potential information. The rate of mutation that changes the number of repeats at microsatellite loci can be as much as six orders of magnitude greater than for single nucleotide substitutions in noncoding nuclear sequences (Hancock 1999), which results in a proliferation of markers for many lineages within a population. Attempts to tap into the information contained in the set of abundant microsatellite mutations that are often encountered within populations, and to simultaneously account for the homoplasy (independent origins of alleles with identical numbers of repeats) expected with these kinds of mutations, have led to the development of several F_{ST} analogs intended specifically for use with microsatellite data (e.g., R_{ST}: Slatkin 1995, Rousset 1996). These measure the degree of population subdivision assuming a specific model of microsatellite evolution and can be used to estimate migration rates. It should be noted that use of these F_{ST} analogs to estimate dispersal rates typically assume the basic island model of migration (Feldman et al. 1999), and increasing the sophistication of the model of molecular evolution cannot overcome inaccuracies that might arise from an oversimplified model of migration and gene flow.

Relatedness and Assignment Tests

It is becoming increasingly common to use population genetic data to infer the degree of relatedness between and among individuals or the natal subpopulation of particular individuals. Relatedness is essentially estimated by comparing observed genetic similarity to that expected from a random sample of the population (Queller & Goodnight 1989), and it is often used to compare groups like resident males to resident females for the purpose of describing social structure and inferring patterns of sex-biased dispersal. Assignment tests aim to identify the place (or subpopulation) of birth for individuals. They rely on the degree of geographical structure in the population genetic data, because individuals born into locally distinctive gene pools carry the genetic "fingerprint" of their natal site in a statistical sense (Waser & Strobeck 1998). Both of these methods can be effective at elucidating current patterns of dispersal without the influence of historical patterns, which might have been considerably different and can strongly influence inferences from equilibrium models such as F_{ST} and its analogues.

The Different Patterns and Processes of mtDNA versus Nuclear DNA Diversity

The reason that subpopulation sizes and numbers of migrants appear so prominently in the various models relating dispersal to population genetics is that genetic drift becomes an increasingly important cause of subpopulation divergence as dispersal and gene flow wane. The same could be said of natural selection, although we will continue to focus on neutral divergence. This raises an important point regarding the interpretation of results obtained using different types of genetic markers, because different markers actually have different population sizes even though they might be observed in the same set of individuals. For example, primate individuals normally have two copies of autosomal nuclear loci (although some sequences or functional genes occur at many loci in the nuclear genome). Males have only a single copy of Y and X chromosome loci, females have two copies of X chromosome loci, and all individuals have a single copy of mtDNA loci. To further complicate matters, these different partitions of the genome have distinctly different patterns of inheritance. Thus, data from these different parts of a primate's genome reflect the histories of some-what different evolutionary lineages, even though today we can find them in the same individual. We focus here on the distinction between mtDNA and autosomal nuclear markers because the vast majority of population genetic data produced to date represents these two categories.

Not only are there half as many copies of an mtDNA marker in a primate population as there are of an autosomal nuclear marker, but mtDNA is generally inherited only from the mother (Melnick et al. 1992). As a result, while males all have mtDNA, mutations that may arise in males' mtDNA are not passed on to their offspring, and thus they do not generally influence the evolution of mtDNA in their population or species. The sizes of mtDNA marker populations are effectively one-quarter those of autosomal nuclear markers in an ideal population (a randomly mating group with a $1:1$ sex ratio, or a Poisson distribution of reproductive success for asexual genomes, like mtDNA). Thus, mtDNA haplotype frequencies drift with four times greater amplitude than do autosomal allele frequencies under this hypothetical population structure. This is usually described as a $4:1$ ratio in the effective population sizes (N_e) of autosomal and mtDNA markers, respectively (Birky et al. 1989, Moore 1995). N_e has numerous technical and context-specific definitions, but it can be generally described as a measure of the strength of genetic drift. In other words, the N_e of a population is equivalent to the size of an ideal population that loses genetic variation due to drift at the same rate. It is important to note that no primate species conforms perfectly to the ideal population model and at least some differ in significant ways.

Female philopatry (low female dispersal) and polygyny (higher reproductive variance among males than females) are two departures from the ideal population model common among primates that increase the N_e of mtDNA relative to autosomal markers, potentially even to the point of being greater (Hoelzer 1997). The effect of philopatry is most obvious when considering the familiar deme-structured models, such as the island model. If we impose deme structure on a panmictic population and make migration between demes rare, large differences among the demes evolve. If there are many demes, the diversity of alleles in the whole system (the sum of all demes) erodes very slowly as they become spatially organized. In fact, in this model drift happens primarily within demes. It happens at the whole system level only to the extent that the system is connected by gene flow. In the complete absence of migration, individual demes become fixed for particular alleles due to drift, but after this point no allele is ever lost from the set of all demes, indicating that drift has no effect at the global scale.

In natural systems that manifest mutation, the amount of genetic variation in a system is commonly used to estimate N_e because systems that lose genetic variants rapidly tend to contain little variation at any particular point in time. Consequently, the N_e of the whole system increases under the island model as the rate of gene flow decreases (Wright 1943). When dispersal of females becomes reduced compared with the dispersal of males, the N_e of mitochondrial markers increases relative to the N_e of autosomal markers, and the amount of neutral genetic variation in the system increases for the mitochondrial genome relative to the nuclear genome. In contrast, the amount of genetic diversity found locally diminishes with increasing philopatry, because the relatedness among neighboring individuals increases. Deme-structured models become family group models as the rate of migration among demes becomes small. In effect, increasing the spatial viscosity of gene flow via dispersal results in both reduced genetic variation locally and increased genetic divergence globally. This apparent contradiction vanishes when one understands that reducing the rate of dispersal between demes allows different subpopulations to exhibit more distinctive gene pools (i.e., they are less like mere replicates of one another).

The marker-specific effect of polygamy can be understood by considering its effects on the relative numbers of markers in different inheritance categories that are transmitted during reproduction. If each reproductively successful male produced an equal number of offspring with each of four females, and the females were completely faithful to their mate, then the number of distinct mtDNA genomes passed from one generation to the next would equal the number of copies passed of any autosomal marker, thus perfectly countering the effects of mtDNA haploidy and maternal-only transmission described above. Of course, the difference in N_e of markers in different inheritance categories is tuned in a continuous fashion by the degree of polygyny or polyandry in a system.

Whichever type of marker has the smaller N_e on a local scale (i.e., within demes) in a particular system will tend to show a greater degree of population subdivision and yield lower estimates of dispersal, because smaller populations retain the homogenizing influence of migration for a shorter period of time before genetic drift counteracts its effects. The important thing to remember is that these are estimates of subdivision and dispersal *for specific markers*, which can differ from one marker to another because they may be "tracking" different sets of genetic lineages with different N_es, dispersal distances and/or rates, and/or mutation rates. The marker of choice in any given study should ideally reflect the organismal population of interest, but the practical availability of particular types of markers often plays a large role.

Given our understanding of the population genetic effects of differing modes of inheritance, contrasting distributions and degrees of genetic variation for different marker types should allow us to test various hypotheses about evolutionary genetic consequences of primate social structure (Melnick & Hoelzer 1992, 1993). In most cases, differing population genetic structures for different genetic markers should not be interpreted as contradictions in need of resolution. Rather, they should be seen as complementary sources of information, allowing us to develop a more complete understanding of primate sociality.

Review of Empirical Studies

Table 5.1 contains some examples of studies that used genetic markers to assess current patterns of dispersal and gene flow among primate populations. Few studies actually exist in the primate literature that attempt to describe dispersal processes among populations, and

Table 5.1. Empirical Studies of Primate Species That Use Genetic Markers to Infer Patterns of Dispersal

Species (Reference)	Social System	Genetic Marker	Type of Analysis	Results Pertaining to Dispersal
Strepsirhines				
Weasel sportive lemur (*Lepilemur mustelinus*; Tomiuk et al. 1997)	Solitary	Allozymes	*F* statistics	Recent habitat fragmentation has not resulted in genetic differentiation of populations yet (low F_{ST}), but high inbreeding (high F_{IS}) suggests low dispersal.
Grey mouse lemur (*Microcebus murinus*; Radespiel et al. 2001)	Dispersed social organization. Solitary foragers but at night form "female sleeping associations."	Microsatellites	Relatedness	Degree of relatedness between females of the same sleeping groups was significantly higher than between groups. Closely related females were caught in closer proximity than male-female dyads. Closely related males were caught farthest from each other. Overall males had significantly less relatedness within the populations than females. These findings support the behavioral hypothesis of female philopatry and male dispersal.
Platyrrhines				
Wild common marmoset (*Callithrix jacchus*; Nievergelt et al. 1999)	Small groups—reproduction restricted to one or maybe two females	Microsatellites	Relatedness	Within-group relatedness is significantly higher than the average intergroup relatedness. Mean relatedness between neighboring groups is higher than between remote groups, indicating either individual dispersing limited to neighboring groups, or the fission of social groups into two adjacent groups. Relatedness between adult males from the same groups is significantly lower than between adult females, indicating female philopatry and male dispersal.

(*continued*)

Table 5.1. Continued

Species (Reference)	Social System	Genetic Marker	Type of Analysis	Results Pertaining to Dispersal
Saddle-back tamarin (*Saguinus fuscicollis*; Peres et al. 1996)	Multimale-multifemale groups	mtDNA	Phylogeography	Samples were taken on both sides of the Juruá river, representing at least 2 of the 14 subspecies recognized in this species: the dark *S. f. fuscicollis*, and the white *S. f. melanoleucus*. Both groups had distinct mtDNA haplotypes, but toward the headwater section of the river intermediate color morphs and haplotypes from both subspecies were found. This suggests that the frequency of gene flow increases toward the headwater region of major rivers, supporting the "riverine barrier hypothesis."
Red howler monkeys (*Alouatta seniculus*; Pope 1998)	Territorial social groups of 2–5 adult females, 1–3 adult males, and their offspring	Allozymes	F statistics, relatedness	In growing, low-density populations, relatedness within groups is essentially random. High-density, socially mature groups had higher relatedness values. This supports the observations that unrelated female individuals evicted from their natal groups form new groups, and mature groups include an alpha female and her daughters. F_{ST} values also increased with population density.
Mantled howler monkeys (*Alouatta palliata*; Ellsworth 2000)	Territorial social groups of 2–15 females, 1–4 males, and 0–7 infants	Microsatellites	F statistics, relatedness	Very low, but statistically significant, levels of genetic subdivision were detected among social groups. This supports the observation that dispersal in this system is common for both males (79%) and females (96%).

Catarrhines

Vervet monkeys (*Cercopithecus aethiops*; Shinada & Shotake 1997; Shimada 2000)	Multimale-multifemale	Allozymes, mtDNA	F statistics, mtDNA polymorphism	F_{ST} values show little or no genetic differentiation among groups. A lower than expected proportion of polymorphic loci is interpreted as a consequence of repeated population bottlenecks under dry and cold climate in the past and successive population expansions afterward. When adult males are excluded, different mtDNA haplogroups are rather discretely distributed among different groups along the Awash river, supporting female philopatry and male dispersal.
Hamadryas baboons (*Papio hamadryas*; Hapke et al. 2001)	Multilevel social system. The smallest unit consists of one male, his mates, and offspring. Several units form a clan, and several clans form a band.	mtDNA	Phylogeography, ANOVA, F statistics	A phylogenetic tree shows well-resolved clades, but no geographic structure. The analysis of molecular variance shows more variation within than between local subpopulations. There is a slight correlation between geographic and genetic distance. These data support female-mediated gene flow, but movement of bands is constrained by geographic distance.
Rhesus monkey (*Macaca mulatta*; Melnick & Hoelzer 1992; Tosi et al. 2002)	Multimale-multifemale social groups. Males are dominant to females but are peripheral to the group and change groups every few years.	Allozymes, mtDNA, Y chromosome	F statistics, phylogeny	Allozyme data show that nearly 99% of the nuclear gene diversity can be found in any population of rhesus monkey surveyed in a study in Pakistan. Furthermore, only 9% of the total species diversity could be apportioned to interregional differences across the species' range. This pattern contrasts markedly with mtDNA variation, which shows the opposite pattern. These data are consistent with the observation of extreme female philopatry and male dispersal. Y chromosome sequence data show evidence of contemporary hybridization between rhesus and long-

(*continued*)

Table 5.1. Continued

Species (Reference)	Social System	Genetic Marker	Type of Analysis	Results Pertaining to Dispersal
				tailed macaques (*M. fascicularis*) in central Indochina. Rhesus Y chromosome haplotypes are infiltrating deep into the territory of long-tailed macaques, a pattern not reflected in mtDNA markers or morphology.
Long-tailed macaques (*Macaca fascicularis*; de Ruiter & Geffen 1998)	Multimale-multifemale groups	Microsatellites	Relatedness	Consistent with female philopatry and male dispersal, relatedness was highest among mother-offspring and father-offspring dyads, and lowest among adult male-male dyads within groups. Relatedness within the high-ranking matriline was higher than the relatedness in the lower ranking matrilines. A likely explanation is that the alpha male in this group mated significantly more with high-ranking than with low-ranking females. The high-ranking matriline was more closely related to segments of the neighboring groups than the lower ranking matrilines were among each other. This is the likely the result of gene flow among high-ranking matrilines of neighboring groups.
Chimpanzees (*Pan troglodytes*; Morin et al. 1993)	Multimale-multifemale communities in which a core of related males patrol the boundaries. Females live a slightly more solitary life.	Microsatellites, mtDNA	Relatedness, mtDNA polymorphism	Males are typically philopatric and cooperate to defend community ranges. Females may emigrate as adolescents or temporarily as reproductively active adults. Males are more related to each other than were the females within the group. Furthermore, the high level of mtDNA polymorphism is consistent with high levels of female-mediated gene flow.

even fewer that try to quantify such parameters as number of migrants per generation or average dispersal distances using genetic information. The most likely reason for this is the difficulty in obtaining sufficient numbers of genetic samples from different populations that together with accurate demographic and behavioral information can provide the necessary data to understand the dynamics of primate dispersal.

Some Impacts of Dispersal and Genetic Diversity

The Average Degrees of Relatedness Within Groups and Potential Effects on Social Behavior

Local Diversity

Dispersal and local genetic diversity are positively correlated. This has important implications for social behavior, because kinship is expected to influence both the degree of cooperation and optimal mate choices. The relative lack of local genetic diversity when dispersal is low reflects a high degree of relatedness among neighbors, which facilitates kin selection (Hamilton 1964, Silk 2002), kin-biased reciprocation, and kin-biased mutualism (Chapais 2001), mechanisms for the evolution of cooperative social networks. While little compelling empirical evidence discriminates among the potential behavioral mechanisms of social structure evolution, it should be recognized that forms of interaction initially motivated by kin-based mechanisms might entrain social cultures, thus diminishing the degree of kin bias observed in groups of mixed relatedness. On the other hand, mating between close relatives is often selected against due to inbreeding depression exhibited by the offspring of consanguineous matings (Perrin & Goudet 2001), which could mean costly searches for suitable mates. It has often been suggested that inbreeding avoidance is likely to be accomplished by the evolution of dispersal patterns that make neighbors of distantly related potential mates (Melnick et al. 1984, Perrin & Goudet 2001). However, when dispersal and local genetic diversity are both high, the potential for individuals to benefit from social networking (the development of social relationships) is largely limited to non–kin-based mechanisms. Clearly such mechanisms exist (e.g., Trivers 1971). Our point is that the potential for these mechanisms exists whether or not interactions among adult kin are common, so when adult kin are separated through dispersal, the diversity of processes available for the building of social relationships is reduced. While we are not surprised to find non–kin-based social structures in such systems (e.g., howler monkeys), it also seems clear that the degree of sociality and the duration of social relationships are typically greater where philopatry keeps adult kin in closer proximity (e.g., female macaques).

While dispersal by all juveniles effectively mixes the population, inbreeding depression can be avoided by most, while retaining the potential advantages of social networking among kin if only one gender disperses (Chesser & Ryman 1986, Perrin & Goudet 2001). This, in fact, seems to be a common situation, especially in "permanent" social groups (Pusey 1987). It is the males who disperse further and more frequently than the females in most mammals (Chepko-Sade & Halpin 1987). Macaques, for example, epitomize this pattern (Pusey & Packer 1987, Clutton-Brock 1989, Melnick & Hoelzer 1996). When this occurs, the patterns of social interaction among female macaques are often highly structured by kinship (see chapter 7). The females are born into a society in which their status is generally destined to be similar to their mothers' and in which affiliative, tolerant, and supportive interactions with other females are often highly biased toward maternal kin. The primary way in which low-ranking female macaques can improve their social standing, and disperse at the same time, is through group fission, in which females comprising the lowest ranking matrilines leave to form a new social group in a nearby area (Furaya 1968, 1969; Chepko-Sade & Sade 1979; Dittus 1988; Purdhomme 1991; Menard & Vallet 1993; Li et al. 1996; Kuester & Paul 1997; Hsu & Lin 2001). In contrast, the social network among dispersing males, and between sexes, is less constrained by relatedness (Witt et al. 1981, van Noordwijk & van Schaik 1988, Sprague et al. 1996).

Geographically Structured Diversity

Genetic divergence among subpopulations results from increasingly distant shared ancestries among individuals from different groups, and therefore should lead to increasingly recognizable differences in phenotype. In addition to genetically induced differences, cultural differences are likely to evolve when exchanges of individuals between groups is rare (e.g., Lynch & Baker 1994). The distinctiveness of individual phenotypes and cultural norms could also contribute to a positive feedback mechanism (e.g., reinforcement: Servedio 2000) driving subpopulation divergence to the point of speciation. Even if positive feedback does not occur, recognizable subpopulation membership is likely to affect the nature of social interactions between individuals with different group identities.

Potential for Evolutionary Change

Local Diversity

Subpopulations are different from species in that their evolutionary paths are ultimately constrained to be similar to the paths of other subpopulations, whereas species boundaries represent relatively discrete breaks in evolutionary cohesion. Therefore, the evolutionary potential of a subpopulation depends more strongly on the potential of the whole system than on its own features, such as the amount of local genetic diversity. If local diversity is high due to historically great dispersal, the potential for locally independent evolution (e.g., adaptation to local conditions) is severely compromised by the loss of natives and the influx of individuals from other areas. If local diversity is low due to historically weak dispersal, it might evoke concern about vulnerability to extinction due to inbreeding depression (see above) or reduced potential for adaptation to a changing environment (Houle 1992). However, if the connectedness of the system as a whole has remained stationary in time, then such a subpopulation has historically relied on immigration as its primary source of adaptive variation, rather than local mutation or availability of a local reservoir of variation. This notion is recognized in plans to impose immigration on increasingly isolated subpopulations found to contain little genetic variation (e.g., Hedrick 1995). Such efforts are essentially designed to maintain stationary metapopulation processes in the face of deteriorating environmental conditions (Woodruff 2001).

Geographically Structured Diversity

Genetic diversity at the regional level of organization greatly facilitates the potential for local adaptation (and drift) and speciation, which canalizes a gene pool boundary and uncouples the population genetics of previously linked subpopulations. Thus, widespread species characterized by low but persistent dispersal would seem to hold the greatest evolutionary potential. This circumstance is relatively rare, because species with weak dispersal abilities seldom achieve widespread distributions. Nevertheless, there is a range of evolutionary potential across species that relates to their specific combinations of dispersal, range, and degree of geographical structuring of genetic variation.

Conservation of Primate Populations

Local Diversity

Low levels of local genetic diversity have been a concern to conservation biologists for three reasons: (1) deficiency in evolutionary potential, (2) the possibility of inbreeding depression, and (3) a desire to avoid losing gene pool distinctiveness and possibly local adaptedness. The first two reasons are related through N_e. Low levels of genetic diversity reflect low N_e, which means that the local population would have a long waiting time before mutation would make available new adaptive variants (Elena et al. 1996, Wahl & Krakauer 2000) when needed. Low N_e also means that the strength of genetic drift is great, which could cause fixations of deleterious alleles (subpopulation level inbreeding depression), further reducing N_e in an extinction vortex ("mutational meltdown": Lynch et al. 1995). Both of these problems can be ameliorated by reestablishing gene flow with other subpopulations, if there are any, which increases local N_e (e.g., Hedrick 1995). Unfortunately, this solution is counterproductive in terms of conserving the distinctiveness of these populations. One reason often stated for protecting distinctive gene pools, regardless of their diversities, is that they offer unique foundations on which future evolution can be built (Waples 1991; Moritz 1994, 1999), but it is difficult to justify conserving uniqueness of very small populations when their evolutionary potential is so limited. Unlike the first two reasons listed above, no theoretical basis has been developed that supports the argument for preserving uniqueness alone. At some level, this argument is almost certainly true (e.g., the extinction of the dinosaurs, excepting birds, probably made for the evolution of a very different biota than would otherwise have occurred), but it is far from clear how this concern should be balanced with population genetic or demographic issues in a conservation context. The possibility of local adaptation is difficult to demonstrate empirically, and its conservation or facilitation also depends on the N_e of local subpopulations (see Localized, Partial Barriers to Dispersal above). The complex relationship between local genetic diversity and conservation biology makes the development of general principles an elusive goal. In practice, population genetic data must be considered in the context of case-specific demography and ecology, not to mention political realities (Taylor & Dizon 1999).

Geographically Structured Diversity

Population subdivision is appreciated as a useful condition for genetic management of captive breeding populations of endangered species because preventing gene flow among the captive populations maintains the greatest amount of genetic diversity under management (Lacy 1987). The set of captive populations is a primary target of concern in this context; so individual subpopulations may be allowed to suffer for the maximization of species-wide genetic diversity. Gene flow could be imposed among captive subpopulations to invigorate groups suffering the effects of inbreeding depression, but this would facilitate loss of allelic variants from the global captive breeding program. The value of applying a subdivided captive breeding design is realized if reintroduction into the wild is attempted. Releasing animals from different captive subpopulations, should this be desirable and possible, will maximize the infusion of genetic diversity into the wild. Concern with the roles of divergent subpopulations as interactive elements in a diversified system in this context provides an interesting contrast with proposals to maintain isolation of genetically distinctive subpopulations in the wild into perpetuity (Waples 1991; Moritz 1994, 1999). The stability of metapopulations can depend on connectedness and number of subpopulations. So loss of some subpopulations from the system by habitat destruction or imposed isolation, even population sinks, can potentially destabilize the whole system (e.g., Foppen et al. 2000, Frouz & Kindlmann 2001, Matter 2001, Silva et al. 2001). It is also important to note that complete isolation of a subpopulation from a metapopulation is equivalent to local habitat destruction in its effects on dynamics of the metapopulation. Therefore, there are potential conservation benefits at the metapopulation level in reconnecting a subpopulation to the network. It is often the difficult task of conservation workers to balance the preservation of uniqueness at the subpopulation level (e.g., Andayani et al. 2001) with the robustness of metapopulation dynamics (Perez-Sweeney et al. 2002).

Conclusion

The theory of spatial population genetics is less mature than is suggested by the immense literature and widespread use of estimation methods based on the island model of migration, because the island model does not actually take spatial constraints to dispersal into account. Estimates of population parameters like migration rate based on this model should be considered cautiously until the ranges of parameter values for which such estimates will reasonably approximate the true values have been elucidated. Indeed, the same argument should hold for inferences drawn from any equilibrium, mean-field model of spatial evolution. Computational approaches to analyzing idiosyncratic and asymmetrical spatial architectures, such as the ones implemented in the LAMARC software package (http://evolution. genetics.washington.edu/lamarc.html), are promising, but it is not yet clear how widely useful they will be. On the other hand, methods for inferring relatedness and natal sites are more straightforward. These methods should be favored when the questions and data in a study lend themselves to such analyses.

Population genetics has been profitably applied to the study of primate dispersal and social structure, although not frequently. It is likely that the difficulties involved with obtaining sufficient numbers of samples within local groups, sufficient numbers of groups, and sufficient coverage at regional scales of geography have significantly impeded this research program. Additional studies of the population genetic structure of primate species, including a greater diversity of primate taxa, will be required before we will begin to see any general empirical patterns that might exist.

Despite the limited data from primate populations, it is clear that dispersal patterns, the geographical structure of species' gene pools, and social structure are highly interrelated through the evolutionary process. In addition, these factors influence the potential for both kin-based and more generalized social structures, which can result in the evolution of social behaviors that feed back to the further evolution of dispersal. Conservation of such evolving, nonequilibrium systems is almost a contradiction in terms, but we can try to minimize the chance that human activity tips the balance toward extinction. However, even this goal can be compromised by ambiguity over the target of conservation due to the tradeoff between the demographic stability obtained through connectedness (migration) of the metapopulation and the uniqueness of local subpopulations.

Acknowledgments   The authors would like to thank the editors, Mary Peacock, and an anonymous reviewer for helpful comments on drafts of this chapter.

References

Andayani, N., Morales, J. C., Forstner, M. R. J., Supriatna, J., & Melnick, D. J. 2001. Genetic variability in mitochondrial DNA of the Javan gibbon (*Hylobates moloch*): implications for the conservation of a critically endangered species. *Conserv. Biol.*, 15, 770–775.

Barton, N. H. & Whitlock, M. C. 1997. The evolution of metapopulations. In: *Metapopulation Biology: Ecology, Genetics, and Evolution* (Ed. by I. A. Hanski & M. E. Gilpin), pp. 183–210. San Diego, CA: Academic Press.

Beerli, P. & Felsenstein, J. 1999. Maximum-likelihood estimation of migration rates and effective population numbers in two populations using a coalescent approach. *Genetics*, 152, 763–773.

Beerli, P. & Felsenstein, J. 2001. Maximum likelihood estimation of a migration matrix and effective population sizes in n subpopulations by using a coalescent approach. *Proc. Natl. Acad. Sci. USA*, 98, 4563–4568.

Birky, C. W., Fuerst, P., & Maruyama, T. 1989. Organelle gene diversity under migration, mutation, and driftùequilibrium expectations, approach to equilibrium, effects of heteroplasmic cells, and comparison to nuclear genes. *Genetics*, 121, 613–627.

Brown, J. H. & Lomolino, M. V. 1998. *Biogeography*. Sunderland, MA: Sinauer Associates.

Chapais, B. 2001. Primate nepotism: what is the explanatory value of kin selection? *Int. J. Primatol.*, 22, 203–229.

Chepko-Sade, B. D. & Halpin, Z. T. 1987. *Mammalian Disperal Patterns*. Chicago: University of Chicago Press.

Chepko-Sade, B. D. & Sade, D. S. 1979. Patterns of group splitting within matrilineal kinship groups. *Behav. Ecol. Sociobiol.*, 5, 67–87.

Chesser, R. K. & Ryman, N. 1986. Inbreeding as a strategy in subdivided populations. *Evolution*, 40, 616–624.

Clobert, J., Danchin, E., Dhondt, A. A., & Nichols, J. D. 2001. *Dispersal*. Oxford: Oxford University Press.

Clutton-Brock, T. H. 1989. Female transfer and inbreeding avoidance in social mammals. *Nature*, 337, 70–72.

de Quiroz, K. 1998. The general lineage concept of species, species criteria, and the process of speciation: a conceptual unification and terminological recommendations. In: *Endless Forms* (Ed. by D. J. Howard & S. H. Berlocher), pp. 57–75. New York: Oxford University Press.

de Ruiter, J. R. & Geffen, E. 1998. Relatedness of matrilines, dispersing males and social groups in long-tailed macaques (*Macaca fascicularis*). *Proc. Royal Soc. Lond., B*, 265, 79–87.

Dittus, W. P. 1988. Group fission among wild toque macaques as a consequence of female resource competition and environmental stress. *Anim. Behav.*, 36, 1626–1645.

Ebert, D., Haag, C., Kirkpatrick, M., Riek, M., Hottinger, J. W., & Pajunen, V. I. 2001. A selective advantage to immigrant genes in a *Daphnia* metapopulation. *Science*, 295, 485–488.

Elena, S. F., Cooper, V. S., & Lenski, R. E. 1996. Punctuated evolution caused by selection of rare beneficial mutations. *Science*, 272, 1802–1804.

Ellsworth, J. A. 2000. Molecular evolution, social structure, and phylogeography of the mantled howler monkey (*Alouatta palliata*). PhD dissertation, University of Nevada, Reno.

Feldman, M. W., Kumm, J., & Pritchard, J. 1999. Mutation and migration in models of microsatellite evolution. In: *Microsatellites: Evolution and Applications* (Ed. by D. B. Goldstein & C. Schlotterer), pp. 98–115. New York: Oxford University Press.

Foppen, R. P. B., Chardon, J. P., & Liefveld, W. 2000. Understanding the role of sink patches in source-sink metapopulations: Reed Warbler in an agricultural landscape. *Conserv. Biol.*, 14, 1881–1892.

Frouz, J. & Kindlmann, P. 2001. The role of sink to source re-colonisation in the population dynamics of insects living in unstable habitats: an example of terrestrial chironomids. *Oikos*, 93, 50–58.

Furaya, Y. 1968. On the fission of troops of Japanese monkeys. I. *Primates*, 9, 323–350.

Furaya, Y. 1969. On the fission of troops of Japanese monkeys. II. *Primates*, 10, 47–69.

Goldstein, D. B. & Schlotterer, C. 1999. *Microsatellites: Evolution and Applications*. New York: Oxford University Press.

Haldane, J. B. S. 1931. A mathematical theory of natural selection. VI. Isolation. *Trans. Camb. Phil. Soc.*, 26, 220–230.

Hamilton, W. D. 1964. The genetical evolution of social behavior, I & 2. *J. Theor. Biol.*, 7, 1–52.

Hancock, J. M. 1999. Microsatellites and other simple sequences: genomic context and mutational mechanisms. In: *Microsatellites: Evolution and Applications* (Ed. by D. B. Goldstein & C. Schlotterer,), pp. 1–9. New York: Oxford University Press.

Hanski, I. A. & Gilpin, M. E. 1997. *Metapopulation Biology: Ecology, Genetics, and Evolution*. San Diego, CA: Academic Press.

Hapke, A., Zinner, D., & Zischler, H. 2001. Mitochondrial DNA variation in Eritrean hamadryas baboons (*Papio hamadryas hamadryas*): life history influences population genetic structure. *Behav. Ecol. Sociobiol.*, 50, 483–492.

Hardy, O. J. & Vekemans, X. 1999. Isolation by distance in a continuous population: reconciliation between spatial autocorrelation analysis and population genetics models. *Heredity*, 83, 145–154.

Hartl, D. L. & Clarke, A. G. 1997. *Principles of Population Genetics*. Sunderland, MA: Sinauer Associates.

Hedrick, P. W. 1995. Gene flow and genetic restoration: the Florida panther as a case-study. *Conserv. Biol.*, 9, 996–1007.

Hedrick, P. W. 1999. Perspective: highly variable loci and their interpretation in evolution and conservation. *Evolution*, 53, 313–318.

Hedrick, P. W. & Gilpin, M. E. 1997. Genetic effective size of a metapopulation. In: *Metapopulation Biology: Ecology, Genetics, and Evolution* (Ed. by I. A. Hanski & M. E. Gilpin), pp. 165–181. San Diego, CA: Academic Press.

Hoelzer, G. A. 1997. Inferring phylogenies from mtDNA variation: mitochondrial-gene trees versus nuclear-gene trees revisited. *Evolution*, 51, 622–626.

Houle, D. 1992. Comparing evolvability and variability of quantitative traits. *Genetics*, 130, 185–204.

Hsu, M. J. & Lin, J. F. 2001. Troop size and structure in free-ranging Formosan macaques (*Macaca cyclopis*) at Mt. Longevity, Taiwan. *Zool. Stud.*, 40, 49–60.

Ingvarsson, P. K. & Whitlock, M. C. 2000. Heterosis increases the effective migration rate. *Proc. Royal Soc. Lond., B,* 267, 1321–1326.

Kimura, M. 1953. The "stepping-stone" model of a population. *Annu. Rep. Natl. Inst. Genetics Japan, Mishima-shi*, 3, 63.

Kimura, M. 1968. Genetic variability maintained in a finite population due to mutational production of neutral and nearly neutral isoalleles. *Genet. Res.*, 11, 247–269.

Kimura, M. & Weiss, G. H. 1964. The steppingstone model of population structure and the decrease of genetic correlation with distance. *Genetics*, 49, 561–576.

Kuester, J. & Paul, A. 1997. Group fission in Barbary macaques (*Macaca sylvanus*) at Affenberg Salem. *Int. J. Primatol.*, 18, 941–966.

Lacy, R. C. 1987. Loss of genetic diversity from managed populations: interacting effects of drift, mutation, immigration, selection, and population subdivision. *Conserv. Biol.*, 1, 143–158.

Li, J. H., Wang, Q. S., & Han, D. M. 1996. Fission in a free-ranging Tibetan macaque troop at Huangshan Mountain, China. *Chin. Sci. Bull.*, 41, 1377–1381.

Lynch, A. & Baker, A. J. 1994. A population memetics approach to cultural-evolution in chaffinch songùdifferentiation among populations. *Evolution*, 48, 351–359.

Lynch, M., Conery, J., & Burger, R. 1995. Mutation accumulation and the extinction of small populations. *Am. Nat.*, 146, 489–518.

Malecot, G. 1948. *The Mathematics of Heredity*. San Francisco: W. H. Freeman.

Matter, S. F. 2001. Synchrony, extinction, and dynamics of spatially segregated, heterogeneous populations. *Ecological Modelling*, 141, 217–226.

Melnick, D. J. & Hoelzer, G. A. 1992. Differences in male and female macaque dispersal lead to contrasting distributions of nuclear and mitochondrial-DNA variation. *Int. J. Primatol.*, 13, 379–393.

Melnick, D. J. & Hoelzer, G. A. 1993. What is mtDNA good for in the study of primate evolution? *Evol. Anthropol.*, 2, 2–10.

Melnick, D. J. & Hoelzer, G. A. 1996. The population genetic consequences of macaque social organisation and behaviour. In: *Evolution and Ecology of Macaque Societies* (Ed. by J. E. Fa & D. G. Lindburg,), pp. 413–443. Cambridge: Cambridge University Press.

Melnick, D. J., Hoelzer, G. A., & Honeycutt, R. L. 1992. Mitochondrial DNA: its uses in anthropological research. In: *Molecular Applications in Biological Anthropology* (Ed. by E. J. Devor), pp. 179–233. Cambridge: Cambridge University Press.

Melnick, D. J., Pearl, M. C., & Richard, A. F. 1984. Male migration and inbreeding avoidance in wild rhesus monkeys. *Am. J. Primatol.*, 7, 229–243.

Menard, N. & Vallet, D. 1993. Dynamics of fission in a wild barbary macaque group (*Macaca sylvanus*). *Int. J. Primatol.*, 14, 479–500.

Moore, W. S. 1995. Inferring phylogenies from mtDNA variationùmitochondrial-gene trees versus nuclear-gene trees. *Evolution*, 49, 718–726.

Morin, P. A., Wallis, J., Moore, J. J., Chakraborty, R., & Woodruff, D. S. 1993. Noninvasive sampling and DNA amplification for paternity exclusion, community structure, and phylogeography in wild chimpanzees. *Primates*, 34, 347–356.

Moritz, C. 1994. Defining evolutionarily-significant-units for conservation. *Trends Ecol. Evol.*, 9, 373–375.

Moritz, C. 1999. Conservation units and translocations: strategies for conserving evolutionary processes. *Hereditas*, 130, 217–228.

Nievergelt, C. M., Digby, L. J., Ramakrishnan, U., & Woodruff, D. S. 2000. Genetic analysis of group composition and breeding system in a wild common marmoset (*Callithrix jacchus*) population. *Int. J. Primatol.*, 21, 1–20.

Peres, C. A., Patton, J. L., & da Silva, M. N. F. 1996. Riverine barriers and gene flow in Amazonian saddle-back tamarins. *Folia Primatol.*, 67, 113–124.

Perez-Sweeney, B., Valladares-Padua, C., & Melnick, D. J. 2002. Using genetics for black lion tamarin (*Leontopithecus chrysopygus*) metapopulation management. *Annu. Meeting Soc. Conserv. Biol.* July, 2002. Canterbury, UK.

Perrin, N. & Goudet, J. 2001. Inbreeding, kinship, and the evolution of natal dispersal. In: *Dispersal* (Ed. by J. Clobert, E. Danchin, A. A. Dhondt, & J. D. Nichols), pp. 121–142. New York: Oxford University Press.

Pope, T. R. 1998. Effects of demographic change on group kin structure and gene dynamics of populations of red howling monkeys. *J. Mammal.*, 79, 692–712.

Prudhomme, J. 1991. Group fission in a semifree-ranging population of barbary macaques (*Macaca sylvanus*). *Primates*, 32, 9–22.

Pusey, A. E. 1987. Sex biased dispersal and inbreeding avoidance in birds and mammals. *Trends Ecol. Evol.*, 2, 295–299.

Pusey, A. E. & Packer, C. 1987. Dispersal and philopatry. In: *Primate Societies* (Ed. by B. B. Smuts, D. L. Cheney, R. M. Seyfarth, R. W. Wrangham, & T. T. Struhsaker), pp. 250–266. Chicago: University of Chicago Press.

Queller, D. C. & Goodnight, K. F. 1989. Estimating relatedness using genetic markers. *Evolution*, 43, 258û275.

Radespiel, U., Sarikaya, Z., Zimmerman, E., & Bruford, M. W. 2001. Sociogenetic structure in a free-living nocturnal primate population: sex-specific differences in the grey mouse lemur (*Microcebus murinus*). *Behav. Ecol. Sociobiol.*, 50, 493–502.

Richards, C. M. 2000. Inbreeding depression and genetic rescue in a plant metapopulation. *Am. Nat.*, 155, 383–394.

Rousset, F. 1996. Equilibrium values of measures of population subdivision for stepwise mutation processes. *Genetics*, 142, 1357–1362.

Rousset, F. 1997. Genetic differentiation and estimation of gene flow from F-statistics under isolation by distance. *Genetics*, 145, 1219–1228.

Sawyer, S. 1977. Rates of consolidation in a selectively neutral migration model. *Ann. Probab.*, 5, 486–493.

Sawyer, S. A. 1976. Results for the stepping-stone model for migration in population genetics. *Ann. Probab.*, 4, 699–728.

Schluter, D., Boughman, J. W., & Rundle, H. D. 2001. Parallel speciation with allopatry. *Trends Ecol. Evol.*, 16, 283–284.

Servedio, M. R. 2000. Reinforcement and the genetics of non-random mating. *Evolution*, 54, 21–29.

Shinada, M. K. 2000. Geographic distribution of mitochondrial DNA variations among grivet (*Cercopithecus aethiops aethiops*) populations in central Ethiopia. *Int. J. Primatol.*, 21, 113–129.

Shinada, M. K. & Shotake, T. 1997. Genetic variation of blood proteins within and between local populations of grivet monkey (*Cercopithecus aethiops aethiops*) in central Ethiopia. *Primates*, 38, 399–414.

Silk, J. B. 2002. Kin selection in primate groups. *Int. J. Primatol.*, 23, 849–876.

Silva, J. A. L., de Castro, M. L., & Justo, D. A. R. 2001. Stability in a metapopulation model with density-dependent dispersal. Bulletin of *Mathematical Biology*, 3, 485–505.

Slatkin, M. 1985a. Gene flow in natural-populations. *Annu. Rev. Ecol. Sys.*, 16, 393–430.

Slatkin, M. 1985b. Rare alleles as indicators of gene flow. *Evolution*, 39, 53–65.

Slatkin, M. 1995. A measure of population subdivision based on microsatellite allele frequencies. *Genetics*, 139, 457–462.

Slatkin, M. & Takahata, N. 1985. The average frequency of private alleles in a partially isolated population. *Theor. Pop. Biol.*, 28, 314–331.

Sprague, D. S., Suzuki, S., & Tsukahara, T. 1996. Variation in social mechanisms by which males attained the alpha rank among Japanese macaques. In: *Evolution and Ecology of Macaque Societies* (Ed. by J. E. Fa & D. G. Lindburg), pp. 444–458. Cambridge: Cambridge University Press.

Taylor, B. L. & Dizon, A. E. 1999. First policy then science: why a management unit based solely on genetic criteria cannot work. *Mol. Ecol.*, 8, S11-S16.

Thompson, S. K. 1992. *Sampling*. New York: Wiley.

Tomiuk, J., Bachmann, L., Leipoldt, M., Ganzhorn, J. U., Reis, R., Weis, M., & Loeschcke, V. 1997. Genetic diversity of *Lepilemur mustelinus ruficaudatus*, a nocturnal lemur of Madagascar. *Conserv. Biol.*, 11, 491–497.

Tosi, A. J., Morales, J. C., & Melnick, D. J. 2002. Y-chromosome and mitochondrial markers in *Macaca fascicularis* indicate introgression with Indochinese *M. mulatta* and a biogeographic barrier in the Isthmus of Kra. *Int. J. Primatol.*, 23, 161–178.

Trivers, R. L. 1971. The evolution of reciprocal altruism. *Q. Rev. Biol.*, 46, 35–57.

van Noordwijk, M. A. & van Schaik, C. P. 1988. Male migration and rank acquisition in wild long-tailed macaques (*Macaca fascicularis*). *Anim. Behav.*, 33, 849–861.

Wahl, L. M. & Krakauer, D. C. 2000. Models of experimental evolution: the role of genetic chance and selective necessity. *Genetics*, 156, 1437–1448.

Waples, R. S. 1991. *Definition of "Species" Under the Endangered Species Act: Application to Pacific Salmon*. Seattle, WA: National Marine Fisheries Service.

Waser, P. M. & Strobeck, C. 1998. Genetic signatures of interpopulation dispersal. *Trends Ecol. Evol.*, 13, 43–44.

Whitlock, M. C. 2001. Dispersal and the genetic properties of metapopulations. In: *Dispersal* (Ed. by J. Clobert, E. Danchin, A. A. Dhondt, & J. D. Nichols), pp. 273–282. New York: Oxford University Press.

Whitlock, M. C., Ingvarsson, P. K., & Hatfield, T. 2000. Local drift load and the heterosis of interconnected populations. *Heredity*, 84, 452–457.

Whitlock, M. C. & McCauley, D. E. 1999. Indirect measures of gene flow and migration: F-ST not equal 1/(4Nm + 1). *Heredity*, 82, 117–125.

Witt, R., Schmidt, C., & Schmitt, J. 1981. Social rank and Darwinian fitness in a multimale group of Barbary macaques (*Macaca sylvanus*, Linnaeus, 1758): dominance reversals and male reproductive success. *Folia Primatol.*, 36, 201–211.

Woodruff, D. S. 2001. Declines of biomes and biotas and the future of evolution. *Proc. Natl. Acad. Sci. USA*, 98, 5471–5476.

Wright, S. 1931. Evolution in Mendelian populations. *Genetics*, 16, 97–159.

Wright, S. 1943. Isolation by distance. *Genetics*, 28, 114–138.

6

The Effects of Demographic Variation on Kinship Structure and Behavior in Cercopithecines

David A. Hill

Demographic processes have a pervasive influence on many aspects of social behavior and social structure in group-living animals. The influence is mediated through the effects of demographic processes on group size and composition. This point was made by Altmann and Altmann (1979), who also noted that social behavior in turn influences demographic processes, thereby closing the loop to produce a cycle of effects. In assessing the effects of demographic processes on kinship structure and behavior, we are primarily concerned with the first two links of this cycle: the impact of demographic processes on group size and composition, and ways in which group size and composition influence social behavior and social relationships.

Group size and composition are determined by the interplay of seven processes: birth, death, maturation, emigration, immigration, group fission, and group fusion (Altmann & Altmann 1979). These processes do not exert even, predictable pressure on a group, but fluctuate in response to changes in the environment. Environmental factors that influence these processes include both the availability of resources, such as food and water, and also aspects of the social environment, such as the degree of intra- and intergroup competition for resources. Some of the temporal fluctuations in the environment follow regular cyclic patterns, such as seasonal variation in patterns of food availability. Other changes, such as periods of drought or food shortages following bush fires, occur at irregular, and therefore unpredictable, intervals.

In addition to environmental factors, group composition is also influenced by chance events. These may include, for example, years when very few females in the group give birth, or all give birth to offspring of the same sex, due entirely to coincidence rather than to any environmental factor. Dunbar (1979) has shown that relatively minor changes in the pattern of births can cause large disruptions with lasting effects on the demographic profile of a group. Thus, group size and composition are dynamic and, in species like primates with relatively slow rates of reproduction, it is probably rare for populations to reach stability in terms of age distribution (Dunbar 1979, 1988).

Variation in group size and composition influences social behavior and social relationships through its consequences for the number and nature of potential social partners. The upper limits of the number and variety of social relationships an individual animal could have are set by the availability of potential partners in its social group. While group size is obviously a key factor determining the number of potential social partners available to form relationships with, group composition, in terms of the age and sex of its members, influences the variety of kinds of relationships that an individual may have. Consequently, the potential for the complexity of social structure, as defined by the nature and patterning of social relationships (Hinde 1976), is constrained by group size and composition. As these factors are determined by demographic processes and life history variables, their combined effects have far-reaching consequences for individuals, in terms of their opportunities to interact and form relationships with others.

There are also specific consequences of demographic processes for kinship structure within a group, which in turn govern the potential for the occurrence and distribution of kin-related behavior. For example, the combined effects of the mean interval at which a female gives birth and the infant mortality rate will determine how likely an infant is to have other immature siblings. Similarly, the age at first birth, interbirth interval, and mortality rates will determine the extent to which there is overlap between generations and consequently the potential for grandparent-offspring interactions.

An association between kinship and patterns of primate social behavior was first noted by Japanese primatologists in the mid-1950s (Kawai 1958a, b; Kawamura 1958). Their early studies revealed that females that were closely related through the maternal line had similar dominance ranks, and that juvenile females appeared to inherit dominance rank from their mothers. In the five decades that followed, kinship emerged as a major explanatory principle for a diverse range of primate social behaviors and aspects of social structure. In addition to dominance rank, kinship has been shown to be associated with the amount of time individuals spend together, the exchange of affiliative behaviors such as grooming and allomothering, time spent feeding together, and support in agonistic interactions (reviewed by Gouzoules & Gouzoules 1987, Silk 2001, chapter 7).

The pioneering work of the early Japanese primatologists took a largely sociological perspective. They were primarily concerned with the ways in which groups of Japanese macaques (*Macaca fuscata*) were organized, and in the nature of relationships between individuals, rather than with evolutionary explanations for the patterns of social behavior they observed (Imanishi 1960). However, since the broad acceptance by biologists of the concepts of kin selection and inclusive fitness (Hamilton 1964, Maynard-Smith 1964), there has been a strong emphasis on evolutionary explanations of kin-related behavior in primates. Indeed, it is almost assumed that any pattern of behavior that can be explained in terms of kinship must have evolved as a result of kin selection. As Chapais (2001) pointed out, such assumptions may not always be justified. In many cases, there is little hard evidence for kin selection, and there is a need to consider alternative explanations more fully. Nevertheless, strong evidence exists for an association between kinship and the patterning of primate social behavior, particularly so for social behavior between close kin (Kapsalis & Berman 1996).

The potential for kinship to exert an influence on social behavior and social structure is clearly constrained by the extent to which kin-related individuals coexist in the same group.

It is therefore important to understand the ways in which demographic processes (such as dispersal, philopatry, and group fission) and demographic variables (such as group size and interbirth interval) might influence kinship structure within a group. This chapter considers the potential for demographic variation to influence the manifestation of kinship effects, concentrating for the most part on studies of rhesus macaques (*M. mulatta*), Japanese macaques, and savannah baboons (*Papio* spp.).

There are two reasons for adopting this narrow taxonomic focus. First, the association between kinship and patterns of social behavior has been studied longer and more extensively in these species than in most others. Second, a risk is associated with taking a broader taxonomic view; namely, introducing confounding differences that are due to phylogenetic rather than demographic variation. For example, even within the genus *Macaca* there is considerable variation in important key aspects of social behavior, such as male dispersal and the degree of nepotism (e.g., Caldecott 1986, Thierry et al. 2000). However, studies of rhesus macaques, Japanese macaques, and savannah baboons have tended to show a high degree of similarity in the social behavior of these species.

Data on Demography and Kinship

To assess the influence of demographic variation on kinship structure and behavior, data at two different levels of focus are required. Analysis of demographic trends should be based on data for one or more large populations. By contrast, determination of kinship by direct observation and description of social relationships between kin both require detailed data at the level of small groups of individuals. Ideally, both sets of data should be derived simultaneously from the same population or populations. This is rarely feasible in practice, however, for primates living under natural conditions (i.e., nonprovisioned, indigenous populations living in relatively undisturbed habitats), because the requirements of data collection are quite different in each case.

Demography originated as the statistical study of the structure and distribution of human populations, and demographic studies of modern and recent human populations are frequently able to draw upon vast and detailed databases obtained in government censuses (e.g., Coleman & Salt 1996, Woods 2000). This allows an assessment of patterns and trends at different geographic scales, for different sections of society and across different periods in history. By contrast, attempts to study demographic variation in nonhuman primates have to make do with data sets that are much smaller and often quite fragmentary. Field studies of primates have typically focused on single groups or occasionally on two or three groups. Few field studies have collected accurate data on group size and composition for more than a few groups. Those that have surveyed a larger subset of a population have involved species living in habitats where visibility is relatively good, such as baboons in savannah and montane grasslands (e.g., Altmann et al. 1985, Henzi & Lycett 1995). These studies essentially provide cross-sectional data on the size and composition of many groups. There may also be detailed data on the life histories of known individuals in one or two groups, but no studies exist in which demographic data and long-term data on kinship are available for many groups of baboons in a naturally occurring population.

Studies of multiple groups of macaques in naturally occurring populations have also been performed, but most have been on a smaller scale than the baboon studies cited above. A population of long-tailed macaques (*M. fascicularis*) at Ketambe in Sumatra was studied for 11 years. Behavioral and demographic data were collected for six groups for part of this period (e.g., van Schaik & van Noordwijk 1986), and for three groups for the entire 11 years (de Ruiter & Geffen 1998, van Noordwijk 1999). Similarly, the behavior and demography of seven groups of rhesus macaques were studied for 42 months in a disturbed forest habitat in Pakistan (Melnick et al. 1984).

More large-scale censuses of some natural populations of Japanese macaques have been attempted (e.g., Yoshihiro et al. 1999), although none has been comparable in accuracy or detail to the baboon studies cited above. This is largely a reflection of the difficulties involved in making accurate counts of multiple groups of unhabituated primates in a forest habitat. The problem is exacerbated by the fact that much of the remaining forest in Japan is in extremely mountainous terrain, making large-scale coordinated censuses a logistic nightmare.

An alternative approach is to combine demographic data from diverse studies of individual groups of the same species (e.g., Takasaki 1981, Takasaki & Masui 1984). This has the benefit of providing data for multiple groups over a much wider geographical area than it would be practical to survey in a single census. There are potential problems with this approach, however. For instance, while rhesus macaques have been studied at various other sites within their distribution, data for groups living under conditions that have not been radically influenced by human activity are scarce. Throughout much of the species' distribution, rhesus monkeys live commensally with humans and receive at least some of their food from them. Most studies of indigenous groups have involved temple monkeys (e.g., Teas et al. 1980) or other groups that regularly interact with people (e.g., Malik & Southwick 1988). This commensal lifestyle of rhesus macaques is not a recent development, and it could be argued that it represents part of the range of conditions under which the species naturally occurs (Richard et al. 1989). However, commensalism takes a variety of forms, so that while some groups may have access to a food source that is much more concentrated and regularly replenished than any natural food resources, other groups may be dependent on scavenging from a variety of relatively meager sources. As abundance of food and concentration of the food supply may affect both demographic and behavioral characteristics of groups (discussed more fully below), combined demographic data from diverse studies of commensal groups would be extremely difficult to interpret.

The situation for Japanese macaques is somewhat different. While groups have been provisioned by researchers and in monkey parks in various parts of Japan (Asquith 1989), this species does not live commensally with humans in the way that rhesus macaques do. Raiding of crops is a growing problem in many parts of Japan, but several populations still receive little if any of their food from human sources. Analyses have been made of demographic data collected in diverse studies of this species (e.g., Takasaki 1981, Takasaki & Masui 1984), but in the vast majority of cases, complementary data on kinship were not available.

One drawback associated with combining data from multiple studies is that the aims of those studies and the methodologies used can vary considerably, and this will inevitably influence the kind and comparability of the data they present. On the whole, this problem is likely to have greater impact on behavioral data than on demographic data. However, even some key measures for demographic studies, such as age and sex categories and group membership, may be open to different interpretations by different observers.

Firm data on maternal kinship are less prone to problems of interpretation; however, collecting sufficient data to determine reasonably complete genealogies is problematic for different reasons. While meaningful data on the demography of a population can be collected over a short period of time, logistic considerations allowing, the accumulation of accurate data on kinship by direct observation requires very long-term study. Individual animals must be followed from birth to parenthood. Macaques are long-lived animals with an interbirth interval of two to three years under natural conditions, and infant mortality may be as high as 25% within the first year (Takahata et al. 1998). So determination of kin relatedness among adults by direct observation requires many years of continuous study. Consequently, there has been only a small number of studies of natural groups of macaques for which kinship was accurately known (Yamagiwa & Hill 1998), and none in which kinship was known for a sizeable population comprising many social groups.

Demography and Kinship in Provisioned Groups

Most of the data on kinship structure and behavior of Japanese and rhesus macaques have come from studies of provisioned groups (reviewed in Asquith 1989, Hill 1999). There are good reasons for this. Provisioned groups tend to be less mobile and easier to observe than natural groups. They are also less prone to extinction, which can bring many years of careful data collection on kinship to a sudden halt (Takahata et al. 1994). These factors have made it much more practical to collect the very long-term data required to establish kinship in provisioned situations than in the wild. In the case of rhesus macaques, studies at artificially established and maintained colonies have made a huge contribution to our knowledge of social behavior and kinship. The most productive of these has been the Cayo Santiago colony in Puerto Rico, where the macaques range freely over a small island and have all of their nutritional needs provided for them (for an account of the history of the colony see Rawlins & Kessler 1986). Japanese macaques have also been studied in ex situ artificial colonies (e.g., Fedigan 1991), but the vast majority of data for this species come from studies of groups that were provided with food in their natural habitat. At most of the established study sites in Japan, natural populations were initially fed by researchers to improve visibility (Sugiyama 1965). Provisioning subsequently became established as a regular routine at these sites (Asquith 1989).

While studies of provisioned groups have provided much of the data on kinship and social behavior of rhesus and Japanese macaques, determining the relevance of their findings to the natural situation can be problematic. This is particularly pertinent when considering the effects of demography, as provisioning has profound effects on a variety of demographic parameters. The effects of a regular supply of artificial food on the demography of macaque populations have been well documented (e.g., Mori 1979, Sugiyama & Ohsawa 1982, Fukuda 1988, Asquith 1989), but it is only quite recently that adequate data from wild populations have been available that allow meaningful quantitative comparisons to be made (Takahata et al. 1998). Females in provisioned groups of Japanese macaques tend to bear their first offspring a year or two earlier than those in natural populations, and birth rate is enhanced, at about 0.5 births per female per year in provisioned groups, compared with 0.27 to 0.35 births per female per year in natural populations (reviewed in Takahata et al. 1998). Infant mortality is also usually much lower in provisioned than in natural groups, although this is not always the case (Fukuda 1988). These changes combine to produce rapid population growth, which is reflected in greatly increased group size. Under natural conditions groups usually have fewer than 50 animals, although they may occasionally exceed 70 animals (Takasaki 1981). However, provisioned groups frequently grow to more than 100 members and may become much larger (data for Japanese macaques reviewed in Yamagiwa & Hill 1998). In 1989, one of the three groups of provisioned Japanese macaques at Takasakiyama included more than 1,000 animals (personal observation).

While the fact that provisioning influences demographic parameters has long been recognized, there has been little concern about its potential influence on social behavior. It has been widely assumed that, while the frequency of some behaviors and activity budgets might be altered by provisioning (e.g., Altmann & Muruthi 1988), patterns of behavior and the quality of social relationships observed would be essentially unchanged. For example, in a review of kinship in primates, Gouzoules and Gouzoules (1987, p. 299) wrote, "Although . . . much of the data on primate kin correlated behavior come from studies of provisioned populations, no evidence suggests that the patterning of behavior in these populations has been altered by feeding." However, strong indications now exist that the effects of provisioning do influence the quality of social relationships observed (Hill & Okayasu 1995, 1996; Hill 1999) and consequently have an impact on the social structure (sensu Hinde 1976) of the group.

Only one population of nonprovisioned Japanese macaques has been studied long enough for matrilineal kinship to be known for almost all females in some study groups. This was on the west coast of Yakushima, an island at the southern limit of the distribution of Japanese macaques, where the monkeys are an endemic subspecies, *Macaca fuscata yakui* (Maruhashi 1982). Unfortunately, the original study groups became extinct, and there has not yet been a stage at which kinship structure and behavior could be assessed in relation to demographic variables. Thus, the data that would be required to make a rigorous and systematic assessment of the effects of demographic variation on kinship structure and behavior of a nonprovisioned population are not currently available, even for such well-studied species as rhesus and Japanese macaques. The alternative taken here is to consider ways in which demographic variation is likely to influence kinship structure and its probable consequences for patterns of behavior and social relationships. As studies of provisioned groups have contributed so much to our current understanding of social behavior in these species, frequent reference will be made to data for provisioned groups. But it is important to keep in mind that data for provisioned groups may not be truly representative of the natural condition, either in terms of demography or social structure (Mori 1979, Sugiyama & Ohsawa 1982, Altmann & Muruthi 1988, Asquith 1989, Hill & Okayasu 1996, Hill 1999), so extrapolation to the natural situation requires caution. The influence of provisioning can also extend beyond the bounds of individual groups. A good example of this is the process of dispersal, which, in addition to its effects on group composition, also makes an important contribution to dynamics at the population level.

Dispersal

Most birds and mammals that live in social groups show some form of dispersal (Greenwood 1980, Pusey 1987). The most common pattern is for individuals of one sex to disperse from the group they are born in, and to attempt to join, or form, another. As dispersal generally precedes the first mating by the dispersing individual, and as breeding between close kin may have deleterious effects on the survival of offspring (Ralls & Ballou 1982, Pusey 1987, Alberts & Altmann 1995), avoidance of inbreeding has been proposed as the ultimate function of dispersal in primates (e.g., Wade 1979), although alternative explanations have also been put forward (Moore & Ali 1984). The consequences of dispersal for the evolution of kin-based groups are considered in detail elsewhere in this volume (chapter 4). The discussion here is therefore limited to a few points that help to set the context for the interplay between demography and kinship in macaques and savannah baboons.

As a primary mechanism influencing the composition of groups, dispersal is clearly of key importance in determining kinship structure and consequently opportunities for kin-related patterns of behavior to develop within groups (Moore 1992). In most cercopithecine primate species, it is the males that routinely disperse (reviewed in Pusey & Packer 1987), and data for natural populations of Japanese and rhesus macaques suggest that all males disperse, usually before reaching full adulthood (Melnick et al. 1984, Suzuki et al. 1998). Even in provisioned populations of rhesus and Japanese macaques, where both opportunities for dispersal and choice of other groups to disperse to can be very limited, most males disperse. In some cases, males from provisioned groups of Japanese macaques may cross extensive areas of cultivation and settlement to join another group (e.g., Huffman 1991). Under these extreme conditions of isolation, however, some males do remain in the group they are born in. In most cases, males who remain are the sons of high-ranking females. These individuals frequently rise to gain high rank themselves, and this has been interpreted as resulting from nepotistic support from their high-ranking female kin (Meikle & Vessey 1981). Rise in rank and mating by natal males has also been reported by one study of savannah baboons (Bulger & Hamilton 1988, Hamilton & Bulger 1990), but it is not clear how widespread the phenomenon is in this population, or whether it occurs at all in other populations. To date there are no records of male rhesus or Japanese macaques staying in their natal group in nonprovisioned populations, and indications are that males in such populations are highly mobile (Melnick et al. 1984, Furuichi 1985, Suzuki et al. 1998), much more so than males in provisioned groups. Indications are that, if it occurs at all in these species, cases of natal males achieving high rank as a result of nepotistic support under natural conditions must be extremely rare.

Studies of patterns of transfer in provisioned colonies of rhesus macaques have also suggested that males may transfer with kin, or preferentially transfer to groups in which there are already male kin (Meikle & Vessey 1981). The fate of dispersing animals can be difficult to determine in natural populations, but what data exist for cercopithecine primates give a somewhat conflicting picture. Patterns of transfer of males in a natural population of seven groups of rhesus monkeys in Pakistan showed no departure from a random model and no indication of kin-related males moving together (Melnick et al. 1984). On the other hand, young male vervets (*Cercopithecus aethiops*) and long-tailed macaques transferring for the first time appear to be attracted by the presence of male peers in adjacent groups, some of whom may be kin (Cheney & Seyfarth 1983, van Noordwijk & van Schaik 1985).

However, two factors make it extremely unlikely that kinship between immigrant males is an influential factor in the social structure of nonprovisioned groups of macaques and baboons. First, the longer interbirth interval and lower infant survival rate that are characteristic of nonprovisioned populations of baboons and macaques (Altmann & Altmann 1979) drastically reduces the occurrence of young males having similar-aged male kin in their natal group around the time of their departure from it.

Second, in relatively undisturbed populations, males remain in a group for only a limited period (an average of two to three years in Japanese macaques: Furuichi 1985, Suzuki et al. 1998) before moving on to another, and they continue to move from group to group throughout their lives (Melnick et al. 1984, Furuichi 1985, Suzuki et al. 1998). Transfer from the natal group is often to an adjacent group, but subsequent transfers usually take males further afield (Suzuki et al. 1998).

Females routinely disperse from the natal group in many species of prosimians, New World monkeys, and apes (Strier 1994). Female dispersal may also be more widespread among other primate taxa than preliminary data may have suggested (Moore & Ali 1984). Nevertheless, female dispersal appears to be exceptional in rhesus and Japanese macaques. Most documented cases have involved provisioned groups and have been associated with some major disturbance, such as a drastic reduction in food supply (Burton & Fukuda 1981, Fukuda 1988). Similarly, it is rare for two groups of females to unite to form a single group. A few cases of "fusion" have been recorded in a natural population of Japanese macaques. However, each case involved the two or three remaining members of a group that had been brought to the brink of extinction by a gradual decline in numbers joining a neighboring group (Takahata et al. 1994, Sugiura et al. 2002).

To summarize, available data suggest that, in natural populations of rhesus and Japanese macaques, dispersal from the natal group is routine among males and that dispersal is effectively limited to males. Consequently, social groups of these species consist of a stable core of kin-related adult females and their immature offspring, and a number of more transient young and adult males who are immigrants, and who normally have no kin in the group, apart from any offspring they may have fathered. Although cases of fusion by females have been recorded, each has involved one or two females joining a larger group, and these rare events probably have a minimal impact on kinship structure in the population as a whole. As a result, all females in a naturally formed group will normally belong to a single matriline, although the mean degree of kin relatedness between females will decrease with new generations. So in considering the ways in which demographic variation influences kinship structure and behavior of these species, the primary focus will be the kin network of females and their immature offspring that forms the stable core of the group.

Effects of Group Size on Social Complexity

As noted above, most study populations of rhesus and Japanese macaques have been provisioned, which typically results in an increase in population density and group size. This suggests that population growth and group size are limited by food, but even with a plentiful food supply and under the same conditions, group size can vary enormously. For example, Itani (1975) reported that the three groups in the provisioned Takasakiyama population of Japanese macaques numbered 517, 150, and 73 in 1962. The relevance of such variation under the influence of provisioning is questionable, but group size also varies considerably in nonprovisioned populations of rhesus and Japanese macaques. For example, in the population of rhesus macaques studied by Alison Richards and her students in Pakistan, the 292 animals were living in seven social groups, which ranged in size from 18 to 65 animals

(Melnick et al. 1984). Similarly, seven natural groups of Japanese macaques living in contiguous forest in Yakushima ranged in size from 13 to 47 animals (Maruhashi 1982). Furthermore, both species are found in a variety of habitat types, across which variation may be even greater. Japanese macaques have a scattered distribution in remaining areas of forest from the northern tip of Honshu to the southern island of Yakushima. Within this distribution, the size of nonprovisioned groups can range from 11 to over 100, although groups at the top end of this range are characteristic of highly disturbed habitats (Takasaki 1981), and around 70 animals is probably a more realistic maximum group size for undisturbed conditions.

In social groups in which individuals form differentiated relationships with several others, the potential for social complexity must increase with group size, because the larger the group, the greater the number and diversity of potential partners that an individual could have a social relationship with. Clearly, if the number of an individual's relationships increased monotonously with group size, and if time allocated to social behavior remained constant, the time available for interacting with each social partner would decrease. In this case, in larger groups individuals would be expected to have a greater number of weaker relationships.

Observations of female grooming networks suggest that this is not the case. In large groups, individuals tend to have relationships with a subset of other females, rather than with all females to diminishing levels (Silk et al. 1999). In a comparison of social dynamics in two groups of savannah baboons, Sambrook et al. (1995) found that the size of grooming network among females in each of the groups was very similar, in spite of one study group being twice the size of the other. Similarly, Henzi, Lycett, and Weingrill (1997) found a mean grooming clique size of 7.4 females in five groups of mountain baboons. In groups with seven or fewer females, each female would have close relationships with a subset of a few others, but would attempt to groom with all others to some extent. However, once the number of females exceeded seven, clear cliques would become differentiated within the group.

It is now well established that, in species with female philopatry, the strongest relationships, in terms of time spent together and frequency of affiliative behaviors, involve the closest female kin (reviewed in Gouzoules & Gouzoules 1987, Kapsalis & Berman 1996). Similarly, the composition of cliques or subgroups of females in groups large enough to have them is closely linked with matrilineal kinship (Silk et al. 1999). As a group steadily increases in size, with successive generations the mean degree of relatedness between individuals in the kin network decreases (Berman et al. 1997). Kinship effects are largely limited to close kin, and more distant relatives are treated in the same way as nonkin (Kapsalis & Berman 1996; Chapais et al. 1997, 2001). Thus, as a group becomes very large, social complexity will take the form of multiple clusters of subgroups of closely related kin.

The fact that these clusters do, nevertheless, belong to the same group has important consequences for the social context. It means, for example, that infants are more likely to encounter females and other immature individuals who are not their close kin in a large group than in a small one. This was demonstrated by Berman et al. (1997), who investigated the effects of changes in group size on the social relationships of infant rhesus macaques in the Cayo Santiago colony. Their analyses were based on data from one social group at different stages of a period of expansion, daughter groups produced by two fissions of the original study group, and comparative data from an unrelated study group. They found marked differences in the social milieux experienced by infants in relation to group size, and associated changes in the mother's behavior. In larger groups, infants had a greater number of individuals within 5 m of them, but the proportion of those individuals that were close kin decreased as group size increased. Also as group size increased, infants spent less time away from their mothers, and the mothers' role in maintaining proximity to their infants increased. Berman et al. interpreted this as reflecting greater concern of the mothers as their infants' exposure to others who were not close kin increased. Interestingly, the number of individuals within 60 cm of infants did not increase significantly with group size, perhaps reflecting the mothers' intolerance of the close proximity of those who are not close kin.

Fission

Fission is the process by which a large group splits to form smaller groups. Apart from natural disasters, it is the only event with the potential to bring about a rapid and major change in kinship structure of a group. Fission is a common phenomenon in both provisioned and natural populations of rhesus and Japanese macaques (provisioned: Furuya 1968, 1969, Koyama 1970, Missakian 1973, Chepko-Sade & Sade 1979; natural: Maruhashi 1982, Melnick & Kidd 1983, Yamagiwa 1985, Oi 1988). A key aspect of the process is that the split tends to occur along kinship lines, such that each smaller "daughter" group consists of females that are more closely related to one another than they are to females in the other daughter groups. This characteristic of fission is apparent in both natural (Melnick & Kidd 1983, Oi 1988) and maintained populations (e.g., Koyama 1970, Missakian 1973) of both species.

The formation of persistent subgroups of females within an expanding group may represent an early step toward group fission (Henzi, Lycett, & Piper 1997, Henzi, Lycett, & Weingrill 1997). Divisions between subgroups will be perpetuated and strengthened by the tendency observed by Berman et al. (1997) for rhesus mothers to try to limit their infants' exposure to individuals who are not close kin. This further weakening of links between subgroups may increase the chances of fission, with subgroups of relatively close kin forming the daughter groups. Apart from reducing group size, fission along family lines also results in an increase in the mean degree of relatedness between individuals in group, and mothers put less effort into trying to maintain proximity to their infants (Berman et al. 1997).

Fission occurs less frequently in provisioned populations than under natural conditions. At Arashiyama, only three cases of fission were observed over a period of 29 years, two of them occurring in the same year (Koyama et al. 1992). Conditions vary enormously between provisioned sites in terms of the amount of food provided, the extent to which the movements of the macaques are limited by physical barriers, the availability of natural plant foods, and whether or not the population is subject to culling. For example, after each of the fissions at Arashiyama, one of the daughter groups was captured and removed (Koyama et al. 1992). However, it is clear from other sites that provisioned groups can persist for very long periods without fissioning (e.g., Koshima: Mori et al. 1989). By contrast, in the natural population on the west coast of Yakushima, the frequency of fission was roughly twice that at Arashiyama, with members of the original study group experiencing three cases of fission in 15 years (Maruhashi 1982, Takahata et al. 1994).

As mentioned above, provisioning leads to population growth and this, in combination with a reduced rate of fission, results in very large group sizes. The reasons why fission should occur more rarely in provisioned groups are not entirely clear, although the provisioned food source itself may hold the answer. Under natural conditions, a group forages throughout its home range and members of the group disperse during foraging (Furuichi 1983). Home ranges of adjacent groups may overlap, and there are rarely empty ranges with no occupants. After each of multiple fissions in the natural population in Yakushima, the daughter groups tended to occupy part of the range of the original group (Maruhashi 1982). The early stages of fissions may be apparent in foraging subgroups of kin, which come together less and less frequently. This pattern may be inhibited in provisioned groups, where individuals are repeatedly brought together at the major food resource.

The mean interval between group fissions will have important ramifications for the kinship structure within the group. The longer the interval between fissions, the lower the mean degree of kin relatedness between females will become. So in provisioned groups increasingly distant kin, who would have been separated by fission, remain in the same group. Furthermore, as older matriarchs die, the closest living links between subsets of kin will become increasingly distantly related, until those kin groups can be thought of as separate matrilines. Studies of provisioned macaques frequently refer to multiple matrilines within a single group (e.g., Chepko-Sade & Sade 1979), and in the case of artificially formed colonies, females who come together to form the original groups may be unrelated (e.g., Sade 1965). But in the case of naturally occurring groups that were provisioned in situ, this is very unlikely to occur, and all females within a group are kin, albeit very distant kin in some cases.

Provisioned naturally occurring groups may represent an extreme situation, but one that serves to illustrate a potential effect of demographic variation on the pattern of kinship within a group. As the interval between fission events increases, so too will the potential for an increasingly complex social structure, consisting of multiple subgroups of closely related kin, as outlined in the previous section.

Demographic Influences on Dominance

The context in which Kawai (1958a, b) and Kawamura (1958) first recognized an association between kinship and patterns of social behavior was female dominance rank. They noted that a clear dominance hierarchy was apparent among female Japanese macaques, and that close female kin had similar dominance ranks. Since then, numerous studies have demonstrated a remarkable degree of order and stability in the dominance relations of female rhesus and Japanese macaques (e.g., Koyama 1967, Missakian 1972, Mori et al. 1989). For the most part, the distinctive pattern found in female dominance rank in these two species can be described by two simple rules. The first is that a mother normally remains dominant to her daughters throughout her life. Although there are occasional exceptions to this, most can be explained in terms of group instability (e.g., Chikazawa et al. 1979). The second rule concerns the process by which juvenile females acquire dominance rank. In both species, the position of juveniles in the dominance hierarchy may be difficult to discern, but for females it typically becomes clear around, or just before, adolescence. At this time, a young female normally acquires a dominance rank immediately below that of her mother. In other words, she becomes dominant to all females who are subordinate to her mother, including her own older sisters, if she has any. This process of young females coming to outrank their older sisters is commonly known as youngest ascendancy.

Since it was first described by Kawamura (1958), the process has been confirmed by numerous studies of provisioned rhesus and Japanese macaques at various sites (Koyama 1967, Missakian 1972, Sade 1972, Datta 1988, Mori et al. 1989, Takahata 1991). Kawamura noted that there were exceptions to the rule, and subsequent studies have also reported cases where youngest ascendancy did not occur, and younger sisters remained subordinate to one or more of their elders (e.g., Mori et al. 1989, Takahata 1991). But these exceptions tended to involve a small minority of females. On the whole, in the provisioned groups of rhesus and Japanese macaques where it has been studied most, the phenomenon appears to be remarkably robust (Hill & Okayasu 1996).

Both the maintenance of maternal dominance and the process of youngest ascendancy have been reported for other primate species that form matrifocal, multimale groups (e.g., Horrocks & Hunte 1983). However, some studies have found that exceptions to the "rules" of rank acquisition may be much more common than they are in rhesus or Japanese macaques (Silk et al. 1981, Paul & Kuester 1987, Hausfater et al. 1982). In attempting to account for these exceptions, several researchers have noted apparent links between demographic variables and patterns of female dominance (Chikazawa et al. 1979; Datta 1989; Hausfater et al. 1981, 1987; Paul & Kuester 1987). Relative physical strength of individual females appears to play a part in dominance acquisition and maintenance under some conditions. For example, among provisioned Barbary macaques (*M. sylvanus*), Paul and Kuester (1987) found that mothers that were outranked by their daughters tended to be much older than the majority of mothers who remained dominant. However, there is also evidence that the support of allies is a key factor in the maintenance of maternal dominance. For example, dominance reversals between mothers and daughters have also been associated with the formation of a new group of rhesus macaques in captivity (Chikazawa et al. 1979). In this case, the mothers involved were not elderly, and were larger than their daughters. The researchers attributed the reversals to the mother's reluctance to fight her daughter, and to the lack of an extensive network of kin from which a mother could normally expect support. The crucial role of the support of others in the maintenance of female dominance relations has also been demonstrated experimentally in captive Japanese macaques (Chapais 1988, 1991).

The process of youngest ascendancy also depends on the support of allies (Datta 1988). At adolescence, when young females rise in rank above their older sisters, they have not yet reached full body size and so are smaller, and presumably less powerful, than the individuals they become dominant to. The rise in rank depends on the support of female kin, primarily that of the mother. Kawamura (1958) first noted the tendency for mothers to give preferential support to their youngest and most vulnerable offspring. However, some evidence shows that the support of individuals other than the mother can also be instrumental in establishing a young female's dominance over her older sisters. Confirmation that a mother's support is not always crucial to youngest ascendancy comes from a 29-year study of changes in the dominance rank of female Japanese macaques at Koshima. Mori et al. (1989) found that, in more than half of all cases where a mother died before her daughter reached adolescence, youngest ascendancy still occurred. Furthermore, a study of the acquisition of dominance in rhesus monkeys at Cayo Santiago demonstrates the proximate mechanism by which youngest ascendancy can take place after the mother's death (Datta 1988). Although the majority of interventions on behalf of the younger sibling in disputes between sisters were made by the mother, about a third of all interventions were made by other female kin.

The availability of potential allies within the network of female kin is influenced by various demographic parameters, such as mean interbirth interval, infant mortality rate, and the rate of group fissions. Thus, the role that kinship can play in the acquisition and maintenance of female dominance rank is determined, to a large extent, by demographic variation. Datta and Beauchamp (1991) developed a model that simulated the effects of demographic variation on the probability that a female would have strong potential allies at key periods of her life for dominance acquisition and maintenance. They used the model to compare two populations. The first was a rapidly expanding population living under favorable conditions, typical of provisioned populations of rhesus and Japanese macaques. In this population, the interbirth interval was short, mortality relatively low, and females reached sexual maturity early. The second was a population living under much harsher conditions, and basically in decline, with long interbirth intervals, relatively high mortality, and late attainment of sexual maturity. The model demonstrated that, in terms of availability of suitable allies, older mothers would be expected to maintain their dominant status, and young females would be expected to rise in rank above their older sisters, in the former population but not in the latter.

It is important to note that all records of the routine occurrence of youngest ascendancy have come from provisioned populations. To some extent, this might be explained by Datta and Beauchamp's model in terms of demographic characteristics of those populations. On the other hand, it raises the possibility that other factors associated with provisioning are contributing to the phenomenon. It must be said that there are very few wild groups for which the requisite data on kinship are available. In one nonprovisioned group of Japanese macaques for which kinship data were available, Hill and Okayasu (1995, 1996) noted that youngest ascendancy was not observed in any of four pairs of sisters. This was in spite of the fact that the mother was alive and undeniably dominant to her daughters when the younger sister reached adolescence in each case. They propose that a key difference between this situation and the provisioned one is in levels of aggression. In provisioned groups, aggression, especially that associated with the concentrated food resource, is frequent and often intense. Under such conditions, aggression may involve a real threat of injury, causing a mother to routinely support her youngest and most vulnerable daughter. Under natural conditions, competition and aggression are much less frequent (Furuichi 1985, Hill & Okayasu 1995), and polyadic aggression in general and a mother's agonistic support of her daughters in particular are very rarely observed (Hill & Okayasu 1995).

Whether Datta and Beachamp's (1991) model provides a valid explanation for variation in the occurrence of youngest ascendancy or not, it does give a clear picture of the ways in which demographic parameters can influence the availability of allies. The rapidly expanding populations typical of provisioned groups of macaques are likely to be a very rare occurrence under natural conditions. Consequently, extensive matrilines in which females have multiple daughters separated in age by intervals of only a few years must be thought of as a product of provisioning, rather than a situation likely to occur with any regularity in natural populations. At the same time, less extreme variation does occur in natural populations. This variation will influence the availability of kin-related partners in the way that the model demonstrates and so, ultimately, the social structure of the group.

Conclusion

Both rhesus and Japanese macaques have been studied under free-ranging conditions for about half a century. Kinship has been recognized as a key factor in the social behavior of both species for most of this period. It seems odd, then, to have to conclude that the amassed data are inadequate to allow a methodical assessment of the influence of demographic variation on kinship structure and behavior. A major reason is that almost all of the data on kinship come from provisioned groups. Studies of provisioned groups have provided important insights into many aspects of social behavior and kinship. At the same time, results from such studies must be interpreted with caution when extrapolating to the natural situation, as they can produce potentially misleading conclusions. This problem becomes acute when considering the role of demographic variation in shaping kinship structure and behavior, because provisioning has such profound effects on both the size and the composition of groups, as well as on major processes, such as fission, that would normally shape kin networks.

Another major hurdle has been that most data come from single-group studies. There are, of course, very good reasons for this group-centered approach, not least being the logistic considerations and practical limitations involved in doing fieldwork at remote sites. The majority of primate species live in forest habitats, where visibility is poor. Study groups can be difficult to habituate to the presence of observers, and even well-habituated groups can be difficult to locate on a regular basis. Synchronous observational studies of multiple groups require multiple fieldworkers, making the work expensive and demanding in terms of project management. So an approach focused largely on individual groups may be unavoidable, but it is important to be aware of its inherent limitations.

The main problem is that it is impossible to assess the extent to which patterns of behavior, or social structure, observed in the study group can be generalized to the population that the group belongs to. In other words, the extent to which findings represent general principles, as opposed to idiosyncratic characteristics of the group in question, cannot be established. The usual way around this has been to identify common features recorded by independent studies of the same species, and to assume that any ubiquitous characteristics can be regarded as species typical, or at least typical of the study population. By this approach, a picture gradually emerges of species-typical social behavior and social structure. Accumulating evidence in this way cannot be considered a rigorous scientific technique, as it does not involve any true hypothesis testing. Nevertheless, the species-typical model is falsifiable, in that each new study is, in a sense, testing the prediction that the pattern will be repeated. This kind of approach may be useful for identifying characteristics of social systems that are sufficiently robust that they remain apparent regardless of variations in environmental conditions, or of natural variations that occur in group size or composition. It is less well suited to studying the extent to which characteristics vary among groups of the same species.

Perhaps the greatest difficulty facing studies of kinship in natural populations is the time required to establish relatedness between adult group members. Essentially it is necessary to observe a group for 15 to 20 years before the majority of female kin relationships are known rather than inferred. With the rapid development of techniques for DNA analysis, especially analysis of mitochondrial DNA, new possibilities are emerging. Studies of matrilineal relatedness of groups and subgroups that would formerly have required such long-term observation can now be made on the basis of molecular analyses (e.g., Hashimoto et al. 1996, de Ruiter & Geffen 1998, chapters 2 and 3). If techniques can be refined to the point that the matrilineal kinship of individuals can be reliably discerned, a systematic assessment of the effects demographic variation on kinship structure and behavior would at last become possible.

Acknowledgments   I am indebted to Carol Berman and Bernard Chapais for their patience and constructive comments, and to Anne E. Main for her untiring support and encouragement.

References

Alberts, S. C. & Altmann, J. 1995. Balancing costs and opportunities: dispersal in male baboons. *Am. Nat.*, 145, 279–306.

Altmann, J., Hausfater, G., & Altmann, S. A. 1985. Demography of Amboseli baboons, 1963–1983. *Am. J. Primatol.*, 8, 113–125.

Altmann, J. & Muruthi, P. 1988. Differences in daily life between semiprovisioned and wild-feeding baboons. *Am. J. Primatol.*, 15, 213–221.

Altmann, S. A. & Altmann, J. 1979. Demographic constraints on behavior and social organization. In: *Primate Ecology and Human Origins* (Ed. by I. S. Bernstein & E. O. Smith), pp. 47–64. New York: Garland STPM Press.

Asquith, P. J. 1989. Provisioning and the study of free-ranging primates: history, effects, and prospects. *Ybk. Phys. Anthropol.*, 32, 129–158.

Berman, C. M., Rasmussen, K. L. R., & Suomi, S. J. 1997. Group size, infant development and social networks in free-ranging rhesus monkeys. *Anim. Behav.*, 53, 405–421.

Bulger, J. & Hamilton, W. J. 1988. Inbreeding and reproductive success in a natural chacma baboon, *Papio cynocephalus ursinus*, population. *Anim. Behav.*, 36, 574–578.

Burton, J. J. & Fukuda, F. 1981. On female mobility: the case of the Yugawara-T group of *Macaca fuscata*. *J. Hum. Evol.*, 10, 381–386.

Caldecott, J. O. 1986. Mating patterns, societies and the ecogeography of macaques. *Anim. Behav.*, 34, 208–220.

Chapais, B. 1988. Rank maintenance in female Japanese macaques: experimental evidence for social dependency. *Behaviour*, 104, 41–59.

Chapais, B. 1991. Matrilineal dominance in Japanese macaques: the contribution of an experimental approach. In: *The Monkeys of Arashiyama* (Ed. by L. M. Fedigan & P. J. Asquith), pp. 251–273. Albany, NY: SUNY Press.

Chapais, B. 2001. Primate nepotism: what is the explanatory value of kin selection? *Int. J. Primatol.*, 22, 203–229.

Chapais, B., Gauthier, C., Prud'homme, J., & Vasey, P. 1997. Relatedness threshold for nepotism in Japanese macaques. *Anim. Behav.*, 53, 1089–1101.

Chapais B., Savard, L., & Gauthier, C. 2001. Kin selection and the distribution of altruism in relation to degree of kinship in Japanese macaques. *Behav. Ecol. Sociobiol.*, 49, 493–502.

Cheney, D. L. & Seyfarth, R. M. 1983. Nonrandom dispersal in free-ranging vervet monkeys: social and genetic consequences. *Am. Nat.*, 122, 392–412.

Chepko-Sade, B. D. & Sade, D. S. 1979. Patterns of group splitting within matrilineal kinship groups: a study of social group structure in *Macaca mulatta* (Cercopithecidae: Primates). *Behav. Ecol. Sociobiol.*, 5, 67–87.

Chikazawa, D., Gordon, T. P., Bean, C. A., & Bernstein, I. S. 1979. Mother-daughter dominance reversals in rhesus monkeys (*Macaca mulatta*). *Primates*, 20, 301–305.

Coleman, D. & Salt, J. (eds.) 1996. *Ethnicity in the 1991 Census. Volume 1. Demographic Characteristics of the Ethnic Minority Populations*. London: H.M.S.O.

Datta, S. 1988. The acquisition of dominance among free-ranging rhesus monkey siblings. *Anim. Behav.*, 36, 754–772.

Datta, S. B. 1989. Demographic influences on dominance structure among female primates. In: *Comparative Socioecology: The Behavioural Ecology of Humans and Other Mammals* (Ed. by V. Standen & R. A. Foley), pp. 265–284. Oxford: Blackwell Scientific Publications.

Datta, S. B. & Beauchamp, G. 1991. Effects of group demography on dominance relationships among female primates.1. Mother-daughter and sister-sister relations. *Am. Nat.*, 138, 201–226.

de Ruiter, J. R. & Geffen, E. 1998. Relatedness of matrilines, dispersing males and social groups in long-tailed macaques (*Macaca fascicularis*). *Proc. Royal Soc. Lond., B*, 265, 79–87.

Dunbar, R. I. M. 1979. Population demography, social organization and mating strategies. In: *Primate Ecology and Human Origins* (Ed. by I. S. Bernstein & E. O. Smith), pp. 65–88. New York: Garland STPM Press.

Dunbar, R. I. M. 1988. *Primate Social Systems*. London: Croom Helm.

Fedigan, L. M. 1991. History of the Arashiyama West Japanese macaques in Texas. In: *The Monkeys of Arashiyama* (Ed by L. M. Fedigan & P. J. Asquith), pp. 54–73. Albany, NY: SUNY Press.

Fukuda, F. 1988. Influence of artificial food supply on population parameters and dispersal in the Hakone T troop of Japanese macaques. *Primates*, 29, 477–492.

Furuichi, T. 1983. Interindividual distance and influence of dominance on feeding in a natural Japanese macaque troop. *Primates*, 24, 445–455.

Furuichi, T. 1985. Inter-male associations in a wild Japanese macaque troop on Yakushima Island, Japan. *Primates*, 26, 219–237.

Furuya, Y. 1968. On the fission of troops of Japanese monkeys. I. Five fissions and social changes between 1955 and 1966 in the Gagyusan troop. *Primates*, 9, 323–349.

Furuya, Y. 1969. On the fission of troops of Japanese monkeys. II. General view of troop fission of Japanese monkeys. *Primates*, 10, 47–69.

Gouzoules, S. & Gouzoules, H. 1987. Kinship. In: *Primate Societies* (Ed. by B. B. Smuts, D. L. Cheney, R. M. Seyfarth, R. W. Wrangham, & T. T. Struhsaker), pp. 299–305. Chicago: University of Chicago Press.

Greenwood, P. J. 1980. Mating system, philopatry and dispersal in birds and mammals. *Anim. Behav.*, 28, 1140–1162.

Hamilton, W. D. 1964. The genetical evolution of social behaviour. *J. Theor. Biol.*, 7, 1–51.

Hamilton, W. J. & Bulger, J. B. 1990. Natal male baboon rank rises and successful challenges to resident alpha-males. *Behav. Ecol. Sociobiol.*, 26, 357–362.

Hashimoto, C., Furuichi, T., & Takenaka, O. 1996. Matrilineal kin relationships and social behavior of wild bonobos (*Pan paniscus*): sequencing the D-loop region of mitochondrial DNA. *Primates*, 37, 305–318.

Hausfater, G., Altmann, J., & Altmann, S. 1982. Long-term consistency of dominance relations among female baboons (*Papio cynocephalus*). *Science*, 217, 752–755.

Hausfater, G., Cairns, S. J., & Levin, R. N. 1987. Variability and stability in the rank relations of nonhuman primate females: analysis by computer simulation. *Am. J. Primatol.*, 12, 55–70.

Hausfater, G., Saunders, C. D., & Chapman, M. 1981. Some applications of computer models to the study of primate mating and social systems. In: *Natural Selection and Social Behavior: Recent Research and New Theory* (Ed. by R. D. Alexander & D. W. Tinkle), pp. 345–360. New York: Chiron Press.

Henzi, S. P. & Lycett, J. E. 1995. Population-structure, demography, and dynamics of mountain baboons—an interim-report. *Am. J. Primatol.*, 35, 155–163.

Henzi, S. P., Lycett, J. E., & Piper, S. E. 1997. Fission and troop size in a mountain baboon population. *Anim. Behav.*, 53, 525–535.

Henzi, S. P., Lycett, J. E., & Weingrill, T. 1997. Cohort size and the allocation of social effort by female mountain baboons. *Anim. Behav.*, 54, 1235–1243.

Hill, D. A. 1999. Effects of provisioning on the social behaviour of Japanese and rhesus macaques: implications for socioecology. *Primates*, 40, 187–198.

Hill, D. A. & Okayasu, N. 1995. Absence of "youngest ascendancy" in the dominance relations of sisters in wild Japanese macaques (*Macaca fuscata yakui*). *Behaviour*, 132, 367–379.

Hill, D. A. & Okayasu, N. 1996. Determinants of dominance among female macaques: demography, nepotism and danger. In: *Evolution and Ecology of Macaque Societies*, (Ed. by J. Fa & D. Lindburg), pp. 459–472. Cambridge: Cambridge University Press.

Hinde, R. A. 1976. Interactions, relationships and social structure. *Man*, 11, 1–17.

Horrocks, J. A. & Hunte, W. 1983. Rank relations in vervet sisters: a critique of the role of reproductive value. *Am. Nat.*, 122, 417–421.

Huffman, M. A. 1991. History of the Arashiyama Japanese macaques in Kyoto, Japan. In: *The Monkeys of Arashiyama* (Ed. by L. M. Fedigan & P. J. Asquith), pp. 22–53. Albany, NY: SUNY Press.

Imanishi, K. 1960. Social organization of subhuman primates in their natural habitat. *Curr. Anthropol.*, 1, 393–407.

Itani, J. 1975. Twenty years with Mount Takasaki monkeys. In: *Primate Utilization and Conservation* (Ed. by G. Bermant & D. G. Lindburg), pp. 101–125. New York: Wiley.

Kapsalis, E. & Berman, C. M. 1996. Models of affiliative relationships among free-ranging rhesus monkeys, *Macaca mulatta*. I. Criteria for kinship. *Behaviour*, 133, 1209–1234.

Kawai, M. 1958a. On the system of social ranks in a natural troop of Japanese monkeys. (I). Basic rank and dependent rank. *Primates*, 1, 111–130.

Kawai, M. 1958b. On the system of social ranks in a natural troop of Japanese monkeys. (II). Ranking order as observed among the monkeys on or near the test box. *Primates*, 1, 131–148.

Kawamura, S. 1958. Matriarchal social ranks in the Minoo-B troop—a study of the rank system of Japanese monkeys. *Primates*, 1, 149–156.

Koyama, N. 1967. On dominance rank and kinship of a wild Japanese monkey troop in Arashiyama. *Primates*, 8, 189–216.

Koyama, N. 1970. Changes in dominance rank and division of a wild Japanese monkey troop in Arashiyama. *Primates*, 11, 335–390.

Koyama, N., Takahata, Y., Huffman, M. A., Norikoshi, K., & Suzuki, H. 1992. Reproductive parameters of female Japanese macaques: thirty years data from the Arashiyama troops, Japan. *Primates*, 33, 33–47.

Malik, I. & Southwick, C. H. 1988. Feeding behavior and activity patterns of rhesus monkeys (*Macaca mulatta*) at Tughlaqabad, India. In: *Ecology and Behavior of Food-Enhanced Primate Groups* (Ed. by J. E. Fa & C. H. Southwick), pp. 95–111. New York: Alan R. Liss.

Maruhashi, T. 1982. An ecological study of troop fissions of Japanese monkeys (*Macaca fuscata yakui*) on Yakushima Island, Japan. *Primates*, 23, 317–337.

Maynard-Smith, J. 1964. Group selection and kin selection. *Nature*, 201, 1145–1147.

Meikle, D. B. & Vessey, S. H. 1981. Nepotism among rhesus-monkey brothers. *Nature*, 294, 160–161.

Melnick, D. J. & Kidd, K. K. 1983. The genetic consequences of social group fission in a wild population of rhesus monkeys (*Macaca mulatta*). *Behav. Ecol. Sociobiol.*, 12, 229–236.

Melnick, D. J., Pearl, M. C., & Richard, A. F. 1984. Male migration and inbreeding avoidance in wild rhesus monkeys. *Am. J. Primatol.*, 7, 229–243.

Missakian, E. A. 1972. Genealogical and cross-genealogical dominance relations in a group of free-ranging rhesus monkeys (*Macaca mulatta*) on Cayo Santiago. *Primates*, 13, 169–180.

Missakian, E. A. 1973. The timing of fission among free-ranging rhesus monkeys. *Am. J. Phys. Anthropol.*, 38, 621–624.

Moore, J. 1992. Dispersal, nepotism, and primate social-behavior. *Int. J. Primatol.*, 13, 361–378.

Moore, J. & Ali, R. 1984. Are dispersal and inbreeding avoidance related? *Anim. Behav.*, 32, 94–112.

Mori, A. 1979. Analysis of population changes by measurement of body weight in the Koshima troop of Japanese monkeys. *Primates*, 20, 371–397.

Mori, A., Watanabe, K., & Yamaguchi, N. 1989. Longitudinal changes of dominance rank among the females of the Koshima group of Japanese monkeys. *Primates*, 30, 147–173.

Oi, T. 1988. Sociological study on the troop fission of wild Japanese monkeys (*Macaca fuscata yakui*) on Yakushima Island. *Primates*, 29, 1–19.

Paul, A. & Kuester, J. 1987. Dominance, kinship and reproductive value in female Barbary macaques (*Macaca sylvanus*) at Affenberg Salem. *Behav. Ecol. Sociobiol.*, 21, 323–331.

Pusey, A. E. 1987. Sex-biased dispersal and inbreeding avoidance in birds and mammals. *Trends Ecol. Evol.*, 2, 295–299.

Pusey, A. E. & Packer, C. 1987. Dispersal and philopatry. In: *Primate Societies* (Ed. by B. B. Smuts, D. L. Cheney, R. M. Seyfarth, R. W. Wrangham, & T. T. Struhsaker), pp. 250–266. Chicago: University of Chicago Press.

Ralls, K. & Ballou, J. 1982. Effects of inbreeding on infant mortality in captive primates. *Int. J. Primatol.*, 3, 491–505.

Rawlins, R. G. & Kessler, M. J. 1986. The history of the Cayo Santiago colony. In: *The Cayo Santiago Macaques* (Ed. by R. G. Rawlins & M. J. Kessler), pp. 13–45. Albany, NY: SUNY Press.

Richard, A. F., Goldstein, S. J., & Dewar, R. E. 1989. Weed macaques: the evolutionary implications of macaque feeding ecology. *Int. J. Primatol.*, 10, 569–594.

Sade, D. S. 1965. Some aspects of parent-offspring and sibling relations in a group of rhesus monkeys, with a discussion of grooming. *Am. J. Phys. Anthropol.*, 23, 1–17.

Sade, D. S. 1972. Sociometrics of *Macaca mulatta*. I. Linkages and cliques in grooming matrices. *Folia Primatol.*, 18, 196–223.

Sambrook, T. D., Whiten, A., & Strum, S. C. 1995. Priority of access and grooming patterns of females in a large and a small group of olive baboons. *Anim. Behav.*, 50, 1667–1682.

Silk, J. B. 2001. Ties that bond: the role of kinship in primate societies. In: *New Directions in Anthropological Kinship* (Ed. by L. Stone), pp. 71–92. Oxford: Rowman & Littlefield.

Silk, J. B., Samuels, A., & Rodman, P. S. 1981. The influence of kinship, rank, and sex on affiliation and aggression between adult female and immature bonnet macaques (*Macaca radiata*). *Behaviour*, 78, 111–137.

Silk, J. B., Seyfarth, R. M., & Cheney, D. L. 1999. The structure of social relationships among female savanna baboons in Moremi Reserve, Botswana. *Behaviour*, 36, 679–703.

Strier, K. B. 1994. Myth of the typical primate. *Yrk. Phys. Anthropol.*, 37, 233–271.

Sugiura, H., Agetsuma, N., & Suzuki, S. 2002. Troop extinction and female fusion in wild Japanese macaques in Yakushima. *Int. J. Primatol.* 23, 69–84.

Sugiyama, Y. 1965. Short history of the ecological and sociological studies on non-human primates in Japan. *Primates*, 6, 457–460.

Sugiyama, Y. & Ohsawa, H. 1982. Population dynamics of Japanese monkeys with special reference to the effect of artificial feeding. *Folia Primatol.*, 39, 238–263.

Suzuki, S., Hill, D. A., & Sprague, D. S. 1998. Intertroop transfer and dominance rank structure of nonnatal male Japanese macaques in Yakushima, Japan. *Int. J. Primatol.*, 19, 703–722.

Takahata, Y. 1991. Diachronic changes in the dominance relations of adult female Japanese monkeys of the Arashiyama B group. In: *The Monkeys of Arashiyama* (Ed by L. M. Fedigan & P. J. Asquith), pp. 123–139. Albany, NY: SUNY Press.

Takahata, Y., Suzuki, S., Agetsuma, N., Okayasu, N., Sugiura, H., Takahashi, H., Yamagiwa, J., Izawa, K., Furuichi, T., Hill, D. A., Maruhashi, T., Saito, C., Sato, S., & Sprague, D. S. 1998. Reproduction of wild Japanese macaque females of Yakushima and Kinkazan Islands: a preliminary report. *Primates*, 39, 339–349.

Takahata, Y., Suzuki, S., Okayasu, N., & Hill, D. A. 1994. Troop extinction and fusion in wild Japanese macaques of Yakushima Island, Japan. *Am. J. Primatol.*, 33, 317–322.

Takasaki, H. 1981. Troop size, habitat quality, and home range area in Japanese macaques. *Behav. Ecol. Sociobiol.*, 9, 277–281.

Takasaki, H. & Masui, K. 1984. Troop composition data of wild Japanese macaques reviewed by multivariate methods. *Primates*, 25, 308–318.

Teas, J., Richie, T., Taylor, H., & Southwick, C. 1980. Population patterns and behavioral ecology of rhesus monkeys (*Macaca mulatta*) in Nepal. In: *The Macaques: Studies in Ecology, Behavior and Evolution* (Ed. by D. G. Lindburg), pp. 247–262. New York: Van Nostrand Reinhold.

Thierry, B., Iwaniuk, A. N., & Pellis, S. M. 2000. The influence of phylogeny on the social behaviour of macaques (Primates: Cercopithecidae, genus *Macaca*). *Ethology,* 106, 713–728.

van Noordwijk, M. A. 1999. The effects of dominance rank and group size on female lifetime reproductive success in wild long-tailed macaques, *Macaca fascicularis*. *Primates*, 40, 105–130.

van Noordwijk, M. A. & van Schaik, C. P. 1985. Male migration and rank acquisition in wild long-tailed macaques (*Macaca fascicularis*). *Anim. Behav.*, 33, 849–861.

van Schaik, C. P. & van Noordwijk, M. A. 1986. The hidden costs of sociality: intra-group variation in feeding strategies in Sumatran long-tailed macaques (*Macaca fascicularis*). *Behaviour*, 99, 296–315.

Wade, T. D. 1979. Inbreeding, kin selection, and primate social evolution. *Primates*, 20, 355–370.

Woods, R. 2000. *The Demography of Victorian England and Wales.* Cambridge: Cambridge University Press.

Yamagiwa, J. 1985. Socio-sexual factors of troop fission in wild Japanese monkeys (*Macaca fuscata yakui*) on Yakushima Island, Japan. *Primates*, 26, 105–120.

Yamagiwa, J. & Hill, D. A. 1998. Intraspecific variation in the social organization of Japanese macaques: past and present scope of field studies in natural habitats. *Primates*, 39, 257–273.

Yoshihiro, S., Ohtake, M., Matsubara, H., Zamma, K., Hanya, G., Tanimura, Y., Kubota, H., Kubo, R., Arakane, T., Hirata, T., Furukawa, M., Sato, A., & Takahata, Y. 1999. Vertical distribution of wild Yakushima macaques (*Macaca fuscata yakui*) in the western area of Yakushima Island, Japan: preliminary report. *Primates*, 40, 409–415.

Part III

Diversity of Effects
of Kinship on Behavior

A captive slender loris (*Loris tardigradus nordicus*) mother with two infants clinging to her belly and back (which is not usual) shares food with the sire and an older sibling of the twins. Colony of Ruhr University bred from stock from Polonnaruwa, Sri Lanka. Photo by Helga Schulze.

7

Matrilineal Kinship and Primate Behavior

Ellen Kapsalis

Previous reviews of the impact of maternal kinship on primate behavior (Gouzoules 1984, Gouzoules & Gouzoules 1987, Walters 1987, Bernstein 1991, but see Silk 2001) have focused on a limited number of the cercopithecine taxa (i.e., macaques, baboons, and vervets) even though it is generally recognized that the female philopatric "cercopithecine model" is representative of only one phylogenetic branch of the primate order as a whole (Gouzoules & Gouzoules 1987, Strier 1994). This situation has arisen due to the excellent observational conditions for many cercopithecine species culminating in numerous long-term studies of kinship. In contrast, most noncercopithecine primates are generally much more difficult to study due to difficult terrain and/or poor visibility. As a result, relatively few researchers endeavored to take on the challenge of specifically identifying and following individuals longitudinally within social groups of these understudied species until the last decade or so. Hence, our overall knowledge regarding the impact of maternal kinship on primates is biased and limited. However, recent mitochondrial DNA (mtDNA) haplotype testing on wild ape groups has given us the opportunity to test some assumptions regarding the impact of maternal kinship on behavior in some male philopatric primate groups (e.g., bonobos and chimpanzees). Additionally, in the last decade, researchers have begun to explore maternal kinship relationships among female philopatric platyrrhine species previously unstudied due to the paucity of habituated social groups (e.g., *Cebus*).

The first part of this chapter briefly describes the cercopithecine model, specifically in regard to how maternal kinship structures social relationships. This description is a steppingstone to a major focus of this chapter, that is, methodological issues that have emerged from cercopithecine maternal kinship studies. For example, there is a growing awareness that Old World female monkeys behave differentially depending upon the category of kin. I describe this research and discuss the importance of kin discrimination evaluations in behavioral analyses. Another issue concerns the difficulty of teasing apart the roles of maternal kinship and rank distance in analyses. I describe how new statistical techniques have largely resolved this problem.

In the last part of this chapter, I discuss what we currently know regarding how maternal kinship structures social relationships of primates other than cercopithecine species. With the exception of studies of wild great ape groups where genetic relations are known, we still know very little of the impact of maternal kinship on behavior. However, recent work on platyrrhine female philopatric species suggests a pattern similar to the cercopithecine model. These studies are described.

Matrilineal Kinship and Cercopithecine Species

As mentioned above, most studies regarding the impact of maternal kinship in primates have concentrated on cercopithecine taxa and excellent reviews of these species have already been done (Gouzoules 1984, Gouzoules & Gouzoules 1987, Walters 1987, Silk 2001). I will not attempt to reinvent the wheel here but I will give a general overview of what is known for cercopithecines, and then focus on some associated methodological issues.

Macaques, baboons, and vervets were the first species that were individually identified and longitudinally followed (e.g., Kawai 1965; Kawamura 1965; Sade 1965; Hausfater 1975; Seyfarth 1980; Cheney & Seyfarth 1980; Seyfarth & Cheney 1980, 1984). Because of this, complete maternal genealogies for some social groups were recognized relatively early in primatological behavioral research. For example, our knowledge of matrilineal genetic lines of Cayo Santiago free-ranging rhesus monkeys (*Macaca mulatta*) goes back to 1956 (Sade et al. 1985, Rawlins & Kessler 1986). Researchers found that these species are female philopatric; males typically transfer from their natal groups upon maturity, whereas females in these species remain in their natal groups throughout their lives. Thus, females represent the stable and enduring aspect of the social group (Kawai 1965; Kawamura 1965; Sade 1965, 1967, 1972a, b; Missikian 1972; Hausfater 1975; Dunbar & Dunbar 1977). The only exceptions to this are group fissioning (e.g., Furuya 1969, Koyama 1970, Nash 1976, Chepko-Sade & Sade 1979, Chepko-Sade & Olivier 1979, Dittus 1988) and rare female transfer (e.g., Ali 1981, Kapsalis & Johnson 1999).

Since cercopthecine females spend their lives together, extensive networks of matrilineal kin (i.e., matrilines) have the potential to develop and associate with each other. Early field workers studying Japanese macaques (*M. fuscata*) and rhesus monkeys were the first to observe the tendency for females to form close relationships with matrilineal kin (Koford 1963; Yamada 1963; Kawai 1965; Sade 1965, 1972a, b; Missikian 1972; Oki & Maeda 1973; Mori 1975; Kurland 1977). Since these early studies, a plethora of field, captive, and laboratory studies have found maternal kinship to be one of the most influential factors structuring affiliative behavior including spatial proximity, grooming, conflict intervention, alliances, reconciliation, and cofeeding among cercopithecine individuals; in particular, differential patterns of grooming and alliance formation have been the behaviors most frequently evaluated (reviewed by Gouzoules 1984; Gouzoules & Gouzoules 1987; Walters 1987; Kapsalis & Berman 1996a, b; Aureli & de Waal 2000; Silk 2001). These kin-biased relationships begin in infancy; daughters develop close and enduring affiliative social relationships with their mothers and other maternal kin as infants that continue into adulthood (reviewed by Pereira & Fairbanks 1993, Fairbanks 2000, chapter 14).

Maternal kinship not only structures affiliative relationships among cercopithecine females, their dominance relationships are profoundly affected by matrilineal kinship as well. Dominance among females is generally socially "inherited" rather than the result of individual prowess or personality. Adult daughters acquire dominance ranks just below those of their mothers (Kawamura 1965, Missakian 1972, Sade 1972, Chapais 1992, Pereira 1992) through a process involving conflict interventions by close female relatives during juvenile female dominance interactions and aggressive responses by these protectors toward their opponents (e.g., Kawai 1965, Kawamura 1965, Koyama 1967, Missikian 1972, Sade 1972, Cheney 1977, de Waal 1977, Lee & Oliver 1979, Berman 1980, Walters 1980, Datta 1983, Horrocks & Hunte 1983, Lee 1983, Paul & Kuester 1987, Chapais 1988, Pereira 1989, Chapais & Gauthier 1993; see chapter 14).

A widespread and robust phenomenon in dominance acquisition of provisioned rhesus and Japanese macaques includes "youngest ascendancy" (Datta 1988; see Hill & Okayasu 1996 for a detailed treatment) in which adult daughters rank in inverse order of their age directly beneath their mother. The most probable mechanism by which this occurs is intervention by mothers (Kawamura 1965) and other close kin (Sade 1972a, Datta 1988, Mori et al. 1989) on behalf of youngest daughters against older daughters during sibling disputes. Schulman and Chapais (1980) presented a model to explain youngest ascendancy based on the assumption that during sibling disputes, mothers should preferentially give support to the daughter that has the highest reproductive value. Since the peak of reproductive value coincides with the age at which the youngest daughter reaches sexual maturity, the reasoning is that mothers benefit by supporting their youngest sexually mature daughter against older ones, and that this results in the youngest daughter rising in rank over her older sisters. An alternative explanation assumes that mothers and daughters are competitors for dominance rank status (Horrocks & Hunte 1983), with mothers striving to prevent alliances among daughters to ensure their own higher rank position over their daughters. By supporting the youngest daughter over her other daughters, the mother creates a situation whereby the most dominant daughter is dependent upon her mother for support and unlikely to seek alliances with her lower ranking sisters (Horrocks & Hunte 1983).

No matter what the underlying explanation, the interplay of a female's close association with kin and the patterning of dominance relations described above results in the development of cohesive cliques of related females occupying contiguous ranks (Sade 1972a). Over generations, this pattern of rank acquisition creates "corporate" matrilineal dominance hierarchies (cf. Silk 2001) in which females within a matriline are linearly ranked, and an entire matriline ranks above or below other matrilines within a social group. These dominance hierarchies are typically very stable for years (Bramblett et al. 1982, Hausfater et al. 1982, Samuels et al. 1987, Silk 1988, Kapsalis & Berman 1996b, Isbell & Young 1993, Isbell & Pruetz 1998), although very disruptive reversals have been documented (Altmann & Altmann 1979, Samuels & Henrickson 1983, Ehardt & Bernstein 1986, Samuels et al. 1987). Kinship may be largely responsible for the stability of dominance hierarchies; few challenges against higher ranking females may occur since the potential for maternally related allies to intervene against an opponent always exists (Silk 2001). Further, the nature of corporate dominance hierarchies is such that changes in physical abilities such as illness, injury, or agedness in one individual are not likely to dramatically affect the hierarchical structure as a whole (Silk 2001).

Maternal kinship plays a role in structuring juvenile male relationships as well. As with juvenile females, affiliative behavior is differentially allocated to young males on the basis of maternal kinship in terms of proximity, grooming, conflict intervention, alliances, reconciliation, and cofeeding (see reviews by Pereira & Fairbanks 1993, Fairbanks 2000, chapter 14). Males associate less with matrilineal kin than females as they mature, and since they typically transfer to nonnatal groups, the impact of matrilineal kinship on their social relationships is limited thereafter. However, some evidence suggests that males are more likely to transfer into nonnatal groups where maternally related adult male relatives already reside (Cheney & Seyfarth 1983). While not a common occurrence, some males stay within their natal group after adulthood. These males are generally sons of high-ranking mothers and tend to occupy high rank during their tenure in their natal groups. Berard (1989, 1999) has suggested that the high dominance ranks that these males have is in large part due to the support of female relatives that natal males receive during aggressive interactions with nonnatal males.

Interspecific Differences Between Macaques

The cercopithecine model described above, which typifies provisioned rhesus and Japanese macaques, might lead one to assume that all macaques are highly kin biased and have kin-based dominance hierarchies. However, other macaques such as bonnet macaques (Silk et al. 1981), Barbary macaques (Paul & Kuester 1987), Tonkean macaques (Thierry 1985, 1990; Demaria & Thierry 2001), and stump-tailed macaques (Butovskaya 1993) show variation in certain features of both affiliative and dominance behaviors for female macaques. For example, these species are exceptional in that females show little or no kin bias in affiliative interactions and may not follow the rule of youngest sister ascendancy (Paul & Kuester 1987; Hill & Okayasu 1995, 1996). Females may rise in rank above their mothers and, in some cases, rank above females dominant to their mothers (Silk et al. 1981).

Thierry (1990) suggested that interspecies differences among macaques may be explained in terms of all macaque species varying along a continuum from "despotic" (e.g., rhesus and Japanese macaques) to "egalitarian" (e.g., Tonkean [*M. tonkeana*], bonnet [*M. radiata*], Barbary [*M. sylvanus*], and stump-tailed [*M. arctoides*] macaques) with a corresponding suite of behaviors. For example, in a comparative analysis of rhesus, long-tailed (*M. fascicularis*) and Tonkean macaques, Thierry (1985) found rhesus to have behaviors he described as despotic, showing little tolerance and frequent and often severe aggression toward nonkin. Kinship profoundly influenced social and dominance relationships (as described above) and reconciliation behaviors happened infrequently. Conversely, for species described as egalitarian (i.e., Tonkean macaques), kinship was less influential in structuring social relationships; affiliative interactions among nonkin were much more frequent than for rhesus. Little severe aggression was observed, aggression was less unidirectional in nature, and conciliatory behaviors were typical for this species. Long-tailed macaques were intermediate between rhesus and Tonkean macaques in terms of aggression and other behaviors toward nonkin. Other studies have also shown interspecific differences for reconciliation as described by Thierry (e.g., de Waal & Ren 1988, Butovskaya 1993, Chaffin et al. 1995, Petit et al. 1997, Demaria and Thierry 2001). However, Thierry (1985) and others (Schino et al. 1988, Butovskaya et al. 1996, Castles et al. 1996, Demaria & Thierry 2001) caution that these interspecific comparisons are based on only one group for each species and may not take into account intraspecific variation.

Hill and Okayasu (1995, 1996) focused on intraspecific variation in the occurrence of youngest ascendancy in Japanese macaques and suggested an alternative ecological explanation. Noting that this phenomenon occurred only among provisioned groups of Japanese macaques under their study, they show that the number and frequency of females in close proximity at food resources clumped in space and time is extremely high compared to groups that feed only on natural food sources. They suggest that this closer proximity results in much higher rates of aggression among provisioned groups than among nonprovisioned social groups. This increased aggression leads mothers to routinely support the most vulnerable daughter (i.e., the youngest) over her older daughters, and also allows the youngest daughter to threaten older sisters with impunity in the presence of their mother.

Maternal Kin Recognition

As the previous section reveals, nepotism plays varying roles in structuring cercopithecine female-female social relations depending upon the species under study. Regardless of its impact, for it to occur at all, relatives must be able to distinguish kin from nonkin. Primate studies of maternal kin recognition have primarily focused on provisioned cercopithecine social groups in which complete maternal kin relationships were known. Since promiscuous mating is the norm for these species and males typically emigrate from natal groups upon maturity, degrees of relatedness are generally calculated between individuals through maternal lines, assuming no inbreeding or sharing of paternal genes. Within large cercopithecine social groups with extended genealogies, females may live with 12 or more different types of direct and collateral types of kin relations.

While the exact mechanisms are not known (Bernstein 1991, Tomasello & Call 1997; for synthesis, see chapter 13), matrilineal kin recognition seems to be based upon a period of long and intimate association or familiarity that begins with the mother-offspring relationship. This long-term relationship provides a variety of cues for each individual to recognize the others (Cheney & Seyfarth 1980, Nakamichi & Yoshida 1986, Pereira 1986), which in turn may allow other individuals to recognize other categories of maternal kin (e.g., Sackett & Fredrickson 1987, Kuester et al. 1994, Martin 1997). Most of the available evidence for cercopithecines suggests that individuals do not recognize paternal kin (Fredrickson & Sackett 1984; Bernstein 1991; Paul et al. 1992, 1996; Kuester et al. 1994; Alberts 1999; but see Smith 2000; Widdig et al. 2001). In contrast, excellent experimental field studies indicate that primates act as if they recognize maternally related individuals within their social milieu (e.g., Cheney & Seyfarth 1980, 1986, 1989, 1990). Indeed, researchers investigating kin recognition by the use of field playback experiments and laboratory studies have found that monkeys behave not only as if they recognize their own kin relations but also those between other group members (e.g., Cheney & Seyfarth 1980, 1982, 1990; Gouzoules et al. 1984; Dasser 1988a, b; see chapter 15). In addition, nonexperimental studies of redirected aggression and reconciliation (York & Rowell 1988, Cheney & Seyfarth 1989, Judge 1991, Das et al. 1997) indicate that relatives of monkeys not originally involved in disputes are targeted afterward in both affiliative and aggressive ways, suggesting third-party discrimination of kin relationships.

Discrimination of Different Categories of Maternal Kin

Over the years, some field and laboratory observers have reported that the distributions of many affiliative behaviors among cercopithecines decrease steeply and asymptotically with degree of relatedness, for example, $r = .5$, mother-daughter; $r = .25$, sisters, grandmother-granddaughter; $r = .125$, aunt-niece; $r = .063$, first cousins (Kurland 1977; Massey 1977; Kaplan 1978; Berman 1982, 1983; Glick et al. 1986; Singh et al. 1992; Kapsalis & Berman 1996a). However, studies analyzing the effects of maternal kinship on cercopithecine social relations have been inconsistent in their operational definitions of maternal kinship. Probably due to incomplete knowledge of matrilineal genetic relations, some studies have analyzed the effects of matrilineal kinship as simply related or unrelated (e.g., Silk et al. 1981; Silk 1982; Bernstein & Ehardt 1985, 1986; de Waal & Luttrell 1986; Thierry et al. 1990; de Waal 1991; Bernstein et al. 1993; Call et al. 1996). Others have analyzed kinship by differentiating degrees of relatedness (Kurland 1977; Massey 1977, 1979; Kaplan 1978; Berman 1982, 1983; Seyfarth & Cheney 1984; Glick et al. 1986; Hunte & Horrocks 1987; Lopez-Vergara et al. 1989; Singh et al. 1992; Rendall et al. 1996; Chapais et al. 1997, 2001; Bélisle & Chapais 2001). Among those that have used a graded system to define kin (e.g., levels or degrees of relatedness), some definitions have been relatively narrow, including relatives sharing an $r > .125$ (Kapsalis & Berman 1996b) or more broad, $r > .063$ (e.g., Rendall et al. 1996). Seyfarth and Cheney (1984) did not differentiate degrees of kinship but rather defined two categories, kin ($r > .25$) and nonkin (all other relationships).

Recent work examining the correlation between categories of kin relations and the distribution of various behaviors (proximity, grooming, conflict intervention, tolerated cofeeding, homosexual interactions) indicate that female macaques not only behave as if they discriminate between different levels of relatedness, they seem to have a "relatedness threshold" (cf. Chapais et al. 1997) for kin discrimination and nepotism beyond which they no longer act as if they discriminate between kin and nonkin (Kapsalis & Berman 1996a; Chapais et al. 1997, 2001; Bélisle & Chapais 2001; Chapais 2001).

For example, in our study of affiliative relationships among the multigenerational free-ranging rhesus monkeys of Cayo Santiago, Carol Berman and I found that, for several affiliative measures, a graded model of kinship that defined kin as individuals related by at least $r = .125$ accounted for a larger proportion of the variation than a similarly graded model that defined kin as individuals related by at least $r = .03125$. In other words, females behaved toward relatives beyond this degree of relatedness ($r < .125$) as if they were nonkin (Kapsalis and Berman 1996a). For the same population, Rendall et al. (1996) found that proximity between kin leveled off at coefficients of relatedness of $r = .125$. However, using playback experiments to test vocal recognition of contact calls, Rendall et al. (1996) found that biased responses to calls of kin encompassed broader categories of maternal kin relations ($r > .063$).

Chapais and his associates have conducted a number of experiments to examine kin discrimination in a captive social group of Japanese macaques (Chapais et al. 1997, 2001; Bélisle & Chapais 2001; Chapais 2001). They have found a consistent pattern for three very different types of behavioral interactions among females (homosexual interactions, conflict interventions, and tolerated cofeeding). Specifically, for each behavior, they found relatedness thresholds for preferential treatment of kin at $r = .25$ (grandmother-granddaughter and sister dyads). Females with genetic relationships beginning at $r = .125$ (aunt-niece dyads) and greater interacted much like nonkin. Recently, Chapais et al. (2001) expanded their research to include conflict intervention of adult females in maternally related juvenile male dominance interactions. They found that adult conflict intervention extended to $r = .125$ but only among directly related kin, that is, mothers, grandmothers, and great-grandmothers. Collateral kin intervened only up to $r = .25$ (siblings). They suggested that there may be more dominance competition among collateral than direct kin, and hence a narrower deployment of assistance to the former.

In addition to the theoretical implications, it is becoming increasingly important to understand how primates distribute nepotistic acts among relatives for various methodological reasons. First, results using a simple discrete model of kinship (related vs. unrelated) to simultaneously evaluate kin and rank effects of behaviors can be misleading. In our study, a regression model using discrete criteria for kinship accounted for smaller proportions of variation in affiliative interactions among adult female rhesus monkeys than two other models that used graded criteria for kinship. Second, we found that the choice of criteria for kinship affected the apparent importance of rank distance. The importance of rank as an organizing principle was spuriously masked when gradations of kinship were extended beyond a coefficient of $r > .125$ (Kapsalis & Berman 1996a). This finding suggests that the manner of defining kinship can dramatically affect interpretation of the relative importance of dominance in structuring female-female relations.

Third, evaluating altruistic acts toward different categories of kin reveals that not only degree of relatedness but also the type of relatedness (direct or collateral) may change the quality of kin social relationships under study (Chapais et al. 2001). More research is needed to assess (a) if similar patterns are observed in other species; (b) if differences in the quality of direct and collateral kin relationships are due to higher rates of interaction among direct rather than collateral kin; and/or (c) are the result of dominance competition constraining nepotism among collateral kin more than with direct kin (Chapais et al. 2001).

Fourth, researchers typically invoke kin selection to explain kin biases in all beneficent behaviors. However, most tests of kin selection have been based on kin/nonkin comparisons (but see Schaub 1996, Chapais et al. 2001). By knowing how individuals discriminate between categories of kin, researchers can better assess the extent to which altruistic behaviors are being distributed in a manner that corresponds to Hamilton's rule or according to other principles, for example, whether altruistic acts are allocated proportionally according to degree of relatedness (Altmann 1979, Weigel 1981, Reiss 1984, Chapais 2001, Silk 2001, chapter 16) and/or whether other evolutionary or nonevolutionary mechanisms are at work (Chapais 2001).

For example, the "limits of nepotism" hypothesis proposed by Sherman (1980, 1981) posits that certain categories of related individuals that have rarely coexisted historically should not be expected to distinguish one another from unrelated individuals. Given that large multigenerational groups (typical of several cercopithecine provisioned populations under study) are unlikely to be found under more natural conditions (e.g., Altmann & Altmann 1979), we might expect that beyond certain categories of relatedness, females would not discriminate behaviorally kin from nonkin. Additionally, demographic and ecological circumstances may alter absolute and relative numbers of available female social partners within a social group from year to year, and between social groups of a particular species (Dunbar 1991, 1992; Sambrook et al. 1995; Kapsalis & Berman 1996b; Henzi, Lycett, & Piper 1997; Henzi, Lycett, & Weingrill 1997; Silk et al. 1999). These differences in availability alone may be responsible for the quantity and quality of nepotistic behaviors to kin (chapter 16). Now that there is an explicit awareness that primates behaviorally discriminate between categories of kin, future research can better evaluate the relative contribution of both evolutionary and nonevolutionary forces.

The Relative Roles of Kin and Rank Distance

Another methodological issue concerns the difficulty of distinguishing between the roles of matrilineal kinship and rank distance, that is, the disparity between the respective dominance position of two monkeys (Demaria & Thierry 1990, de Waal 1991) in corporate matrilineal dominance hierarchies typified by rhesus and Japanese macaques. Early on, researchers found that apart from maternal kinship, affiliative relationships as exhibited through grooming patterns also occurred among unrelated cercopithecine females (e.g., Bramblett 1970, Rhine 1972, Sade 1972, Kummer 1975, Seyfarth 1976). Hence, although maternal kinship is a strong influence in structuring social relationships among cercopthecine females, other organizing principles must structure affiliative relations among unrelated females. Seyfarth (1977) proposed an influential model of rank-based attractiveness to explain the structuring of affiliative grooming, specifically among closely ranked unrelated females. However, since closeness in rank also correlates positively with closeness in kinship, it was difficult to assess whether a bias in affiliative behavior toward a close-ranking individual actually reflected a preference for a closely ranked or closely related partner.

Distinguishing between the roles of maternal kinship and rank distance has been difficult not only because the two variables tend to be related, but also because appropriate methods for analyzing social interaction matrices have not been available. Earlier analyses generally used parametric regression/correlation tests. These tests make the assumption that the data points are independent of each other. However, in many ethological studies, the data points tend to be interdependent. Hence, the use of these parametric tests is inappropriate; the P values obtained using these tests overestimate the relationship between the dependent and independent variables. Only in the past decade or so have appropriate statistical methods for analyzing social interaction matrices became available (along with corresponding software, e.g., MATSQUAR & Matman: Hemelrijk 1990a, de Vries et al. 1993). These new matrix association/correlation methods (e.g., the Mantel, Kr, and Zr tests) are increasingly being used for behavioral data that are not independent (e.g., Schnell et al. 1985; de Waal & Luttrell 1986; Smouse et al. 1986; Hemelrijk 1990a, b; de Waal 1991; de Vries 1993; Sambrook et al. 1995; Silk et al. 1999; Watts 2000a, b; Mitani et al. 2002) since they do not overestimate the true significance of the relationship between dependent and independent variables. Specifically, matrix permutation methods use observed values to generate a sampling distribution against which to assess the significance of a sample correlation statistic (Dow et al. 1987, Hemelrijk 1990a, de Vries et al. 1993).

Since the problem of teasing apart the respective roles of maternal kinship and rank distance on behavior has been resolved, rank has been found to independently structure affiliative behavior in a number of studies (e.g., Chapais et al. 1991, 1994; de Waal 1991; Sambrook et al. 1995; Kapsalis & Berman 1996a, b; Chapais & St. Pierre 1997; Silk et al. 1999). For example, research on captive rhesus reveals that independent of maternal kinship, rank distance is positively correlated with proximity and cofeeding (de Waal 1991). Experimental research on Japanese macaques indicates that kin-based alliances are the basis of a female's initial rank; however, rank-based alliances could later dominate as the major criterion for alliance formation (Chapais et al. 1991, 1994; Chapais & St. Pierre 1997). Kapsalis and Berman (1996a, b) found independent rank effects for affiliative behaviors for three out of four years, and theorized that this variation might be due to changes of group composition of kin and nonkin individuals. Similarly, for wild baboons, strong kin effects but the lack of rank effects in grooming distributions in wild female savanna baboons has been attributed to group size (Sambrook et al. 1995; Silk et al. 1999). These studies reveal that maternal kinship consistently remains a strong correlate of affiliative relationships among cercopithecines; however, changes in demography seem to strongly influence rank-related behavior of primates (e.g., de Waal 1976, Altmann & Altmann 1979, Dunbar 1987, Berman 1988; see chapter 16).

The Impact of Matrilineal Kinship on Noncercopithecine Primates

Capuchins

Since the late 1980s, different species of *Cebus* have been studied longitudinally by a number of researchers (e.g., de Ruiter 1986; O'Brien 1988, 1990, 1991, 1993a, b; Rose 1988, 1994; O'Brien & Robinson 1991, 1993; Fedigan 1993; Fedigan & Rose 1995; Rose & Fedigan 1995; Fedigan et al. 1996; Perry 1996a, b, 1997; Manson 1999; Manson et al. 1999). These species are of particular interest for primate kinship studies since capuchin social structure seems similar to cercopithecine species in several salient features. For example, these species are classified as generally female philopatric (e.g., O'Brien 1990, Moore 1993, Isbell & Van Vuren 1996), although female transfer does occur more commonly than for cercopithecine species (Manson et al. 1999). Both sexes are highly social, and grooming interactions and alliance formation occur commonly among females (Rose 1988, 1994; O'Brien 1991, 1993a, b; O'Brien & Robinson 1991, 1993; Fedigan 1993; Perry 1996a, b, 1997; Manson 1999; Manson et al. 1999). Female dominance hierarchies are found for two wild *Cebus* species, wedged-capped (*Cebus olivaceus*) and white-faced capuchins (*Cebus capucinus*) (O'Brien & Robinson 1991; O'Brien 1993a, b; Perry 1996a; Manson et al. 1999).

Unfortunately, our knowledge of the role of maternal kinship in these wild populations of *Cebus* is still very limited. At this writing, adult and juvenile maternal kin relationships were known for only three groups of wedge-capped capuchins (O'Brien 1993a, b; O'Brien & Robinson 1993). For white-faced capuchin groups at two different sites, maternal kin relationships were known for only three adult dyads. Another difficulty is the vast differences in group size and composition between most cercopithecine social groups and the *Cebus* currently under study. *Cebus* groups tend be small relative to many cercopithecine social groups (Dunbar 1988, Chapais 1995) with the exception of some vervet social groups (Isbell & Young 1993, Isbell & Pruetz 1998), with the average number of adult females being about 7.5 for wedge-capped capuchins and 5.2 for white-faced capuchins. Cercopithecine kinship data have often come from much larger provisioned social groups with many more female residents. Hence, current and future comparisons between cercopithecine females and these platyrrhine females must take this issue into consideration (Manson et al. 1999). Overall, less than 75 adult females have been studied longitudinally in the field for the most studied *Cebus* species. Given the small numbers of capuchins overall that are currently under study, it is still very early to make any kind of assessment regarding the impact of maternal kinship on behavior, although some intriguing suggestive preliminary data have emerged.

O'Brien and Robinson's (O'Brien 1988, 1990, 1991, 1993a, b; O'Brien & Robinson 1991, 1993) research on two social groups of wedge-capped capuchins have found female dominance hierarchies that in part are based on matrilineal social inheritance. These dominance hierarchies are similar to those of cercopithecines in being stable over years; however, social inheritance of dominance rank seems to be restricted to the highest ranking juveniles (O'Brien 1993a, b; O'Brien & Robinson 1993). They have also found positive correlations between grooming and coalition formation (O'Brien 1993a, b; O'Brien & Robinson 1993) typical of cercopithecine females; however, grooming relationships are not necessarily strongest among mother-daughter dyads, as one might expect. Instead, adjacently ranked unrelated females are more likely to associate with each other.

For white-faced capuchins at two different sites, female dominance hierarchies exist as well; however wide variations in stability occur at different sites, with some groups showing years of stability while others exhibit intense dominance struggles among females (Manson et al. 1999). Few kin relationships among adult females are known for this species. However, the current data show that at one site, mother-daughter pairs are closely ranked and groom frequently, while at another the one known related adult female-female dyad is distantly ranked, and one partner intervened at times against her relative during juvenile conflicts (Perry unpublished data, cited in Manson et al. 1999).

Additionally, adult female white-faced capuchins exhibit extreme variation between sites in patterns of grooming and coalition formation. At Lomas Barbudal, female-female dyads that groomed more also formed more coalitions against other females, and more closely ranked dyads groomed at higher rates (Perry 1996a, Manson et al. 1999). In contrast, at the Santa Rosa site, there was no tendency for closely ranked females to groom each other more than distantly ranked females (Manson et al. 1999).

These data thus far suggest that maternal kinship plays a weaker role in structuring social relationships among *Cebus* females than it does for the corporate cercopithecine model. One reason for this may be that while *Cebus* are generally female philopatric, some female transfer between groups does occur, at least for white-faced capuchins. Manson et al. (1999) found that female immigration into white-faced capuchin social groups was more common than previously believed, and hence average maternal genetic relatedness within groups was probably lower for white-faced capuchin females than for cercopithecine females. With an expanded knowledge of kin relationships in white-faced capuchins, more longitudinal research, and more *Cebus* groups studied, more detailed analyses will be available that will clarify the similarities and differences between platyrrhine and cercopithecine female philopatric species.

Mountain Gorillas

Most data regarding the impact of maternal kinship on gorillas come from the Virungas mountain gorilla populations that have been studied since 1967 at the Karisoke Research Centre, Rwanda. Both males and females disperse from their natal groups (Stewart & Harcourt 1987); however, there is a great deal of variation for females. Some females reproduce in their natal groups, others transfer to groups that contain kin or transfer with kin, and some transfer to groups with no kin at all (Harcourt 1978, Stewart & Harcourt 1987, Watts 1990). Both natal and secondary transfer is common for females, and, while the data are limited, it appears that sisters may transfer together or join groups that contain females from their original natal group (Stewart & Harcourt 1987, Watts 1996). Hence, in spite of female dispersal, females typically spend more than 70% of their reproductive years with female relatives (Watts 1996). Male dispersal from natal groups may be less common than for females (Robbins 1995), although the data are limited at present (Watts 1996).

The strongest adult social relationships within mountain gorilla groups occur between unrelated females and males, particularly for females who have no female kin within the social groups. Most adult females spend more of their time associating with males than with other females (Harcourt 1979; Stewart & Harcourt 1987; Watts 1992, 1994a). However, when related adult gorilla females do live together, maternal kinship structures social relationships for females in that social group. Affiliative relationships in terms of proximity, grooming, and alliance formation occur mostly among female kin and rarely among unrelated females (Watts 1985, 1991, 1994a, 1996; Harcourt & Stewart 1987, 1989; Stewart & Harcourt 1987).

However, mountain gorilla females differ from the cercopithecine model in that they do not usually form linear dominance hierarchies and there is no social inheritance of rank (Watts 1994b, 1996). Most aggressive interactions between females are bidirectional and undecided. Moreover, maternal support of daughters does not consistently ensure daughters' success in aggressive interactions. Further, adult males frequently intervene to end conflicts between females (Harcourt 1979, Watts 1991). Hence agonistic relationships are generally unresolved, leading to egalitarian female agonistic relationships (Watts 1994a). This pattern of aggression between females along with male intervention in female disputes may be responsible for the lack of nepotistic dominance hierarchies for females (Harcourt & Stewart 1987, 1989). An ultimate explanation for the difference between cercopithecine and mountain gorilla female dominance relationships may be found in terms of feeding ecology. Watts (1996) suggests that contest competition for food and cooperation between female relatives has minor nutritional importance for mountain female gorillas in comparison to cercopithecines. Hence, linear dominance hierarchies are not crucial for female mountain gorillas' reproductive success.

There are few affiliative interactions among adult mountain gorilla males. This is because the majority of mountain gorilla groups only contain one adult male (Harcourt et al. 1981, Weber & Vedder 1983) and adult males typically become increasingly intolerant of maturing males (Harcourt 1979, Watts 1992, Robbins 1996). Those adult males within multimale groups typically have low rates of affiliative interaction, and although the data are limited, there is some evidence that paternal kinship rather than maternal kinship is more influential in structuring male-male social relationships. That is, older males may be more tolerant of sons and full brothers than of more distant kin (Harcourt 1979; Harcourt & Stewart 1981, 1987; Yamagiwa 1986; Watts & Pusey 1993; Sicotte 1994; Robbins 1995, 1996). Most coalitionary support between male dyads against extragroup males probably involves father-son pairs (Yamagiwa 1986; Stewart & Harcourt 1987). Maternal kinship probably plays a limited role in structuring male-male relationships since (1) male dispersal occurs in gorillas and (2) long interbirth intervals limit the number of maternally related adult males within the same group (Robbins 1996).

Chimpanzees

Chimpanzee social structure differs greatly from the cercopithecine model in its fluidity. These great apes live in communities, and individuals often form temporary associations called parties that vary in size and composition (Nishida 1968, Goodall 1986, Boesch & Boesch-Achermann 2000). Unlike cercopithecine monkeys, chimpanzee females usually leave their natal groups upon maturity while males stay in their natal groups for life (Nishida & Hosaka 1996), although some adult females at Gombe continue to associate with natal members and maintain social relationships with subadult and adult daughters (Goodall 1986). Genetic analyses verify that male chimpanzees within social groups are more closely related maternally to each other than females (Morin et al. 1994, Goldberg & Wrangham 1997, Mitani et al. 2000, Vigilant et al. 2001). For example, at Gombe, the average degree of relatedness among males is very high at $r = .25$ (Morin et al. 1994).

Given this arrangement, it has long been assumed that maternal kinship would play a strong role in patterns of affiliation among male chimpanzees much in the way it would for female cercopithecines. In fact, strong social relationships between male chimpanzees are the norm, as observed through patterns of association, proximity, grooming, coalitions, meat sharing, and territorial behavior (Nishida 1968, 1983; Simpson 1973; Riss & Goodall 1977; Goodall et al. 1979; Wrangham & Smuts 1980; Nishida et al. 1992; Nishida & Hosaka 1996; Watts 1998, 2000a, b; Newton-Fisher 1999; Wrangham 1999; Boesch & Boesch-Achermann 2000; Mitani & Watts 2001; Watts & Mitani 2001). For example, coalitionary support by presumed maternal brothers at Gombe against higher ranking individuals allowed several males to elevate their rank status (Goodall 1986). In contrast, chimpanzee females generally maintain no permanent associations beyond their own offspring.

However, recent mtDNA haplotype sharing studies of one community of chimpanzees in Ngobo, Uganda, reveal that maternal kinship is not strongly associated with male-male patterns of association, proximity, grooming, alliances, meat sharing, and boundary patrols. Instead of maternal kin, males preferentially formed coalitions, shared meat, and patrolled territorial boundaries with members of their same age cohort and rank status (Goldberg & Wrangham 1997, Mitani et al. 2000, Mitani et al. 2002). One possible explanation is that demographic constraints (i.e., long interbirth intervals) limit the number of adult brothers available to effectively assist each other to improve their social status within communities. On the other hand, familiar unrelated males belonging to the same age cohort and rank status may be more inclined to establish effective relationships insofar as they share similar needs, access to resources, and ability to exchange social benefits (Mitani et al. 2002).

Bonobos

Male bonobos also remain in their natal communities while female transfer is common (Kano 1982, Furuichi 1989, Hohmann et al. 1999, Hashimoto et al. 1996). Behavioral studies indicate and molecular analyses of the genetic relatedness between community members validate that (1) the highest association rates occur between adult females and their adult sons (Hasegawa 1987, Furuichi 1989, Ihobe 1992, Kano 1992, Furuichi & Ihobe 1994, Furuichi 1997, Hohmann et al. 1999), and (2) the longest term associations are found exclusively among mother-son dyads (Hohmann et al. 1999). Additionally, a mother's presence in the social unit-group seems to be influential in her son's gaining high dominance rank

(Hasegawa 1987; Fuiriuchi 1989; Ihobe 1992; Kano 1992, 1996; Muroyama & Sugiyama 1994; White 1996). High association rates between maternally related adult brothers have also been confirmed through genetic analyses (Hohmann et al. 1999).

On the other hand, genetic studies indicate that bonobo females do not selectively transfer into social units that contain their sisters (Hashimoto et al. 1996), nor choose matrilineally related females they find in the new group as association partners (Hashimoto et al. 1996, Hohmann et al. 1999). While bonobo females do form close associations with each other within communities, these associations are usually of much shorter duration than adult mother-adult son associations and are not based on kinship (Furuichi 1987, White 1988, Kano 1992, Hashimoto et al. 1996, Parish 1996, Hohmann et al. 1999).

Conclusion

It is difficult to overstate the impact of maternal kinship on cercopithecine females. Maternal kinship systemically structures both affiliative and agonistic interactions within social groups to form strong affiliative relationships among related females and the corporate matrilineal dominance hierarchies typical of the cercopithecine model. In the last decade, researchers have begun to examine the influence of maternal kinship on cercopithecines in a more molecular manner. These studies reveal that primates may discriminate between different categories of maternal kin in their distribution of nepotistic behaviors and that there may be a "kinship threshold" beyond which individuals treat related animals as they would non-kin. More in-depth research is needed to evaluate if not only the degree of relatedness but also the type of relatedness (direct or collateral) changes the quality of kin relationships.

Relatively few other noncercopithecine species have been studied intensively enough to make any conclusive statements regarding the impact of maternal kinship overall for nonhuman primates. However, thus far, maternal kinship seems to play a far smaller role in structuring social relationships for noncercopithecine than for Old World monkeys. Among the non-cercopithecine species, capuchin monkeys promise to be the best comparative primate species for the cercopithecine model since they are generally female philopatric, have differentiated social relationships, and have some form of female dominance hierarchies. However, larger sample sizes and a more complete knowledge of *Cebus* maternal relationships are needed before true comparative analyses can be conducted.

With mtDNA genetic analyses, recent research has reevaluated the assumption that maternal kinship is a strong influence structuring male-male relationships in chimpanzees and bonobos. While these males do stay in their natal communities, maternal kinship plays a relatively weak role in structuring their interactions with each other. It appears that demographic constraints (i.e., the lack of available age-appropriate male relatives) restrict the role maternal kinship plays among these primates.

Overall, this review reveals that our knowledge regarding the influence of maternal kinship on nonhuman primates is still very biased and limited. That is, we know a lot about a relatively small number of species (i.e., Old World monkeys). However, we do have the advantage of having a very rich body of knowledge, research protocols, and hypotheses from studying cercopithecines as a basis from which to study the effect of kinship on the rest of the primate order. Moreover, genetic analyses are now available to quickly ascertain maternal relationships within an entire social group or population in an understudied species.

So, within the next decade, I believe we can anticipate a truer representation of the existing primate order in terms of kinship studies.

References

Alberts, S. C. 1999. Paternal kin discrimination in wild baboons. *Proc. Royal Soc. Lond., B*, 266, 1501–1506.

Ali, R. 1981. The ecology and social behavior of the Agastyamali bonnet macaque (*Macaca radiata diluta*). PhD dissertation, University of Bristol, Bristol, UK.

Altmann, J. 1979. Altruistic behaviour: the fallacy of kin deployment. *Anim. Behav.*, 27, 958–962.

Altmann, S. A. & Altmann, J. 1979. Demographic constraints on behavior and social organization. In: *Primate Ecology and Human Origins: Ecological Influences on Social Organization* (Ed. by I. S. Bernstein & E. O. Smith), pp. 47–63. New York: Garland STPM Press.

Aureli, F. & de Waal, F. B. M. 2000. *Natural Conflict Resolution*. Berkeley, CA: University of California Press.

Bélisle, P. & Chapais, B. 2001. Tolerated co-feeding in relation to degree of kinship in Japanese macaques. *Behaviour*, 138, 487–509.

Berard, J. 1989. Life histories of male Cayo Santiago macaques. *Puerto Rico Health Sci. J.*, 8, 61–64.

Berard, J. 1999. A four-year study of the association between male dominance rank, residency status, and reproductive activity in rhesus macaques (*Macaca mulatta*). *Primates*, 40, 159–175.

Berman, C. M. 1980. Early agonistic experience and rank acquisition among free-ranging infant rhesus monkeys. *Int. J. Primatol.*, 1, 153–170.

Berman, C. M. 1982. The ontogeny of social relationships with group companions among free-ranging infant rhesus monkeys. I. Social networks and differentiation. *Anim. Behav.*, 30, 149–162.

Berman, C. M. 1983. Early differences in relationships between infants and other group members based on mothers' status: their possible relationship to peer-peer rank acquisition. In: *Primate Social Relationships: An Integrated Approach* (Ed. by R. A. Hinde), pp. 154–156. Sunderland, MA: Sinauer Associates.

Berman, C. M. 1988. Demography and mother-infant relationships: implications for group structure. In: *Ecology and Behavior of Food-Enhanced Primate Groups* (Ed. by J. E. Fa & C. H. Southwick), pp. 269–296. New York: Alan R. Liss.

Bernstein, I. S. 1991. The correlation between kinship and behaviour in non-human primates. In: *Kin Recognition* (Ed. by P. G. Hepper), pp. 6–29. Cambridge: Cambridge University Press.

Bernstein, I. S. & Ehardt, C. L. 1985. Agonistic aiding: kinship, rank, age, and sex influences . *Am. J. Primatol.*, 8, 37–52.

Bernstein, I. S. & Ehardt, C. L. 1986. The influence of kinship and socialization on aggressive behaviour in rhesus monkeys (*Macaca mulatta*). *Anim. Behav.*, 34, 739–747.

Bernstein, I. S., Judge, P. G., & Ruehlmann, T. E. 1993. Kinship, association, and social relationships in rhesus monkeys (*Macaca mulatta*). *Am. J. Primatol.*, 31, 41–53.

Boesch, C. & Boesch-Achermann, H. 2000. *The Chimpanzees of the Tai Forest: Behavioral Ecology and Evolution*. New York: Oxford University Press.

Bramblett, C. A. 1970. Coalitions among gelada baboons. *Primates*, 11, 327–333.

Bramblett, C. A., Bramblett, S. S., Bishop, D. A., & Coelho, A. M. Jr. 1982. Longitudinal stability in adult status hierarchies among vervet monkeys (*Cercopithecus aethiops*). *Am. J. Primatol.*, 2, 43–51.

Butovskaya, M. 1993. Kinship and different dominance styles in groups of three species of the genus *Macaca* (*M. arctoides, M. mulatta, M. fascicularis*). *Folia Primatol.*, 60, 210–224.

Butovskaya, M., Kozintsev, A., & Welker, W. 1996. Conflict and reconciliation in two groups of crab-eating monkeys differing in social status by birth. *Primates*, 37, 261–270.

Call, J., Judge, P. G., & de Waal, F. B. M. 1996. Influence of kinship and spatial density on reconciliation and grooming in rhesus monkeys. *Am. J. Primatol.*, 39, 35–45.

Castles, J., Aureli, F., & de Waal, F. B. M. 1996. Variation in conciliation tendency and relationship quality across groups of pigtail macaques. *Anim. Behav.*, 52, 389–403.

Chaffin, C. L., Friedlen, K., & de Waal, F. B. M. 1995. Dominance style of Japanese macaques compared with rhesus and stumptail macaques. *Am. J. Primatol.*, 35, 103–116.

Chapais, B. 1988. Rank maintenance in female Japanese macaques: experimental evidence for social dependency. *Behaviour*, 104, 41–59.

Chapais, B. 1992. The role of alliances in social inheritance of rank among female primates. In: *Coalitions and Alliances in Humans and Other Animals* (Ed. by A. H. Harcourt & F. B. M. de Waal), pp. 29–59. New York: Oxford University Press.

Chapais, B. 1995. Alliances as a means of competition in primates: evolutionary, developmental, and cognitive aspects. *Ybk. Phys. Anthropol.*, 28, 115–136.

Chapais, B. 2001. Primate nepotism: what is the explanatory value of kin selection? *Int. J. Primatol.*, 22, 203–229.

Chapais, B. & Gauthier, C. 1993. Early agonistic experience and the onset of matrilineal rank acquisition in Japanese macaques. In: *Juvenile Primates: Life History, Development, and Behavior* (Ed. by M. E. Pereira & L. A. Fairbanks), pp. 245–258. New York: Oxford University Press.

Chapais, B., Gauthier, C., Prud'homme, J., & Vasey, P. 1997. Relatedness threshold for nepotism in Japanese macaques. *Anim. Behav.*, 53, 1089–1101.

Chapais, B., Girard, M., & Primi, G. 1991. Non-kin alliances, and the stability of matrilineal dominance relations in Japanese macaques. *Anim. Behav.*, 41, 481–491.

Chapais, B., Prud'homme, J., & Teijeiro, S. 1994. Dominance competition among siblings in Japanese macaques: constraints on nepotism. *Anim. Behav.*, 48, 1335–1347.

Chapais, B., Savard, L., & Gauthier, C. 2001. Kin selection and the distribution of altruism in relation to degree of kinship in Japanese macaques (*Macaca fuscata*). *Behav. Ecol. Sociobiol.*, 49, 493–502.

Chapais, B. & St. Pierre, C.-E. G. 1997. Kinship bonds are not necessary for maintaining matrilineal rank in captive Japanese macaques. *Int. J. Primatol.*, 18, 375–385.

Cheney, D. L. 1977. The acquisition of rank and the development of reciprocal alliances among free-ranging immature baboons. *Behav. Ecol. Sociobiol.*, 2, 303–318.

Cheney, D. L. & Seyfarth, R. M. 1980. Vocal recognition in free-ranging vervet monkeys. *Anim. Behav.*, 28, 362–367.

Cheney, D. L. & Seyfarth, R. M. 1982. Recognition of individuals within and between groups of free-ranging vervet monkeys. *Am. Nat.*, 22, 519–529.

Cheney, D. L. & Seyfarth, R. M. 1983. Nonrandom dispersal in free-ranging vervet monkeys: social and genetic consequences. *Am. Nat.*, 122, 392–412.

Cheney, D. L. & Seyfarth, R. M. 1986. The recognition of social alliances by vervet monkeys. *Anim. Behav.*, 34, 1722–1731.

Cheney, D. L. & Seyfarth, R. M. 1989. Redirected aggression and reconciliation among vervet monkeys, *Cercopithecus aethiops. Behaviour*, 110, 258–275.

Cheney, D. L. & Seyfarth, R.M. 1990. *How Monkeys See the World: Inside the Mind of Another Species*. Chicago: University of Chicago Press.

Chepko-Sade, B. D. & Olivier, T. J. 1979. Coefficient of genetic relationship and the probability of intragenealogical fission in *Macaca mulatta*. *Behav. Ecol. Sociobiol.*, 5, 263–278.

Chepko-Sade, B. D. & Sade, D. S. 1979. Patterns of group splitting within matrilineal kinship groups: a study of social group structure in *Macaca mulatta* (Cercopithecidae: Primates). *Behav. Ecol. Sociobiol.*, 5, 67–86.

Das, M., Penke, Z., & van Hooff, J. A. R. A. M. 1997. Affiliation between aggressors and third parties following conflicts in long-tailed macaques (*Macaca fascicularis*). *Int. J. Primatol.*, 18, 159–181.

Dasser, V. 1988. Mapping social concepts in monkeys. In: *Machiavellian Intelligence* (Ed. by R. W. Byrne & A. Whiten), pp. 85–93. Oxford: Clarendon Press.

Dasser, V. 1988. A social concept in Java monkeys. *Anim. Behav.*, 36, 225–230.

Datta, S. B. 1983. Relative power and the acquisition of rank. In: *Primate Social Relationships: An Integrated Approach* (Ed. by R. A. Hinde), pp. 93–103. Sunderland, MA: Sinauer Associates.

Datta, S. B. 1988. The acquisition of dominance among free-ranging rhesus monkey siblings. Anim. Behav., 36, 754–772.

Demaria, C. & Thierry, B. 1990. Formal biting in stumptailed macaques (*Macaca arctoides*). *Am. J. Primatol.*, 20, 133–140.

Demaria, C. & Thierry, B. 2001. A comparative study of reconciliation in rhesus and Tonkean macaques. *Behaviour*, 138, 397–410.

de Ruiter, J. R. 1986. The influence of group size on predator scanning and foraging behaviour of wedgecapped capuchin monkeys (*Cebus olivaceus*). *Behaviour*, 98, 240–258.

de Vries, H. 1993. The rowwise correlation between two proximity matrices and the partial rowwise correlation. *Psychometrika*, 58, 53–69.

de Vries, H., Netto, W. J., & Hanegraaf, P. L. H. 1993. Matman: a program for the analysis of sociometric matrices and behavioural transition matrices. *Behaviour*, 125, 157–175.

de Waal, F. B. M. 1977. The organization of agonistic relations within two captive groups of Java monkeys (*Macaca fascicularis*). *Z. Tierpsychol.*, 44, 225–282.

de Waal, F. B. M. 1991. Rank distance as a central feature of rhesus monkey organization: a sociometric analysis. *Anim. Behav.*, 41, 383–395.

de Waal, F. B. M. & Luttrell, L. M. 1986. The similarity principle underlying social bonding among female rhesus monkeys. *Folia Primatol.*, 46, 215–234.

de Waal, F. B. M. & Ren, R.M. 1988. Comparison of the reconciliation behavior of stumptail and rhesus macaques. *Ethology*, 78, 129–142.

Dittus, W. P. J. 1988. Group fission among wild toque macaques as a consequence of female resource competition and environmental stress. *Anim. Behav.*, 36, 1626–1645.

Dow, M. M., Cheverud, J. M., & Friedlaender, J. S. 1987. Partial correlation of distance matrices in studies of population structure. *Am. J. Primatol.*, 72, 343–352.

Dunbar, R. I. M. 1987. Demography and reproduction. In: *Primate Societies* (Ed. by B. B. Smuts, D. L. Cheney, R. M. Seyfarth, R. W. Wrangham, & T. T. Struhsaker), pp. 240–249. Chicago: University of Chicago Press.

Dunbar, R. I. M. 1988. *Primate Social Systems*. Ithaca, NY: Cornell University Press.

Dunbar, R. I. M. 1991. Functional significance of social grooming in primates. *Folia Primatol.*, 57, 121–131.

Dunbar, R. I. M. 1992. Time: a hidden constraint on the behavioural ecology of baboons. *Behav. Ecol. Sociobiol.*, 31, 35–49.

Dunbar, R. I. M. & Dunbar, E. P. 1977. Dominance and reproductive success among female gelada baboons. *Nature*, 266, 351–352.

Ehardt, C. L. & Bernstein, I. S. 1986. Matrilineal overthrows in rhesus monkey groups. *Int. J. Primatol.*, 7, 157–181.

Fairbanks, L. A. 2000. Maternal investment throughout the life span in old world monkeys. In: *Old World Monkeys* (Ed. by P. F Whitehead & C. J. Jolly), pp. 341–367. Cambridge: Cambridge University Press.

Fedigan, L. 1993. Sex differences and intersexual relations in adult white-faced capuchins (*Cebus capucinus*). *Int. J. Primatol.*, 14, 853–877.

Fedigan, L. M. & Rose, L. M. 1995. Interbirth interval variation in three sympatric species of neotropical monkey. *Am. J. Primatol.*, 37, 9–24.

Fedigan, L. M., Rose, L. M., & Avila, R. M. 1996. See how they grow: tracking capuchin monkey (*Cebus capucinus*) populations in a regenerating Costa Rican dry forest. In: *Adaptive Radiations of Neotropical Primates* (Ed. by M. A. Norconk, A. L. Rosenberger, & P. A. Garber), pp. 289–307. New York: Plenum Press.

Fredrickson, W. T. & Sackett, G. P. 1984. Kin preferences in primates (*Macaca nemestrina*): relatedness or familarity? *J. Comp. Psychol.*, 98, 29–34.

Furuichi, T. 1987. Sexual swelling, receptivity, and grouping of wild pygmy chimpanzee females at Wamba, Zaire. *Primates*, 28, 309–318.

Furuichi, T. 1989. Social interactions and the life history of female *Pan paniscus* in Wamba, Zaire. *Int. J. Primatol.*, 10, 173–197.

Furuichi, T. 1997. Agonistic interactions and matrifocal dominance rank of wild bonobos (*Pan paniscus*) at Wamba. *Int. J. Primatol.*, 18, 855–875.

Furuichi, T. & Ihobe, H. 1994. Variation in male relationships in bonobos and chimpanzees. *Behaviour*, 130, 211–228.

Furuya, Y. 1969. On the fission of troops of Japanese monkeys. II. General view of troop fission of Japanese monkeys. *Primates*, 10, 47–69.

Glick, B. B., Eaton, G. G., Johnson, D. F., & Worlein, J. 1986. Social behavior of infant and mother Japanese macaques (*Macaca fuscata*): effects of kinship, partner sex, and infant sex. *Int. J. Primatol.*, 7, 139–155.

Goldberg, T. L. & Wrangham, R. W. 1997. Genetic correlates of social behaviour in wild chimpanzees: evidence from mitochondrial DNA. *Anim. Behav.*, 54, 559–570.

Goodall, J. 1986. *The Chimpanzees of Gombe: Patterns of Behavior*. Cambridge, MA: Harvard University Press.

Goodall, J., Bandora, A., Bergmann, E., Busse, C., Matama, H., Mpongo, E., Pierce, A., & Riss, D. 1979. Intercommunity interactions in the chimpanzee population of the Gombe National Park. In: *Perspectives on Human Evolution. Vol. 5: The Great Apes* (Ed. by D. A. Hamburg & E. R. McCown), pp. 13–53. Menlo Park, CA: Benjamin/Cummings.

Gouzoules, S. 1984. Primate mating systems, kin associations, and cooperative behavior: evidence for kin recognition? *Ybk. Phys. Anthropol.*, 27, 99–134.

Gouzoules, S. & Gouzoules, H. 1987. Kinship. In: *Primate Socieites* (Ed. by B. B. Smuts, D. L. Cheney, R. M. Seyfarth, R. W. Wrangham, & T. T. Struhsaker), pp. 299–305. Chicago: University of Chicago Press.

Gouzoules, S., Gouzoules, H., & Marler, P. 1984. Rhesus monkey (*Macaca mulatta*) screams: representational signalling in the recruitment of agonistic aid. *Anim. Behav.*, 32, 182–193.

Harcourt, A. H. 1978. Strategies of emigration and transfer by primates, with particular reference to gorillas. *Z. Tierpsychol.*, 48, 401–420.

Harcourt, A. H. 1979. Social relationships between adult male and female mountain gorillas in the wild. *Anim. Behav.*, 27, 325–342.

Harcourt, A. H. & Stewart, K. J. 1981. Gorilla male relationships: can differences during immaturity lead to contrasting reproductive tactics in adulthood? *Anim. Behav.*, 29, 206–210.

Harcourt, A. H. & Stewart, K. J. 1987. The influence of help in contests on dominance rank in primates: hints from gorillas. *Anim. Behav.*, 35, 182–190.

Harcourt, A. H & Stewart, K. J. 1989. Functions of alliances in contests within wild gorilla groups. *Behaviour*, 109, 176–190.

Harcourt, A. H., Stewart, K. J., & Fossey, D. 1981. Gorilla reproduction in the wild. In: *Reproductive Biology of the Great Apes: Comparative and Biomedical Perspectives* (Ed. by C. E. Graham), pp. 265–279. New York: Academic Press.

Hasegawa, T. 1987. Copulatory behavior of wild chimpanzees. *Reichorui Kenkyu*, 3, 156.

Hashimoto, C., Furuichi, T., & Takenaka, O. 1996. Matrilineal kin relationship and social behavior of wild bonobos (*Pan paniscus*): sequencing the D-loop region of mitochondrial DNA. *Primates*, 37, 305–318.

Hausfater, G. 1975. Dominance and reproduction in baboons (*Papio cynocephalus*). *Contrib. Primatol.*, 7, 1–150.

Hausfater, G., Altmann, J., & Altmann, S. 1982. Long-term consistency of dominance relations among female baboons (*Papio cynocephalus*). *Science*, 217, 752–755.

Hemelrijk, C. K. 1990a. A matrix partial correlation test used in investigations of reciprocity and other social interaction patterns at group level. *J. Theor. Biol.*, 143, 405–420.

Hemelrijk, C. K. 1990b. Models of, and tests for, reciprocity, unidirectionality and other social interaction patterns at a group level. *Anim. Behav.*, 39, 1013–1029.

Henzi, S. P., Lycett, J. E., & Piper, S. E. 1997. Fission and troop size in a mountain baboon population. *Anim. Behav.*, 53, 525–535.

Henzi, S. P., Lycett, J. E., & Weingrill, T. 1997. Cohort size and the allocation of social effort by female mountain baboons. *Anim. Behav.*, 54, 1235–1243.

Hill, D. A. & Okayasu, N. 1995. Absence of "youngest ascendancy" in the dominance relations of sisters in wild Japanese macaques (*Macaca fuscata yakui*). *Behaviour*, 132, 267–279.

Hill, D. A. & Okayasu, N. 1996. Determinants of dominance among female macaques. In: *Evolution and Ecology of Macaques Societies* (Ed. by J. E. Fa & D. G. Lindburg), pp. 265–279. Cambridge: Cambridge University Press.

Hohmann, G., Gerloff, U., Tautz, D., & Fruth, B. 1999. Social bonds and genetic ties: kinship, association and affiliation in a community of bonobos (*Pan paniscus*). *Behaviour*, 136, 1219–1235.

Horrocks, J. & Hunte, W. 1983. Maternal rank and offspring rank in vervet monkeys: an appraisal of the mechanisms of rank acquisition. *Anim. Behav.*, 31, 772–782.

Hunte, W. & Horrocks, J. 1987. Kin and non-kin interventions in the aggressive disputes of vervet monkeys. *Behav. Ecol. Sociobiol.*, 20, 257–263.

Ihobe, H. 1992. Male-male relationships among wild bonobos (*Pan paniscus*) at Wamba, Republic of Zaire. *Primates*, 33, 163–179.

Isbell, L. A. & Pruetz, J. D. 1998. Differences between vervets (*Cercopithecus aethiops*) and patas monkeys (*Erythrocebus patas*) in agonistic interactions between adult females. *Int. J. Primatol.*, 19, 837–855.

Isbell, L. A. & Van Vuren, D. 1996. Differential costs of locational and social dispersal and their consequences for female group-living primates. *Behaviour*, 133, 1–36.

Isbell, L. A. & Young, T. P. 1993. Social and ecological influences on activity budgets of vervet monkeys, and their implications for group living. *Behav. Ecol. Sociobiol.*, 32, 377–385.

Judge, P.G. 1991. Dynamic and triadic reconciliation in pigtail macaques (*Macaca nemestrina*). *Am. J. Primatol.*, 23, 225–237.

Kano, T. 1982. The social group of pygmy chimpanzees (*Pan paniscus*) of Wamba. *Primates*, 23, 171–188.

Kano, T. 1992. *The Last Ape: Pygmy Chimpanzee Behavior and Ecology*. Stanford, CA: Stanford University Press.

Kano, T. 1996. Male rank order and copulation rate in a unit-group of bonobos at Wamba, Zaire. In: *Great Ape Societies* (Ed. by W. C. McGrew, L. F. Marchant, & T. Nishida), pp. 135–145. Cambridge: Cambridge University Press.

Kaplan, J. R. 1978. Fight interference and altruism in rhesus monkeys. *Am. J. Phys. Anthropol.*, 49, 241–249.

Kapsalis, E. & Berman, C. M. 1996a. Models of affiliative relationships among free-ranging rhesus monkeys (*Macaca mulatta*). I. Criteria for kinship. *Behaviour,* 133, 1209–1234.

Kapsalis, E. & Berman, C. M. 1996b. Models of affiliative relationships among free-ranging rhesus monkeys (*Macaca mulatta*). II. Testing predictions for three hypothesized organizing principles. *Behaviour*, 133, 1235–1263.

Kapsalis, E. & Johnson, R. L. 1999. Female homosexual behavior among free-ranging rhesus macaques (*Macaca mulatta*). *Am. J. Primatol.*, 49, 68–69.

Kawai, M. 1965. On the system of social ranks in a natural troop of Japanese monkeys. 1. Basic rank and dependent rank. In: *Japanese Monkeys* (Ed. by K. Imanishi & S. A. Altmann), pp. 66–86. Atlanta, GA: Emory University Press.

Kawamura, S. 1965. Matriarchial social ranks in the Minoo-B troop: a study of the rank system of Japanese monkeys. In: *Japanese Monkeys* (Ed. by K. Imanishi & S. A. Altmann), pp. 105–112. Atlanta, GA: Emory University Press.

Koford, C. B. 1963. Group relations in an island colony of rhesus monkeys. In: *Primate Social Behavior: An Enduring Problem. Selected Readings* (Ed. by C. H. Southwick), pp. 136–152. Princeton, NJ: D. Van Nostrand.

Koyama, N. 1967. On dominance rank and kinship of a wild Japanese monkey troop in Arashiyama. *Primates*, 8, 189–216.

Koyama, N. 1970. Changes in dominance rank and division of a wild Japanese monkey troop in Arashiyama. *Primates*, 11, 335–390.

Kuester, J., Paul, A., & Arnemann, J. 1994. Kinship, familiarity and mating avoidance in Barbary macaques, *Macaca sylvanus. Anim. Behav.*, 48, 1183–1194.

Kummer, H. 1975. Rules of dyad and group formation among captive gelada baboons (*Theropithecus gelada*). In: *Proceedings from the Symposia of the Fifth Congress of the International Primate Society* (Ed. by S. Kondo, M. Kawai, A. Ehara, & S. Kawamura), pp. 129–159. Tokyo: Japan Science Press.

Kurland, J. A. 1977. Kin selection in the Japanese monkey. *Contrib. Primatol.*, 12, 1–145.

Lee, P. C. 1983. Ecological influences on relationships and social structure. In: *Primate Social Relationships: An Integrated Approach* (Ed. by R. A. Hinde), pp. 225–229. Sunderland, MA: Sinauer Associates.

Lee, P. C. & Oliver, J. 1979. Competition, dominance and the acquisition of rank in juvenile yellow baboons (*Papio cynocephalus*). *Anim. Behav.*, 27, 576–585.

Lopez-Vergara, L., Santillan-Doherty, A. M., Mayagoita, L., & Mondragon-Ceballos, R. 1989. Self and social grooming in stump-tailed macaques: effects of kin presence or absence within the social group. *Behav. Proc.*, 18, 99–106.

Manson, J. 1999. Infant handling in wild *Cebus capucinus*: testing bonds between females? *Anim. Behav.*, 57, 911–921.

Manson, J. H., Rose, L. M., Perry, S., & Gros-Louis, J. 1999. Dynamics of female-female relationships in wild *Cebus capucinus*: data from two Costa Rican sites. *Int. J. Primatol.*, 20, 679–706.

Martin, D. A. 1997. Kinship bias: a function of familiarity in pigtailed macaques (*Macaca nemestrina*). PhD dissertation, University of Georgia, Athens.

Massey, A. 1977. Agonistic aids and kinship in a group of pigtail macaques. *Behav. Ecol. Sociobiol.*, 2, 31–40.

Massey, A. 1979. Reply to Kurland and Gaulin. *Behav. Ecol. Sociobiol.*, 8, 81–83.

Missikian, E. A. 1972. Genealogical and cross-genealogical dominance relationships in a group of free ranging rhesus monkeys (*Macaca mulatta*) on Cayo Santiago. *Primates*, 13, 169–180.

Mitani, J. C., Merriwether, D. A., & Zhang, C. B. 2000. Male affiliation, cooperation and kinship in wild chimpanzees. *Anim. Behav.*, 59, 885–893.

Mitani, J. C. & Watts, D. P. 2001. Why do chimpanzees hunt and share meat? *Anim. Behav.*, 61, 915–924.

Mitani, J. C., Watts, D. P., Pepper, J. W., & Merriwether, D. A. 2002. Demographic and social constraints on male chimpanzee behaviour. *Anim. Behav.*, 64, 727–737.

Moore, J. 1993. Inbreeding and outbreeding in primates: what's wrong with "the dispersing sex"? In: *The Natural History of Inbreeding and Outbreeding* (Ed. by N. W. Thornhill), pp. 392–426. Chicago: University of Chicago Press.

Mori, A. 1975. Intratroop spacing mechanism of the wild Japanese monkeys of the Koshima troop. In: *Contemporary Primatology* (Ed. by S. Kondo, M. Kawai, & A. Ehara), pp. 423–427. Basel: S. Karger.

Mori, A., Watanabe, K., & Yamaguchi, N. 1989. Longitudinal changes of dominance rank among the females of the Koshima group of Japanese monkeys. *Primates*, 147–173.

Morin, P. A., Wallis, J., Moore, J. J., & Woodruff, D. S. 1994. Paternity exclusion in a community of wild chimpanzees using hypervariable simple sequence repeats. *Mol. Ecol.*, 3, 469–477.

Muroyama, Y. & Sugiyama, Y. 1994. Grooming relationships in two species of chimpanzees. In: *Chimpanzee Cultures* (Ed. by R. W. Wrangham, W. C. McGrew, F. B. M. de Waal & P. G. Heltne), pp. 169–180. Cambridge, MA: Harvard University Press.

Nakamichi, M. & Yoshida, A. 1986. Discrimination of mother by infant among Japanese macaques (*Macaca fuscata*). *Int. J. Primatol.*, 7, 481–489.

Nash, L. T. 1976. Troop fission in free-ranging baboons in the Gombe Stream National Park, Tanzania. *Am. J. Phys. Anthropol.*, 44, 63–77.

Newton-Fisher, N. E. 1999. Termite eating and food sharing by male chimpanzees in the Budongo Forest, Uganda. *Afr. J. Ecol.*, 37, 369–371.

Nishida, T. 1968. The social group of wild chimpanzees in the Mahali mountains. *Primates*, 9, 167–224.

Nishida, T. 1983. Alpha status and agonistic alliance in wild chimpanzees (*Pan troglodytes schweinfurthii*). *Primates*, 24, 318–336.

Nishida, T., Hasegawa, T., Hayaki, H., Takahata, Y., & Uehara, S. 1992. Meat-sharing as a coalition strategy by an alpha male chimpanzee? In: *Topics in Primatology. Vol. 1: Human Origins* (Ed. by T. Nishida, W. C. McGrew, & P. Marler), pp. 159–174. Tokyo: University of Tokyo Press.

Nishida, T. & Hosaka, K. 1996. Coalition strategies among adult male chimpanzees of the Mahale mountains, Tanzania. In: *Great Ape Societies* (Ed. by W. C. McGrew, L. F. Marchant, & T. Nishida), pp. 114–134. Cambridge: Cambridge University Press.

O'Brien, T. G. 1988. Parasitic nursing behavior in the wedge-capped capuchin monkey (*Cebus olivaceus*). *Am. J. Primatol.*, 16, 341–344.

O'Brien, T. G. 1990. Determinants and consequences of social structure in a neotropical primate, *Cebus olivaceus.* PhD dissertation, University of Florida, Gainesville.

O'Brien, T. G. 1991. Female-male social interactions in wedge-capped capuchin monkeys: benefits and costs of group living. *Anim. Behav.,* 41, 555–567.

O'Brien, T. G. 1993a. Allogrooming behaviour among adult female wedge-capped capuchin monkeys. *Anim. Behav.,* 46, 499–510.

O'Brien, T. G. 1993b. Asymmetries in grooming interactions between juvenile and adult female wedge-capped capuchin monkeys. *Anim. Behav.,* 46, 929–938.

O'Brien, T. G. & Robinson, J. G. 1991. Allomaternal care by female wedge-capped capuchin monkeys: effects of age, rank and relatedness. *Behaviour,* 119, 30–50.

O'Brien, T. G. & Robinson, J. G. 1993. Stability of social relationships in female wedge-capped capuchin monkeys. In: *Juvenile Primates: Life History, Development, and Behavior* (Ed. by M. E. Pereira & L. A. Fairbanks), pp. 197–210. New York: Oxford University Press.

Oki, J. & Maeda, Y. 1973. Grooming as a regulator of behavior in Japanese macaques. In: *Behavioral Regulators of Behavior in Primates* (Ed. by C. R. Carpenter), pp. 149–163. Lewisburg, PA: Bucknell University Press.

Parish, A. R. 1996. Female relationships in bonobos (*Pan paniscus*): evidence for bonding, cooperation, and female dominance in a male-philopatric species. *Hum. Nat.,* 7, 61–96.

Paul, A. & Kuester, J. 1987. Dominance, kinship and reproductive value in female Barbary macaques (*Macaca sylvanus*) at Affenberg Salem. *Behav. Ecol. Sociobiol.,* 21, 323–331.

Paul, A., Kuester, J., & Arnemann, J. 1992. DNA fingerprinting reveals that infant care by male barbary macaques (*Macaca sylvanus*) is not paternal investment. *Folia Primatol.,* 58, 93–98.

Paul, A., Kuester, J., & Arnemann, J. 1996. The sociobiology of male-infant interactions in Barbary macaques, *Macaca sylvanus. Anim. Behav.,* 155–170.

Pereira, M. E. 1986. Maternal recognition of juvenile offspring coo vocalizations in Japanese macaques. *Anim. Behav.,* 34, 935–937.

Pereira, M. E. 1989. Agonistic interactions of juvenile savanna baboons. II. Agonistic support and rank acquisition. *Ethology,* 80, 152–171.

Pereira, M. E. 1992. The development of dominance relations before puberty in Cercopithecine societies. In: *Aggression and Peacefulness in Humans and Other Primates* (Ed. by J. Silverberg & J. P. Gray), pp. 117–149. New York: Oxford University Press.

Pereira, M. E. & Fairbanks, L. A. (eds.) 1993. *Juvenile Primates: Life History, Development, and Behavior.* New York: Oxford University Press.

Perry, S. 1996a. Female-female social relationships in wild white-faced capuchin monkeys, *Cebus capucinus. Am. J. Primatol.,* 440, 167–182.

Perry, S. E. 1996b. Intergroup encounters in wild white-faced capuchins (*Cebus capucinus*). *Int. J. Primatol.,* 17, 309–330.

Perry, S. 1997. Male-female social relationships in wild white-faced capuchins (*Cebus capucinus*). *Behavior,* 134, 477–510.

Petit, O., Abegg, C., & Thierry, B. 1997. A comparative study of aggression and conciliation in three cercopithecine monkeys (*Macaca fuscata, macaca nigra, Papio papio*). *Behaviour,* 134, 415–432.

Rawlins, R. G. & Kessler, M. J. 1986. Cayo Santiago bibliography (1938–1985). In: *The Cayo Santiago Macaques: History, Behavior and Biology* (Ed. by R. G. Rawlins & M. J. Kessler), pp. 283–300. Albany, NY: SUNY Press.

Reiss, M. J. 1984. Kin selection, social grooming and removal of ectoparasites: a theoretical discussion. *Primates,* 25, 185–191.

Rendall, D., Rodman, P. S., & Emond, R. E. 1996. Vocal recognition of individuals and kin in free-ranging rhesus monkeys. *Anim. Behav.*, 51, 1007–1015.

Rhine, R. J. 1972. Changes in the social structure of two groups of stumptail macaques (*Macaca arctoides*). *Primates*, 13, 181–194.

Riss, D. & Goodall, J. 1977. The recent rise to the alpha-rank in a population of free-living chimpanzees. *Folia Primatol.*, 27, 134–151.

Robbins, M. M. 1995. A demographic analysis of male life history and social structure of mountain gorillas. *Behaviour*, 132, 21–47.

Robbins, M. M. 1996. Male-male interactions in heterosexual and all-male wild mountain gorilla groups. *Ethology*, 102, 942–965.

Rose, L. M. 1988. Behavioral ecology of white-faced capuchins (*Cebus capucinus*) in Costa Rica. PhD dissertation, Washington University, St. Louis.

Rose, L. M. 1994. Benefits and costs of resident males to females in white-faced capuchins, *Cebus capucinus*. *Am. J. Primatol.*, 32, 235–248.

Rose, L. M. & Fedigan, L. M. 1995. Vigilance in white-faced capuchins, *Cebus capucinus*, in Costa Rica. *Anim. Behav.*, 49, 63–70.

Sackett, G. P. & Fredrickson, W. T. 1987. Social preferences by pigtailed macaques: familiarity versus degree and type of kinship. *Anim. Behav.*, 35, 603–606.

Sade, D. S. 1965. Some aspects of parent-offspring and sibling relations in a group of rhesus monkeys, with a discussion of grooming. *Am. J. Phys. Anthropol.*, 23, 1–18.

Sade, D. S. 1967. Determinants of dominance in a group of free-ranging rhesus monkeys. In: *Social Communication Among Primates* (Ed. by S. A. Altmann), pp. 99–114. Chicago: University of Chicago Press.

Sade, D. S. 1972a. A longitudinal study of social behavior of rhesus monkeys. In: *The Functional and Evolutionary Biology of Primates* (Ed. by R. H. Tuttle), pp. 378–379. Chicago: Aldine.

Sade, D. S. 1972b. Sociometrics of *Macaca mulatta*. 1. Linkages and cliques in grooming matrices. *Folia Primatol.*, 18, 196–223.

Sade, D. S., Chepko-Sade, B. D., Schneider, J. M., Roberts, S. S., & Richtsmeier, J. T. 1985. *Basic Demographic Observations on Free-Ranging Rhesus Monkeys*. New Haven, CT: Human Relations Area Files.

Sambrook, T. D., Whiten, A., & Strum, S. C. 1995. Priority of access and grooming patterns of females in a large and a small group of olive baboons. *Anim. Behav.*, 50, 1667–1682.

Samuels, A. & Henrickson, R. V. 1983. Outbreak of severe aggression in captive *Macaca mulatta*. *Am. J. Primatol.*, 5, 277–281.

Samuels, A., Silk, J. B., & Altmann, J. 1987. Continuity and change in dominance relations among female baboons. *Anim. Behav.*, 35, 785–793.

Schaub, H. 1996. Testing kin altruism in long-tailed macaques (*Macaca fascicularis*) in a food-sharing experiment. *Int. J. Primatol.*, 17, 445–467.

Schino, G., Aureli, F., & Troisi, A. 1988. Equivalence between measures of allogrooming: an empirical comparison in three species of macaques. *Folia Primatol.*, 51, 214–219.

Schnell, G., Watt, D., & Douglas, M. 1985. Statistical comparison of proximity matrices: applications in animal bevaviour. *Anim. Behav.*, 33, 239–253.

Schulman, S. R. & Chapais, B. 1980. Reproductive value and rank relations among macaque sisters. *Am. Nat.*, 115, 580–593.

Seyfarth, R. M. 1976. Social relationships among adult female baboons. *Anim. Behav.*, 24, 917–938.

Seyfarth, R. M. 1977. A model of social grooming among adult female monkeys. *J. Theor. Biol.*, 65, 671–698.

Seyfarth, R. M. 1980. The distribution of grooming and related behaviours among adult female vervet monkeys. *Anim. Behav.*, 28, 798–813.

Seyfarth, R. M. & Cheney, D. L. 1980. The ontogeny of vervet monkey alarm calling behavior: a preliminary report. *Z. Tierpsychol.*, 54, 37–56.

Seyfarth, R. M. & Cheney, D. L. 1984. Grooming, alliances and reciprocal altruism in vervet monkeys. *Nature*, 308, 541–543.

Sherman, P. W. 1980. The limits of ground squirrel nepotism. In: *Sociobiology: Beyond Nature/Nurture?* (Ed. by G. W. Barlow & J. Silverberg), pp. 504–544. Boulder, CO: Westview Press.

Sherman, P. W. 1981. Kinship, demography and Belding's ground squirrel nepotisim. *Behav. Ecol. Sociobiol.*, 8, 251–260.

Sicotte, P. 1994. Effect of male competition on male-female relationships in bi-male groups of mountain gorillas. *Ethology*, 97, 47–64.

Silk, J. B. 1982. Altruism among female *Macaca radiata*: explanations and analysis of patterns of grooming and coalition formation. *Behaviour*, 79, 162–188.

Silk, J. B. 1988. Social mechanisms of population regulation in a captive group of bonnet macaques (*Macaca radiata*). *Am. J. Primatol.*, 14, 111–124.

Silk, J. B. 2001. Ties that bond: the role of kinship in priimate societies. In: *New Directions in Anthropological Kinship* (Ed. by L. Stone), pp. 71–92. Oxford: Rowman Littlefield.

Silk, J. B., Samuels, A., & Rodman, P. S. 1981. The influence of kinship, rank, and sex on affiliation and aggression between adult female and immature bonnet macaques (*Macaca radiata*). *Behaviour*, 78, 111–137.

Silk, J. B., Seyfarth, R. M., & Cheney, D. L. 1999. The structure of social relationships among female savanna baboons in Moremi Reserve, Botswana. *Behaviour*, 136, 679–703.

Simpson, M. J. A. 1973. The social grooming of male chimpanzees: a study of eleven free-living males in the Gombe Stream National Park, Tanzania. In: *Comparative Ecology and Behavior of Primates* (Ed. by R. P. Michaell & J. H. Crook), pp. 41–505. New York: Academic Press.

Singh, M., D'Souza, L., & Singh, M. 1992. Hierarchy, kinship and social interaction among Japanese monkeys (*Macaca fuscata*). *J. Biosci.*, 17, 15–27.

Smith, K. 2000. Paternal kin matter: the distribution of social behavior among wild adult female baboons. PhD dissertation, University of Chicago, Chicago.

Smouse, P. E., Long, J. C., & Sokal, R. R. 1986. Multiple regression and correlation extensions of the Mantel test of matrix correspondence. *Sys. Zool.*, 35, 627–632.

Stewart, K. J. & Harcourt, A. H. 1987. Gorillas: variation in female relationships. In: *Primate Societies* (Ed. by B. B. Smuts, D. L. Cheney, R. M. Seyfarth, R. W. Wrangham, & T. T. Struhsaker), pp. 155–164. Chicago: University of Chicago Press.

Strier, K. B. 1994. Myth of the typical primate. *Ybk. Phys. Anthropol.*, 37, 233–271.

Thierry, B. 1985. Patterns of agonistic interactions in three species of macaques (*Macaca mulatta, M. fascicularis, M. tonkeana*). *Aggress. Behav.*, 11, 223–233.

Thierry, B. 1990. Feedback loop between kinship and dominance: the macaque model. *J. Theor. Biol.*, 145, 511–521.

Thierry, B., Gauthier, C., & Peignot, P. 1990. Social grooming in Tonkean macaques. *Int. J. Primatol.*, 11, 357–375.

Tomasello, M. & Call, J. 1997. *Primate Cognition*. New York: Oxford University Press.

Vigilant, L., Hofreiter, M., Siedel, H., & Boesch, C. 2001. Paternity and relatedness in wild chimpanzee communities. *Proc. Natl. Acad. Sci. USA*, 98, 12890–12895.

Walters, J. 1980. Interventions and the development of dominance relationships in female baboons. *Folia Primatol.*, 34, 61–89.

Walters, J. R. 1987. Kin recognition in nonhuman primates. In: *Kin Recognition in Animals* (Ed. by D. J. C. Fletcher & C. D. Michener), pp. 359–393. New York: Wiley.

Watts, D. P. 1985. Relations between group size and composition and feeding competition in mountain gorilla groups. *Anim. Behav.*, 33, 72–85.

Watts, D. P. 1990. Ecology of gorillas and its relation to female transfer in mountain gorillas. *Int. J. Primatol.*, 11, 21–45.

Watts, D. P. 1991. Harassment of immigrant female mountain gorillas by resident females. *Ethology*, 89, 135–153.

Watts, D. P. 1992. Social relationships of immigrant and resident female mountain gorillas. I. Male-female relationships. *Am. J. Primatol.*, 28, 159–181.

Watts, D. P. 1994a. Social relationships of immigrant and resident female mountain gorillas. II. Relatedness, residence, and relationships between females. *Am. J. Primatol.*, 32, 13–30.

Watts, D. P. 1994b. Agonistic relationships between female mountain gorillas (*Gorilla gorilla beringei*). *Behav. Ecol. Sociobiol.*, 34, 347–358.

Watts, D. P. 1996. Comparative socio-ecology of gorillas. In: *Great Ape Societies* (Ed. by W. C. McGrew, L. F. Marchant, & T. Nishida), pp. 16–28. Cambridge: Cambridge University Press.

Watts, D. P. 1998. Coalitionary mate guarding by male chimpanzees at Ngogo, Kibale National Park, Uganda. *Behav. Ecol. Sociobiol.*, 44, 43–55.

Watts, D. P. 2000a. Grooming between male chimpanzees at Ngogo, Kibale National Park. I. Partner number and diversity and grooming reciprocity. *Int. J. Primatol.*, 21, 189–210.

Watts, D. P. 2000b. Grooming between male chimpanzees at Ngogo, Kibale National Park. II. Influence of male rank and possible competition for partners. *Int. J. Primatol.*, 21, 211–238.

Watts, D. P. & Mitani, J. C. 2001. Male chimpanzee boundary patrolling and social bonds. *Am. J. Phys. Anthropol.*, Suppl. 32, 160–161.

Watts, D. P. & Pusey, A. E. 1993. Behavior of juvenile and adolescent great apes. In: *Juvenile Primates: Life History, Development, and Behavior* (Ed. by M. E. Pereira & L. A. Fairbanks), pp. 148–167. New York: Oxford University Press.

Weber, A. W. & Vedder, A. 1983. Population dynamics of the Virunga gorillas: 1959–1978. *Biol. Conserv.*, 26, 341–366.

Weigel, R. M. 1981. The distribution of altruism among kin: a mathematical model. *Am. Nat.*, 118, 191–201.

White, F. J. 1988. Party composition and dynamics in *Pan paniscus*. *Int. J. Primatol.*, 9, 179–193.

White, F. J. 1996. *Pan paniscus* 1973 to 1996: twenty-three years of field research. *Evol. Anthropol.*, 5, 11–17.

Widdig, A., Nuernberg, P., Krawczak, M., Streich, W. J., & Bercovitch, F. B. 2001. Paternal relatedness and age proximity regulate social relationships among adult female rhesus macaques. *Proc. Natl. Acad. Sci. USA*, 98, 13769–13773.

Wrangham, R. W. 1999. Evolution of coalitionary killing. *Ybk. Phys. Anthropol.*, 42, 1–30.

Wrangham, R. W. & Smuts, B. B. 1980. Sex differences in the behavioural ecology of chimpanzees in the Gombe National Park, Tanzania. *J. Fert. Reprod.*, Suppl. 28, 13–31.

Yamada, M. 1963. A study of blood-relationship in the natural society of the Japanese macaque—An analysis of co-feeding, grooming, and playmate relationships in Minoo-B-troop. *Primates*, 4, 43–65.

Yamagiwa, J. 1986. On the display behaviors of gorillas and chimpanzees. *Reichorui Kenkyu*, 2, 148.

York, A. D. & Rowell, T. E. 1988. Reconciliation following aggression in patas monkeys, *Erythrocebus patas*. *Anim. Behav.*, 36, 502–509.

8

Patrilineal Kinship and Primate Behavior

Karen B. Strier

U nderstanding the impact of patrilineal kinship on behavior in primates is complicated by a variety of factors, beginning with the limited abilities of primates—and their human observers—to recognize patrilineal kin when they occur. Few primates mate monogamously when other potential partners are available, and inferring paternity from observed copulations is difficult, even when the timing of fertilizations can be estimated from physical, hormonal, or behavioral cues (e.g., de Ruiter et al. 1994). The challenges of identifying patrilineal kinship relationships beyond those involving paternity are further compounded by incomplete genealogies and imperfect observation conditions, especially but not exclusively in the wild.

Genetic analyses of wild primates have the power to address a wide range of evolutionary questions, including those that pertain to the degree of relatedness and social interactions among patrilineal kin. These data are still available only for a small number of species, but the emerging findings provide intriguing insights into the conditions that affect the availability of patrilineal kin. For example, although high-ranking males in hierarchical societies sire more offspring than low-ranking males, the degree to which fertilizations are skewed varies widely (table 8.1). The paternal sibships that arise from this reproductive skew have measurable effects on the genetic structure of groups and populations and on the genealogical relationships among their members (Altmann et al. 1996). Different levels of reproductive skew will therefore lead to differences in the opportunities for paternally related partners to interact (Altmann & Altmann 1979, van Hooff & van Schaik 1994).

In the absence of genetic data, efforts to evaluate the behavioral impact of patrilineal kinship have focused on males that remain in their natal groups for life. Traditionally regarded as rare in primates, male philopatry is now known to occur in a diversity of anthropoid species (Moore 1984, 1992; Strier 1994). Indeed, although male philopatry is not found in the prosimians (Kappeler 1997) and is rare in cercopithecines (Di Fiore & Rendall 1994),

Table 8.1. Reproductive Skew and Male Tenure at Rank

Species	Percent Paternity Attributed to Alpha Male	Alpha Male's Tenure	Number of Infants Sampled
Red howler monkeys[a]	100%	7.5 years (mean)	12 in 4 multimale troops
Long-tailed macaques[b]	50–90%	1–5 years (range)	21 in 3 groups
Japanese macaques[c]	22% of all; 67% of six infants sired by group males	<2 years	9 in 1 group
Yellow baboons[d]	81%	4 years (max)	27 in 2 groups
Bonobos[e]	50–70% (among two high-ranking males)	?	10 in 1 community
Chimpanzees[f]	35.7% alpha male; 50% for alpha + one high-ranking male	8 years?	14 in 1 community

[a]Pope 1990, 1992, 1998.
[b]de Ruiter et al. 1994, de Ruiter & Geffen 1998.
[c]Soltis et al. 2001.
[d]Altmann et al. 1996.
[e]Gerloff et al. 1999.
[f]Constable et al. 2001.

it may be the ancestral pattern in both New World monkeys (Strier 1999a, Pope 2000) and apes (Wrangham 1987, Fuentes 2000). Yet even in species with male philopatry, investigations into patterns of social interactions have emphasized male behavior toward maternal relatives instead of toward paternally and patrilineally related kin (Furuichi & Ihobe 1994, Hashimoto et al. 1996, Goldberg & Wrangham 1997, Mitani et al. 2000, Strier et al. 2002). The question of nepotism among paternally related and patrilineal kin in these societies, like that among paternally related kin in matrilocal societies, requires paternity data to resolve (Alberts 1999, Widdig et al. 2001, Buchan et al. 2003).

There are at least two sets of contexts in which patrilineal kin might be expected to interact differently with one another than with nonkin. One context involves inbreeding avoidance, which relies upon mechanisms such as dispersal, which reduces opportunities for patrilineal kin to breed, or kin recognition, which may reduce the attractiveness of paternal siblings as mates (Alberts 1999). A second context involves nepotism, which is typically equated with altruism among kin and assumes that individual fitness costs to the altruist are offset by the inclusive fitness benefits gained by helping a relative. Nepotism is difficult to distinguish from mutualism, however, if benefits accrue to each participant without obvious fitness costs (Chapais 2001). For example, coalitions among male chimpanzees are mutualistic if each of the males achieves greater access to females by participating in the coalition than he would by himself (Watts 1998). These coalitions could be considered examples of "mutualistic nepotism" if males preferentially choose patrilineal kin over nonkin as coalition partners (Chapais 2001), but whether they routinely do this is not currently known.

In this chapter I review what is—and is not yet—known about the impact of patrilineal kinship on the behavior of wild primates. I begin by decoupling patrilineal kinship from male philopatry, or patrilocality as it is more appropriately called in the ethnographic literature, and then briefly discuss the mechanisms by which patrilineal kin recognition may occur. Next, I review some of the principle conditions that affect access to patrilineal kin. Dispersal, mating, and demographic patterns interact to determine the availability of patrilineal kin of both sexes as social partners, and therefore the potential for interactions among them to occur. Finally, I examine some examples of social cooperation among patrilineally related primates that may be better described by the more conservative criteria of mutualistic nepotism than by nepotism per se.

One persistent question that emerges from this review is how an absence of nepotism among patrilineal kin can be confidently distinguished from constraints on the abilities of primates to recognize patrilineal kin. A second set of questions for future studies to address pertains to the influence of maternally as well as paternally related kin on male cooperation with one another against groups of unrelated males, and the effects of paternal sibships on cooperative relationships among both males and females within groups.

Assumptions of Kinship Terminology

Kinship terminology is one of the most well developed areas of social anthropology (Fox 1967), but surprisingly few primatologists or behavioral ecologists adhere to the definitions established by ethnographers when describing kinship relationships or their behavioral correlates in animals. These are more than discipline-specific semantic differences, however, because they reflect divergent assumptions about the meanings of kinship and the abilities of human and nonhuman primates to identify kin (Sade 1991).

In human societies, kinship systems characterize not only genealogical descent and residence patterns, but also the cultural rules that dictate inheritance of land and other resources within descent groups and the distribution of alliance and marriage partners among descent groups (Fox 1975). Human kinship terminology distinguishes descent through males (i.e., patrilines and patrilineages) and females (i.e., matrilines and matrilineages) from residence patterns, which are patrilocal when females reside in the groups where their husbands' families live, and matrilocal when males reside in the groups where their wives' families live. These distinctions are necessary to make sense out of the broader kinship networks that define alliances and the exchange of marriage partners within and among groups of humans. Because humans maintain their association with descent groups through self- and group-level identification, their kinship systems do not always map directly on to their extent of consanguinity in the way that studies of kinship in primates and other animals assume. Nonetheless, the distinctions between descent and residence patterns have important implications for understanding whether patrilineal kinship affects primate behavior.

In primates, by contrast, such extended kinship networks and their implications for behavior are rarely considered except when describing the distribution of genetic variation among groups in populations (e.g., Melnick & Hoelzer 1992; Pope 1992, 1998; de Ruiter & Geffen 1998). Instead, we tend to extrapolate the kinship structures of primate groups from their dispersal regimes and to evaluate interactions among relatives based on predictions about behavior derived from kin selection and inclusive fitness theory (Hamilton 1964).

What matters from the evolutionary perspective of kin selection and inclusive fitness theory is the degree of relatedness between individuals, or r. Relatedness is usually calculated independently of the sex of the ancestor through whom it is traced. On average, individuals that share one parent (i.e., half siblings) share 25% of their genes with one another, as well as with each of their grandparents, and 12.5% of their genes with their uncles and aunts, and so on. In theory, kin with similar degrees of relatedness should give and receive similar levels of altruism, provided that the fitness costs to actors and benefits to recipients are the same. However, the "profitability" of altruism among different categories of kin will differ depending on variables such as sex, age, and competitive abilities, as well as on the abilities of primates to discriminate among them (Berman & Kapsalis 1999, Chapais et al. 2001).

In human kinship analyses, the sex of the common ancestor determines whether individuals belong to the same or different patrilineages and matrilineages. Thus, while a father's son and daughter are both members of his patriline, only his son's offspring will also belong. Patrilineally related males share the same Y chromosome, but although daughters belong to their fathers' patrilines, they do not pass their patrilineage on to their offspring (figure 8.1). Grandchilden share the same proportion of their genes with each of their four grandparents ($r = .25$), but they are only patrilineally related to their fathers' fathers and patrilaterally related to other kin on their fathers' side of the family (table 8.2).

Making the ethnographic distinction between genealogical kinship and residence patterns more explicit calls attention to two obvious but often overlooked facts, both of which have implications for how interactions among patrilineal kin are interpreted by primatologists. The first is that offspring of both sexes sired by different fathers in patrilocal societies can be less closely related to one another than offspring sired by the same father in matrilocal societies. The second is that patrilocality permits males to co-reside with their mothers and maternally related brothers for life (van Hooff & van Schaik 1994), just as matrilocality permits paternally related females, who may also be matrilineally related, to coreside.

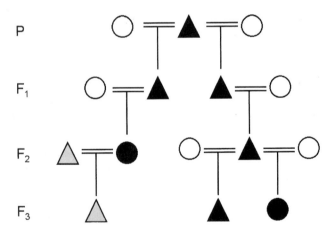

Figure 8.1. Patrilineal kinship. Triangles signify males; circles signify females. Black symbols represent individuals of both sexes that are patrilineally related through the ancestral male shown (P). Gray symbols represent a different patriline (ancestry not shown). Although F_2 female is a member of her father's (F_1) patriline, and therefore patrilineally related to her father's brother's son (and both of his offspring, which are paternal siblings), F_2 female's own son (F_3) is not a member of her patrilineage.

Table 8.2. Terminology for Describing Human Kinship Relationships

Term	Description
Patrilineal kin	
Patriline	If ego is a male, his patriline includes his direct male ancestors (e.g., his father, his father's father, etc.) and his direct male descendants (e.g., his son, his son's son, etc.), and his daughters (but not their offspring). If ego is a female, her patriline includes only her direct male ancestors (e.g., her father, father's father, etc.). Her descendants belong to their fathers' patrilines.
Patrilineage	All males related through male lines, and their daughters. Collateral kin include the aunts, uncles, nieces, nephews, and cousins, etc. of males in the patriline.
Patrilateral kin	All kin (of both sexes) on the father's side of the family.
Patrilocal	When sons remain in their own birth groups after marriage.

While $r = .25$ between paternal half siblings in any kind of society, lifelong coresidence among more distant patrilineal kin ($r < .25$) is only likely to occur when males are patrilocal. Yet the relatedness among such distant kin may have little, if any, direct impact on behavior. Chapais (2001) identified $r = .25$ as the "relatedness threshold for altruism," or RTA, based on evidence that Japanese macaques (*Macaca fuscata*) do not recognize familiar but more distantly related kin. In a more extensive study, Chapais et al. (2001) found evidence of altruism on behalf of lineal kin when $r = .125$, such as that between great-grandmothers and their great-grandsons, while the unequivocal RTA for collateral kin, such as half siblings, was higher ($r = .25$). Whether the differences in RTAs for lineal and collateral kin reflect different limits on kin recognition or on the profitability of altruism for different categories of kin has not yet been resolved, but their implications for interpreting the potential impact of patrilineal kinship on the behavior of primates should not be ignored. For example, if the relatedness of a male to his paternal uncle ($r = .125$), or that between the sons of paternally related brothers ($r = .063$) falls below the RTA, then there is no basis for assuming that such distantly related patrilineal kin in patrilocal societies will treat one another as such.

Although cooperation among patrilocal males in coalitions against unrelated males is often attributed to their patrilineal kinship, the possible influence of maternally related males on patterns of cooperation in between-group interactions cannot be dismissed. Patrilocality creates unique opportunities for lifelong interactions among close, maternally related males, which may be more likely to recognize one another as kin than they are to recognize more distantly related patrilineal kin. Decoupling assumptions about kinship and residence patterns emphasizes the dual importance of looking beyond patrilocal societies for evidence of differential behavior among both male and female patrilineal kin, as well as distinguishing between patterns of interactions among patrilineal and matrilineal kin in patrilocal societies.

Mechanisms of Patrilineal Kin Recognition

Familiarity appears to be a necessary requirement for kin recognition, even among closely related kin (Bernstein 1999, Chapais 2001). For example, in captive studies of pigtailed macaques (*M. nemestrina*) (Fredrikson & Sackett 1984) and yellow baboons (*Papio cynocephalus*) (Erhart et al. 1997), interactions among unfamiliar paternal half siblings did not differ from those among nonkin. In the wild, however, cohorts of paternal siblings will usually grow up in the same groups together and will therefore be familiar to one another for at least part, if not all, of their lives. In the matrilocal societies of rhesus macaques (*M. mulatta*), for example, paternal sisters appear to maintain significantly stronger affiliations with one another than with nonkin (Widdig et al. 2001). Conversely, males in matrilocal societies can disperse from their natal groups with familiar peers. Among Peruvian squirrel monkeys (*Saimiri sciureus*), for example, relationships among dispersing male cohorts, which may also be paternal brothers, can persist over multiple migration events involving successive groups (Mitchell 1994).

While familiarity is a necessary condition for patrilineal kin recognition in primates, the mechanisms by which familiar kin and nonkin are distinguished are still poorly understood. Phenotypic matching has been proposed to account for the lower levels of affiliation and sexual behavior among paternally related yellow baboon consorts compared to nonkin consorts (Alberts 1999), and for the strength of affiliations among paternally related female rhesus macaques (Widdig et al. 2001). In both studies, paternal kin that were close to one another in age were more discriminating than those with greater age differences between them. Alberts (1999) suggested that age proximity might be a social cue responsible for the lower probabilities of sexual consortships among baboon paternal siblings born within two years of one another. Similarly, Widdig et al. (2001) suggest that female rhesus macaques recognize paternal kin that are close to them in age by matching behavioral cues associated with similar personality traits, reminiscent of the anecdotal descriptions of compatible personalities among male chimpanzee (*Pan troglodytes*) coalition partners at Mahale (Nishida & Hosaka 1996).

In addition to age proximity and its associated behavioral cues, phenotypic matching may also be restricted to certain categories of kin. For example, the ability of chimpanzees to match the physical phenotypes of unfamiliar mother-son pairs, but not mother-daughter or father-daughter pairs, might reflect past selection for mechanisms of kin recognition that reduce the risks of inbreeding between mothers and sons (Parr & de Waal 1999). Indeed, despite the opportunities patrilocality creates, matings between mothers and their sexually mature sons are rare in wild populations of chimpanzees (Constable et al. 2001), bonobos (*Pan paniscus*) (Gerloff et al. 1999), and muriquis (*Brachyteles arachnoides*) (Strier 1997a). By contrast, mechanisms for recognizing father-daughter kinship might be less important because father-daughter matings should be precluded in patrilocal societies when daughters transfer, and in matrilocal societies by the secondary dispersal of fathers.

Evidence that infant hanuman langurs (*Presbytis entellus*) are exclusively defended by genetic fathers and males that were residents in multimale groups when the infants were conceived (Borries et al. 1999) is consistent with the idea that mechanisms of paternity recognition may operate on shorter time frames (e.g., from conception to infancy) than those necessary to encompass the age at which females reach sexual maturity. Genetic data also demonstrate that adult male yellow baboons differentially support their juvenile offspring, relying on a combination of phenotype matching and their consort time with females to identify their offspring (Buchan et al. 2003).

While phenotypic matching may be a mechanism involved in recognizing one's familiar kin (Alberts 1999, Widdig et al. 2001) or mother-son kinship in unfamiliar individuals (e.g.,

Parr & de Waal 1999), primates that live in social groups also have unique sets of experiences that can influence their interactions with kin. For example, Berman & Kapsalis (1999) found that similarities in the kin-biased affiliative networks of rhesus macaque infants and their mothers reflect similarities in their respective levels of social risk, as measured by the number of other individuals in close proximity. Infants adjusted their own affiliations with kin in response to their own social experiences, independently of their interactions with their mothers. Such "socially biased independent learning," which incorporates both maternal transmission through an infant's association with her mother, and independent learning (Berman & Kapsalis 1999) could also be involved in the development of biases toward patrilineal kin. However, the conditions under which a mother and her infant's patrilineal kin are the same and available to them as social partners may be limited.

Conditions Affecting Access to Patrilineal Kin

In addition to understanding the limits of patrilineal kin recognition, it is important to consider the conditions that influence the availability of patrilineal kin as potential social partners (Chapais 2001). For example, dispersal patterns dictate the duration of associations among same- and opposite-sexed kin. Mating patterns determine the probability of shared paternity among all members of a cohort, including males and females. Demographic patterns impact the size of cohorts and kin groups, as well as the age differences among maternal siblings. Interactions among these social and demographic variables affect access to paternally related and more extended patrilineal kin (figure 8.2).

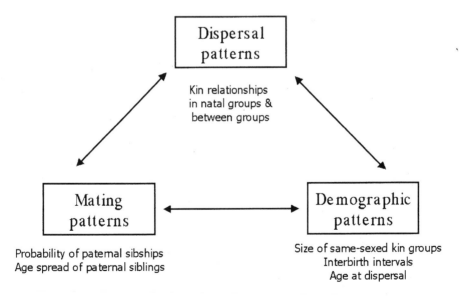

Figure 8.2. Mechanisms that affect the availability of patrilineal kin.

Dispersal Regimes

Dispersal patterns affect the genetic composition of primate groups, as well as the kinds of kin present and the duration of their associations with one another. Nonrandom dispersal patterns have implications for the genetic composition among groups within populations (Cheney & Seyfarth 1983, Melnick & Hoelzer 1992) and can vary from cohort to cohort or group to group in the same population over time (e.g., de Ruiter & Geffen 1998). For example, in expanding populations of red howler monkeys (*Alouatta seniculus*), both males and females typically disperse from their natal groups. But in saturated habitats, where dispersal opportunities are limited, the daughters of high-ranking mothers are retained with the result that the female membership of troops can evolve into extended matrilines (Crockett 1996, Pope 1998). Because alpha male red howler monkeys can monopolize fertilizations (Pope 1990), cohorts of offspring born in these troops can be paternally as well as matrilineally related to one another.

Males in patrilocal societies may be either paternal or maternal half siblings, or related to one another through more extended patrilineal relationships. Extended matrilineal relationships, such as those between a male and his sisters' sons, will be precluded in patrilocal societies if females routinely transfer out of their natal groups (e.g., bonobos: Gerloff et al. 1999). However, in the Kasakela chimpanzee community at Gombe, four of nine adult females sampled in a recent genetic study were known to have reproduced in their natal communities (Constable et al. 2001), and the proportion of natal females that have remained and reproduced in this community may be even higher. As a result, at least some of the males in at least some patrilocal societies coreside with their distant matrilineal as well as patrilineal male kin.

Limited dispersal opportunities caused by habitat fragmentation have been associated with cases of bisexual philopatry (e.g., muriquis: Strier 1991, 1999b), and the effects of altered habitats on population densities and the relative strength of within- versus between-group competition have been associated with cases of bisexual dispersal in societies in which dispersal is ordinarily sex biased (e.g., chimpanzees: Sugiyama 1999; langurs: Sterck 1998, Sterck et al. 1997; mantled howler monkeys (*Alouatta palliata*): Jones 1999). Limited dispersal opportunities can increase the levels of local mate competition among kin and thus reduce the profitability of altruism among them (e.g., Taylor 1992, Wilson et al. 1992, West et al. 2001), although the benefits of cooperating with kin against other groups may offset or outweigh the fitness costs.

In patrilocal societies, males remain in their natal groups, with their fathers and extended patrilineal kin for life. Females in patrilocal societies rarely transfer more than once, so sons usually also end up residing with their mothers and maternally related brothers (Moore 1992). Immigrant female chimpanzees (Goodall 1986, Constable et al. 2001), bonobos (Gerloff et al. 1999), and muriquis (Strier 1999b, Strier & Ziegler 2000) have maintained their memberships over the multiple decades that some populations of these species have been studied. In matrilocal societies, by contrast, males tend to maintain their group memberships for much shorter durations and to disperse on multiple occasions. For example, the tenure of male Peruvian squirrel monkeys rarely exceeds two annual breeding seasons (Boinski & Mitchell 1994) and the median tenure of nonnatal males in yellow baboon troops is 24 months (Alberts & Altmann 1995).

While patrilocal males have the opportunity to interact with both their maternal and paternal brothers for life (figure 8.3), dispersing males will only do so if they disperse nonrandomly into the same groups as one another (van Hooff & van Schaik, 1994). How close they are in age, which is minimally determined by interbirth intervals, will determine the degree to which maternal brothers overlap with one another in their natal groups, and thus whether they are potentially recognizable to one another as kin.

Males born into matrilocal societies automatically sever their associations with their mothers and other female kin when they disperse. In yellow baboons, however, nearly 50% of natal males become sexually active prior to dispersing (Alberts & Altmann 1995) and they do not avoid sexual consortships with their paternal sisters (Alberts 1999). Associations among matrilineally related male kin and paternal brothers can also persist outside of their natal groups because in many species (but curiously, not yellow baboons) males do not disperse from their natal groups alone.

Dispersing with cohorts would presumably reduce the vulnerability of both males and females to predators, but access to familiar kin may be more important for males that need allies to successfully join extant groups (e.g., ring-tailed lemurs [*Lemur catta*]: Sussman 1992; vervet monkeys [*Cercopithecus aethiops*]: Cheney & Seyfarth 1983; macaques: van Noordwijk & van Schaik 1985; Peruvian squirrel monkeys: Mitchell 1994) or to defend the

Figure 8.3. Patrilocal male muriquis sitting together and keeping tabs on each other as they monitor a receptive female (not shown). Extreme scramble competition among males in their egalitarian society may reduce the probability of paternal sibships among age cohorts, but increase the probability of paternal sibships across ages compared to the paternal sibships that arise through contest competition in the hierarchical societies of most other primates. Photo by K. B. Strier.

new groups they establish (e.g., red howler monkeys: Pope 1990). Cohorts of dispersing males may subsequently split up when males transfer secondarily in search of more favorable sex ratios (Sussman 1992) and correspondingly better reproductive opportunities (Cheney & Seyfarth 1983, van Noordwijk & van Schaik 1985), or they may persist over successive dispersal events (Mitchell 1994). In both cases, they demonstrate the potential for extended associations among male kin to persist beyond their natal group.

Male cohorts that disperse together are more likely to be paternal brothers than maternal brothers, if they are related to one another at all. The likelihood of being paternal brothers is a function of the degree to which single males can monopolize fertilizations (van Hooff & van Schaik 1994). Unless they are twins, maternal brothers will belong to different cohorts that are separated in age by at least one interbirth interval. Males may follow their older maternal brothers into the same groups (Cheney & Seyfarth 1983, van Noordwijk & van Schaik 1985), or even transfer into their fathers' natal groups, where their fathers' mothers and paternally related sisters reside (figure 8.4). Nonetheless, the absence of prior familiarity among a dispersing male and females belonging to his father's matriline should preclude them from recognizing one another as kin.

In wild long-tailed macaques (*M. fascicularis*), de Ruiter & Geffen (1998) found an average degree of relatedness among paternal half siblings of .35 (versus .25 for maternal half sibs) because many of them were related not only through the same high-ranking father, but also through the matrilineal relationships among their mothers. The relatedness of dispersing pairs of male long-tailed macaques from the same natal groups could be as high as

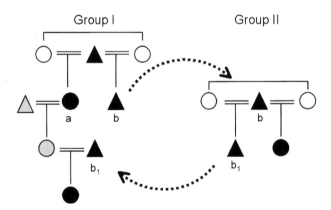

Figure 8.4. Discontinuous local patriline resulting from nonrandom dispersal patterns of males in matrifocal societies. Brackets connecting open circles indicate matrilineal kinship among females (e.g., sisters, half sisters, etc.) in their respective matrilocal groups. In Group I, female (a) and male (b) are both patrilateral half siblings *and* matrilateral cousins. Other symbols are the same as in figure 8.1. Dotted arrows indicate male transfers (male b and his son, b_1) from their respective natal groups. Note that extended patrilineal and matrilateral kin can end up as coresidents in matrilocal societies if a son (e.g., b_1) disperses into his father's natal group.

.43 and .58, in contrast to relatedness among maternal half siblings, which rarely exceeded .25 because the fathers of maternal half sibs were usually unrelated to one another.

Paternal brothers may maintain their associations with one another, but it is not clear whether they preferentially disperse with one another or with more distant matrilineally related males that may be more familiar to them through the relationships among their mothers. For example, kin-biased affiliative networks established through socially biased independent learning could influence the identities of male dispersal partners (Berman & Kapsalis 1999). In matrilocal societies, the availability of both paternally related males as potential dispersal partners and paternally related females as lifelong affiliates will be determined by the degree of reproductive skew among breeding males and the length of time these males maintain their status as breeders.

Effects of Mating Systems

When an alpha male can monopolize a disproportionate percentage of fertilizations, many, if not all, of the offspring born during his breeding tenure will be paternally related half siblings (Altmann & Altmann 1979, van Hooff & van Schaik 1994). Paternal sibships can occur independently of whether a society is matrilocal or patrilocal, but the limited sample of paternity studies suggests that levels of reproductive skew may be lower in patrilocal societies than in matrilocal societies and societies in which both sexes disperse (table 8.1).

Genetic data show that alpha male red howler monkeys sire 100% of the offspring produced in multimale troops (Pope 1990) during the 7.5 years of their tenure as alpha males (Pope 1992). As a result of their reproductive monopolies, all infants born in a howler monkey troop during an alpha male's reign are paternal half siblings, and they may be even more closely related if their mothers are also members of the same matrilines (Pope 1998). In other primate species for which paternity data are available, the proportion of paternities attributed to alpha or high-ranking males is more variable than it appears to be in red howler monkeys (table 8.1).

The degree to which fertilizations are skewed can reflect differences in group size and the number of potential breeding males and females present (e.g., de Ruiter et al. 1994, de Ruiter & Geffen 1998). In yellow baboons at Amboseli National Park, Kenya, for example, 81% of paternities during a four-year period could be assigned to the alpha males of two groups, despite the presence of 20 adult males in this sample (Altmann et al. 1996). In wild Japanese macaques, the number of females mating influenced the alpha male's ability to monopolize them from both low-ranking males in the group and from extragroup males (Soltis et al. 2001).

The fertilization success of alpha male chimpanzees is lower than that of yellow baboons, long-tailed macaques, and red howler monkeys. Only 5 of 14 paternities, or less than 36% of the paternities in the Kasakala community at Gombe, could be assigned to the males that held alpha status at the times that these conceptions occurred (Constable et al. 2001). Adding the two infants sired by one high-ranking male to the five infants sired by alpha males raises the percentage of paternities by alpha and high-ranking males to 50%. Interestingly, this percentage is similar to the 50% to 70% of 10 paternities accounted for by two high-ranking male bonobos out of the eight possible fathers at Lomako (Gerloff et al. 1999). Indeed, despite the greater levels of scramble competition associated with access to mates in bonobos compared to chimpanzees (Furuichi & Ihobe 1994), high-ranking males in these species appear similar to one another in their fertilization success.

Female chimpanzee grouping patterns, which are more fluid than those of female bonobos, might make it more difficult for high-ranking male chimpanzees to monopolize fertilizations than expected based on their more rigid dominance hierarchies. Paternity data from a larger sample of primates representing other dispersal and grouping patterns will be necessary before legitimate comparisons of levels of reproductive skew can be made.

Demography and Life Histories

Both the distribution of fertilizations and the relatedness among males, and the number of fertile females and their relatedness to one another, have implications for the size and genetic relatedness among age cohorts. The length of interbirth intervals relative to the duration of male breeding monopolies affects the probability that successive infants born to the same mother are full or half siblings (table 8.3). For example, in both yellow baboons and red howler monkeys, interbirth intervals average 2 years, whereas the tenures of alpha males are up to 4 years and 7.5 years, respectively (Pope 1992, Altmann et al. 1996). Successive sons born to a yellow baboon or red howler monkey mother could share the same father and therefore be both paternally and maternally related to one another. In chimpanzees and bonobos, by contrast, longer interbirth intervals (4–5 years) relative to male tenures at high rank and the lower proportion of fertilizations males monopolize make the probability of shared paternity among maternal siblings less likely. Indeed, there was no evidence that any of the maternally related siblings in either the Kasakela chimpanzees or the Lomako bonobos had been sired by the same fathers, despite the lack of male dispersal in these patrilocal communities (Gerloff et al. 1999, Constable et al. 2001).

Infant sex ratios and survivorship further influence the probability that same-sex paternally related or patrilineal kin will be available as social partners in either natal groups in patrilocal societies or at the time at which males disperse (Strier 2000a). In patrilocal societies, the potential for interactions among males reflects, at least in part, the number of surviving males that were born and matured in the group. Similarly, when males disperse,

Table 8.3. Interbirth Intervals Relative
to Alpha Males' Tenures[a]

Species	Interbirth Interval	Alpha Male Tenure
Red howler monkeys[a]	~2 years	7.5 years
Long-tailed macaques[b]	~2 years	1–5 years
Yellow baboons[c]	~2 years	up to 4 years
Bonobos[d]	~4–5 years	?
Chimpanzees[e]	~5 years	up to 8 years?

[a]Pope 1990, 1992, 1998.
[b]de Ruiter et al. 1994, de Ruiter & Geffen 1998.
[c]Altmann et al. 1996.
[d]Gerloff et al. 1999.
[e]Constable et al. 2001.

the opportunities for them to do so with patrilineally as well as matrilineally related male kin will be determined by the number of males in their natal cohorts that survive to dispersal age. The presence of male age cohorts with which to disperse does not always result in cohort dispersal, as exemplified by the solitary dispersals of male yellow baboons (Alberts & Altmann 1995). Yet access to same-sex peers is obviously a necessary condition for cohort dispersal to occur.

Apart from predictive models of sex ratio allocation under conditions of local resource competition (e.g., Clark 1978, Silk 1983) or enhancement (e.g., Gowaty & Lennartz 1985), stochastic demographic and ecological processes can impact infant sex ratios and survivorship in wild populations (Strier 2000b). Changes in the availability of paternally related or patrilineal kin at different times in an individual's lifetime can affect the availability of social allies, including coalition partners, and thus influence the levels of kin-biased behavior expressed (Chapais 2001).

Evidence of Nepotism Among Patrilineal Kin

Primates do not appear to recognize their own unfamiliar kin (Fredrikson & Sackett 1984, Sackett & Fredrikson 1987, Erhart et al. 1997), and their behavior toward distant familiar kin ($r < .125$ for lineal kin) is indistinguishable from that toward nonkin (Chapais 2001). It is therefore appropriate to restrict considerations of nepotism to the possible kinds of relationships among familiar individuals whose relatedness falls within the minimum RTA for direct ($r > .125$) and collateral ($r > .25$) kin (Chapais et al. 2001). These relationships include only patrilineal kin from father–offspring ($r = .50$) to great-grandfather–great-grandoffspring ($r = .125$) and paternally related siblings of both sexes ($r = .25$). It is important to reiterate that other patrilineal kin in patrilocal societies, such as sons and their fathers' paternal brothers, will be more distantly related ($r < .125$) than the unequivocal RTA for collateral kin. Even if their fathers' and uncles' mothers are the same, and $r = .25$, the greater relatedness among them is not a result of their patrilineal kinship.

Males in Patrilocal Societies

Patrilocality provides an established group into which natal males can mature. Males in patrilocal societies do not need allies to join extant groups or establish new groups, but they may rely on coalitions with one another in conflicts against other groups of patrilocal males and against other males in their natal groups. Patrilocal males also have unique opportunities to associate and interact with one another, such as in the grooming dyads of male spider monkeys (*Ateles geofroyi*) (Ahumada 1992), the nearest neighbors of male woolly monkeys (*Lagothrix lagothricha*) (Stevenson 1988) and muriquis (Strier 1997b), and the hunting parties of male chimpanzees (Mitani & Watts 2001), although they do not appear to associate preferentially with known maternal brothers (e.g., muriquis: Strier et al. 2002) or with males with similar mtDNA haplotypes (e.g., chimpanzees: Goldberg & Wrangham 1997; Mitani et al. 2000, 2002).

Cooperation Against Other Groups

The cooperation male chimpanzees exhibit during hostile encounters with males from neighboring communities is consistent with predictions based on kin selection because on average, males in chimpanzee communities are more closely related to one another than to males from other communities (Morin et al. 1994, Vigilant et al. 2001). Nonetheless, recent findings that both high- and low-ranking males participate similarly in these interactions (Wilson et al. 2001) and that male participation correlates with mating success and social bonds in the unusually large community of males at Ngogo in Kibale National Park, Uganda (Watts & Mitani 2001) imply that their cooperation is self-interested.

Similar examples of patrilocal male primates cooperating with one another in hostile intergroup encounters occur in spider monkeys (*Ateles* spp.) (Symington 1987, 1990), muriquis (Strier 1990, 1999a), and Costa Rican squirrel monkeys (*Saimiri orstedii*) (Boinski 1994). Male cooperation in between-group competition may also occur when male golden lion tamarins (*Leontopithecus rosalia*) (Baker et al. 1993) and other callitrichids (Garber 1997), red howler monkeys (Pope 1990), and mountain gorillas (*Gorilla gorilla*) (Robbins 1995, Watts 1996) remain in their natal groups instead of dispersing, and assist their fathers in resisting threats or incursions from extragroup males. In primates such as callitrichids, red howler monkeys, and mountain gorillas, however, patrilocality appears to be a facultative response to constraints on dispersal in saturated habitats, and cooperation among paternally related males is mutualistic because it is beneficial to both the fathers and their sons. For example, alpha male mountain gorillas maintain longer tenures in multimale groups than in unimale groups (Robbins 1995), and coalitions of paternally related red howler monkey males are more successful in defending their troops than single males or than coalitions involving unrelated partners (Pope 1990). The subordinate males in these coalitions may have limited, or no, sexual access to females, but their chances of inheriting breeding opportunities from their fathers may be greater than their chances of acquiring breeding opportunities by dispersing on their own.

Much of the variation in the degree of cooperation in intergroup encounters by males in patrilocal societies can be explained by ecological and demographic factors. For example, intergroup encounters among muriqui groups living at low densities in large, undisturbed forests are rare compared to those living at high densities in disturbed forests (Moraes et al. 1998, Strier 2000b). Similarly, more concentrated food resources may permit the relatively small chimpanzee communities at the Taï National Park, Ivory Coast to exploit core feeding areas in their respective territories, and thereby minimize the frequency of intercommunity encounters compared to East African chimpanzees (Boesch & Boesch-Achermann 2000, Herbinger et al. 2001). The more frequent territorial patrols and larger size of patrol parties of the Ngogo chimpanzees have been attributed to the larger size and greater number of males in this community compared to those at Taï and Gombe (Watts & Mitani 2001).

The only male primates that participate in lethal raids associated with coalitionary killings of extracommunity males are East African chimpanzees (Wrangham 1999). Although patrilocal male bonobos sometimes cooperate during agonistic intercommunity encounters, they do so much less frequently and with less persistence than chimpanzees (Furuichi & Ihobe 1994). Efforts to explain these differences between male bonobos and chimpanzees have tended to focus on the more prominent role females play in mediating male social and sexual interactions in bonobo societies (Furuichi & Ihobe 1994, Furuichi 1997), but the effects of community demography and corresponding differences in their communities' kinship structures might also be involved. For example, the smaller number of males relative to females, as well as greater scramble competition, could lead to lower fitness benefits from cooperation or lower levels of patrilineal kinship among bonobos (Gerloff et al. 1999) than among East African chimpanzees.

The greater diversity of mtDNA haplotypes found in the Lomako bonobos (Gerloff et al. 1999) may also result in lower levels of matrilineal kinship among patrilocal males compared to chimpanzees such as those at Gombe, where not all natal females emigrate. Indeed, although male chimpanzees with similar mtDNA haplotypes at Kibale National Park, Uganda, may associate more frequently with one another than expected by chance (Goldberg & Wrangham 1997), they are not more likely to be coalition partners on a communitywide basis (Mitani et al. 2000, 2002). Thus, while matrilineal kinship may not influence interactions among patrilocal males within their communities, its contribution in facilitating mutualistically beneficial cooperation against males in other communities cannot be ruled out.

Cooperation Within Groups

The lower proportion of fertilizations achieved by high-ranking males in patrilocal societies of chimpanzees and bonobos compared to high-ranking males in the matrilocal societies of yellow baboons and long-tailed macaques (table 8.1) may reflect constraints on competition among patrilineally related male kin (van Hooff & van Schaik 1994, van Hooff 2000) as well as differences in the abilities of individual males to monopolize females. Yet with the exception of northern muriquis, in which there is no evidence of agonistically based hierarchies among males at any time of year (Strier 1990, 1992; Strier et al. 2002), patrilocal and patrilineally related male primates compete with one another to varying degrees for status and the priority of access to females that high status brings.

Coalitions of males in within-group competition typically involve participants that are close to one another in age and rank (e.g., chimpanzees: Nishida & Hosaka 1996, Watts 1998, Mitani et al. 2002; Costa Rican squirrel monkeys: Boinski 1994), where the possibility, if not high probability, of paternal relatedness among them exists. Indeed, the limitations on coalitions among maternally related male chimpanzees (Goldberg & Wrangham 1997, Mitani et al. 2000) may reflect age differentials resulting from interbirth intervals that reduce their value to one another as allies. Coalition partners that are close to one another in age will make better allies if they are similar in competitive ability and also paternally related siblings (van Hooff & van Schaik 1994).

Among the unusually large male community of Ngogo chimpanzees, coalitions of two to three males cooperate with one another to monopolize access to periovulatory females (Watts 1998). In contrast to male baboons, among whom only one partner in a successful coalition consorts with the female (Noë 1992), Ngogo chimpanzees form coalitions when doing so increases each participant's copulation success over what it would be without the other males' help (Watts 1998). Whether coalitions involving paternally related males are more stable or more successful than those among more distant patrilineal kin or nonkin in this community cannot yet be assessed because the necessary paternity data are not yet available. Nonetheless, the variance in male mating success, even among coalition partners, suggests that fertilizations of Ngogo chimpanzees may be at least as skewed in favor of high-ranking males as they are in other chimpanzee communities with fewer competing males (Constable et al. 2001, Vigilant et al. 2001).

The absence of male coalitions in within-group competition in the patrilocal societies of bonobos (Furuichi & Ihobe 1994, Furuichi 1997) or muriquis (Strier et al. 2000) may be a consequence of the greater influence of females or greater difficulties males face in detecting female ovulation in these species (Gerloff et al. 1999, Strier et al. 2002). Bonobo males rely primarily on their mothers instead of one another for support in agonistic interactions (Furuichi 1997), while in the egalitarian societies of muriquis, males appear to increase their mating success by maintaining close spatial associations that permit them to monitor one another and share in their mating efforts (Strier 1997b, Strier et al. 2002). Under such extreme scramble competition in muriquis, the probability of paternal sibships among their age cohorts should be lower, while that across their age cohorts may be more common than it is when access to females involves contest competition and fertilizations are monopolizable.

Patrilineally Related Females in Matrilocal Societies

Maternal and matrilineal kinship are typically invoked to explain the cooperation of matrilocal females in both between- and within-group competition. However, as recent studies indicate, paternal kinship, particularly among age cohorts, plays a role in mediating not only social relationships among females (Widdig et al. 2001) but also the probability and nature of sexual relationships between females and males (Alberts 1999) in the matrilocal societies of rhesus macaques and yellow baboons, respectively.

Yet many basic questions remain about the effects of paternal kinship on the tradeoffs between female cooperation and competition, and the role of paternal kin in cooperation against other groups of females. For example, same-aged paternal sisters might be more or less likely to support one another over more distantly related members of their respective matrilines depending on the fitness costs and benefits and the relative strength of paternal versus matrilineal kin biases (Chapais et al. 2001). While Widdig et al. (2001) found the average affiliation index (a measure of proximity, grooming, and approach scores) among same-aged paternal sisters to be significantly higher than that among nonkin, it was still significantly lower than that among maternal sisters (1.745 ± 0.739 vs. 10.771 ± 1.039, respectively) in rhesus macaques. What the magnitude of this difference implies for cooperation among different categories of kin in within- versus between-group competition is a topic for future research to explore.

Relationships between paternal sisters in matrilocal societies also differ fundamentally from those between maternal brothers in patrilocal societies, because only paternal siblings can be age peers. One might therefore predict that the greatest opportunities for paternal kinship to impact the behavior of female primates would be in those matrilocal societies in which female relationships are egalitarian and cooperation among matrilineal kin is minimal (Sterck et al. 1997).

Dispersing Males

To associate with females, dispersing males must join extant groups when females are matrilocal, or establish new groups with females when females also disperse from their natal groups. Males that disperse with peers from their natal groups will be familiar to one another, but whether they are paternally related will be determined by the mating patterns that prevailed when they were conceived.

The best examples of dispersing males maintaining long-term coalitions in other groups come from societies, such as those of red howler monkeys, in which alpha males account for most, if not all, fertilizations during their tenures. In red howler monkeys, coalitions of males from the same natal troops endured longer and were more successful at taking over troops and repelling takeover attempts than coalitions of males from different natal troops (Pope 1990). Because alpha male red howler monkeys can maintain exclusive breeding monopolies, natal males that form coalitions to establish new troops are minimally paternal half siblings ($r = .25$), even without the potential matrilineal kinship they inherit through their mothers (Pope 1998).

Among Peruvian squirrel monkeys, one or two males monopolize most copulations during the annual four-week breeding season (Mitchell 1994). Although dispersal should result in lower levels of relatedness among males in Peruvian squirrel monkey troops than in patrilocal Costa Rican squirrel monkeys (Boinski & Mitchell 1994), similarities in the mating monopolies of alpha males in both species should result in a high probability of paternal sibships among age cohorts in both species (Boinski 1987). Indeed, matrilineal kinship among matrilocal female Peruvian squirrel monkeys may result in even closer genetic relatedness among paternal siblings than occurs among age cohorts in the patrilocal societies of Costa Rican squirrel monkeys. Such close kinship could contribute to the longevity of migration alliances in Peruvian squirrel monkeys, which involve two to four males that are close to one another in both age and rank (Mitchell 1994).

Like male chimpanzee and red howler monkey coalitions, the migration alliances of male squirrel monkeys appear to be mutually beneficial to the participants (Mitchell 1994). Alliance partners cooperate with one another in agonistic interactions that involve genital displays toward other males, and all but the lowest ranking males are more successful at eliciting subordinate responses when they display together than alone. Cooperative genital displays toward females, which are dominant over males in this species, are less likely to succeed than those of solitary males, suggesting that it is the benefits of cooperating in male-male competition instead of in attracting females that accounts for the long duration of male squirrel monkey alliances. However, whether paternal kinship also contributes to the longevity or success of squirrel monkey migration alliances, as it appears to affect red howler monkey coalitions, is not yet known.

Young male ring-tailed lemurs, vervet monkeys, and long-tailed macaques also disperse from their natal groups with peers, but these associations usually dissolve when older males transfer groups again (Cheney & Seyfarth 1983, van Noordwijk & van Schaik 1985, Sussman 1992, Sauther et al. 1999). Males in these species forfeit their associations with familiar males that may also be paternally related kin when they secondarily disperse on their own in search of reproductive opportunities.

Dispersal partners will benefit mutually if doing so increases the males' success at joining or taking over extant groups, or establishing new groups, over what it would be if they dispersed alone, but whether familiar dispersal partners choose one another based on recognized kinship or some other criteria is not clear. Similarly, although the friendly interactions during intergroup encounters involving male white-fronted capuchin monkeys (*Cebus albifrons*) (Perry 1996) and gibbons (Reichard & Sommer 1997, Brockelman et al. 1998, Oka &

Takenaka 2001) have been attributed to familiarity that may facilitate their subsequent transfers into adjacent groups, it is not known whether they preferentially join groups based on the presence of recognizable patrilineal kin. Data on the paternal relatedness of dispersing males and the relative success of different kinds of migration alliances are necessary to determine whether males prefer patrilineal kin over other available dispersal partners that are similarly familiar to them.

Conclusion and Future Directions

Understanding the full impact of patrilineal kinship on the behavior of primates requires a more comprehensive integration of behavioral and genetic data than current knowledge permits. While the genetic determination of paternity is a necessary starting point, data on more distant genealogical relationships are also needed to evaluate social relationships among other categories of patrilineal kin in wild populations. Analyses of behavioral data must also be more sensitive than they have been to the social and demographic factors that affect the availability of paternally related and patrilineal kin in both patrilocal and nonpatrilocal societies alike (e.g., Alberts 1999, Mitani et al. 2002). Moreover, the degree to which both male and female primates can adjust their behavior to increase (or decrease) their access to patrilineal kin in different kinds of societies are questions that have not yet been explored.

There is clear evidence that in many species of primates, males cooperate with one another in social contexts both within and beyond their natal groups. Coalitions among males in between-group conflicts are more common (e.g., chimpanzees, spider monkeys, muriquis, Costa Rican squirrel monkeys) and more successful (e.g., mountain gorillas, red howler monkeys, Peruvian squirrel monkeys) when the participants are kin, but because they may be related to one another through their mothers as well as fathers, the effect of patrilineal versus matrilineal kinship on their behavior has not yet been teased apart. In contrast to participation in coalitions against other groups, which include a mixture of males of different ages and ranks, males that cooperate in coalitions for mate guarding (e.g., chimpanzees) or cooperative agonistic displays against other males in their group (e.g., Peruvian squirrel monkeys) tend to be more similar to one another in age. Thus, the probability of paternal relatedness among males that cooperate in within-group competition may be higher than it is among males that cooperate in between-group competition. The significantly stronger affiliations among paternal sisters that are close to one another in age than among unrelated female rhesus macaques suggest the importance of further investigations into mechanisms of phenotypic matching for personality traits (Widdig et al. 2001) and socially biased independent learning (Berman & Kapsalis 1999) in both male and female primates living in different social environments.

Decoupling patrilocality from patrilineal kinship emphasizes the potential roles of matrilineal kinship among males in patrilocal societies and paternal sibships among males and females in both patrilocal and matrilocal societies. Mating and demographic patterns interact with dispersal patterns to affect the availability of different kinds of kin at different times, and thus set constraints on the potential of patrilineal kinship to impact behavior.

Acknowledgments I am grateful to John Moore for his help with human kinship terminology, John Mitani and Fred Bercovitch for stimulating discussions on the topic of paternal sibships, and Bernard Chapais and Carol Berman for inviting my contribution to their volume. Carol Berman, Bernard Chapais, John Mitani, and John Moore provided many helpful suggestions on an earlier version of this manuscript.

References

Ahumada, J. A. 1992. Grooming behavior of spider monkeys (*Ateles geoffroyi*) on Barro Colorado Island, Panama. *Int. J. Primatol.*, 13, 33–49.

Alberts, S. C. 1999. Paternal kin discrimination in wild baboons. *Proc. Royal Soc. Lond., B*, 266, 1501–1506.

Alberts, S. C. & Altmann, J. 1995. Balancing costs and opportunities: dispersal in male baboons. *Am. Nat.*, 145, 279–306.

Altmann, J., Alberts, S. C., Haines, S. A., Dubach, J., Muruthi, P., Coote, T., Geffen, E., Cheesman, D. J., Mututua, R. S., Saiyalele, S. N., Wayne, R. K., Lacy, R. C., & Bruford, M. W. 1996. Behavior predicts genetic structure in a wild primate group. *Proc. Natl. Acad. Sci. USA*, 93, 5797–5801.

Altmann, S. A. & Altmann, J. 1979. Demographic constraints on behavior and social organization. In *Primate Ecology and Human Origins* (Ed. by I. S. Bernstein & E. O. Smith), pp. 47–64. New York: Garland Press.

Baker, A. J., Dietz, J. M., & Kleiman, D. G. 1993. Behavioural evidence for monopolization of paternity in multi-male groups of golden lion tamarins. *Anim. Behav.*, 46, 1091–1103.

Berman, C. M. & Kapsalis, E. 1999. Development of kin bias among rhesus monkeys: maternal transmission or individual learning? *Anim. Behav.*, 58, 883–894.

Bernstein, I. S. 1999. Kinship and the behavior of nonhuman primates. In: *The Nonhuman Primates* (Ed. by P. Dolhinow & A. Fuentes), pp. 202–205. Mountain View, CA: Mayfield.

Boesch, C. & Boesch-Achermann, H. 2000. *The Chimpanzees of the Taï Forest*. New York: Oxford University Press.

Boinski, S. 1987. Mating patterns in squirrel monkeys (*Saimiri oerstedi*). *Behav. Ecol. Sociobiol.*, 21, 13–21.

Boinski, S. 1994. Affiliation patterns among male Costa Rican squirrel monkeys. *Behaviour*, 130, 191–209.

Boinski, S. & Mitchell, C. L. 1994. Male residence and association patterns in Costa Rican squirrel monkeys (*Saimiri oerstedi*). *Am. J. Primatol.*, 34, 157–169.

Borries, C., Launhardt, K., Epplen, C., Epplen, J. T., & Winkler, P. 1999. Males as infant protectors in hanuman langurs (*Presbytis entellus*) living in multi-male groups—defence pattern, paternity, and sexual behaviour. *Behav. Ecol. Sociobiol.*, 46, 350–356.

Brockelman, W. Y., Reichard, U., Treesucon, U., & Raemakers, J. J. 1998. Dispersal, pair formation and social structure in gibbons (*Hylobates lar*). *Behav. Ecol. Sociobiol.*, 42, 329–339.

Buchan, J. C., Alberts, S. A., Silk, J. B., & Altmann, J. 2003. True paternal care in a multi-male primate society. *Nature*, 425, 179–181.

Chapais, B. 2001. Primate nepotism: what is the explanatory value of kin selection? *Int. J. Primatol.*, 22, 203–229.

Chapais, B., Savard, L., & Gautheir, C. 2001. Kin selection and the distribution of altruism in relation to degree of kinship in Japanese macaques (*Macaca fuscata*). *Behav. Ecol. Sociobiol.*, 49, 493–502.

Cheney, D. L. & Seyfarth, R. M. 1983. Non-random dispersal in free-ranging vervet monkeys: social and genetic consequences. *Am. Nat.*, 122, 392–412.

Clark, A. B. 1978. Sex ratio and local resource competition in a prosimian primate. *Science*, 201, 163–165.

Constable, J. L., Ashley, M. V., Goodall, J., & Pusey, A. E. 2001. Noninvasive paternity assignment in Gombe chimpanzees. *Mol. Ecol.*, 10, 1279–1300.

Crockett, C. M. 1996. The relation between red howler monkey (*Alouatta seniculus*) troop size and population growth in two habitats. In: *Adaptive Radiations of Neotropical Primates* (Ed. by M. A. Norconk, A. L. Rosenberger, & P. A. Garber), pp. 489–510. New York: Plenum Press.

de Ruiter, J. R. & Geffen, E. 1998. Relatedness of matrilines, dispersing males and social groups in long-tailed macaques (*Macaca fascicularis*). *Proc. Royal Soc. Lond., B*, 265, 79–87.

de Ruiter, J. R., van Hooff, J. A. R. A. M., & Scheffrahn, W. 1994. Social and genetic aspects of paternity in wild long-tailed macaques. *Behaviour*, 129, 203–224.

Di Fiore, A. & Rendall, D. 1994. Evolution of social organization: a reappraisal for primates by using phylogenetic methods. *Proc. Natl. Acad. Sci. USA*, 91, 9941–9945.

Erhart, E. M., Coelho, A. M. J., & Bramblett, C. A. 1997. Kin recognition by paternal half-siblings in captive *Papio cynocephalus*. *Am. J. Primatol.*, 43, 147–157.

Fox, R. 1967. *Kinship and Marriage*. Baltimore, MA: Penguin Books.

Fox, R. 1975. Primate kin and human kinship. In: *Biosocial Anthropology* (Ed. by R. Fox), pp. 9–35. New York: Halsted Press.

Fredrikson, W. T. & Sackett, G. 1984. Kin preferences in primates, *Macaca nemestrina*: relatedness or familiarity? *J. Comp. Psychol.*, 98, 29–34.

Fuentes, A. 2000. Hylobatid communities: changing views on pair bonding and social organization in hominoids. *Ybk. Phys. Anthropol.*, 43, 33–60.

Furuichi, T. 1997. Agonistic interactions and matrifocal dominance rank of wild bonobos (*Pan paniscus*) at Wamba. *Int. J. Primatol.*, 18, 855–875.

Furuichi, T. & Ihobe, H. 1994. Variation in male relationships in bonobos and chimpanzees. *Behavior*, 130, 212–228.

Garber, P. A. 1997. One for all and breeding for one: cooperation and competition as a tamarin reproductive strategy. *Evol. Anthropol.*, 5, 187–199.

Gerloff, U., Hartung, B., Fruth, B., Hohmann, G., & Tautz, D. 1999. Intracommunity relationships, dispersal pattern and paternity success in a wild living community of Bonobos (*Pan paniscus*) determined from DNA analyses of faecal samples. *Proc. Royal Soc. Lond., B*, 266, 1189–1195.

Goldberg, T. L. & Wrangham, R. W. 1997. Genetic correlates of social behaviour in wild chimpanzees: evidence from mitochondrial DNA. *Anim. Behav.*, 54, 559–570.

Goodall, J. 1986. *The Chimpanzees of Gombe: Patterns of Behavior*. Cambridge, MA: Harvard University Press.

Gowaty, P. A. & Lennartz, M. R. 1985. Sex ratios of nestling and fledgling red cockaded woodpeckers (*Picoides borealis*) favor males. *Am. Nat.*, 126, 347–353.

Hamilton, W. D. 1964. The genetical evolution of social behaviour. I and II. *J. Theor. Biol.*, 7, 1–52.

Hashimoto, C., Furuichi, T., & Takenaka, O. 1996. Matrilineal kin relationship and social behavior of wild bonobos (*Pan paniscus*): sequencing the D-loop region of mitochondrial DNA. *Primates*, 37, 305–318.

Herbinger, I., Boesch, C., & Rothe, H. 2001. Territory characteristics among three neighboring chimpanzee communities in the Taï National Park, Côte d'Ivoire. *Int. J. Primatol.*, 22, 143–167.

Jones, C. 1999. Why both sexes leave: effects of habitat fragmentation on dispersal behavior. *Endangered Species UPDATE*, 15, 70–73.

Kappeler, P. M. 1997. Determinants of primate social organization: comparative evidence and new insights from Malagasy lemurs. *Biol. Rev.*, 72, 111–151.

Melnick, D. J. & Hoelzer, G. A. 1992. Differences in male and female macaque dispersal lead to contrasting distributions of nuclear and mitochondrial DNA variation. *Int. J. Primatol.*, 13, 379–393.

Mitani, J. C., Merriwether, D. A., & Zhang, C. 2000. Male affiliation, cooperation and kinship in wild chimpanzees. *Anim. Behav.*, 59, 885–893.

Mitani, J. C. & Watts, D. P. 2001. Why do chimpanzees hunt and share meat? *Anim. Behav.*, 61, 915–924.

Mitani, J. C., Watts, D. P., Pepper, J. W., & Merriwether, D. A. 2002. Demographic and social constraints on male chimpanzee behaviour. *Anim. Behav.*, 64, 727–737.

Mitchell, C. L. 1994. Migration alliances and coalitions among adult male South American squirrel monkeys (*Saimiri sciureus*). *Behaviour*, 130, 169–190.

Moore, J. 1984. Female transfer in primates. *Int. J. Primatol.*, 5, 537–589.

Moore, J. 1992. Dispersal, nepotism, and primate social behavior. *Int. J. Primatol.*, 13, 361–378.

Moraes, P. L. R., Carvalho Jr., O., & Strier, K. B. 1998. Population variation in patch and party size in muriquis (*Brachyteles arachnoides*). *Int. J. Primatol.*, 19, 325–337.

Morin, P. A., Moore, J. J., Chakraborty, R., Jin, L., Goodall, J., & Woodruff, D. S. 1994. Kin selection, social structure, gene flow, and the evolution of chimpanzees. *Science*, 265, 1193–1201.

Nishida, T. & Hosaka, K. 1996. Coalition strategies among adult male chimpanzees of the Mahale Mountains, Tanzania. In: *Great Ape Societies* (Ed. by W. C. McGrew, L. F. Marchant, & T. Nishida), pp. 114–134. New York: Cambridge University Press.

Noë, R. 1992. Alliance formation among male baboons: shopping for profitable partners. In: *Coalitions and Alliances* (Ed. by A. H. Harcourt & F. B. M. de Waal), pp. 285–322. Oxford: Oxford University Press.

Oka, T. & Takenaka, O. 2001. Wild gibbons' parentage tested by non-invasive DNA sampling and PCR-amplified polymorphic microsatellites. *Primates*, 42, 67–73.

Parr, L. A. & de Waal, F. B. M. 1999. Visual kin recognition in chimpanzees. *Nature*, 399, 647–648.

Perry, S. 1996. Intergroup encounters in wild white-faced capuchins (*Cebus capucinus*). *Int. J. Primatol.*, 17, 309–330.

Pope, T. R. 1990. The reproductive consequences of male cooperation in the red howler monkey: paternity exclusion in multi-male and single-male troops using genetic markers. *Behav. Ecol. Sociobiol.*, 27, 439–446.

Pope, T. R. 1992. The influence of dispersal patterns and mating system on genetic differentiation within and between populations of the red howler monkey (*Alouatta seniculus*). *Evolution*, 46, 1112–1128.

Pope, T. R. 1998. Effects of demographic change on group kin structure and gene dynamics of populations of red howling monkeys. *J. Mammal.*, 79, 692–712.

Pope, T. R. 2000. Reproductive success increases with degree of kinship in cooperative coalitions of female red howler monkeys (*Alouatta seniculus*). *Behav. Ecol. Sociobiol.*, 48, 253–267.

Reichard, U. & Sommer, V. 1997. Group encounters in wild gibbons (*Hylobates lar*): agonism, affiliation, and the concept of infanticide. *Behaviour*, 134, 1135–1174.

Robbins, M. M. 1995. A demographic analysis of male life history and social structure of mountain gorillas. *Behaviour*, 132, 21–47.

Sackett, G. P. & Fredrikson, W. T. 1987. Social preferences by pigtailed macaques: familiarity versus degree and type of kinship. *Anim. Behav.*, 35, 603–607.

Sade, D. S. 1991. Kinship. In: *Understanding Behavior: What Primate Studies Tell Us*

About Human Behavior (Ed. by J. D. Loy & C. V. Peters), pp. 229–241. New York: Oxford University Press.

Sauther, M. L., Sussman, R. W., & Gould, L. 1999. The socioecology of ringtailed lemur: thirty-five years of research. *Evol. Anthropol.*, 8, 120–132.

Silk, J. B. 1983. Local resource competition and facultative adjustment of sex ratios in relation to competitive abilities. *Am. Nat.*, 108, 203–213.

Soltis, J., Thomsen, R., & Takenaka, O. 2001. The interaction of male and female reproductive strategies and paternity in wild Japanese macaques, *Macaca fuscata*. *Anim. Behav.*, 62, 485–494.

Sterck, E. H. M. 1998. Female dispersal, social organization and infanticide in langurs: are they linked to human disturbance? *Am. J. Primatol.*, 44, 235–254.

Sterck, E. H. M., Watts, D. P., & van Schaik, C. P. 1997. The evolution of female social relationships in nonhuman primates. *Behav. Ecol. Sociobiol.*, 41, 291–309.

Stevenson, P. R. 1988. Proximal spacing between individuals in a group of woolly monkeys (*Lagothrix lagotricha*) in Tingua National Park, Colombia. *Int. J. Primatol.*, 19, 299–311.

Strier, K. B. 1990. New World primates, new frontiers: insights from the wooly spider monkey, or muriqui (*Brachyteles arachnoides*). *Int. J. Primatol.*, 11, 7–19.

Strier, K. B. 1991. Demography and conservation in an endangered primate, *Brachyteles arachnoides*. *Conserv. Biol.*, 5, 214–218.

Strier, K. B. 1992. Causes and consequences of nonaggression in woolly spider monkeys. In: *Aggression and Peacefulness in Humans and Other Primates* (Ed. by J. Silverberg & J. P. Gray), pp. 100–116. New York: Oxford University Press.

Strier, K. B. 1994. Brotherhoods among *atelins*. *Behaviour*, 130, 151–167.

Strier, K. B. 1997a. Mate preferences of wild muriqui monkeys (*Brachyteles arachnoides*): reproductive and social correlates. *Folia Primatol.*, 68, 120–133.

Strier, K. B. 1997b. Subtle cues of social relations in male muriqui monkeys (*Brachyteles arachnoides*). In: *New World Primates: Evolution, Ecology and Behavior* (Ed. by W. G. Kinzey), pp. 109–118. New York: Aldine de Gruyter.

Strier, K. B. 1999a. Why is female kin bonding so rare: comparative sociality of New World primates. In: *Primate Socioecology* (Ed. by P. C. Lee), pp. 300–319. Cambridge: Cambridge University Press.

Strier, K. B. 1999b. Predicting primate responses to "stochastic" demographic events. *Primates*, 40, 131–142.

Strier, K. B. 2000a. From binding brotherhoods to short-term sovereignty: the dilemma of male Cebidae. In: *Primate Males: Causes and Consequences of Variation in Group Composition* (Ed. by P. M. Kappeler), pp. 72–83. Cambridge: Cambridge University Press.

Strier, K. B. 2000b. Population viabilities and conservation implications for muriquis (*Brachyteles arachnoides*) in Brazil's Atlantic forest. *Biotropica*, 32, 903–913.

Strier, K. B., Carvalho, D. S., & Bejar, N. O. 2000. Prescription for peacefulness. In: *Natural Conflict Resolution* (Ed. by F. Aureli & F. B. M. de Waal), pp. 315–317. Berkeley, CA: University of California Press.

Strier, K. B., Dib, L. T., & Figueira, J. E. C. 2002. Social dynamics of male muriquis (*Brachyteles arachnoides hypoxanthus*). *Behaviour*, 139, 315–342.

Strier, K. B. & Ziegler, T. E. 2000. Lack of pubertal influences on female dispersal in muriqui monkeys, *Brachyteles arachnoides*. *Anim. Behav.*, 59, 849–860.

Sugiyama, Y. 1999. Socioecological factors of male chimpanzee migration at Bossou, Guinea. *Primates*, 40, 61–68.

Sussman, R. W. 1992. Male life history and intergroup mobility among ringtailed lemurs (*Lemur catta*). *Int. J. Primatol.*, 13, 395–413.

Symington, M. M. 1987. Sex ratio and maternal rank in wild spider monkeys: when daughters disperse. *Behav. Ecol. Sociobiol.*, 20, 333–335.

Symington, M. M. 1990. Fission-fusion social organization in *Ateles* and *Pan*. *Int. J. Primatol.*, 11, 47–61.

Taylor, P. D. 1992. Altruism in viscous populations—an inclusive fitness model. *Evol. Ecol.*, 6, 352–356.

van Hooff, J. A. R. A. M. 2000. Relationships among non-human primate males: a deductive framework. In: *Primate Males: Causes and Consequences of Variation in Group Composition* (Ed. by P. M. Kappeler), pp. 183–191. Cambridge: Cambridge University Press.

van Hooff, J. A. R. A. M. & van Schaik, C. P. 1994. Male bonds: affiliative relationships among nonhuman primate males. *Behaviour*, 130, 309–337.

van Noordwijk, M. A. & van Schaik, C. P. 1985. Male migration and rank acquisition in wild long-tailed macaques (*Macaca fascicularis*). *Anim. Behav.*, 33, 849–861.

Vigilant, L., Hofreiter, M., Siedel, H., & Boesch, C. 2001. Paternity and relatedness in wild chimpanzee communities. *Proc. Natl. Acad. Sci. USA*, 98, 12890–12895.

Watts, D. P. 1996. Comparative socio-ecology of gorillas. In: *Great Ape Societies* (Ed. by W. C. McGrew, L. F. Marchant, & T. Nishida), pp. 16–28. New York: Cambridge University Press.

Watts, D. P. 1998. Coalitionary mate guarding by male chimpanzees at Ngogo, Kibale National Park, Uganda. *Behav. Ecol. Sociobiol.*, 44, 43–55.

Watts, D. P. & Mitani, J. C. 2001. Boundary patrols and intergroup encounters in wild chimpanzees. *Behaviour*, 138, 299–327.

West, S. A., Murray, M. G., Machado, C. A., Griffin, A. S., & Herre, E. A. 2001. Testing Hamilton's rule with competition between relatives. *Nature*, 409, 510–513.

Widdig, A., Nürnberg, P., Krawczak, M., Streich, W. J., & Bercovitch, F. M. 2001. Paternal relatedness and age proximity regulate social relationships among adult female rhesus macaques. *Proc. Natl. Acad. Sci. USA*, 98, 13769–13773.

Wilson, D. S., Pollock, G. B., & Dugatkin, L. A. 1992. Can altruism evolve in purely viscous populations? *Evol. Ecol.*, 6, 331–341.

Wilson, M. L., Hauser, M. D., & Wrangham, R. W. 2001. Does participation in intergroup conflict depend on numerical assessment, range location, or rank for wild chimpanzees? *Anim. Behav.*, 61, 1203–1216.

Wrangham, R. W. 1987. The significance of African apes for reconstructing human social evolution. In: *The Evolution of Human Behavior: Primate Models* (Ed. by W. G. Kinzey), pp. 51–71. Albany, NY: SUNY Press.

Wrangham, R. W. 1999. Evolution of coalitionary killing. *Ybk. Phys. Anthropol.*, 42, 1–30.

9

Kinship and Behavior Among Nongregarious Nocturnal Prosimians: What Do We Really Know?

Leanne T. Nash

When I was first invited to participate in this volume, my initial reaction was: "A chapter on what we know about kinship among the nocturnal primates . . . hmm, now *that* will be a short chapter!" In this review, I include tarsiers and all the nocturnal strepsirhine primates (lemurs, lorises, galagos, pottos). Until recently, I tended to tell my students that what we currently know about dispersal patterns among nocturnal primates, and therefore what is known or mainly inferred about kinship relationships among individuals, is similar to the state of our information about diurnal primates in the mid-1970s. At that time there were few long-term field studies of individually recognized animals, so we knew very little. However, a few very recent studies, which I review below, have catapulted our knowledge into the twenty-first century, as new genetic techniques have been combined with older and very hard won methods of gathering information on sociality in these species. This is truly an exciting time to be thinking about questions of kinship and behavior among these species. One of the main goals in writing this review is to encourage more young primatologists to take on the challenges (which are many) of studying nocturnal primates. After three decades of studying and appreciating these primates, I can attest that the questions *can* be tackled, and it can even be fun doing it.

In this chapter, first I briefly discuss the kinds of questions we might ask about kinship and review some common generalizations about kinship effects on the behavior of nocturnal primates. Many of these generalizations are based on very small sample sizes. I critically review the most extensive studies among galagos, pottos, lorises, tarsiers, and nocturnal lemurs that bear on the relationship of kinship and behavior and, finally, suggest areas for future research.

What We Want to Know

The questions we ask about the relationship between kinship and behavior in nocturnal primates are the same as those we ask about diurnal gregarious primate species, for example, paternal and maternal kinship effects on behavior, dispersal and kinship, and mechanisms of kin recognition. These are well reviewed in the other chapters in this volume (e.g., chapters 4, 6, 7, 8, and 13). Within primate studies, it is important to remember that nongregarious does not mean nonsocial. This message needs to be repeated often. Nongregarious mammals do not forage together, but they must be social to reproduce. Though nongregarious, the primates included here do maintain year-round social networks and social contacts. Sociality is expressed in many ways. In nocturnal primates, it is primarily studied by examining (1) range overlap of individuals, (2) congregations at sleeping places (often nests or tree holes), and (3) direct social interactions (Harcourt & Nash 1986, Nash & Harcourt 1986, Bearder 1987, Sterling et al. 2000, Sterling & Radespiel 2000). These social interactions may be face-to-face, including grooming, mating, aggression, and potentially cooperative interactions (e.g., alliances in defense of a nest site, nursing another female's offspring) but among nocturnal primates, they prominently include vocalizations and scent-marking behaviors, that is, communication at a distance in time and in space. These modalities of communication are more challenging to decode, and analyses require special expertise. In addition to these descriptions of the social systems (mainly sharing sleeping sites) and spacing systems of behavior (revealed through range overlap), ideally we want to know about the mating systems (Sterling & Richard 1995), but these are much more difficult to determine, as copulations are rarely observed. Inferences from genetic data are very revealing (see below).

Until the advent of new genetic methods (e.g., mtDNA and microsatellite data), the key to understanding the role of kinship in behavior among gregarious diurnal primates was data from studies spanning several generations with identified individuals. In principle, given the rapid life history schedule in nocturnal primates (Harvey et al. 1987, Strier 2003), we should be able to gather equivalent data on them in about half the number of years of continuous study that has been required in monkeys and apes. In fact, however, as is detailed below, there have been few such studies. Where we do see cooperative behaviors and potential paternal investment, is kin selection operating? Does co-sleeping in the same nest lead to improved thermoregulation or predator avoidance (Radespiel et al. 1998)? Are nests a limited resource shared by kin? Does the learning of new foods take place by observing closely related social companions (Nash 1993)? Genetic data and lifetime data on reproductive success are needed. Again, we need sufficiently longitudinal studies across multiple generations. We also need to be cautious in making conclusions about kinship and behavior, as not all nepotistic behaviors may evolve by the mechanism of kin selection (Chapais 2001), and the effects of dispersal on kin structure in a local area are not always as expected (Burland et al. 2002).

If kin selection is operating, how does it interact with other factors in producing the behavioral variations we find among nocturnal prosimians? For example, Clark found that there might be differentiated mother–offspring relationships among twin *Galago crassicaudatus* and marked temperamental differences between twins. She wondered if it related to an advantage of "diversifying" offspring to maximize maternal reproductive success, either through the nature of the mother-infant interactions during development or due to offspring having different fathers (Clark 1982a, Clark & Ehlinger 1987). If choosing mates such that successive offspring (or litter mates) are not of the same father increases a female's fitness (e.g., Spencer et al. 1998), how does this impact kin-selected behavior? Are there alternative but equivalent reproductive tactics among males? Is kinship involved in which are chosen? Among *Galago moholi*, males may show two morphological classes, possibly analogous to flanged and unflanged male orangutans (Bearder & Martin 1980a, Bearder 1987, Nash 1993, Pullen et al. 2000, Maggioncalda & Sapolsky 2002). It remains to be seen if there is a genetic basis to the difference and to examine the impact such differences might have on kin-selected behaviors. Though life history may be generally fast in nocturnal primates, how does life history variation among species, for example in litter size or timing of reproductive maturity relative to seasonality of resources, impact the possibility of kin-related behaviors (Kappeler 1996)? What are the mechanisms of kin recognition in these species that primarily use vocal and olfactory communication?

What We Think We Know

Most impressions in the secondary literature about kinship in nongregarious species have been based on patterns of dispersal suggested for these species within a social organization, usually suggested to be found in all nocturnal primates, that is variously described as "noyau" (population nucleus) or "dispersed polygyny." These terms carry the implication that all species are socially alike. One of the first studies of nocturnal primate social behavior was that of Bearder on *Galago moholi* (formerly *Galago senegalensis*), and his influential review indicated that females of this species formed "matriarchies."

> Maturing daughters . . . adopt territorial behaviors toward strange females and young, and their movements tend to become separate from those of the mother. However, they are often unable to establish separate territories because of the presence of aggressive female neighbors. They continue to have regular amicable contacts with the mother . . . and related females may rear offspring in the same nest. If, . . . an adjacent territory becomes available . . . either the daughter or the mother will make an abrupt shift to the new area, and their relationship soon becomes one of mutual intolerance. . . . Most males make a prereproductive migration away from the area of their birth. (Bearder 1987: 18)

Even though Bearder then went on to discuss other social patterns, this image was very powerful. Similarly, in my own work with Harcourt, we suggested similar matriarchies may be present. Concerning *Galagoides zanzibaricus*, we wrote "[male ranges] were shared with one or two (probably related) females (and their offspring) with which the male slept. . . . Young males dispersed from their natal ranges while females did not" (Harcourt & Nash 1986: 353). For *Otolemur garnettii*, we stated, "Males dispersed further and at a younger age than did females" (Nash & Harcourt 1986: 364).

This has led to the following sorts of statements about nocturnal primates in secondary sources such as textbooks and review articles. "Females are likely to stay with their mothers . . . and it is likely that nesting groups consist of individuals of the same matriline . . . " (Sussman 2000: 92). "In *all* bushbabies, adult females have relatively small ranges that overlap those of related females, while adult males have considerably larger ranges that incorporate those of a number of females. This distribution is partly due to the fact that males leave the mother's territory upon reaching sexual maturity, but females do not. Therefore, the small social groups . . . are matrilineal" (Falk 2000: 70–71, emphasis added). About *Galagoides demidoff,* "Social structure: . . . Matriarchies are present" (Rowe 1996: 22). And last, from my own work, "In those species where dispersal patterns are known, males disperse more often, farther, and, in some cases, earlier than females. . . . However, among these species some females also disperse. . . . Mothers, sisters, aunts, and nieces may compete for ranges" (Nash 1993: 123).

What We Really Know

The title of this chapter and many of the quotations above might be taken to imply that all the species included in this discussion are socially similar. What we clearly do know, after three decades of effort, is that among the nongregarious, solitary foraging, nocturnal primates, several social patterns are present that seem to correspond with social system differences among diurnal primates. Based mainly on ranging patterns and sleeping associations, Bearder (1987) described five different patterns: (1) male ranges are larger than female with "matriarchies" present; (2) male ranges are larger than female without "matriarchies"; (3) male and female ranges coincide, but they are nongregarious and there are no apparent matriarchies; (4) male and female ranges seem to show little overlap, and presence of matriarchies is unknown; and (5) a male, female, and associated immatures forage gregariously, but the presence of matriarchies is unknown.

More recent studies have shown that not all of the nocturnal primates fall easily into these categories (Bearder 1999). Müller and Thalmann (2000), following Sterling and Richard (1995), differentiated between the spacing system (the most easily obtainable data on the spatiotemporal distribution of individuals), the social system (based on relationships and interactions that are difficult to observe, i.e., scent and vocal communication, as well as cosleeping), and the mating system (reproductive interactions best identified through paternity testing). After eliminating the mating system, due to the usual lack of genetic information, they then identify four possible patterns of social organization, using the following terms: (1) monogamy (i.e., ranges of one male and one female coincide), (2) multimale/multifemale system (i.e., the range of a male overlaps that of several females while a female's range overlaps that of several males), (3) harem (i.e., the range of one male overlaps those of several females but a female's range overlaps only one male, sometimes called dispersed polygyny), and (4) polyandry (i.e., the home range of one female overlaps those of several males). They further point out that among all the "solitary" mammals, we need to discriminate between "spatial social organization," with no social contact outside the breeding season, and "dispersed social organization," whereby individuals forage alone but exhibit a continuing social network. All the nongregarious primates exhibit a dispersed social organization as far as is known, so Müller and Thalmann conclude that dispersed social organization is the ancestral condition of primate social organization, indicating that social networks must have been present already among the first primates. Of the four dispersed social organization patterns, nocturnal primates exhibit dispersed monogamy, dispersed multimale/multifemale, and dispersed harem patterns. All three types are represented among both the "lorisiform" species and the Malagasy lemurs (Müller & Thalmann 2000). However, a dispersed harem system is rare and to date is only known from *Galago alleni* and, possibly, *Lepilemur leucopus*. There is probably similar diversity within the tarsiers (Niemitz 1984, Crompton & Andau 1986, Gursky 2000a). There are still many species for which no data are currently available.

What is the proportion of nocturnal nongregarious primate species for which good (or even *some*) data about social behavior are available? That depends on how many species are recognized. In all the major clades of nocturnal primates, extensive taxonomic revision is ongoing. This has led to considerable splitting of existing species, debate about correct taxonomy, and discovery of new species (e.g., Harcourt 1984, Nash et al. 1989, Schmid & Kappeler 1994, Bearder et al. 1995, Honess 1996, Rowe 1996, Zimmermann et al. 1998, Bearder 1999, Groves 2001, Masters & Brothers 2002). Using Groves' (2001) taxonomy, there are 64 species that could have been considered in this review. Among those, some field data are available on social behavior for less than half, about 25. However, even among the better studied species, sample sizes are often very limited for data upon which to make inferences about natal dispersal and kinship.

The reasons for the proliferation of species are many and beyond the scope of this review (Bearder 1999, Zimmermann et al. 2000). However, the consequence of these taxonomic revisions for comparative studies is that it is very easy, when looking at older literature, to lump animals that are now considered to be separate species into an analysis as single species and mask variability (Clark 1988). For example, most descriptions of the behavior of *Galago senegalensis* in secondary sources (e.g., Fleagle 1988) are actually based on data from South Africa on what is now recognized as a different species, *Galago moholi* (Bearder & Doyle 1974a; Bearder & Martin 1980a, b). To be fair, this was changed in Fleagle's 1999 edition and newer texts are catching up (e.g., Falk 2000).

Galagos and Lorises: Galagonidae, Loridae

The galagos were among the first nocturnal primates for which good field data were collected (Bearder 1974, Charles-Dominique 1977, Bearder & Martin 1980a, Clark 1985). These studies, from western and southern Africa, all suggested that males dispersed futher from their mother's range than did females in *G. moholi, Galago demidoff* (=*Galagoides demidoff*), and *Otolemur crassicaudatus* (=*Galago crassicaudatus*). In these species, young adult females often shared ranges with their mother and/or sisters. Charles-Dominique (1977) described the pattern as the formation of "matriarchies." A pattern of galago female natal residency and male dispersal was critical to Clark's (1978) "local resource competition" hypothesis of sex ratio adjustment. Later studies, in East Africa, of *Galago zanzibaricus* (=*Galagoides zanzibaricus*) and *Otolemur garnettii* (=*Galago garnettii*) suggested a similar pattern for these species (Harcourt & Nash 1986, Nash & Harcourt 1986). However, table 9.1 illustrates that these conclusions were based on studies that rarely exceeded two years in length and were also based on very small sample sizes for known maturing, and thus potentially dispersing, males or females. In spite of the limited data, the notion that all nocturnal prosimians show dispersed polygyny and matriarchies persists in the literature to the point that it has been called "the myth of the typical solitary primate" (Kappeler 1997a: 122). In fact, these studies have also suggested that there is quite a lot of variation in social organization in nocturnal prosimians. For example, using Müller and Thalmann's (2000) definitions, *G. alleni* has a dispersed harem organization, *G. zanzibaricus* a dispersed mo-

Table 9.1. Nature of Field Data Available on Dispersal in the Better Studied Galagos (Galagonidae)[a]

Species and Source [b]	Site	Study Duration	Type [c]	Ranging Data Type	Dispersal Observed	(N)f [d]	(N)m [d]	Age at Sexual Maturity; Litter Size
Otolemur crassicaudatus (1)	Tzaneen, South Africa	15 months	c	Trap-retrap	y	1	1	Sexual maturity, 18–24 months; twins and triplets
O. crassicaudatus (2)	Thompson Farm (Wallacedale), Louis Trichard, South Africa	16 months	i?	Trap, mark	y	7	24	See above
O. crassicaudatus (3)	Same as above	Parts of each season (summer, winter) in one year (across about an 18-month period)	i	Trap, mark, some radio tracking	y	?	?	See above
O. garnettii (4)	Gedi, Kenya	21 months	c	Trap-retrap, some radio tracking	y	6	2	Wild: first mating, 20 months, both sexes; singletons
O. garnettii (5)	Diani, Kenya	23 months	c	Radio tracking, some trap-retrap	y	1	6	See above
Euoticus elegantulus (6)	Makokou, Gabon	42 months over 7 visits within a 104-month period	i	Ad lib	n			Sexual maturity: 10 months; 1 singleton
Galago senegalensis (7)	Mutara, Kenya	18 days	c	Survey, unmarked Ss	n			Sexual maturity: about 1 year
G. moholi (8)	Mosdene, near Naboonspruit, South Africa	24 months	c	Radio tracking, trap-retrap	y	5–6?	10?	First birth: 10–12 months; twins common

Table 9.1. Continued

Species and Source [b]	Site	Study Duration	Type [c]	Ranging Data Type	Dispersal Observed	(N)f [d]	(N)m [d]	Age at Sexual Maturity; Litter Size
G. moholi (9)	Same as above	Parts of each season (summer, winter) in one year (across an 18-month period of trapping)	i	Trap, mark, some radio tracking	y	3+?	?	See above
G. moholi (10)	Nylsvley Nature Reserve, South Africa	Span of 20 months in periods of 9, then 9 months with 2 months break	c	Radio tracking, trap-retrap	y	5	11	See above
G. alleni (11)	Makokou, Gabon	42 months over 7 visits within a 104-month period	i	Radio tracking, trap-retrap	y	?	2	Sexual maturity 8–10 months; singleton?
G. zanzibaricus (12)	Gedi, Kenya	21 months	c	Trap-retrap, some radio tracking	y	6	3	Wild: female's first birth, 12 months; singletons
G. zanzibaricus (13)	Diani, Kenya	23 months	c	Radio tracking, some trap-retrap	y	4	7	See above
G. demidoff (14)	Makokou, Gabon	42 months over 7 visits within a 104-month period	i	Radio tracking, trap-retrap	y	?	?	Sexual maturity 8–10 months; singletons

[a]For context, the following species of galagos from Groves (2001) lack any pertinent data: G. thomasi, G. murinus, G. orinus, G. rondoensis, G. udzungwensis, G. nyasae, G. granti, G. gabonensis, G. cameronensis, G. matschiei, G. gallarum, E. pallidus, O. monteiri, and three undescribed species in the "orinus" and "zanzibaricus" groups.

[b]Sources: (1) Bearder 1974; (2) Crompton 1983, 1984; (3) Harcourt 1980, 1986; (4) Nash & Harcourt 1986; (5) Nash & Harcourt 1986; (6) Charles-Dominique 1977; (7) Izard & Nash 1988, Nash & Whitten 1989; (8) Bearder & Doyle 1974b, Bearder & Martin 1980a, Bearder 1987; (9) Harcourt 1980, 1986; (10) Pullen 2000, Pullen et al. 2000; (11) Charles-Dominique 1977; (12) Harcourt & Nash 1986; (13) Harcourt & Nash 1986; (14) Charles-Dominique 1977, as well as Honess 1996. Only the study of G. moholi by Pullen (2000, 2002; Pullen et al. 2000) incorporated any genetic information.

[c]Study type: continuous, c; intermittent, i.

[d]Numbers of individuals of each sex (f, female; m, male) for which some inference about dispersal could be made. Often difficult to determine in the sources.

nogamous organization, and the remainder of the species mentioned above might have a dispersed multimale organization. This would suggest considerable variation in kinship structure among these species.

Available research provides few indications of which behaviors might be potentially kin selected. In many of these species, animals share a sleeping place—a potentially limited resource (e.g., *G. alleni, Euoticus elegantulus, G. demidoff, G. zanzibaricus, G. moholi, O. garnettii,* and *O. crassicaudatus* each form sleeping groups). These groups are most likely composed of mothers, their infants, and possibly other kin such as possible fathers and sibs. However, this remains to be well established beyond mothers and dependent young. Subadult and nonmother adult females may stay in proximity to a parked infant while its mother forages (Clark 1985). *G. zanzibaricus* will join together in mobbing a potential predator (Nash, personal observation). Captive *G. senegalensis* females (sisters or mother and daughter) who have infants at the same time will nurse each other's young (Nash, personal observation). Young animals pay close attention to the feeding choices of adults (Charles-Dominique 1977, Nash, personal observation). Learning of potentially safe food from kin may be critical; a young, hand-reared Demidoff's bushbaby died after eating an irritant moth that, presumably, its natural mother would not have eaten (Charles-Dominique 1977). Young Senegal galagos (figure 9.1) were more successful at "food taking" attempts directed at older sibs and their mother than at adult males (Nash 1991). Juvenile *G. demidoff* may follow the "A" (larger, dominant) male associated with their mother and have amicable

Figure 9.1. A captive lesser galago (*Galago senegalensis*) juvenile (center) watches adults eat and attempts to take food. Arizona State University galago colony. Photo by Leanne T. Nash.

enough relationships to be trapped in the same trap (Charles-Dominique 1977). It is unknown if these represent father-young associations. No clear paternal behavior was noted for *G. zanzibaricus* (Harcourt & Nash 1986), despite its possible monogamy.

The only behavioral genetic field study published to date is on *G. moholi* (Pullen et al. 2000, Pullen & Dixson 2002). The division of adult males into large and socially dominant individuals (A males) and smaller, socially subordinate ones (B males) did not clearly differentiate paternity success. This finding implicated the importance of scramble (sperm) competition among males for mating in this species. The study did not extensively address dispersal and relatedness among females or female-female interactions.

Even less is known about the pottos of Africa and the lorises of Asia. *Perodicticus potto* do not show matriarchies in Gabon (Charles-Dominique 1977) but may live in male-female pairs (based on ranging but rare cosleeping) in Cameroon (Pimley et al. 2002). In Gabon, Charles-Dominique suggested that a female might abandon her range for her daughter, but this was based on only one case. Two of the three young maturing males he observed moved out of their natal range. The young male who stayed was a special case in which the adult male sharing a range with the young male's mother had died *and* this range locale was relatively isolated from the main forest block, making the young male's dispersal less easy. Pimley's work is not yet sufficiently long term to provide information on dispersal.

The best data to date on lorises is for the slender loris, *Loris tardigradus lydekkerianus* (Nekaris 2000, 2002; Radhakrishna 2001; Radhakrishna & Singh 2002). These studies were done at the same site over a 21-month period. The ranging and sleeping systems suggest a dispersed multimale social organization. Only three juvenile females matured to subadult status during the study. In two cases the daughter left the maternal range and in the third, the adult female, as in the one potto case, left her range and the daughter stayed. Sleeping in "family groups" and food sharing with offspring (Part III figure), as in galagos, has been observed in captive animals (Schulze & Meier 1995).

Tarsiers: Tarsiidae

Gursky's studies of *Tarsius spectrum* provide intriguing suggestions of behaviors that appear kin related (Gursky 1994, 1997, 2000a, b, c). The modal grouping pattern is a male-female pair and dependent young that share a range and sleeping sites. For a nocturnal primate, there is considerable contact during the night and Gursky has questioned if the species should be described as nongregarious. The degree of sociality is higher when food supplies are more abundant. Subadult females offer considerable allocare to infants, presumed sibs, from within their group, and male adults and subadults also provide some care. Possible male parental investment is suggested, as males also patrol group boundaries and may actively defend against predators. In one case, three males from neighboring groups and a female mobbed and actually attacked a snake that had predated a tarsier. Since some male and female young adults ($n = 3$) who established new territories did so near their parents' range, Gursky speculated that this response on the part of neighbors might be kin based. In a larger data set, males were found to disperse further from their natal range than did females, and established pairs showed relatively high site fidelity across periods of two to five years (Gursky, personal communication). This suggests that male or female kin may continue to interact over long periods, but the influences on dispersal by either sex need more study. Gursky's recent experimental studies with predator models suggest that grouping patterns among probable mates and offspring are sensitive to predation threat as well as food supply (Gursky 2001, 2002a). The addition of genetic data to these sorts of observations would be most informative.

Tarsius bancanus has also been studied in the field, but the studies have either been done in semicaptive animals in cages in the forest (Niemitz 1984) or have not been of sufficient duration to examine natal dispersal patterns (Crompton & Andau 1987). Since the latter study indicates that the social system of *T. bancanus* may not be as pair based as that of *T. spectrum*, comparative genetic data on kinship and behavior would be of interest.

Malagasy Lemurs: Daubentoniidae, Megaladapidae (=Lepilemuridae), Indridae

Daubentonia madagascariensis, the aye-aye, is the most peculiar of all primates. Aye-ayes have been the subject of a study by Sterling lasting two years and of other shorter studies at two sites (Sterling 1993, 1994; Ancrenaz et al. 1994; Andriamasimanana 1994; Sterling & Richard 1995). Among the Malagasy lemurs, they are the only species that does not show highly seasonal breeding, probably due to their specialized diet of foods that are available year round. Males' ranges are very large relative to females' ranges. Females' ranges do not overlap each other, but males' ranges overlap with those of several other males, and with the ranges of several females. During an estrous cycle, females advertise estrus with scent marking and vocalization, and they mate with more than one male. The social organization is best described as dispersed multimale. In contrast to *O. crassicaudatus*, where encounters between animals are frequent and prolonged (Clark 1985), aye-ayes' affiliative interactions are brief. However, occasional foraging associations are seen between an adult male and other adult or young males. It is unknown if these animals were related. In Sterling's study, none of the females had associated young, but in one of the shorter studies, an older juvenile female abruptly disappeared from her natal range at the time her mother resumed estrous cycling (Andriamasimanana 1994). Thus, matriarchies may not be present in the aye-aye spacing system, but much more information is needed.

No studies of *Lepilemur* have been done that exceeded one year. Field studies of these small-bodied folivores have suggested that *L. edwardsi* live in pairs or in dispersed family groups, *L. ruficaudatus* may live in pairs, and *L. leucopus* may be variable in pattern at different sites or at different seasons (Charles-Dominique & Hladik 1971, Russell 1978, Ganzhorn & Kappeler 1996, Warren & Crompton 1997, Nash 1998, Thalmann 2001). Methodological limitations frustrate comparisons, as these animals can be difficult to capture, and field data are based on small numbers of each sex. No good data on dispersal of maturing offspring are available. These field studies agree that several animals may share a nest site and may occasionally forage in the same tree. For example, I have observed a male foraging in a tree with a juvenile with no other adult present. However, kin relationships among such individuals remain unknown.

Although nocturnal, the woolly lemurs, *Avahi occidentalis* and *A. laniger,* are described as showing "gregarious monogamy" (Müller & Thalmann 2000) and thus might be considered outside the scope of this chapter. Even though they are nocturnal, they forage in a more cohesive pattern reminiscent of *Aotus* (Harcourt 1991, Warren & Crompton 1997, Thalmann 2001). Thalmann (2001) argues that, since they feed selectively on relatively rare resources, females may select males who share knowledge of and can help them defend localized, repeatedly exploited, resources. Thus, males might be offering paternal investment in young, but no detailed studies of male–young interactions, long-term studies of dispersal, or genetic data are available.

Malagasy Lemurs: Cheirogaleidae

From the perspective of understanding kinship and behavior among nocturnal prosimians, some of the most detailed data have come from studies of *Cheirogaleus medius*, *Microcebus murinus*, *Mirza coquereli,* and *Phaner furcifer*. *Cheirogaleus medius* has been the subject of two field studies in western Madagascar, at Kirindy (Fietz 1999a, b; Fietz & Ganzhorn 1999; Fietz et al. 2000), and northwestern Madagascar (Müller 1998, 1999a, b) at Ampijoroa. These studies are reviewed by Müller and Thalmann (2002). The social organization is dispersed monogamy, with a male, female, and dependent young from several successive years sharing a range and sleeping sites (figure 9.2). Young may stay in the natal range for up to three years after birth and provide allocare to subsequent offspring. Males and females both nest guard and provide warmth in the nest to newborns. This species exhibits several months of torpor in a shared nest site during the cool dry season of low food availability. Male parental investment may also take the form of territorial defense, as males

Figure 9.2. A western fat-tailed dwarf lemur (*Cheirogaleus medius*) couple leaving its sleeping site at dusk at Ampijoroa Forestry Station in northwestern Madagascar. Note the radio collar on the animal on the right. Photo by Urs Thalmann.

emerge from torpor before females do so and (re)establish their territories. The animals need to put on sufficient fat to get them through the season of torpor. New indications are that the plants that they consume (fruit, flowers, and nectar high in sugars) may be rare and show low species diversity (Müller & Thalmann, 2002). Thus, territorial defense by the male may be critical to his mate's reproductive success. "Floater" males may also exist in the population.

Microsatellite genetic data (Fietz et al. 2000) now indicate that yearlings and infants living together were siblings (actual offspring of the social mother) but that there had also been successful extrapair copulations (EPCs). The social father was excluded from paternity in 44% of the cases. None of the EPCs were attributable to floaters, but instead came from males siring offspring with their social mates as well as with other females from other social pairs. If males do not disperse far from the natal range, as some data suggest, it may be that a male who raises the offspring of another male might be raising the offspring of a closely related male. Better long-term data on dispersal and genetic relatedness are still needed, even though this species is relatively well studied.

Microcebus murinus has also been the subject of intensive behavioral study by research teams working at Kirindi in western Madagascar (Ganzhorn & Schmid 1998; Schmid 1998; Schmid & Kappeler 1998; Fietz 1999c; Schmid & Speakman 2000; Eberle & Kappeler 2002a, b; Wimmer et al. 2002) and at Ampijoroa in northwestern Madagascar (Radespiel et al. 1998, Radespiel 2000, Radespiel, Ehresmann, et al. 2001, Radespiel, Sarikaya, et al. 2001, Schmelting et al. 2001). Both studies incorporated behavioral and genetic data. The Kirindy study involved trap-retrap of marked individuals and some radio tracking over a four-year period. Genetic data (microsatellites and mtDNA) were also analyzed for 85 individuals. At Ampijoroa, similar methods were used over a three-year period and microsatellite genotypes were determined for 161 individuals (see below).

In both study areas, females' ranges often overlapped extensively, and females shared tree hole sleeping sites with each other (even when they had infants) to a greater extent than with males. In many other nocturnal primates, for example galagos, females with newborns avoided sleeping with others, including previous offspring, and became aggressive to them (Clark 1985, Nash personal observation). Males rarely slept in a nest together without females. Thus *M. murinus* is characterized by dispersed multimale organization.

Females fattened prior to the cool, dry season and entered torpor on a daily basis during this season of low food availability. Estimates at Kirindy suggest that females may have saved up to 30% of their daily energy budget by exhibiting this behavioral and physiological strategy. Appropriately insulated tree holes were important to allow females to enter daily torpor, and such tree holes may have been in limited supply. These animals were also subject to high predation, and protected sleeping holes may have been an important resource for that reason (Goodman et al. 1993). Recent work filming mothers and infants in a shared sleeping hole showed that females sometimes nursed each other's young (Eberle & Kappeler 2002a). Males showed no evidence of anything that might be called parental investment behavior toward young and did not defend territories that they shared with females. Both scramble (sperm) competition and contest competition were implicated in the way males competed for matings.

The genetic data from Kirindy using mtDNA has shown that the females who share a nest are indeed more closely related to each other than to other such sets of females. In addition, members of several such nesting groups that share mtDNA haplotypes have ranges that are clustered in space. The data from Ampijoroa show a similar pattern. These are the first data definitely showing female philopatry, defined on a genetic basis, in a nocturnal primate. The genetic data indicate that at least some males disperse more from the natal area than do females, but more study is needed of male and female dispersal behavior to fully understand the genetic data.

Captive work also raises issues of kinship and behavior. For example, litters can be fathered by more than one male (Radespiel et al. 2002). Captive work has also shown that female mouse lemurs have highly variable estrus cycle lengths. Variability is due more to individual differences than to the effects of housing, parity, or age (Radespiel & Zimmermann 2001). Given the potential advantages of social nesting and synchronized births, as well as the high relatedness of conesting females in the wild, the genetic control of cycle length needs more investigation (Radespiel & Zimmermann 2001).

Recent studies of *Mirza coquereli* over a four-year period, encompassing 88 marked individuals, have also incorporated both ranging behavior and genetic data (Kappeler 1997a, Wimmer et al. 2000, Kappeler et al. 2002). Although earlier studies had suggested pair bonding outside of the mating season (Pagès 1978, 1980), seasonal increases in male range size and testes size led Kappeler and coworkers to predict a multimale, multifemale mating system in this species. Their genetic studies revealed that most infants in an annual cohort within the study population had different fathers, even some sets of twins. mtDNA data suggest that females sharing the same haplotype share ranges but that males do not show that pattern. A few cases of facultative dispersal by both sexes were also indicated by the genetic data. Given that females do not show cooperative range defense, the advantages of female philopatry remain unclear. The animals mainly eat insects, which might not be a good candidate for shared feeding information, but sharing information about safe sleeping places may be another possibility. To date, other advantages of sharing sleeping sites and ranges have not been investigated to the extent they have in *Microcebus murinus*.

Phaner furcifer lives in pairs that sleep together and forage relatively cohesively, although they may be alone for about 80% of the night. They stay in contact with frequent vocal duets (Charles-Dominique & Petter 1980). They have small testes for their size, suggesting low sperm competition, and there is evidence of cooperative antipredator behavior involving a pair and their possible offspring (Schülke 2001). However, microsatellite data on eight pairs indicate that the social father may not always be the biological father (Schülke et al. 2001). This presents another example, as in *Cheirogaleus medius*, where more information on dispersal patterns, social interactions, and genetic paternity will allow testing of hypotheses about sexual selection, male parental investment, kinship, and behavior.

While other cheirogalids have been the subject of limited fieldwork (Harcourt 1987, Wright & Martin 1995, Atsalis 2000), none of it has been of sufficient duration to provide insights into social organization or kinship effects on behavior. Atsalis's (2000) work on *M. rufus* in the eastern rainforest at Ranomafana suggested that females were the more stable part of the population. Nevertheless, similarities and differences between the behavior of the well-studied *M. murinus* and its newly discovered congeners, *Microcebus ravelobensis* in the northwest (Zimmermann et al. 1998) and *Microcebus myoxinus* in the west (Ortmann & Heldmaier 1997, Schwab 2000), suggests that comparative work on these species will be informative.

What We Want to Know Now and What It Will Require

A major issue that this comparative review reveals is that we need better data on just how gregarious these "nongregarious" nocturnal species may be. Excluding mother-infant associations, slender lorises spend an average of 13%, with a maximum of 36%, of their nighttime activity in close proximity to a conspecific (Nekaris 2000). Similarly, Gursky has shown that spectral tarsiers may spend about 28% of their nightly activity budget within 10 m or less (including in contact) with another adult group member (Gursky 2002b). We need far more comparative work to understand the ecological and social reasons for species differences among the nocturnal primates in degree of gregariousness and in their social patterning (Waser 1981, Kappeler 1997b, Müller & Thalmann 2000).

It is clear that we need both more studies that incorporate genetic data and more long-term studies. Even the studies that incorporate genetic information leave unresolved questions that cannot be answered without information on dispersal and social interactions over several generations. The absence of such longitudinal data, given the research efforts to date, speaks to the logistic and funding challenges such studies meet. The tantalizing hints of cooperation among adults, including behaviors that appear to be parental effort by males, need to be tested. These studies also reveal that there may be confusion over the terms "matriarchy" and "matriline" such that they need to be defined carefully and used in a consistent manner. Wimmer et al. (2002) suggest that *Microcebus murinus* mothers and daughters sharing a sleeping site should be called family units (not matrilines) and that members of several family units whose ranges are clustered in space and who share a common female ancestor should be called a matriline. Males may then disperse away from their natal ranges that were within the ranges of such a cluster of females. Incorporation of mt-DNA, microsatellite, and Y chromosome data may help sort out kinship through the male and female lines.

Studies of mechanisms of kin recognition are also needed. It is likely, as in most primates, that familiarity through association with the mother is important (Gouzoules 1984, chapter 13). Given the importance of smell and vocalizations to these species, more studies of the correlates of genetic differences and differences in the vocal or olfactory "message" would be interesting (figure 9.3). *Otolemur garnettii, O. crassicaudatus,* and *Galago senegalensis* are capable of discriminating an individual's sex and female reproductive condition through olfaction. The greater galagos, and possibly *G. senegalensis,* can also discriminate between the two species and probably between individuals based solely on olfaction (Clark 1982b, c; Carpetis & Nash 1983). In rodents, the disparate haplotypes of the major histocompatability complex (MHC) affect mate choice and can be perceived by smell (Penn & Potts 1999, Schaefer et al. 2001). Aspects of vocalizations have been used extensively to distinguish species among nocturnal prosimians (Bearder et al. 1995, Bearder 1999). Recent work has shown that features of the vocalization of grey mouse lemurs are more strongly correlated with genetic differences than with morphological differences (Zimmermann & Hafen 2001). Although vocalization differences may be "dialectlike" among demes and influenced by learning (Hafen et al. 1998), no study of a nocturnal prosimian has yet looked for vocalization similarities at the level of the individual that might aid in kin recognition.

We also need to investigate the "boundaries" of kin recognition and differential treatment of kin. Work in *Lemur catta* and in Japanese macaques suggests that a cutoff point for

Figure 9.3. Individual identity and degree of kin relationship might be perceived through olfaction. Drawing by Laura Bidner.

treating others as kin may be at a minimal degree of relatedness of about $r = .25$ (Chapais et al. 1997; Pereira & Kappeler 1997, personal communication; Belisle & Chapais 2001). This boundary may also be found in other mammals, for example ground squirrels, and may develop through experience with the mother and littermates (Holmes 1994, 1995; Holmes & Mateo 1998). A number of species have seasonal but asynchronous estrus, litters, evidence for sperm competition by males, and varying degrees of potential male parental investment behavior. Given that multiple paternity within a litter has now been documented for two genera (*Microcebus* and *Mirza*, see above), the effects of multiple-paternity litters need to be investigated for adult male-offspring and sibling relationships (Holmes 1986).

Gathering such data will be challenging. For any species, the logistics of keeping a longitudinal study going are difficult. The technology for studying nocturnal species includes radio tracking, headlamps, microchips for long-term individual identification, and expensive camera equipment for filming interactions inside sleeping places. Study sites are usually far "off the grid," and a source of electricity is required for much of this equipment that far exceeds a simple single solar panel to run a computer. This technology is expensive. For example, each radio tracking transmitter (collar) costs $150 to $300 and usually must be considered a disposable item, as one cannot always recover them for refurbishing. Even with radio tracking, to get data on several individuals at the same time, a team approach is required. This not only adds to costs, but many researchers have found it difficult to find local field assistants who are willing to work at night. Also, radio tracking does not solve all problems of making direct observations. One can be directly under an animal, with the radio "pointing in all directions," and not be able to see it (Nash personal observation—many times!).

Capture of nocturnal primates is required for radio tracking and often for acquisition of genetic samples (including feces). It can be relatively easy if animals will come to a banana-baited trap (e.g., *Cheirogaleus*, *Microcebus*, most galagos, *Mirza*). *Avahi*, a folivore that is not attracted to fruit, tends to spend the day sitting in tree forks where it can be darted (Warren & Crompton 1997, Thalmann 2001). However, gummivorous galagos and *Phaner*, highly insectivorous lorises, some pottos and tarsiers, and folivores like *Lepilemur* do not readily come into traps or sleep in tree holes or nests, or both, making darting difficult. Thus, capture ease varies with species and habitats (Charles-Dominique 1977, Warren & Crompton 1997, Nash personal observation, Kar Gupta personal communication). Once captured, there are logistical and ethical issues concerning handling of the animals, as reviewed for nocturnal primates by Jolly et al. (2003). For example, in some studies traps are monitored throughout the night, whereas other researchers put traps out in the evening and return to check the following morning. Since the first few hours of the evening are often the most productive (Jolly et al. in press), animals might spend most of the night trapped. A mother might be kept from her "parked" infant or an infant in a nest, as few of these species carry clinging infants.

Despite the methodological issues, the nocturnal primates might be less different from their diurnal relatives in the effect of kinship on their behavior, and in many facets of their social relationships, than has been supposed. Those who rise to the challenge of investigating these species may be able to test a variety of generalizations about primate social behavior that have been derived from the study of diurnal anthropoids (Kappeler 1996, 1997b, 1999; Kappeler & Heymann 1996).

> Given that nocturnal primates often communicate with olfactory signals that are temporally deferred, that vocalizations sometimes cannot be traced to their source, and that observations conditions are often difficult, how much of the inferred differences between group living diurnal and nocturnal primates is an artifact of sampling difficulties? It is possible that nocturnal primates communicate the same types of information and develop complex relationships similar to gregarious diurnal primates, but we do not have the sensory capacity to record these interactions. The presumed differences may be variations on a theme, with gregarious diurnal primate interactions merely representing a spatially and temporally compressed version of nocturnal primate interactions. (Sterling 1993: 8)

Acknowledgments I want to thank Carol Berman and Bernard Chapais for their almost infinite patience during the writing of this chapter and Michael Pereira for initially involving me in this project as well as for sharing his thoughts on several topics. Alexandra Müller, Sharon Gursky, and Peter Kappeler shared unpublished data and manuscripts. Alexandra Müller, Urs Thalmann, Helga Schultz, and Laura Bidner provided photos and drawings. Carol Berman, Laura Bidner, Bernard Chapais, Sharon Gursky, Sue Howell, Stephanie Meredith, Alexandra Müller, Mike Nash, and Melissa Schaefer provided feedback on the manuscript in various stages (always in an incredibly short time). I also thank Arizona State University for support of my research on captive galagos and the Wenner-Gren Foundation for Anthropological Research, National Science Foundation, National Institutes of Health, EARTHWATCH, National Geographic Society, The Institute for Human Origins at Arizona State University, and Arizona State University Faculty Grants-in-Aid for funding my fieldwork on galagos and *Lepilemur*.

References

Ancrenaz, M., Lackman-Ancrenaz, I., & Mundy, N. 1994. Field observations of aye-ayes (*Daubentonia madagascariensis*) in Madagascar. *Folia Primatol.*, 69, 22–36.

Andriamasimanana, M. 1994. Ecoethological study of free-ranging aye-ayes (*Daubentonia madagascariensis*) in Madagascar. *Folia Primatol.*, 69, 37–45.

Atsalis, S. 2000. Aspects of spatial distribution and group composition of *Microcebus rufus* (Cheirogaleidae; Primates): results from a long-term trap-retrap study in Ranomafana National Park, Madagascar. *Am. J. Primatol.*, 51, 61–78.

Bearder, S. K. 1974. Aspects of the behaviour and ecology of the thick-tailed bushbaby, *Galago crassicaudatus*. PhD dissertation, University of the Witwatersrand, Witwatersrand, S. Africa.

Bearder, S. K. 1987. Lorises, bushbabies, and tarsiers: diverse societies in solitary foragers. In: *Primate Societies* (Ed. by B. B. Smuts, D. L. Cheney, R. M. Seyfarth, R. W. Wrangham, & T. T. Struhsaker), pp. 11–24. Chicago: University of Chicago Press.

Bearder, S. K. 1999. Physical and social diversity among nocturnal primates: a new view based on long term research. *Primates*, 40, 267–282.

Bearder, S. K. & Doyle, G. A. 1974a. Ecology of bushbabies *Galago senegalensis* and *Galago crassicaudatus*, with some notes on their behaviour in the field. In: *Prosimian Primates* (Ed. by R. D. Martin, G. A. Doyle, & A. C. Walker), pp. 109–130. London: Duckworth.

Bearder, S. K. & Doyle, G. A. 1974b. Field and laboratory studies of social organization in bushbabies (*Galago senegalensis*). *J. Hum. Evol.*, 3, 37–50.

Bearder, S. K., Honess, P. E., & Ambrose, L. 1995. Species diversity among galagos with special reference to mate recognition. In: *Creatures of the Dark: The Nocturnal Prosimians* (Ed. by L. Alterman, G. A. Doyle, & M. K. Izard), pp. 331–352. New York: Plenum.

Bearder, S. K. & Martin, R. D. 1980a. The social organization of a nocturnal primate revealed by radio tracking. In: *A Handbook on Biotelemetry and Radio Tracking* (Ed. by C. J. Amlaner & D. W. MacDonald), pp. 633–648. Oxford: Pergamon Press.

Bearder, S. K. & Martin, R. D. 1980b. *Acacia* gum and its use by bushbabies, *Galago senegalensis* (Primates Lorisidae). *Int. J. Primatol.*, 1, 103–128.

Belisle, P. & Chapais, B. 2001. Tolerated co-feeding in relation to degree of kinship in Japanese macaques. *Behaviour*, 138, 487–509.

Burland, T. M., Bennett, N. C., Jarvis, J. U. M., & Faulkes, C. G. 2002. Eusociality in African mole-rats: new insights from patterns of genetic relatedness in the Damaraland mole-rat (*Cryptomys damarensis*). *Proc. Royal Soc. Lond., B*, 269, 1025–1030.

Carpetis, R. G. & Nash, L. T. 1983. The communicative potential of urine and its relation to urine-washing in *Galago senegalensis braccatus*. *Am. J. Primatol.*, 4, 339.

Chapais, B. 2001. Primate nepotism: what is the explanatory value of kin selection? *Int. J. Primatol.*, 22, 203–229.

Chapais, B., Gauthier, C., Prud'homme, J., & Vasey, P. 1997. Relatedness threshold for nepotism in Japanese macaques. *Anim. Behav.*, 53, 1089–1101.

Charles-Dominique, P. 1977. *Ecology and Behaviour of Nocturnal Prosimians*. London: Duckworth.

Charles-Dominique, P. & Hladik, C. M. 1971. Le *Lepilemur* du sud de Madagascar: ecologie, alimentation et vie sociale. *Terre Vie*, 25, 3–66.

Charles-Dominique, P. & Petter, J. J. 1980. Ecology and social life of *Phaner furcifer*. In: *Nocturnal Malagasy Primates: Ecology, Physiology, and Behavior* (Ed. by P. Charles-Dominique, H. M. Cooper, A. Hladik, C. M. Hladik, E. Pagès, G. F. Pariente, A. Petter-Rousseaux, J.-J. Petter, & A. Schilling), pp. 75–95. New York: Academic Press.

Clark, A. B. 1978. Sex ratio and local resource competition in a prosimian primate. *Science*, 201, 163–165.

Clark, A. B. 1982a. Sibling relationships and social development in the greater bushbaby (*Galago crassicaudatus*). Paper presented at the Midwest Regional Animal Behaviour Conference, Urbana, IL.

Clark, A. B. 1982b. Scent marks as social signals in *Galago crassicaudatus*. I. Sex and reproductive status as factors in signals and responses. *J. Chem. Ecol.*, 8, 1133–1151.

Clark, A. B. 1982c. Scent marks as social signals in *Galago crassicaudatus*. II. Discrimination between individuals by scent. *J. Chem. Ecol.*, 8, 1153–1165.

Clark, A. B. 1985. Sociality in a nocturnal "solitary" prosimian: *Galago crassicaudatus*. *Int. J. Primatol.*, 6, 581–600.

Clark, A. B. 1988. Interspecific differences and discrimination of auditory and olfactory signals of *Galago crassicaudatus* and *Galago garnettii*. *Int. J. Primatol.*, 9, 557–572.

Clark, A. B. & Ehlinger, T. J. 1987. Pattern and adaptation in individual behavioural differences. In: *Perspectives in Ethology* (Ed. by P. P. G. Bateson & P. H. Klopfer), pp. 1–47. New York: Plenum.

Crompton, R. H. 1983. Age differences in locomotion of two subtropical Galaginae. *Primates*, 24, 241–259.

Crompton, R. H. 1984. Foraging, habitat structure, and locomotion in two species of *Galago*. In: *Adaptations for Foraging in Nonhuman Primates: Contributions to an Organismal Biology of Prosimians, Monkeys, and Apes* (Ed. by P. S. Rodman & J. G. H. Cant), pp. 73–111. New York: Columbia University Press.

Crompton, R. H. & Andau, P. M. 1986. Locomotion and habitat utilization in free-ranging *Tarsius bancanus*: a preliminary report. *Primates*, 27, 337–355.

Crompton, R. H. & Andau, P. M. 1987. Ranging, activity rhythms, and sociality in free-ranging *Tarsius bancanus*: a preliminary report. *Int. J. Primatol.*, 8, 43–47.

Eberle, M. & Kappeler, P. M. 2002a. Cooperative breeding in grey mouse lemurs (*Microcebus murinus*). Paper presented at the XIXth Congress of the International Primatological Society, August 2002, Beijing, China.

Eberle, M. & Kappeler, P. M. 2002b. Mouse lemurs in space and time: a test of the socioecological model. *Behav. Ecol. Sociobiol.*, 51, 131–139.

Falk, D. 2000. *Primate Diversity*. New York: W.W. Norton & Company.

Fietz, J. 1999a. Demography and floating males in a population of *Cheirogaleus medius*. In: *New Directions in Lemur Studies* (Ed. by B. Rakotosamimanana, H. Rasamimanana, J. U. Ganzhorn, & S. M. Goodman), pp. 159–172. New York: Plenum.

Fietz, J. 1999b. Monogamy as a rule rather than exception in nocturnal lemurs: the case of the fat-tailed dwarf lemur, *Cheirogaleus medius*. *Ethology*, 105, 255–272.

Fietz, J. 1999c. Mating system of *Microcebus murinus*. *Am. J. Primatol.*, 48, 127–133.

Fietz, J. & Ganzhorn, J. 1999. Feeding ecology of the hibernating primate *Cheirogaleus medius*: how does it get so fat? *Oecologia*, 121, 157–164.

Fietz, J., Zischler, H., Schwiegk, C., Tomiuk, J., Dausmann, K. H., & Ganzhorn, J. U. 2000. High rates of extra-pair young in the pair-living fat-tailed dwarf lemur, *Cheirogaleus medius*. *Behav. Ecol. Sociobiol.*, 49, 8–17.

Fleagle, J. G. 1988. *Primate Adaptation & Evolution*. New York: Academic Press.

Fleagle, J. G. 1999. *Primate Adaptation & Evolution*, 2nd ed. New York: Academic Press.

Ganzhorn, J. U. & Kappeler, P. M. 1996. Lemurs of the Kirindy forest. *Primate Rep.*, 46, 257–274.

Ganzhorn, J. U. & Schmid, J. 1998. Different population dynamics of *Microcebus murinus* in primary and secondary deciduous dry forests of Madagascar. *Int. J. Primatol.*, 19, 785–796.

Goodman, S. M., O'Connor, S., & Langrand, O. 1993. A review of predation on lemurs: implications for the evolution of social behavior in small, nocturnal primates. In: *Lemur Social Systems and Their Ecological Basis* (Ed. by P. M. Kappeler & J. Ganzhorn), pp. 51–66. New York: Plenum.

Gouzoules, S. 1984. Primate mating systems, kin associations, and cooperative behavior: evidence for kin recognition? *Ybk. Phys. Anthropol.*, 27, 99–134.

Groves, C. 2001. *Primate Taxonomy*. Washington, DC: Smithsonian Institution Press.

Gursky, S. 1994. Infant care in the spectral tarsier (*Tarsius spectrum*), Sulawesi, Indonesia. *Int. J. Primatol.*, 15, 843–853.

Gursky, S. 1997. Group size and composition in the spectral tarsier, *Tarsier spectrum*: implications for social organization. *Trop. Biodiversity*, 3, 57–58.

Gursky, S. 2000a. Sociality in the spectral tarsier, *Tarsius spectrum*. *Am. J. Primatol.*, 51, 89–101.

Gursky, S. 2000b. Effect of seasonality on the behavior of an insectivorous primate, *Tarsius spectrum*. *Int. J. Primatol.*, 21, 477–496.

Gursky, S. 2000c. Allocare in a nocturnal primate: data on the spectral tarsier, *Tarsius spectrum*. *Folia Primatol.*, 71, 39–54.

Gursky, S. 2001. Determinants of gregariousness in a nocturnal primate. *Am. J. Phys Anthropol.*, Suppl 32, 74.

Gursky, S. 2002a. Predation experiments on an infant spectral tarsier. Paper presented at the XIXth Congress of the International Primatological Society, August 2002, Beijing, China.

Gursky, S. 2002b. Determinants of gregariousness in the spectral tarsier (Prosimian: *Tarsius spectrum*). *J. Zool.*, 256, 401–410.

Hafen, T., Neveu, H., Rumpler, Y., Wilden, I., & Zimmermann, E. 1998. Acoustically dimorphic advertisement calls separate morphologically and genetically homogeneous populations of the grey mouse lemur (*Microcebus murinus*). *Folia Primatol.*, 69 (Suppl. 1), 342–356.

Harcourt, C. S. 1980. Behavioral adaptations in South African galagos. PhD dissertation, University of the Witwatersrand, Witwatersrand, S. Africa.

Harcourt, C. S. 1984. A comparison of Kenyan and South-African galagos—are they the same species? *Int. J. Primatol.*, 5, 345–345.

Harcourt, C. S. 1986. Seasonal variation in the diet of South African galagos. *Int. J. Primatol.*, 7, 491–506.

Harcourt, C. S. 1987. Brief trap/retrap study of the brown mouse lemur (*Microcebus rufus*). *Folia Primatol.*, 49, 209–211.

Harcourt, C. S. 1991. Diet and behaviour of a nocturnal lemur, *Avahi laniger*, in the wild. *J. Zool. Lond.*, 223, 667–674.

Harcourt, C. S. & Nash, L. T. 1986. Social organization of galagos in Kenyan Coastal Forests. I. *Galago zanzibaricus*. *Am. J. Primatol.*, 10, 339–355.

Harvey, P. H., Martin, R. D., & Clutton-Brock, T. H. 1987. Life histories in comparative perspective. In: *Primate Societies* (Ed. by B. B. Smuts, D. L. Cheney, R. M. Seyfarth, & R. W. Wrangham), pp. 181–196. Chicago: University of Chicago Press.

Holmes, W. G. 1986. Identification of paternal half-siblings by captive Belding's squirrels. *Anim. Behav.*, 34, 321–327.

Holmes, W. G. 1994. The development of littermate preferences in juvenile Belding's ground squirrels. *Anim. Behav.*, 48, 1071–1084.

Holmes, W. G. 1995. The ontogeny of littermate preferences in juvenile golden-mantled ground squirrels: effects of rearing and relatedness. *Anim. Behav.*, 50, 309–322.

Holmes, W. G. & Mateo, J. M. 1998. How mothers influence the development of litter-mate preferences in Belding's ground squirrels. *Anim. Behav.*, 55, 1555–1570.

Honess, P. 1996. Speciation among galagos (Primates, Galagidae) in the Tanzanian forests. PhD dissertation, Oxford Brookes University, Oxford, UK.

Izard, M. K. & Nash, L. T. 1988. Contrasting reproductive parameters in *Galago senegalensis braccatus* and *G. s. moholi*. *Int. J. Primatol.*, 9, 519–527.

Jolly, C. J., Philips-Conroy, J. E., & Müller, A. E. 2003. Trapping. In: *Field and Laboratory Methods in Primatology* (Ed. by J. M. Setchell & D. J. Curtis). Cambridge: Cambridge University Press.

Kappeler, P. M. 1996. Causes and consequences of life-history variation among strepsirhine primates. *Am. Nat.*, 148, 868–891.

Kappeler, P. M. 1997a. Intrasexual selection in *Mirza coquereli*: evidence for scramble competition polygyny in a solitary primate. *Behav. Ecol. Sociobiol.*, 41, 115–127.

Kappeler, P. M. 1997b. Determinants of primate social organization: comparative evidence and new insights from Malagasy lemurs. *Biol. Rev.*, 72, 111–151.

Kappeler, P. M. 1999. Lemur social structure and convergence in primate socioecology. In: *Comparative Primate Socioecology* (Ed. by P. C. Lee), pp. 273–299. Cambridge: Cambridge University Press.

Kappeler, P. M. & Heymann, E. W. 1996. Nonconvergence in the evolution of primate life history and socio-ecology. *Biol. J. Linnaean Soc.*, 59, 297–326.

Kappeler, P. M., Wimmer, B., Zinner, D., & Tautz, D. 2002. Hidden matrilineal structure of a solitary lemur: implications for primate evolution. *Proc. Royal Soc. Lond., B*, 269, 1755–1763.

Maggioncalda, A. N. & Sapolsky, R. M. 2002. Disturbing behaviors of the orangutan. *Sci. Am.*, 286, 60–65.

Masters, J. C. & Brothers, D. J. 2002. Lack of congruence between morphological and molecular data in reconstructing the phylogeny of the Galagonidae. *Am. J. Phys. Anthropol.*, 117, 79–93.

Müller, A. E. 1998. A preliminary report on the social organization of *Cheirogaleus medius* (Cheirogaleidae; Primates) in north-west Madagascar. *Folia Primatol.*, 69, 160–166.

Müller, A. E. 1999a. Aspects of social life in the fat-tailed dwarf lemur (*Cheirogaleus medius*): inferences from body weights and trapping data. *Am. J. Primatol.*, 49, 265–280.

Müller, A. E. 1999b. Social organization of the fat-tailed dwarf lemur (*Cheirogaleus medius*) in northwestern Madagascar. In: *New Directions in Lemur Studies* (Ed. by B. Rakotosamimanana, H. Rasamimanana, J. U. Ganzhorn, & S. M. Goodman), pp. 139–156. New York: Plenum.

Müller, A. E. & Thalmann, U. 2000. Origin and evolution of primate social organisation: a reconstruction. *Biol.Rev.*, 75, 405–435.

Müller, A. E. & Thalmann, U. 2002. The biology of the fat-tailed dwarf lemur (*Cheirogaleus medius* E. Geoffroy 1812): new results from the field. *Evol. Anthropol.*, 11, 79–81.

Nash, L. T. 1991. Development of food sharing in captive infant *Galago senegalensis braccatus*. In: *Primatology Today. Proceedings of the XIIIth Congress of the International Primatological Society, Japan* (Ed. by A. Ehara, T. Kumura, O. Takenaka, & M. Iwamoto), pp. 181–182. Amsterdam: Elsevier.

Nash, L. T. 1993. Juveniles in nongregarious primates. In: *Juvenile Primates: Life History, Development and Behavior* (Ed. by L. A. Fairbanks & M. E. Pereira), pp. 119–137. Oxford: Oxford University Press.

Nash, L. T. 1998. Vertical clingers and sleepers: seasonal influences on the activities and substrate use of *Lepilemur leucopus* at Beza Mahafaly Special Reserve, Madagascar. *Folia Primatol.*, 69 (Suppl.1), 204–217.

Nash, L. T., Bearder, S. K., & Olson, T. R. 1989. Synopsis of galago species characteristics. *Int. J. Primatol.*, 10, 57–80.

Nash, L. T. & Harcourt, C. S. 1986. Social organization of galagos in Kenyan coastal forests. II. *Galago garnettii*. *Am. J. Primatol.*, 10, 357–369.

Nash, L. T. & Whitten, P. L. 1989. Preliminary observations on the role of *Acacia* gum chemistry in *Acacia* utilization by *Galago senegalensis* in Kenya. *Am. J. Primatol.*, 17, 27–39.

Nekaris, K. A. I. 2000. The socioecology of the slender loris (*Loris tardigradus lydekkerianus*) in Dindigul, Tamil Nadu, South India. PhD dissertation, Washington University, St. Louis, MO.

Nekaris, K. A. I. 2002. Adult/infant interactions in three subspecies of slender loris (*Loris tardigradus*). Paper presented at the XIXth Congress of the International Primatological Society, August 2002, Beijing, China.

Niemitz, C. 1984. An investigation and review of the territorial behaviour and social organisation of the genus *Tarsius*. In: *Biology of Tarsiers* (Ed. by C. Niemitz), pp. 117–127. New York: Gustav Fischer Verlag.

Ortmann, S. & Heldmaier, G. 1997. Spontaneous daily torpor in Malagasy mouse lemurs. *Naturwissenschaften*, 84, 20–32.

Pagès, E. 1978. Home range, behaviour and tactile communication in a nocturnal Malagasy lemur *Microcebus coquereli*. In: *Recent Advances in Primatology: Evolution* (Ed. by D. J. Chivers & K. A. Joysey), pp. 171–177. New York: Academic Press.

Pagès, E. 1980. Ethoecology of *Microcebus coquereli* during the dry season. In: *Nocturnal Malagasy Primates: Ecology, Physiology, and Behavior* (Ed. by P. Charles-Dominique, H. M. Cooper, A. Hladik, C. M. Hladik, E. Pagès, G. F. Pariente, A. Petter-Rousseaux, J.-J. Petter, & A. Schilling), pp. 97–116. New York: Academic Press.

Penn, D. J. & Potts, W. K. 1999. The evolution of mating preferences and major histocompatibility complex genes. *Am. Nat.*, 153, 145–164.

Pereira, M. E. & Kappeler, P. M. 1997. Divergent systems of agonistic behaviour in lemurid primate. *Behaviour*, 134, 225–274.

Pimley, E., Bearder, S., & Dixson, A. 2002. Patterns of ranging and social interactions in pottos (*Perodicticus potto potto*) in Cameroon. Paper presented at the XIXth Congress of the International Primatological Society, August 2002, Beijing, China.

Pullen, S. L. 2000. Behavioural and genetic studies of the mating system in a nocturnal primate: the lesser galago (*Galago moholi*). PhD dissertation, University of Cambridge, UK.

Pullen, S. L., Bearder, S. K., & Dixson, A. F. 2000. Preliminary observations on sexual behavior and the mating system in free-ranging lesser galagos (*Galago moholi*). *Am. J. Primatol.*, 51, 79–88.

Pullen, S. L. & Dixson, A. 2002. Male mating strategies in the lesser galago (*Galago moholi*) and their reproductive consequences. Paper presented at the XIXth Congress of the International Primatological Society, August 2002, Beijing, China.

Radespiel, U. 2000. Sociality in the gray mouse lemur (*Microcebus murinus*) in northwestern Madagascar. *Am. J. Primatol.*, 51, 21–40.

Radespiel, U., Cepok, S., Zietemann, V., & Zimmermann, E. 1998. Sex-specific usage patterns of sleeping sites in grey mouse lemurs (*Microcebus murinus*) in northwestern Madagascar. *Am. J. Primatol.*, 46, 77–84.

Radespiel, U., Dal Secco, V., Drögemüller, C., Braune, P., Labes, E., & Zimmermann, E.

2002. Sexual selection, multiple mating and paternity in grey mouse lemurs, *Microcebus murinus*. *Anim. Behav.*, 63, 259–268.

Radespiel, U., Ehresmann, P., & Zimmermann, E. 2001. Contest versus scramble competition for mates: the composition and spatial structure of a population of gray mouse lemurs (*Microcebus murinus*) in north-west Madagascar. *Primates*, 42, 207–220.

Radespiel, U., Sarikaya, Z., Zimmermann, E., & Bruford, M. W. 2001. Sociogenetic structure in a free-living nocturnal primate population: sex-specific differences in the grey mouse lemur (*Microcebus murinus*). *Behav. Ecol. Sociobiol.*, 50, 493–502.

Radespiel, U. & Zimmermann, E. 2001. Dynamics of estrous synchrony in captive gray mouse lemurs (*Microcebus murinus*). *Int. J. Primatol.*, 22, 71–90.

Radhakrishna, S. 2001. Reproductive and social behavior of slender loris (*Loris tardigradus lydekkerianus*) in its natural habitat. PhD dissertation, University of Mysore, Manasagangotri, Mysone, India.

Radhakrishna, S. & Singh, M. 2002. Home range and ranging pattern in the slender loris (*Loris tardigradus lydekkerianus*). *Primates*, 43, 237–248.

Rowe, N. 1996. *The Pictorial Guide to the Living Primates*. East Hampton, NY: Pogonias Press.

Russell, R. 1978. The behavior, ecology and environmental physiology of a nocturnal primate, *Lepilemur mustelinus* (Strepsirhini, Lemuriformes, Lepilemuridae). PhD dissertation, Duke University, Durham, NC.

Schaefer, M. L., Young, D. A., & Restrepo, D. 2001. Olfactory fingerprints for major histocompatibility complex-determined body odors. *J. Neurosci.*, 21, 2481–2487.

Schmelting, B., Berke, O., Radespiel, U., & Zimmermann, E. 2001. Sexual selection, male mating strategies and reproductive success in a promiscuous, nocturnal Malagasy primate, the grey mouse lemur (*Microcebus murinus*). *Primate Rep.*, 60–1, 38–39.

Schmid, J. 1998. Tree holes used for resting by gray mouse lemurs (*Microcebus murinus*) in Madagascar: insulation capacities and energetic consequences. *Int. J. Primatol*, 19, 797–810.

Schmid, J. & Kappeler, P. M. 1994. Sympatric mouse lemurs (*Microcebus* sp.) in western Madagascar. *Folia Primatol.*, 63, 162–170.

Schmid, J. & Kappeler, P. M. 1998. Fluctuating sexual dimorphism and differential hibernation by sex in a primate, the gray mouse lemur (*Microcebus murinus*). *Behav. Ecol. Sociobiol.*, 43, 125–132.

Schmid, J. & Speakman, J. R. 2000. Daily energy expenditure of the grey mouse lemur (*Microcebus murinus*): a small primate that uses torpor. *J. Comp. Physiol.*, B170, 633–641.

Schülke, O. 2001. Social anti-predator behaviour in a nocturnal lemur. *Folia Primatol.*, 72, 332–334.

Schülke, O., Zischler, H., & Kappeler, P. M. 2001. The potential for sexual selection in a pair-living nocturnal primate. *Primate Rep.*, Special issue, 60–1, 39.

Schulze, H. & Meier, B. 1995. Behavior of captive *Loris tardigradus nordicus*: a qualitative description, including some information about morphological bases of behavior. In: *Creatures of the Dark: The Nocturnal Prosimians* (Ed. by L. Alterman, G. A. Doyle, & M. K. Izard), pp. 221–250. New York: Plenum.

Schwab, D. 2000. A preliminary study of the social and mating system of pygmy mouse lemurs (*Microcebus myoxinus*). *Am. J. Primatol.*, 51, 41–60.

Spencer, P. B. S., Horsup, A. B., & Marsh, H. D. 1998. Enhancement of reproductive success through mate choice in a social rock-wallaby, *Petrogale assimilis* (Macropodidae) as revealed by microsatellite markers. *Behav. Ecol. Sociobiol.*, 43, 1–9.

Sterling, E. J. 1993. Patterns of range use and social organization in aye-ayes (*Daubentonia madagascariensis*) on Nosy Mangabe. In: *Lemur Social Systems and Their Ecological Basis* (Ed. by P. M. Kappeler & J. Ganzhorn), pp. 1–10. New York: Plenum.

Sterling, E. J. 1994. Evidence for nonseasonal reproduction in wild aye-ayes (*Daubentonia madagascariensis*). *Folia Primatol.*, 62, 46–53.

Sterling, E. J., Nguyen, N., & Fashing, P. J. 2000. Spatial patterning in nocturnal prosimians: a review of methods and relevance to studies of sociality. *Am. J. Primatol.*, 51, 3–19.

Sterling, E. J. & Radespiel, U. 2000. Advances in the studies of sociality in nocturnal prosimians. *Am. J. Primatol.*, 51, 1–2.

Sterling, E. J. & Richard, A. F. 1995. Social organization in the aye-aye (*Daubentonia madagascariensis*) and the perceived distinctiveness of nocturnal primates. In: *Creatures of the Dark* (Ed. by L. Alterman, G. A. Doyle, & K. M. Izard), pp. 439–451. New York: Plenum.

Strier, K. B. 2003. *Primate Behavioral Ecology*. Boston: Allyn & Bacon.

Sussman, R. W. 2000. *Primate Ecology and Social Structure, Vol. 1: Lorises, Lemurs and Tarsiers*. Heights, MA: Pearson Custom Publishing.

Thalmann, U. 2001. Food resource characteristics in two nocturnal lemurs with different social behavior: *Avahi occidentalis* and *Lepilemur edwardsi*. *Int. J. Primatol.*, 22, 287–324.

Warren, R. D. & Crompton, R. H. 1997. A comparative study of the ranging behaviour, activity rhythms and sociality of *Lepilemur edwardsi* (Primates, Lepilemuridae) and *Avahi occidentalis* (Primates, Indriidae) at Ampijoroa, Madagascar. *J. Zool. Lond.*, 243, 397–415.

Waser, P. 1981. Sociality or territorial defense? The influence of resource renewal. *Behav. Ecol. Sociobiol.*, 8, 231–237.

Wimmer, B., Tautz, D., & Kappeler, P. M. 2000. Genetic and social structure of a *Mirza coquereli* population. *Folia Primatol.*, 71, 216.

Wimmer, B., Tautz, D., & Kappeler, P. M. 2002. The genetic population structure of the gray mouse lemur (*Microcebus murinus*), a basal primate from Madagascar. *Behav. Ecol. Sociobiol.*, 52, 166–175.

Wright, P. C. & Martin, L. B. 1995. Predation, pollination and torpor in two nocturnal prosimians: *Cheirogaleus major* and *Microcebus rufus* in the rain forest of Madagascar. In: *Creatures of the Dark: The Nocturnal Prosimians* (Ed. by L. Alterman, G. A. Doyle, & M. K. Izard), pp. 45–60. New York: Plenum Press.

Zimmermann, E., Cepok, S., Rakotoarison, N., Zietemann, V., & Radespiel, U. 1998. Sympatric mouse lemurs in north-west Madagascar: a new rufous mouse lemur species (*Microcebus ravelobensis*). *Folia Primatol.*, 69, 106–114.

Zimmermann, E. & Hafen, T. G. 2001. Colony specificity in a social call of mouse lemurs (*Microcebus* sp.). *Am. J. Primatol.*, 54, 129–142.

Zimmermann, E., Masters, J. C., & Rumpler, Y. 2000. Introduction to diversity and speciation in the Prosimii. *Int. J. Primatol.*, 21, 789–791.

10

Kinship Structure and Reproductive Skew in Cooperatively Breeding Primates

James M. Dietz

The question of how and why reproduction is distributed in social groups has become one of the primary focuses of research on cooperatively breeding animals. Empirical evidence used to develop and support various models predicting allocation of reproduction has come from long-term studies on several species of social insects, birds, and a few mammals, but relatively little from work on primates. Field studies have demonstrated that marmosets and tamarins (Callitrichidae) live in a diverse array of social groupings. The majority of breeding groups contain one reproductive male and female, but polygyny and polyandry have also been reported (e.g., Goldizen 1987, 1988; Baker et al. 1993; Dietz & Baker 1993; Digby & Ferrari 1994; Garber 1994). Breeding groups usually contain adult helpers, but intrasexual aggression is often unidirectional, with socially dominant individuals doing most breeding (Baker et al. 1993, Digby 1995b).

Helpers often remain in their natal groups beyond sexual maturity. Although the degree of relatedness between helpers and offspring varies from very low to very high, all adults in the group carry and feed all infants born in the group (reviewed by Tardif 1997). These characteristics suggest that the callitrichids may serve as an appropriate model system to examine the basis of cooperative breeding in primates and other taxa.

Kinship may affect callitrichid mating patterns in a number of ways. Genetic relatedness between (dominant) breeders and (subordinate) helpers contributes to the amount of inclusive fitness benefits accruing to dominants, hence their readiness to tolerate related subordinates. At the same time, relatedness between subordinate helpers and group offspring influences the amount of inclusive fitness benefits to subordinates, a factor that might act as an incentive for them to stay in their natal group. On the other hand, kinship between breeders and opposite-sexed subordinates might affect mating patterns through the cost of inbreeding, which is hypothesized to effect both the subordinates' decisions to leave or stay and the amount of indirect fitness gains for the dominants. In turn, mating patterns may affect the genetic structure of groups: reproductive skew among males generates paternal sibships and among females generates maternal sibships, in both cases increasing the average relatedness between dominants and subordinates.

In this chapter, I use empirical data from studies of captive and wild callitrichids to test the assumptions of reproductive skew models and to assess the extent to which predictions derived from reproductive skew theory can explain the diversity of mating systems observed in populations of these species. Many of the examples in this chapter are taken from literature on golden lion tamarins (*Leontopithecus rosalia*) (figure 10.1) in Poço das Antas Reserve, first because of the relatively long duration and large sample sizes in this study and second because of the available data on kinship. Kinship among group members is an important factor in various aspects of reproductive skew theory, which are highlighted in this chapter. Where empirical data are lacking, I suggest hypotheses that might be tested in future research.

Reproduction in multigenerational families may be monopolized by a single individual of each sex (high reproductive skew) or distributed more equitably among all adults of each sex (low reproductive skew). Theoretical models attempting to explain reproductive sharing within social groups may be divided into two general categories: optimal skew models (OSM), also known as transactional or reproductive concession models, and incomplete control models (ICM). OSM are based on the premise that a social contract exists between dominant and subordinate individuals of each sex in the group. When a dominant benefits

Figure 10.1. Reproductive female golden lion tamarins (*Leontopithecus rosalia*) produce one or occasionally two litters each year. Modal litter size is two. Photo by James Dietz.

from the presence or actions of a subordinate of the same sex, the dominant may share reproduction as an incentive for the subordinate to remain peaceably in the group. Subordinates are thought to weigh the costs and benefits of remaining as helpers in the current group versus emigrating to attempt reproduction elsewhere.

Optimal skew theory is based on three assumptions: (1) the presence of subordinates results in a net increase in fitness to dominants (i.e., benefits from subordinate's help are greater than costs of retaining the subordinate); (2) subordinates are less likely to leave the group or to challenge the dominant if they share in reproduction (a corollary to this assumption is that subordinates will avoid breeding with close relatives); and (3) dominants are able to control the breeding status of subordinates. ICM, in contrast, assume that dominants are not always able to control reproduction by same-sex subordinates. Division of reproduction is the result of a constant struggle between dominant and subordinate, the outcome of which depends on the costs and benefits to both (Emlen 1995; Mumme 1997; Clutton-Brock 1998a, b; Emlen et al. 1998; Reeve et al. 1998; Reeve & Emlen 2000).

Three parameters enter into basic skew models. The first is x, the probability of a subordinate finding a breeding opportunity and breeding independently. Ecological constraints are those factors that diminish the likelihood that an emigrating individual will succeed in finding such a breeding opportunity. Therefore, x approaches 1 when ecological constraints are low. Callitrichids generally breed on relatively stable territories (Goldizen et al. 1996, Dietz et al. 1997); thus x may be estimated by the number of breeding opportunities in adjacent groups plus the number of discrete unoccupied habitat patches adjacent to the natal group. The other parameters are k, group productivity when a subordinate is retained, and r, the relatedness of the dominant to the subordinate (Reeve et al. 1998). For callitrichids, one estimate of k is the mean change in infant survival to weaning resulting from each additional helper in the group; r could be determined from pedigree or molecular genetic data. If ecological constraints are strong, for example, all available habitat is occupied by breeding groups with intense competition for few breeding vacancies or a high probability of mortality during dispersal, $x < r(k - 1)$, and subordinates are predicted to stay in the group as nonreproductive helpers. However, as the probability of breeding outside the natal group increases, $x > r(k - 1)$, and relatedness of individuals in newly formed groups decreases. Under these circumstances, opposing predictions emerge for OSM and ICM.

Predictions of ICM differ from those of OSM in at least two ways. First, for ICM, the subordinate's share of reproduction in the natal group is the result of a contest with the dominant and is largely unaffected by severity of ecological constraints; whereas for OSM, subordinate reproduction in the natal group varies inversely with severity of ecological constraints limiting independent breeding elsewhere. Second, for ICM, subordinate reproduction is unrelated to, or increases with the subordinate's genetic relatedness to the dominant; whereas for OSM, subordinate reproduction varies inversely with the subordinate's genetic relatedness to the dominant. The basis of this distinction is that relatedness between dominant and subordinate results in indirect fitness gains to the subordinate, regardless of its success in attempts at reproduction. When fitness benefits to dominant and subordinate are symmetrical, the degree of relatedness may have no influence on the relative investment of both individuals and thus on reproductive skew (Clutton-Brock 1998a, Emlen et al. 1998, Reeve et al. 1998).

The Assumptions of Reproductive Skew Models

Helpers Improve the Reproductive Success of Breeders

As ecological constraints diminish, subordinates should be more inclined to leave the natal group and attempt to breed elsewhere. Under these circumstances, concession models predict that dominants may share reproduction with same-sex subordinates as a "staying incentive" (Reeve & Nonacs 1992, Reeve & Ratnieks 1993) for peacefully remaining in the group as an alloparent. Implicit in this prediction is the assumption that helpers augment the fitness of the breeders they assist, either by increasing territory resources and/or decreasing parental workload and thus improving survivorship of the parents, or by increasing the number of surviving offspring produced by the parents. Any cost to the dominant for sharing reproduction (e.g., reduced resources available for its own offspring) should be outweighed by the benefit of retaining the subordinate in the group.

The energetic demands on a breeding female simultaneously rearing two relatively large infants have been hypothesized to explain the evolution of cooperative breeding in the Callitrichidae (Leutenegger 1980, Garber et al. 1984, Terborgh & Goldizen 1985). In most callitrichids, the majority of carrying and provisioning of young infants is done by alloparents (Tardif et al. 1993, Snowdon 1996). Indeed, several lines of evidence from field studies suggest that the presence of helpers increases the number and survivorship of offspring. In eight years of monitoring 19 groups of golden lion tamarins in Poço das Antas Reserve, researchers did not observe one case of a female raising its offspring to maturity without the help of at least one male (Dietz & Baker 1993). In the same study population, care provided by other group members allowed two five-week-old infants to survive the death of their mother (Bales et al. 2001). Infants this age have not been weaned but have started to accept solid food provided by adults (Baker 1991). Also in this population, the number of live births in second litters per year (Bales et al. 2001) and the duration of the reproductive tenure of males (Bales et al. 2000) was significantly related to the number of helpers in the group. The number of offspring was greater in two-male groups than in one-male groups (Baker et al. 1993).

In mustached tamarins (*Saguinus mystax*), the number of adult males in wild-trapped groups was significantly correlated with the number of infants surviving to become juveniles (Garber et al. 1984, Sussman & Garber 1987). Savage, Snowdon et al. (1997) reported that infant survival in wild groups of cotton-top tamarins (*Saguinus oedipus*) did not reach a maximum in groups containing less than three alloparents in addition to both parents. Terborgh and Goldizen (1985) suggested that a reproductive pair of saddle-back tamarins (*Saguinus fuscicollis*) is incapable of raising twin offspring without the aid of at least one additional alloparent (but see Windfelder 2000). Common marmoset (*Callithrix jacchus*) groups containing more than two adult males contained more surviving infants than groups with only one adult male (Koenig 1995). A similar effect of helpers on infant survival was not observed for this species in captivity (Rothe et al. 1993). However, Price (1992) found that captive infant *Saguinus oedipus* in large groups were carried more and received more food than infants in small groups. In summary, the available evidence suggests that helpers do significantly increase the number and survivorship of offspring in free-ranging callitrichid groups. The greater availability of food in captive environments may mask this effect.

The Cost Incurred by Dominants Sharing Reproduction with Same-Sex Subordinates Is Low Relative to the Benefits Provided

Breeding by subordinate females in free-ranging callitrichid groups has been documented in a variety of species (Dawson 1977, Soini 1982, Garber et al. 1984, Hubrecht 1984, Digby 1995b, Goldizen et al. 1996). To my knowledge, the fitness cost to dominants sharing reproduction with subordinates has been assessed in only one field study of callitrichids. In an eight-year study of golden lion tamarins in Poço das Antas Reserve, Dietz and Baker (1993) used demographic data to estimate the costs to reproductive females sharing reproduction with their daughters. In that study, approximately 10% of dominant females shared reproduction with their daughters in the natal group. In most cases, polygynous relationships did not exceed two years. Reproductive success of breeding females, defined as the number of infants surviving to semiannual sampling dates, was significantly greater in monogynous groups (0.829/female) than that of dominants in polygynous groups (0.684/female), suggesting that there is a fitness cost to sharing resources (e.g., helpers and food) with a subordinate. The possibility of a role reversal in the social dominance hierarchy is an additional cost to dominant females and occurred in 7% of polygynous group samples in the Poço das Antas population.

Benefits to the dominant reproductive female included a gain in indirect fitness equal to half the daughter's expected reproductive success in the natal group. Results of a demographic model suggested that the benefits to a dominant female for evicting her daughter slightly outweighed the costs when inbreeding was not a factor. Costs of not expelling the daughter were significantly greater when the daughter was closely related to all adult males in the group and inbreeding was likely to reduce infant survival (Dietz et al. 2000) and thus decrease the dominant's indirect fitness gains. Indeed, the majority of polygynous mating in that population took place when groups contained males unrelated to subordinates (Baker et al. 2002). In summary, the cost to dominant females of sharing reproduction with their daughters in the natal group was relatively low if the group contained a potential mate unrelated to the daughter, and if the daughter was unable to replace her mother as dominant in the group. Under these circumstances, the dominant female would recover the costs of sharing reproduction during the first breeding season after the daughter emigrated and bred outside the natal group.

Observations on *Callithrix jacchus* in northeastern Brazil also support the assumption that dominant females do not incur significant fitness loss by sharing reproduction with subordinates in the natal group (Digby 1995a). In that study, dominant females gave birth to more infants and had higher infant survival than subordinate breeding females. Subordinate females were successful in rearing infants only when such births did not overlap with the dependency period of infants born to the dominant female.

Subordinates Breeding in the Natal Group Are Less Likely to Disperse Than Those Not Breeding in the Natal Group

One assumption of reproductive skew theory is that shared reproduction functions as a "staying incentive" to entice subordinates to remain as helpers in the natal group rather than disperse and breed elsewhere. Data necessary to examine this assumption are scarce for free-ranging callitrichids. Long-term patterns of emigration and reproduction were documented for golden lion tamarins in Poço das Antas Reserve (ca. 18 study groups monitored continuously since 1986). In that population, a small percentage of two-year-old females resident with their mothers were known to conceive, and all confirmed conceptions occurred in groups containing at least one male unrelated to the daughter (Baker et al. 2002). However, the incidence of daughters that became pregnant in their natal groups increased to over 50% for those that were three or four years of age, regardless of whether an unrelated male was present in the group. Although these findings are consistent with the assumption that shared reproduction causes subordinates to delay dispersal, other explanations are also possible. For example, older daughters may become pregnant in their natal groups because dominant females lose the ability to regulate reproduction of older offspring (as predicted by ICM) and/or because the residual reproductive value of subordinates decreases with age.

Approximately 70% of breeding groups of golden lion tamarins contained more than one adult male (Baker et al. 1993, 2002). However, many of these males were sons of the breeding female and thus not likely to breed with her (Abbott 1984, Dietz et al. 2000). Of multimale groups, 46% contained two or three potential breeding males (i.e., nonnatal adult males). Aggression among potential breeding males was rare, but dominance interactions (fights, arch walks, etc.) were unidirectional within most male duos. Dominant males were responsible for 94% of male sexual behavior during the time when the reproductive female was expected to be fertile (Baker et al. 1993). The authors concluded that a single dominant male typically monopolized reproduction in groups in this population. Thus behavioral evidence does not support the assumption that shared reproduction delays male dispersal in this population.

Dominants Can Control the Breeding of Subordinates

Callitrichid primates are typical of cooperatively breeding mammals in that they demonstrate delayed breeding by subordinates (Abbott 1987, French & Inglett 1991). The proximate mechanisms maintaining singular breeding in callitrichid groups have been detailed in a large volume of work done in captivity (reviewed by French 1997). In several callitrichid species, first ovulation is delayed in subordinate females still in their natal groups (e.g., Abbott 1984, Tardif 1984). Delay in the onset of puberty in natal females in captive settings may be due in part to the lack of stimuli necessary to induce reproductive maturation, for example, the presence of unfamiliar males in the group (e.g., Tardif 1984, Ziegler 1987, Savage et al. 1988) or olfactory or behavioral cues related to the presence of a dominant female (see review by Abbott et al. 1990).

In captivity, subordinate females in social groups of cotton-top tamarins are anovulatory while in the presence of their natal parents (French 1984, Ziegler 1987, Savage et al. 1988). However, daughters in a single group of wild cotton-top tamarins showed signs of ovarian activity as evidenced by periodic elevations in fecal progesterone and estrogen metabolites, in spite of residing with their natal mother and father (Savage, Shideler et al. 1997). These findings suggest that levels of physiological inhibition of reproduction in captive environments may not be representative of those in free-ranging callitrichid females.

Unlike other callitrichids, reproduction by lion tamarin daughters in natal family groups and by subordinates in peer groups is not physiologically suppressed in captivity (French et al. 1989). Most subordinate female lion tamarins living in captivity begin to demonstrate normal patterns of ovarian cycles at 16 months of age (French & Inglett 1991, French et al. 1992). Saltzman et al. (1997) demonstrated that the reproductive status of the breeding female may affect the likelihood of a daughter showing ovulatory cycles in the natal family group. While ovulation by subordinate females was common in family groups in which the mother was ovulating but not producing viable offspring, subordinate female ovulation was less common when the mother was regularly conceiving and carrying infants to term.

French et al. (in press) reported that young daughters in intact natal groups of golden lion tamarins in Poço das Antas showed patterns of infertility while in intact natal groups. Older daughters, however, demonstrated elevated fecal concentrations of sex steroids. Subordinate females in groups containing both biological parents were less likely to show elevated steroid concentrations than females in groups with stepparents or no related adults. Although most subordinate females did not conceive while in intact natal groups, a single exception raises the possibility of an additional factor affecting reproduction by subordinate females. One daughter conceived while in her natal group. Her mother, the dominant female in the group, was very old and died shortly after giving birth. This result suggests that reproduction by subordinate females is also sensitive to the age and health of the dominant female and thus the likelihood of inheriting dominant breeding status in the natal group.

In some birds, for example pied kingfisher (Reyer et al. 1986) and Harris's hawk (Mays et al. 1991), titers of sex steroids are low in nonbreeding males related to the breeding females but are high in unrelated males, suggesting that suppression of subordinate male reproduction may be endocrinological where the potential for incest exists. The extent to which male reproductive suppression is mediated through physiological mechanisms is not known for wild populations of cooperatively breeding primates. However, preliminary results (French, Bales, & Dietz unpublished data) show significant variation in fecal testosterone (T) levels among male golden lion tamarins in Poço das Antas. Sons in intact natal groups had concentrations of T that were less than half those of any other class of males, and approximately 25% those of males in monogamous social groups.

In summary, while considerable evidence suggests that reproduction by subordinate callitrichid females is reduced in the presence of dominant females, a variety of factors may affect this relationship. These include the age of the subordinate, the presence of unrelated males, the environment (captive vs. wild), and the health, age, and reproductive status of the dominant female. Behavioral and endocrine data for golden lion tamarins in Poço das Antas suggest that subordinate males do not reproduce. Genetic analysis will be necessary to resolve this question.

Individuals Will Avoid Breeding with Close Relatives

Implicit in skew models is the assumption that deleterious consequences of inbreeding will inhibit mating of close relatives. In a summary of infant mortality rates for inbred and noninbred primates in captivity, Ralls and Ballou (1982) found significantly higher mortality for nearly all populations of 16 species. Studies documenting inbreeding depression in wild primate populations, although few and sometimes equivocal, generally support the conclusion that there is potential for inbreeding depression in primates (Moore 1993, Crnokrak & Roff 1999).

The decision of whether or not to breed with relatives is particularly important for subordinates in cooperatively breeding primate groups. In relatively stable callitrichid populations, subordinates that remain as helpers in their natal groups may wait for several breeding cycles before an unrelated individual of the opposite sex enters the group (Goldizen et al. 1996, Savage et al. 1996, Lazaro-Perea et al. 2000). In theory, the subordinate's decision regarding inbreeding should take into account negative consequences of inbreeding depression and diminishing reproductive value. If ecological constraints are severe, subordinates in intact family groups may reach an age at which their fitness can be improved by mating with a close relative, if the only other option is not mating at all (Bengtsson 1978). As ecological constraints relax, subordinates should disperse and outbreed rather than breed with a close relative.

Dietz et al. (2000) reported that about 10% of all lion tamarin offspring born in Poço das Antas Reserve were inbred. Inbreeding resulted from mating of close relatives in the same group and from individuals dispersing into and breeding in groups containing relatives. In that study, survival of newborn inbred infants was lower than that of noninbred infants and decreased with the severity of inbreeding. Most subordinate females that reached three years of age in their natal group became pregnant (Baker et al. 2002). The question of which males sired offspring of these subordinate females is unresolved. When all resident males were closely related to natal females, natal female pregnancies resulted from either extragroup copulations or within-group incestuous matings. Behavioral evidence regarding this issue was conflicting: both incestuous and extragroup sexual behaviors were documented among natal females in this study (Dietz et al. 2000). Molecular genetic analysis will be necessary to resolve this question.

While pregnancies resulting from incestuous matings appear to be rare in captive golden lion tamarins (French et al. 1989), they have been documented in intact captive groups of golden-headed lion tamarins (*Leontopithecus chrysomelas*: De Vleeschouwer et al. 2001) and other callitrichid species, for example, *Saguinus oedipus* (Price & McGrew 1991), *Callithrix jacchus* (Rothe & Koenig 1991), and *Cebuella pygmaea* (Schroepel 1998).

The Predictions of Skew Models

Predictions from OSM are based on the premise that subordinates and dominants are constantly estimating fitness costs and benefits influenced by the following factors: (1) for subordinates, the relative fitness consequences of dispersing and attempting to breed outside the natal group versus remaining as a helper in the current group; (2) for dominants, the magnitude of any inclusive fitness benefit realized as a result of the subordinate remaining in the group versus the dominant's fitness if the subordinate leaves the group; and (3) for dominants and subordinates, the probability that the subordinate can forcibly displace the dominant (Emlen 1997, Clutton-Brock 1998a). As changes take place in opportunities for independent breeding or current group composition (deaths, emigrations, or immigrations), predictions regarding the degree of distribution of reproduction would also change. In contrast, ICM predict that a subordinate's share of reproduction will be less affected by environmental constraints and genetic relationships than by the subordinate's ability to escape reproductive control by the dominant. Below I examine the extent to which empirical data on callitrichids conform to selected theoretical predictions from these models.

Reproductive Skew Will Be High When the Probability of Breeding Outside the Natal Group Is Low

Among cooperatively breeding birds and mammals, individuals that remain in their natal groups usually do not reproduce. Explanations proposed for why young adults choose not to disperse focus on benefits of staying at home and risks associated with dispersal. As ecological constraints on successful dispersal intensify (e.g., the number of breeding positions or available territories decreases, or competition for them increases), the probability of reproduction outside the natal group is reduced. In theory, multigenerational families form when offspring delay dispersal while waiting for breeding opportunities to become available outside the family territory (Brown 1987; Koenig et al. 1992; Emlen 1994, 1995). In times of plentiful breeding opportunities and/or reduced competition for those opportunities, maturing adults should leave their natal groups and breed elsewhere (Komdeur 1992, Walters et al. 1992, Emlen 1997, Hatchwell & Komdeur 2000).

Data on golden lion tamarins in Poço das Antas generally support the prediction that reproductive skew will be high when ecological constraints are severe. Prior to 1996, all suitable habitat in Poço das Antas Reserve was occupied by breeding groups of golden lion tamarins, and dispersal accounted for 78% of postinfancy losses of natal individuals (Baker 1991, Baker et al. 2002). Although mean age of natal dispersal was not substantially different for males and females, ecological constraints on breeding outside the natal group were greater for females than males. Fifty-three percent of dispersing males joined groups adjacent to their natal group, thus avoiding dangerous prospecting for breeding opportunities in unknown areas. In comparison, only 12% of dispersing females joined adjacent groups (Baker 1991, Baker et al. 2002). Given that females have fewer options for independent breeding, optimal skew models predict that dominants need confer a smaller reproductive staying incentive to females than males, resulting in greater reproductive skew for females.

The available behavioral and demographic data do not support this prediction. While both dominant and subordinate males in most two-male groups in Baker et al. (1993) exhibited sexual behavior with the reproductive female, dominant males were responsible for 94% of mounts, copulations, and episodes of consorting behavior during periods in which the breeding female was expected to be fertile. The authors concluded that despite the frequency of polyandrous group composition and copulations by more than one male, reproduction in this population was typically monopolized by a single dominant male per group (Baker et al. 1993). The results of a demographic model suggested that a subordinate male, even if achieving no current paternity, would typically experience higher lifetime reproductive success by staying in his current group and waiting for the dominant male to die or for a vacancy in a neighboring territory than he would by leaving the group and searching for breeding opportunities elsewhere (Baker et al. 1993).

The majority of golden lion tamarin groups in Poço das Antas contained a single reproductive female, but polygyny was a regular occurrence. Based on data collected between March 1984 and September 1991, Dietz and Baker (1993) classified 10.6% of group samples as polygynous. However, when only groups that contained more than one adult female were considered, the rate of pregnancy polygyny (two females observed to be pregnant in the same group in one breeding season) was much higher, 44% of group seasons (Baker et al. 2002). Reproductive skew was apparently greater in males than in females.

Reproductive Skew Will Be High When Dominant and Subordinate Are Close Relatives

Optimal skew models also predict that subordinates less closely related to dominants receive smaller indirect fitness gains and therefore require a greater staying incentive to remain in the group. Thus reproduction should be more equitably distributed in sibling-sibling associations than in parent-offspring groupings (Keller & Reeve 1994, Reeve & Keller 1995). In contrast with this prediction, mother-daughter sets of golden lion tamarins accounted for all cases of successful rearing by two females within a season (Baker et al. 2002). Sister-sister polygyny was less common, and never resulted in successful polygynous rearing. Duos of unrelated females were least common, and also never resulted in successful rearing (Baker et al. 2002). Some dominant females "allowed" subordinate females to remain and reproduce, while others apparently evicted same-sex subordinates from the group. The basis of these decisions does not appear to conform to predictions of optimal skew theory and may include variations in group membership, ecological conditions, and/or individual nutritional or health status.

Subordinates Are More Likely to Remain in Groups in Which They Provide Care for Close Relatives Than in Groups Containing Unrelated Offspring

Data on the effects of kinship on dispersal by subordinates in callitrichids are available only for the Poço das Antas population of golden lion tamarins. In that study, 22 breeding groups were monitored for 11 to 125 months each. Study groups contained more than one male unrelated to the breeding female in 46% of group months. Coresident males were close relatives in 76% of the 37 cases. In the nine remaining duos, the two males were born in different groups and thus were probably not close relatives. There was not a large difference in mean duration for kin-based versus nonkin duos (kin based: mean 15.1 months, range 0–45 months; nonkin: mean 12.0 months, range 5–26 months). However, a number of the kin-based duos were still intact at the end of the study. An analysis of duo duration that appropriately treated intact duos suggested a longer mean duration for kin-based duos compared to nonkin duos (Baker et al. 2002). As mentioned above, behavioral data suggest that the dominant male monopolizes reproduction in multimale groups, regardless of the degree of relatedness of the males. Thus there is some evidence supporting the prediction that male subordinates are less likely to disperse from groups in which they care for close relatives.

Life history characteristics of golden lion tamarins in Poço das Antas make it difficult to determine the effects of kinship on dispersal by subordinate females. In contrast with males, females rarely immigrated into established breeding groups but often inherited the breeding position upon death of a same-sex dominant (Baker et al. 2002). Thus most female subordinates were closely related to the breeding female and to the infants in the group. Goldizen et al. (1996) documented a similar relationship in their long-term study of *Saguinus fuscicollis*.

Nonreproductive Alloparents Will Provide Most Help to Closely Related Offspring

Kinship is an important factor in optimal skew theory. Subordinates are thought to weigh the costs of staying in a group as a nonreproductive helper against, among other things, inclusive fitness gains accrued by increasing the survival and reproduction of relatives in the group (Hamilton 1964). Following this logic, helpers are predicted to provide more care for closely related offspring than for unrelated offspring.

In captivity, callitrichid alloparents provide care for all group infants, including those to which they are not closely related (Cleveland & Snowdon 1984, Wamboldt et al. 1988, Achenbach & Snowdon 1998). The degree of care varies significantly with helper age but not sex or reproductive status (Tardif et al. 1992, Tardif 1997). In a study of captive pygmy marmosets (*Cebuella pygmaea*), breeding males and male alloparents did not differ in their care of related and unrelated infants, whereas mothers were more responsive to their own infants (Wamboldt et al.1988).

Testing this prediction using wild groups of callitrichids is complicated by the fact that the degree of relatedness among offspring and helpers is usually unknown. In one group of *Saguinus oedipus* in which kinship was known, there were no significant differences in the frequency of carrying by related versus unrelated helpers (Savage 1990). In another field study, Baker (1991) determined the time that golden lion tamarins in seven groups spent carrying infants to which they were more or less closely related. He found that male helpers spent significantly more time carrying closely related infants, but that carrying by female alloparents was not affected by relatedness to infants. He argued that the sex difference was explained by the greater importance for young females to gain parental experience. Tardif (1997) raised two concerns about this interpretation. First, the level of relatedness between alloparents and infants was relatively high in all cases. Second, in five of seven cases the female whose infants were more closely related to the potential alloparents was the dominant female, whereas the female with less closely related infants was subordinate. Social dominance is known to affect infant carrying in common marmosets (Digby & Ferrari 1994) and may contribute to the lack of observed sex bias in carrying by female alloparents.

In summary, although there is considerable individual variation in degree of help provided, callitrichid alloparents generally care for all infants in the group, regardless of kinship. Although there is some evidence of a relationship between helper-offspring kinship and the degree of care provided by wild male golden lion tamarins, data from wild callitrichids are insufficient to address the prediction that helpers will provide more care for closely related infants.

Discussion

The first step in determining if optimal skew or incomplete control models explain cooperative breeding and the diversity of mating systems observed in the Callitrichidae is to ask if empirical observations conform to the assumptions and predictions of the models. Data from field and captive studies generally support the assumption of OSM, that subordinate helpers increase the fitness of dominants in the group. Helpers generally increase infant survival in the wild and infant carrying and feeding in captivity. Where data are available, benefits to dominants "allowing" subordinates to remain in the group appear to outweigh costs. However, additional observations on the effects of helpers in groups containing zero and one alloparent would provide a more critical evaluation of this assumption.

Testing the assumption that subordinate callitrichids are less likely to leave the group or to challenge the dominant if they share in reproduction is problematic for two reasons when considering males. First, the level of reproduction by male callitrichids is unknown for most field studies. Second, field studies for which relevant data are available may not provide a suitable test of the assumption for subordinate males. Behavioral data on golden lion tamarins in Poço das Antas (Baker et al. 1993) and genetic data on three groups of common marmosets in northeastern Brazil (Nievergelt et al. 2000) suggest that subordinate males do not share in reproduction. However, most groups in the former study and all three in the latter study contained several alloparents, and thus reproductive sharing might not be predicted under optimal skew models.

Behavioral evidence does not provide strong support for the assumption that shared reproduction delays dispersal by subordinate male lion tamarins in Poço das Antas. However, one interpretation of the available data is that the staying incentive offered to subordinate males is permission to remain in the group and thus an increased probability of future reproduction relative to the likely outcome following forced dispersal into habitat containing few breeding opportunities. The tendency for kin-based male duos to last longer than unrelated duos in golden lion tamarins (Baker et al. 2002) also suggests that male helpers remain longer in groups containing offspring to which they are related. Although few daughters remained in their natal group to three years of age, most of these reproduced in the natal group, as predicted by OSM. In contrast, excepting a few cases where dominance roles were reversed, subordinate daughters did not breed more than twice in their natal group, a result that seems to support predictions of ICM.

Another assumption of optimal skew models is that dominants are able to control the breeding status of subordinates. Although a variety of factors may affect this relationship, it appears that reproduction by subordinate callitrichid females is reduced in the presence of dominant females. Aggressive interactions among adult male golden lion tamarins were rare but unidirectional, suggesting that enforced social dominance limits reproduction by subordinate males (ICM). Molecular genetic data for common marmosets (Nievergelt et al. 2000) also suggest that subordinate males do not reproduce in established groups. It is unclear whether dominant female golden lion tamarins allow their older daughters to breed (OSM) or have lost the ability to expel them from the group and/or regulate their reproduction (ICM). It is also unclear whether reproductive daughters in intact groups breed incestuously or through extragroup copulations.

There is some evidence that the reproductive suppression seen in established groups does not obtain in newly formed groups. In two recently formed groups of *Callithrix jacchus*, multiple males copulated with females who were likely to have been ovulating at the time (Lazaro-Perea et al. 2000). These authors reported low levels of male-male aggression and no interference with each other's copulations, suggesting cooperative polyandry as the likely mating system. Peres (1986) reported promiscuous matings among two adult males and two adult females in a newly formed group of golden lion tamarins in Poço das Antas. No intrasexual aggression was observed and both females conceived.

A first prediction of OSM is that reproductive skew will be high when the probability of breeding outside the natal group is low. In general, observations on golden lion tamarins in Poço das Antas conform to the prediction. Few breeding opportunities were available for dispersing individuals of either sex, and reproductive skew was high in both sexes (Baker et al. 1993, Dietz & Baker 1993). Constraints on independent breeding were apparently greater for females than males in Poço das Antas. Similar results were reported for saddleback tamarins in Manu National Park (Goldizen et al. 1996). Given fewer options for females to breed independently, OSM predict that dominants need provide smaller staying incentives to females than to males. In contrast with the predicted results of higher reproductive skew in females, reproductive skew was apparently greater for males than for females in both studies.

A second prediction of OSM, that reproductive skew will be high when dominants and subordinates are close relatives, is generally supported by observations on golden lion tamarins. In most cases, dominants and subordinates were close relatives, often brothers or mother and daughter. Reproductive skew was high for both sexes. Behavioral evidence notwithstanding, without molecular genetic analysis it is difficult to test for a difference in skew between duos of related and unrelated males. The observation that kin-based duos of males tended to persist longer than nonkin duos in spite of extreme skew could also be viewed as evidence in support of OSM and would not be expected under ICM. A similar comparison was not possible for females, as virtually all subordinates were closely related to dominant females in that population.

A final prediction of OSM is that subordinate alloparents will provide more care to related offspring than unrelated offspring. In general, captive and free-ranging callitrichid alloparents carry and provision all infants in the group. Few studies have attempted to quantify the amount of care provided to related versus unrelated offspring in wild populations. Baker (1991) found that male golden lion tamarin helpers spent more time carrying closely related infants, as predicted by OSM. However, carrying by female alloparents was not affected by degree of kinship with the infant. Additional research is necessary to adequately address this question.

A Testable Hypothesis Explaining Reproductive Skew in Callitrichids

In summary, although specific aspects of both models have predictive value, neither OSM nor ICM appear to adequately explain the diversity of mating systems and social groupings observed in wild callitrichid populations. One possible explanation for these equivocal findings is that aspects of both ICM and OSM obtain in callitrichid populations, the prevalence of each varying with ecological constraints, sex of subordinates, and the help required by dominants. I hypothesize that dominants in newly formed groups, those that have not produced a first litter, require the most help from subordinates—not only to care for future offspring but also to establish and defend the new territory. Data on newly formed groups are rare, first because this transitional period is often short, and second because stable, established groups are more likely to be habituated to the presence of human observers and thus more easily studied. If dominants are offering reproductive incentives to induce subordinates to remain in the group, I speculate that this will occur at the time of group formation. Available data suggest that within-group promiscuous mating may occur during group formation and dominants do not appear to interfere with copulations by subordinates. The rate of new group formation is a function of population turnover and colonization. Thus populations that have high rates of predation or hunting pressure, or are colonizing new areas, are likely to conform to predictions of OSM.

Populations that are stable and have a low turnover rate are more likely to conform to predictions of ICM. Once formed, callitrichid groups rapidly increase in size through births and/or immigrations. In relatively stable populations, colonized habitat is quickly saturated with breeding groups, and thus ecological constraints on breeding opportunities outside the natal group are typically high. Dominants in established groups effectively inhibit reproduction by most subordinates, through behavioral (males) and/or physiological means (females).

In established groups, there appears to be a contest for control of breeding, at least among females; 7% of the few reproductive subordinate female golden lion tamarins became dominant to their mothers and continued to breed for several years in the natal group. The remainder of breeding subordinates left the group, presumably expelled, after one or two years. It is not clear whether older daughters reproduced because they were able to escape reproductive suppression or because dominants did not expel them from the group. There is some evidence that subordinate females can escape reproductive suppression by females who are old or weak (French et al. in press), providing additional support for ICM.

Once groups are established and ecological constraints limit breeding opportunities for independent breeding, it does not appear that dominant males provide subordinates with reproductive staying incentives. As predicted by ICM, dominant males in established groups use mate guarding to inhibit subordinate reproduction during the relatively short periods when females are fertile. If there is a social contract between males in established groups, it may consist of the dominant male allowing subordinate males to remain in the group helping to raise related offspring while waiting for a breeding vacancy to appear in an adjacent group.

Needs for Future Research

Two factors have been important obstacles to testing predictions of skew models in free-ranging callitrichid populations. The first is a lack of information on genetic relatedness of dominants and subordinates. However, recent advances in molecular genetic techniques may allow resolution of this problem. Microsatellite genotyping used to examine paternity, mating patterns, and population structure in both wild and captive Old World primates (e.g., Morin, Moore et al. 1994; Morin, Wallis et al. 1994; Gerloff et al. 1995, 1999) is now available for similar analyses in some New World primates. In one field study, Nievergelt et al. (2000) used 11 variable microsatellite loci obtained from DNA extracted from hair to establish pedigree relations in three wild groups of common marmosets.

The second general issue is the paucity of long-term field data on callitrichid populations. There have been relatively few field studies on cooperatively breeding primates in which significant numbers of identifiable individuals were observed for time periods necessary to document changes in patterns of dispersal and reproduction under varying ecological conditions. Additional studies in which habitat is not saturated, groups contain few helpers, and turnover rate is high would be useful in providing critical tests of reproductive skew models.

In conclusion, callitrichids provide fertile ground for testing and expanding theory explaining reproductive skew and the evolution of cooperative breeding and family living. Clearly, additional studies on a variety of callitrichid taxa and populations will be necessary to distinguish between optimal skew and incomplete control models—and perhaps alternatives such as parasitism and mutualism (Clutton-Brock 2002). Available data suggest that evolutionary mechanisms maintaining cooperative breeding in the Callitrichidae are complex and diverse. Each additional experiment in a captive environment and each year of field data reveal new layers of complexity in the ecological and genetic factors explaining callitrichid life history decisions.

References

Abbott, D. H. 1984. Behavioral and physiological suppression of fertility in subordinate marmoset monkeys. *Am. J. Primatol.*, 6, 169–186.

Abbott, D. H. 1987. Behaviorally mediated suppression of reproduction in female primates. *J. Zool.*, 213, 455–470.

Abbott, D. H., George, L. M., Barrett, J., Hodges, K. T., O'Byrne, K. T., Sheffield, J. W., Sutherland, I. A., Chambers, G. R., Lunn, S. F., & Ruiz de Elvira, M. C. 1990. Social control of ovulation in marmoset monkeys: a neuroendocrine basis for the study of infertility. In: *Socioendocrinology of Primate Reproduction* (Ed. by T. E. Ziegler & F. B. Bercovitch), pp. 135–158. New York: Wiley-Liss.

Achenbach, G. G. & Snowdon, C. T. 1998. Response to sibling birth in juvenile cotton-top tamarins (*Saguinus oedipus*). *Behaviour*, 135, 845–862.

Baker, A. J. 1991. Evolution of the social system of the golden lion tamarin. PhD dissertation, University of Maryland, College Park, MD.

Baker, A. J., Bales, K., & Dietz, J. M. 2002. Mating system and group dynamics in golden lion tamarins (*Leontopithecus rosalia*). In: *The Lion Tamarins of Brazil* (Ed. by D. G. Kleiman & A. B. Rylands), pp. 188–212. Washington, DC: Smithsonian Institution Press.

Baker, A. J., Dietz, J. M., & Kleiman, D. G. 1993. Behavioural evidence for monopolization of paternity in multi-male groups of golden lion tamarins. *Anim. Behav.*, 46, 1091–1103.

Bales, K., Dietz, J., Baker, A., Miller, K., & Tardif, S. D. 2000. Effects of allocare-givers on fitness of infants and parents in callitrichid primates. *Folia Primatol.*, 71, 27–38.

Bales, K., O'Herron, M., Baker, A. J., & Dietz, J. M. 2001. Sources of variability in numbers of live births in wild golden lion tamarins (*Leontopithecus rosalia*). *Am. J. Primatol.*, 54, 211–221.

Bengtsson, B. O. 1978. Avoiding inbreeding—at what cost? *J. Theor. Biol.*, 73, 439–444.

Brown, J. L. 1987. *Helping and Communal Breeding in Birds*. Princeton, NJ: Princeton University Press.

Cleveland, J. & Snowdon, C. T. 1984. Social development during the first twenty weeks in the cotton-top tamarin (*Saguinus o. oedipus*). *Anim. Behav.*, 32, 432–444.

Clutton-Brock, T. 2002. Behavioral ecology—breeding together: kin selection and mutualism in cooperative vertebrates. *Science*, 296, 69–72.

Clutton-Brock, T. H. 1998a. Reproductive skew, concessions and limited control. *Trends Ecol. Evol.*, 13, 288–292.

Clutton-Brock, T. H. 1998b. Reproductive skew: disentangling concessions from control—reply. *Trends Ecol. Evol.*, 13, 459–459.

Crnokrak, P. & Roff, D. A. 1999. Inbreeding depression in the wild. *Heredity*, 83, 260–270.

Dawson, G. 1977. Composition and stability of social groups of the tamarin, *Saguinus oedipus geoffroyi*, in Panama: ecological and behavioral implications. In: *The Biology and Conservation of the Callithrichidae* (Ed. by D. G. Kleiman), pp. 23–37. Washington, DC: Smithsonian Institution Press.

De Vleeschouwer, K., Van Elsacker, L., & Leus, K. 2001. Multiple breeding females in captive groups of golden-headed lion tamarins (*Leontopithecus chrysomelas*): causes and consequences. *Folia Primatol.*, 72, 1–10.

Dietz, J. M. & Baker, A. J. 1993. Polygyny and female reproductive success in golden lion tamarins, *Leontopithecus rosalia*. *Anim. Behav.*, 46, 1067–1078.

Dietz, J. M., Baker, A. J., & Ballou, J. D. 2000. Demographic evidence of inbreeding depression in wild golden lion tamarins. In: *Genetics, Demography and Viability of Frag-*

mented Populations (Ed. by A. G. Young & G. M. Clarke), pp. 203–211. Cambridge: Cambridge University Press.

Dietz, J. M., Peres, C. A., & Pinder, L. 1997. Foraging ecology and use of space in wild golden lion tamarins (*Leontopithecus rosalia*). *Am. J. Primatol.*, 41, 289–305.

Digby, L. 1995a. Infant care, infanticide, and female reproductive strategies in polygynous groups of common marmosets (*Callithrix jacchus*). *Behav. Ecol. Sociobiol.*, 37, 51–61.

Digby, L. J. 1995b. Social organization in a wild population of *Callithrix jacchus*. 2. Intragroup social behavior. *Primates*, 36, 361–375.

Digby, L. J. & Ferrari, S. F. 1994. Multiple breeding females in free-ranging groups of *Callithrix jacchus*. *Int. J. Primatol.*, 15, 389–397.

Emlen, S. T. 1994. Benefits, constraints and the evolution of the family. *Trends Ecol. Evol.*, 9, 282–285.

Emlen, S. T. 1995. An evolutionary theory of the family. *Proc. Natl. Acad. Sci. USA*, 92, 8092–8099.

Emlen, S. T. 1997. Predicting family dynamics in social vertebrates. In: *Behavioural Ecology: An Evolutionary Approach* (Ed. by J. R. Krebs & N. B. Davies), pp. 228–253. Oxford: Blackwell Science.

Emlen, S. T., Reeve, H. K., & Keller, L. 1998. Reproductive skew: disentangling concessions from control. *Trends Ecol. Evol.*, 13, 458–459.

French, J. A. 1984. The effects of social environment on estrogen excretion, scent marking, and sociosexual behavior in tamarins (*Saguinus oedipus*). *Am. J. Primatol.*, 6, 155–167.

French, J. A. 1997. Proximate regulation of singular breeding in callitrichid primates. In: *Cooperative Breeding in Mammals* (Ed. by N. G. Solomon & J. A. French), pp. 34–75. Cambridge: Cambridge University Press.

French, J. A., Bales, K., Baker, A. J., & Dietz, J. M. In press. Endocrine monitoring of wild dominant and subordinate female golden lion tamarins (*Leontopithecus rosalia*). *Int. J. Primatol.*

French, J. A., deGraw, W. A., Hendricks, S. E., Wegner, F. H., & Bridson, W. E. 1992. Urinary and plasma gonadotropin concentrations in golden lion tamarins (*Leontopithecus rosalia*). *Am. J. Primatol.*, 26, 53–59.

French, J. A. & Inglett, B. J. 1991. Responses to novel social stimuli in callitrichid monkeys: a comparative perspective. In: *Primate Responses to Environmental Change* (Ed. by H. O. Box), pp. 275–294. London: Chapman Hall.

French, J. A., Inglett, B. J., & Dethlefs, T. M. 1989. The reproductive status of nonbreeding group members in captive golden lion tamarin social groups. *Am. J. Primatol.*, 18, 73–86.

Garber, P. A. 1994. Phylogenetic approach to the study of tamarin and marmoset social systems. *Am. J. Primatol.*, 34, 199–219.

Garber, P. A., Moya, L., & Malaga, C. 1984. A preminary field study of the moustached tamarin monkey (*Saguinus mystax*) in northern Peru: questions concerned with the evolution of a communal breeding system. *Folia Primatol.*, 42, 17–32.

Gerloff, U., Hartung, B., Fruth, B., Hohmann, G., & Tautz, D. 1999. Intracommunity relationships, dispersal pattern and paternity success in a wild living community of bonobos (*Pan paniscus*) determined from DNA analysis of faecal samples. *Proc. Royal Soc. Lond., B*, 266, 1189–1195.

Gerloff, U., Schlotterer, C., Rassmann, K., Rambold, I., Hohmann, G., Fruth, B., & Tautz, D. 1995. Amplification of hypervariable simple sequence repeats (microsatellites) from excremental DNA of wild living bonobos (*Pan paniscus*). *Mol. Ecol.*, 4, 515–518.

Goldizen, A. 1987. Tamarins and marmosets: communal care of offspring. In: *Primate Soci-*

eties (Ed. by B. Smuts, D. L. Cheney, R. M. Seyfarth, R. W. Wrangham, & T. T. Struhsaker), pp. 34–43. Chicago: University of Chicago Press.

Goldizen, A. W. 1988. Tamarin and marmoset mating systems: unusual flexibility. *Trends Ecol. Evol.*, 3, 36–40.

Goldizen, A. W., Meldelson, J., Vanvlaardingen, M., & Terborgh, J. 1996. Saddle-back tamarin (*Saguinus fuscicollis*) reproductive strategies: evidence from a thirteen-year study of a marked population. *Am. J. Primatol.*, 38, 57–83.

Hamilton, W. D. 1964. Genetical evolution of social behaviour. I and 2. *J. Theor. Biol.*, 7, 1–52.

Hatchwell, B. J. & Komdeur, J. 2000. Ecological constraints, life history traits and the evolution of cooperative breeding. *Anim. Behav.*, 59, 1079–1086.

Hubrecht, R. C. 1984. Field observations on group size and composition of the common marmoset (*Callithrix jacchus jacchus*), at Tapacura, Brazil. *Primates*, 25, 13–21.

Keller, L. & Reeve, H. K. 1994. Partitioning of reproduction in animal societies. *Trends Ecol. Evol.*, 9, 98–102.

Koenig, A. 1995. Group size, composition and reproductive success in wild common marmosets (*Callithrix jacchus*). *Am. J. Primatol.*, 35, 311–317.

Koenig, W. D., Pitelka, F. A., Carmen, W. J., Mumme, R. L., & Stanback, M. T. 1992. The evolution of delayed dispersal in cooperative breeders. *Q. Rev. Biol.*, 67, 111–150.

Komdeur, J. 1992. Importance of habitat saturation and territory quality for evolution of cooperative breeding in the Seychelles warbler. *Nature*, 360, 768–768.

Lazaro-Perea, C., Castro, C. S. S., Harrison, R., Araujo, A., Arruda, M. F., & Snowdon, C. T. 2000. Behavioral and demographic changes following the loss of the breeding female in cooperatively breeding marmosets. *Behav. Ecol. Sociobiol.*, 48, 137–146.

Leutenegger, W. 1980. Monogamy in callitrichids: a consequence of phyletic dwarfism? *Int. J. Primatol.*, 1, 95–98.

Mays, N. A., Vleck, C. M., & Dawson, J. 1991. Plasma luteinizing hormone, steroid hormones, behavioral role, and nest stage in cooperatively breeding Harris' hawk (*Parabuteo unicinctus*). *Auk*, 108, 619–637.

Moore, J. J. 1993. Inbreeding and outbreeding in primates: what's wrong with the dispersing sex? In: *The Natural History of Inbreeding and Outbreeding* (Ed. by N. W. Thornhill), pp. 392–426. Chicago: University of Chicago Press.

Morin, P. A., Moore, J. J., Chakraborty, R., Jin, L., Goodall, J., & Woodruff, D. S. 1994. Kin selection, social structure, gene flow, and the evolution of chimpanzees. *Science*, 265, 1193–1201.

Morin, P. A., Wallis, J., Moore, J. J., & Woodruff, D. S. 1994. Paternity exclusion in a community of wild chimpanzees using hypervariable simple sequence repeats. *Mol. Ecol.*, 3, 469–477.

Mumme, R. L. 1997. A bird's-eye view of mammalian cooperative breeding. In: *Cooperative Breeding in Mammals* (Ed. by N. G. Solomon & J. A. French), pp. 364–383. Cambridge: Cambridge University Press.

Nievergelt, C. M., Digby, L. J., Ramakrishnan, U., & Woodruff, D. S. 2000. Genetic analysis of group composition and breeding system in a wild common marmoset (*Callithrix jacchus*) population. *Int. J. Primatol.*, 21, 1–20.

Peres, C. 1986. Costs and benefits of territorial defense in golden lion tamarins, *Leontopithecus rosalia*. MSc thesis, University of Florida, Gainesville, FL.

Price, E. C. 1992. The benefits of helpers—effects of group and litter size on infant care in tamarins (*Saguinus oedipus*). *Am. J. Primatol.*, 26, 179–190.

Price, E. C. & McGrew, W. 1991. Departures from monogamy in colonies of captive cotton-top tamarins. *Folia Primatol.*, 57, 16–27.

Ralls, K. & Ballou, J. 1982. Effects of inbreeding on infant mortality in captive primates. *Int. J. Primatol.*, 3, 491–505.

Reeve, H. K. & Emlen, S. T. 2000. Reproductive skew and group size: an n-person staying incentive model. *Behav. Ecol.*, 11, 640–647.

Reeve, H. K., Emlen, S. T., & Keller, L. 1998. Reproductive sharing in animal societies: reproductive incentives or incomplete control by dominant breeders? *Behav. Ecol.*, 9, 267–278.

Reeve, H. K. & Keller, L. 1995. Partitioning of reproduction in mother-daughter versus sibling associations—a test of optimal skew theory. *Am. Nat.*, 145, 119–132.

Reeve, H. K. & Nonacs, P. 1992. Social contracts in wasp societies. *Nature*, 359, 823–825.

Reeve, H. K. & Ratnieks, F. L. W. 1993. Queen-queen conflicts in polygynous societies: mutual tolerance and reproductive skew. In: *Queen Number and Sociality in Insects* (Ed. by L. Keller), pp. 45–85. Oxford: Oxford University Press.

Reyer, H. U., Dittami, J. P., & Hall, M. R. 1986. Avian helpers at the nest: are they psychologically castrated. *Ethology*, 71, 216–228.

Rothe, H. & Koenig, A. 1991. Variability of social organization in captive common marmosets (*Callithrix jacchus*). *Folia Primatol.*, 57, 28–33.

Rothe, H., Koenig, A., & Darms, K. 1993. Infant survival and number of helpers in captive groups of common marmosets (*Callithrix jacchus*). *Am. J. Primatol.*, 30, 131–137.

Saltzman, W., Schultz-Darken, N. J., Severin, J. M., & Abbott, D. H. 1997. Escape from social suppression of sexual behavior and of ovulation in female common marmosets. *Ann. NY Acad. Sci.*, 807, 567–570.

Savage, A. 1990. The reproductive biology of the cotton-top tamarin (*Saguinus oedipus oedipus*) in Colombia. PhD dissertation, University of Wisconsin, Madison.

Savage, A., Giraldo, K. H., Soto, K. H., & Snowdon, C. T. 1996. Demography, group composition, and dispersal in wild cotton-top tamarin (*Saguinus oedipus*) groups. *Am. J. Primatol.*, 38, 85–100.

Savage, A., Shideler, S. E., Soto, L. H., Causado, J., Giraldo, L. H., Lasley, B. L., & Snowdon, C. T. 1997. Reproductive events of wild cotton-top tamarins (*Saguinus oedipus*) in Colombia. *Am. J. Primatol.*, 43, 329–337.

Savage, A., Snowdon, C. T., Giraldo, H., & Soto, K. H. 1997. Parental care patterns and vigilance in wild cotton-top tamarins (*Saguinus oedipus*). In: *Adaptive Radiation of Neotropical Primates* (Ed. by M. A. Norconk, A. Rosenberger, & P. A. Garber), pp. 187–199. New York: Plenum Press.

Savage, A., Ziegler, T. E., & Snowdon, C. T. 1988. Sociosexual development, pair bond formation, and mechanisms of fertility suppression in female cotton-top tamarins (*Saguinus oedipus oedipus*). *Am. J. Primatol.*, 14, 345–359.

Schroepel, M. 1998. Multiple simultaneous breeding females in a pygmy marmoset group (*Cebuella pygmaea*). *Neotrop. Primates*, 6, 1–7.

Snowdon, C. T. 1996. Infant care in cooperatively breeding species. In: *Parental Care: Evolution, Mechanisms, and Adaptive Significance* (Ed. by J. S. Rosenblatt & C. T. Snowdon), pp. 643–689. New York: Academic Press.

Soini, P. 1982. Ecology and population dynamics of the pygmy marmoset, *Cebuella pygmaea*. *Folia Primatol.*, 39, 1–21.

Sussman, R. W. & Garber, P. A. 1987. A new interpretation of the social organization and mating system of the callitrichidae. *Int. J. Primatol.*, 8, 73–92.

Tardif, S. D. 1984. Social influences on sexual maturation of female *Saguinus oedipus oedipus*. *Am. J. Primatol.*, 6, 199–209.

Tardif, S. D. 1997. The bioenergetics of parental behavior and the evolution of alloparental care in marmosets and tamarins. In: *Cooperative Breeding in Mammals* (Ed. by N. G. Solomon & J. A. French), pp. 11–33. Cambridge: Cambridge University Press.

Tardif, S. D., Carson, R. L., & Gangaware, B. L. 1992. Infant-care behavior of non-reproductive helpers in a communal-care primate, the cotton-top tamarin (*Saguinus oedipus*). *Ethology*, 92, 155–167.

Tardif, S. D., Harrison, M. L., & Simek, M. A. 1993. Communal infant care in marmosets and tamarins: relation to energetics, ecology, and social organization. In: *Marmosets and Tamarins: Systematics, Behaviour, and Ecology* (Ed. by A. B. Rylands), pp. 220–234. Oxford: Oxford University Press.

Terborgh, J. & Goldizen, A. W. 1985. On the mating system of the cooperatively breeding saddle-backed tamarin (*Saguinus fuscicollis*). *Behav. Ecol. Sociobiol.*, 16, 293–299.

Walters, J. R., Copeyon, C. K., & Carter, J. H. 1992. Test of the ecological basis of cooperative breeding in red-cockaded woodpeckers. *Auk*, 109, 90–97.

Wamboldt, M. Z., Gelhard, R. E., & Insel, T. R. 1988. Gender differences in caring for infant *Cebuella pygmaea*: the role of infant age and relatedness. *Dev. Psychobiol.*, 21, 187–202.

Windfelder, T. L. 2000. Observations on the birth and subsequent care of twin offspring by a lone pair of wild emperor tamarins (*Saguinus imperator*). *Am. J. Primatol.*, 52, 107–113.

Ziegler, T. E. 1987. Social interactions and determinants of ovulation in tamarins (*Saguinus*). *Int. J. Primatol.*, 8, 457–457.

11

Kinship Structure and Its Impact on Behavior in Multilevel Societies

Fernando Colmenares

The link between kinship structure and social behavior in primate societies has been fairly well documented in several species of Old World monkeys with multimale/multifemale social systems (e.g., macaques [*Macaca* spp.], savanna baboons [*Papio* spp.] and vervets [*Cercopithecus aethiops*]). In these species, dispersal is male biased, and females tend to establish strong and enduring bonds and alliances with maternal kin (Gouzoules & Gouzoules 1987). In contrast, much less is known about the impact of kinship structure on the behavior and decision-making processes of individuals in monkey species with one-male/multifemale social systems, in which the species' typical dispersal pattern can be more variable. For example, both sexes may transfer or female resident patterns may prevail. To complicate matters further, some of these species may form fusion-fission, multilayered societies, in which individuals establish and maintain relationships with others at various levels of nested groupings. For example, hamadryas (*Papio hamadryas*) and geladas (*Theropithecus gelada*) are known to organize themselves into one-male units, clans or teams, bands, and troops or herds (hamadryas: Sigg et al. 1982; Abegglen 1984; Kummer 1968a, 1984; gelada: Dunbar & Dunbar 1975; Kawai 1979; Kawai et al. 1983; Dunbar 1984, 1986, 1988a).

The objective of this chapter is to present an overview of what is known about the social systems of hamadryas and gelada baboons in order to assess the relation between kinship structure and social behavior in one-male social systems where a variety of possibilities can occur: multilevel structures, two-male social systems, female bondedness, cross-sex bonding, and kin-based male associations and alliances. First, I briefly outline the theoretical framework within which the empirical evidence will be analyzed and its implications assessed. Second, I describe the characteristics of the database available on which explanatory principles concerned with the "deep structure" (cf. Hinde 1976) of social systems are to be abstracted and tested. Third, I review what is known about the multilevel group structure and mating patterns of the two species. Fourth, I examine the link between the two species' dispersal patterns and the consequences they have for the kinship structure of the social systems they produce. Fifth, I analyze the nature of the social relationships within and between the sexes to see how they are affected by kinship and how they constrain the individuals' options to associate and cooperate with kin and with unrelated partners. Finally, I discuss the rather incomplete data available on the social systems of these two species in the light of current socioecological theory, emphasizing the economic nature of the rules that seem to underlie the patterns of partner preferences found in each species and the variable value that kin have to males and females in the two social systems.

I conclude by suggesting that similarities and differences between hamadryas and geladas in the patterns of bonding and cooperation may well reflect similarities and differences in the characteristics of the biomarkets where individuals play out their fitness-related strategies, although a possible role for genetically based species differences in the males' contribution to these contrasting social systems is also considered. The "surface structures" of the multilevel societies of hamadryas and geladas differ from one another; however, they are based on the operation of similar principles concerned with the deep structure.

Theoretical Framework

Primate sociality and cooperation are thought to have evolved in response to ecological and social pressures (e.g., Wrangham 1980, 1987; van Schaik 1983, 1989, 1996; Dunbar 1988a, b; Isbell 1991; Sterck et al. 1997). Thus, group living in general and the social relationships that group members establish with each other in particular are seen as strategies whereby partners try to successfully solve the problems of survival (i.e., avoidance of predators and access to food resources) and reproduction (i.e., mating success and infant survival). Living in a group, however, is not a cost-free option. Within primate groups, food and other resources, including mates and social partners, are limited, and this results in increased competition and stress, reproductive suppression and failure, and, ultimately, large variation among group members in reproductive success.

Within the framework of socioecology, concerned with the study of the proximate, functional, and evolutionary causes and consequences of variation in social systems (Crook 1970, Rasmussen 1981, Rubenstein & Wrangham 1986, Dunbar 1988a), several theoretical models have been proposed that have been especially influential in the formulation and testing of specific hypotheses and in setting the agenda of questions generally asked by researchers in this field. Kinship selection theory (Hamilton 1964) claims that the performance of costly behaviors that benefit social partners should be more likely between kin than nonkin, as this would result in increased inclusive fitness (Wrangham 1980, 1982; Gouzoules & Gouzoules 1987; van Schaik 1989; Moore 1992). Nevertheless, cooperation between unrelated individuals, whether based on reciprocity or mutualism, is also expected (Trivers 1971), has been well documented in primate studies (Harcourt & de Waal 1992, Chapais 1995), and has been demonstrated to have a strong impact on the shape of primate social systems (Dunbar 1988a, van Hooff & van Schaik 1992). Sexual selection theory (Trivers 1972) predicts that males and females should be expected to display different social and reproductive strategies, as the factors that constrain the maximizing of their reproductive success vary between the sexes. In general, males should be more competitive, more promiscuous, and less committed to parental care than females (Smuts 1987, Dunbar 1988a). In addition, males might resort to coercion as an alternative strategy to gain more copulations whenever females are choosy and males are not equally attractive to them (Smuts & Smuts 1993, Clutton-Brock & Parker 1995). Of course, the expression of such sex-linked, predicted differences should vary as a function of variation in the particular ecological and social scenarios in which members of either sex have to play out their strategies. Thus, conditions exist under which one might theoretically expect females to be highly competitive and promiscuous and males to be choosy and involved in some form of parental care, and there are empirical data substantiating such predictions (Smuts 1987, Whitten 1987, Small 1993).

Social relationships are thought of as investments (Kummer 1978; Seyfarth 1983; Dunbar 1984, 1988b; Cords 1997; van Hooff 2001) whereby partners attempt to maximize their short-term access to commodities and resources and their lifetime reproductive success. Thus, by cultivating relationships with valuable partners, that is, those who provide vital services (e.g., access to resources, protection against predators or against competitors) whether they are relatives or unrelated, individuals may expect to derive fitness related benefits. However, within a social group, the value of any given group member (and of the services it provides) depends on how many partners are in a position to provide those services (i.e., the level of supply) and on how many individuals are competing for them (i.e., the level of demand). The biological markets theory (Noë & Hammerstein 1995, Noë 2001) predicts that these market forces should be expected to largely determine the strategic options open to group members to establish and service social relationships with fitness consequences (see also Dunbar 1984). It must be pointed out, however, that the organization of a primate social system and hence of the component relationships among the group members is very likely to reflect the operation of time budget and social constraints. In effect, the size and the demographic composition of a group crucially influences the number and quality of the social relationships that any given group member can service, because time for socializing is limited and because the expression of one's partner choice preferences can be constrained by other group members' own partner preferences (Colmenares et al. 2002).

Database

Hamadryas and gelada baboons are both terrestrial, medium-sized African primates of the subfamily Cercopithecinae (Dunbar 1988a). Hamadryas are also called desert baboons because they inhabit the arid regions located on both coasts of the Red Sea, at altitudes from 0 to 2,600 m (Kummer 1995), where food and water resources are dispersed, scarce, and sometimes hard to find and exploit, and predators are not abundant. They are omnivorous, with a marked preference for fruits and seeds. Geladas are also known as mountain baboons because they inhabit areas at altitudes from about 2,000 to 4,000 m. They are adapted to a grazing niche, so the resources they feed on (i.e., the grasses) are evenly distributed and abundant, and they face relatively high predation pressure (Dunbar 1986).

The social organization of the hamadryas and the gelada baboon has been studied both in the wild and in captive settings. As can be seen in table 11.1, unfortunately the field studies carried out on wild populations of hamadryas and gelada baboons have had a number of limitations. First, they have been relatively short term. The study by Kummer and his students and coworkers of Ethiopian hamadryas baboons spanned a fairly long period of time (from May 1971 to February 1977); however, in practice the total time that investigators actually spent in the field sites was only 49 months (out of 69 months), including time lags of up to eight months. Second, the whole sample of study subjects and social units that were individually recognized and intensively studied was rather limited. Observers recognized individually only some of the adult and subadult males and females of their study populations (hamadryas: Kummer 1968a, Abegglen 1984; gelada: Dunbar & Dunbar 1975, Kawai 1979, Dunbar 1984). Finally, and more important, genealogical relationships among the adult members of the study groups were unknown. In some cases, female kin relations were inferred from observations of female grooming clusters and the relative ages of the grooming partners (hamadryas: Kummer 1968a, Abegglen 1984; gelada: Dunbar 1979a, 1980a, 1984) and male kin relations were inferred from morphological resemblance and the relative ages of the males of clans and leader-follower associations (hamadryas: Abegglen 1984; Kummer 1968a, 1995).

The longest study of hamadryas baboons has been carried out in captivity on the large colony housed at the Madrid Zoo (e.g., Colmenares & Gomendio 1988, Colmenares 1992). It started in June 1972, when the colony was established, and has continued ever since. During the past 30 years, long-term data have been collected uninterruptedly on the females' reproductive life history, the membership of social units, and the males' social and reproductive ontogenetic trajectories. All colony individuals are recognized individually and their matrilineal kin relations are also known. At this writing, we have named 340 individuals, of which 312 were born in the Madrid Zoo. As regards paternity, throughout the study we have assumed that a harem's leader male can be considered the putative father of his females' offspring, since all overt copulations occur exclusively between the harem male and his females. We did not make this assumption, however, when the mating access to any given female had been contested by some rival male during the female's follicular phase, and changeovers of varying length had been recorded, or when sneaky copulations by follower males were observed. The Madrid colony of baboons is a largely unmanipulated primate population. Major removals have been done only on three occasions: 1983 to 1985 (Colmenares & Gomendio 1988, Colmenares 1992), March 1995, and July 1999, and unfamiliar animals have been introduced only once, in March 1985 (Colmenares & Gomendio 1988). In many respects, the Madrid colony of hamadryas baboons resembles a wild band (see below). As the 1985 manipulation involved the removal of all adult males and the introduction of three novel adult males, we have considered that the Madrid colony can actually be seen as consisting of two bands: band 1 from June 1972 to April 1985, and band 2 from April 1985 to August 2001 (table 11.1).

Multilevel Group Structure and Mating System

Hamadryas and gelada baboons form multilevel, fusion-fission societies (Kummer 1968a, 1984, 1990; Dunbar & Dunbar 1975; Kawai 1979; Sigg et al. 1982; Kawai et al. 1983; Abegglen 1984; Dunbar 1988a). The basic social and reproductive unit is the harem or one-male unit. It consists of a single breeding male, the so-called harem male or unit leader, and a variable number of sexually mature females (table 11.1). Some such harem units may also contain one or (more rarely) several sexually mature males who do not mate with the unit females, although they may try and sneak some copulations. These males are called follow-

Table 11.1. Field and Enclosure Studies of Social Behavior in Multilevel Societies of Hamadryas and Gelada Baboons[a]

Species	Site	Duration	Period	Troops[b]	Bands[c]	Band Size[d]	Clans[e]	Harems[e]	Harem Size[f]
Hamadryas	Erer-Gota[g] (Ethiopia)	61 mo	Nov 1960–Oct 1961	White Rock	1	80	8	6	1–7
			May 1971–Jul 1972	Ravine Rock	3	52–82	3	2	
			Apr and May 1973	Cone Rock				9	
			Jan 1974	Gota Rock				3	
			Mar 1974–May 1975						
	Enclosure[h] Study		Jan 1976–Feb 1977					3	2
			Mar 1974–May 1975						
	Several sites[i] (Saudi Arabia)	2 mo	Jan–Feb 1980	—	—	—	—	—	1–7
	Zurich Zoo[j] (Switzerland)	?	1955, 1958, and 1959	—	1	15	—	3	2–7
	Madrid Zoo[k] (Spain)	362 mo	Jun 1972–still ongoing	—	2	26–91	4	17	1–9
						29–103	4	26	1–9

Geladas	Sankaber[l] (Ethiopia)	9 mo	Jul 1971–Mar 1972	—	—	2	85–289	—	11	1–10
		9 mo	1974–1975	—	—	2	128–280	—	14	2–8
	Gich[m] (Ethiopia)	9 mo	Jun 1973—Mar 1974	—	—	1	112	—	6	—
	Delta Primate Research Center[n] (U.S.A.)	?	1967	—	—	—	—	—	4	1–3

[a] I excluded field studies and sites where the focus was only or mostly the collection of demographic information (see text). I also excluded enclosure studies where the number of one-male units was very small (<3) or the focus was only marginal to the objectives of this chapter (see text).
[b] Only hamadryas troops from which bands, clans, and/or OMUs for close study were selected.
[c] Only those that were studied intensively.
[d] Only bands from the close study sample.
[e] Only clans and harems that were studied intensively.
[f] Number of breeding females from harems that were studied intensively.
[g] Kummer 1968a, Sigg 1980, Sigg et al. 1982, Abegglen 1984, Stammbach 1987.
[h] Sigg 1980.
[i] Kummer et al. 1985.
[j] Kummer & Kurt 1965, Kummer 1995.
[k] Colmenares 1996, 1992, unpublished data; Colmenares & Gomendio 1988.
[l] Dunbar & Dunbar 1975; Dunbar 1979a, 1980b, 1984.
[m] Kawai 1979.
[n] Kummer 1975.

ers. Several one-male units may aggregate to form higher groupings called clans (in hamadryas) or teams (in geladas). The next higher layer consists of several harem units together with clans or teams, if any, and this is termed a band. Finally, the top level in this hierarchy of nested groupings is termed a troop (in hamadryas) or herd (in geladas), and results when several bands (or band subunits, i.e., harems or clans/teams) associate with one another. In the gelada, bachelor males may also form all-male units. No such type of grouping has been described in hamadryas baboons (see, however, Abegglen 1984: 100).

As shown in table 11.2, the functions of the different grouping units observed in hamadryas and gelada baboons overlap partially. The scarcity of predator-safe sleeping places leads to the formation of the largest grouping units observed in hamadryas baboons, that is, troops, during the night. In contrast, the abundance of food and the high predator pressure faced by geladas favors their forming the largest aggregations of harem units and bands while foraging during the day. On the other hand, the smallest units emerge for foraging in hamadryas baboons and for sleeping in geladas. In both species, the harem is the basic reproductive and social unit, although, especially in the gelada, female grooming cliques and coalitions can also be identified within units, notably when they are large (Dunbar 1984, 1988a).

The social career or life trajectory of the males and of the social units they hold has been described for both species (Kummer 1968a, Dunbar & Dunbar 1975, Mori 1979c, Sigg et

Table 11.2. Mating System, Social Structure, Dispersal Pattern, and Group Cohesion in Hamadryas and Gelada Baboons

Variable	Hamadryas	Gelada
Mating system	Harem defense polygyny and sequential polyandry	Harem defense polygyny and sequential polyandry
Social structure	Multilevel, fusion-fission Patrilineal	Multilevel, fusion-fission Matrilineal
Grouping units and functions		
Harem	Mating, rearing, socializing, and foraging	Mating, rearing, socializing, and sleeping
Clan/team	Foraging	Foraging
Band	Foraging	Foraging
Troop/herd	Sleeping	Foraging
All-male group	Not described	Socializing and foraging
Dispersal and transfer[a]	Both sexes. Females are transferred mostly across OMUs and clans within the bands. Males also transfer across bands.	Only males transfer across bands. Many return and breed at their natal bands.
Cohesion[b]		
Harem	Cross-sex bonding, male herding, and female bonding when harem females are kin	Female bonding among kin
Clan/team	Bonds among male kin	Bonds among female kin

[a]There is some controversy about the pattern of dispersal and transfer of both sexes in the hamadryas, and of the males in the gelada (see text).
[b]The possible effect of male kinship has not yet been thoroughly investigated in the gelada.

al. 1982, Kawai et al. 1983, Abegglen 1984, Dunbar 1984) and may include the following positions: young follower, young leader, prime leader, old leader, and old follower (Colmenares 1992). The most apparent differences between the two species have to do with the early stages: some hamadryas subadult or young adult males start their reproductive careers by herding prepubertal females and establishing initial units, which are small, highly unstable, and often part of a larger established unit typically owned by a prime leader. Gelada males do not follow this path (Dunbar & Dunbar 1975; Mori 1979c, e); instead they first join spatially and socially detached all-male groups, which hamadryas do not form (Kummer 1968a, Abegglen 1984). Later, when they attain sexual maturity, they may attach themselves to a unit as young followers, and by building up bonds with prepubertal or subadult females, they make their way toward eventually setting up their initial or incipient units as young leaders. It is important to emphasize that in both species, not all of the individual males necessarily complete the whole sequence, and that the age at which individual males reach each stage in this life trajectory is strongly demography dependent and therefore may be highly variable within and between bands (hamadryas: Abegglen 1984; Colmenares 1992, unpublished data; gelada: Dunbar & Dunbar 1975, Mori 1979c, Dunbar 1984). In the hamadryas, at least, it is clear that a low socionomic sex ratio leads to more sexually mature males becoming followers and forming queues while they wait for their opportunity to gain access to females. In both species, defeated harem leaders may remain attached to their former units as old followers.

The mating system of both hamadryas and gelada baboons consists of harem defense polygyny and sequential polyandry (Kummer 1984, Dunbar 1984). Table 11.3 shows the degree of polygyny and polyandry exhibited by a selected sample of males and females from the Madrid hamadryas colony. The Madrid males vary in their reproductive success, which is related to the length of their tenure as breeding males and to the intensity of mate competition for females (Colmenares unpublished data). Also, males vary in the reproductive value of the females they mate with. Thus, units of young and old leader males tend to contain not only fewer females than harems of males at the peak of their fighting power, but they also contain a larger proportion of young prepubertal and old postprime females whose fecundity is low or null (Kummer 1968a, Abegglen 1984, Dunbar 1984, Colmenares unpublished data).

Hamadryas and gelada baboon females also exhibit large variation in their reproductive success. In a sample of 11 founding females from Madrid band 1 for whom complete life span data were available, strong differences in both longevity and reproductive output were found (table 11.3). In the gelada, it has been reported that high-ranking females reproduce at a higher rate than low-ranked females (Dunbar & Dunbar 1977; Dunbar 1980a, 1984) and although the total number of offspring tends to increase with increasing unit size (Dunbar 1980a), the breeding rate per female shows a nonlinear relation to unit size (Dunbar & Sharman 1983). A preliminary analysis of a data subset from our long-term records revealed that a female's per capita breeding rate per year is an inverted J-shaped function of unit size, peaking at 7 members and then declining (Colmenares unpublished data). Dunbar (1980a, 1984) has shown that gelada females gain a competitive advantage and reduce harassment from other unit females if they form long-term coalitions with frequent female grooming partners, and that low-ranking females are more likely candidates than high-ranking females are to desert their current unit leader (see also Kummer 1975). We also have evidence that Madrid hamadryas females prefer to attach themselves to powerful pro-

Table 11.3. Reproductive Parameters of a Selected
Sample of Males and Females from the Madrid
Colony of Hamadryas Baboons

Variable	Total	Adjusted
Reproductive success[a]		
Males[b]		
Founding	14.25 (6–29)	1.66 (0.75–2.60)
First Cohorts	9.25 (3–15)	2.14 (0.94–3.52)
Females[c]		
Founding	6.18 (1–12)	0.42 (0.16–0.76)
Longevity[d] (mo)		
Females	178.90 (47–304)	
Polygyny[e]		
Founding	12 (10–13)	
First Cohorts	15.5 (11–20)	
Polyandry[d]	6.90 (2–11)	

[a]Mean number and range in parenthesis of estimated (for males) and recorded (for females) offspring produced. Adjusted values represent mean rates per year.

[b]Only males from band 1. Founding males ($n = 4$): data from June 1972 to December 1983. First cohorts includes the males born in 1974 ($n = 1$) and 1975 ($n = 3$): data from January 1980 to October 1984.

[c]Only females from band 1 who were brought together in 1972 as a subgroup of immatures, who survived at least the first year of colony formation and from whom reproductive data of their complete reproductive life spans, from menarche to death (1972–1998), were available ($n = 11$).

[d]Only data from the sample of 11 founding females (band 1). Longevity refers to length of time from menarche to death. Mean number and ranges in parentheses.

[e]Only data from two samples of band 1 males: founding ($n = 4$, 1972–1983) and 1974 and 1975 cohorts ($n = 4$, 1980–1984). It includes females they mated with, regardless of whether or not they had offspring together. Mean number and range in parentheses.

tector males and that, within the unit, low-ranking females are less loyal than the high-ranked females to their current unit leader (see also Sigg 1980, Abegglen 1984). Sequential polyandry is the result of the periodic natural replacement of the harem-holding males. In the Madrid hamadryas colony, females may associate sequentially over the course of their whole reproductive lifespan with as many as 11 different males (table 11.3).

Hamadryas males do use aggressive and nonaggressive patterns of herding to control their females' spatial, affiliative, and sexual interactions with other unit members and with extragroup individuals (Kummer 1968a, Abegglen 1984, Colmenares & Anaya-Huertas 2001). However, this strategy of mate coercion is rarely used by gelada males (Dunbar & Dunbar 1975, Mori 1979b).

Dispersal Patterns and Kinship Structure

The patterns of dispersal from the natal breeding units and of transfer across the units available in the population and the mating system exhibited by the individuals will determine the kinship structure of the population, that is, the degree of genetic relatedness among individuals within and between units (Melnick & Pearl 1987). The first difficulty that we come across when the multilevel, fusion-fission social systems of hamadryas and gelada baboons are examined has to do with establishing the breeding unit from which dispersal and transfers are to be considered. Moreover, the polygynous mating strategies of hamadryas and gelada males lead to groups in which age cohorts tend to be paternal sibships. The implications of this mating scenario for the kinship structure of groups has only recently begun to be fully appreciated (Altmann 1979, Melnick 1987, Melnick & Pearl 1987, Altmann et al. 1996) and certainly have not yet been thoroughly examined with regard to the social systems of hamadryas and gelada baboons (see, however, Dunbar 1985, Colmenares 1992, Hapke et al. 2001, Swedell & Woolley-Barker 2001).

Although primate socioecologists have come to agree that bands of hamadryas and gelada baboons are ecologically equivalent to the typical *Papio* and *Macaca* troops (Kummer 1968a, 1990; Abegglen 1984; Dunbar 1984, 1986; Stammbach 1987; Barton 2000), the implications of this for an analysis of dispersal and kinship have not been fully considered. On the other hand, the view that hamadryas clans are based on male kinship, whereas gelada teams are based on female kinship (table 11.2), and that this is so because hamadryas baboons are a female transfer species whereas geladas conform to the typical mammalian pattern of male dispersal (Dunbar & Dunbar 1975; Abegglen 1984; Dunbar 1984, 1986; Stammbach 1987; Kummer 1984, 1990) has gained widespread acceptance and has been uncritically incorporated into most classification schemes of primate social organization and philopatry (e.g., Wrangham 1980, Moore 1984, Pusey & Packer 1987, Clutton-Brock 1989, Foley & Lee 1989, van Schaik 1989, Rodseth et al. 1991, Sterck et al. 1997, Barton 2000), despite the limitations of the empirical information available on which such a view is based.

Kummer and his group hold the view that hamadryas females are the transfer sex and that hamadryas harems typically consist of unrelated females (e.g., Sigg et al. 1982, Abegglen 1984, Kummer 1990). However, a careful reading of their publications suggests a more complicated picture (see also Dunbar 1986: 349; 1988a: 301). Sigg et al. (1982) explicitly recognized that individuals of either sex transferred across clans within the same band or even across bands, although they concluded that "the exchange of individuals between social units predominantly occurred *within the band*" (p. 473, italics added). These authors also emphasized that all the study males of known origin returned to their natal clans where they established their harem units, and that the juvenile females tended to remain in their birth clan. Kummer and his colleagues also noted a tendency for females to maintain social bonds with their mothers and other closely related kin and to associate spatially with them whenever they had the chance to do so (Kummer 1968a, Sigg et al. 1982, Abegglen 1984, see also Swedell 2002), although this impression was largely based on inferences about the likely matrilineal kin relations between the study females. On the other hand, Abegglen's (1984) data on the males' dispersal and transfer patterns suggest that wild hamadryas clans are probably based on male patrilineal kinship. According to Abegglen, the ontogeny of hamadryas clans would be based on the long-term association that followers establish with leaders, which eventually leads to the formation of two or more harem units spatially associated with one another by the bonds among the males. The studies conducted by Sugawara (1982) and by Phillips-Conroy et al. (1991, 1992) in the hybrid zone located along the Awash valley, in Ethiopia, where hamadryas baboons crossbreed with olive baboons, show that hamadryas males can and do disperse from their natal bands and move into nonnatal bands (see also Swedell & Woolley-Barker 2001). Recent analyses of mitochondrial DNA samples from Eritrean hamadryas baboons have revealed that individuals of both sexes do transfer across bands (Hapke et al. 2001).

The analyses of data from our long-term study of the Madrid colony of hamadryas baboons indicate that both males and females tend to associate spatially with their kin. In band 1, strong spatial associations between males occurred more often than expected in dyads involving fathers/sons, full and maternal siblings, and cousins/nephews and were significantly less frequent than expected by chance for dyads of paternal siblings and of unrelated males (Colmenares 1992: figure 7). Table 11.4 shows that, during the period 1983–1985, the one-male units from Madrid band 1 contained mother-daughter pairs and paternal sisters more often than expected by chance and comprised maternal sisters and dyads of unrelated females less often than expected by chance. Thus, mother-infant bonds and associations of paternal sisters of the same age cohort tended to be preserved over a long period of time.

Colmenares (1992) reported the emergence of two different types of clans in Madrid band 1. Clans were more cohesive if they consisted of father-son associations and/or maternal (or full) brothers and if they contained extended matrilines (i.e., type A clans). In contrast, the cohesion of clans made of paternal brothers was looser and mainly due to the kinship-based bonds and familiarity among the clan females (i.e., type B clans). In this respect, type B clans would resemble gelada teams (Dunbar & Dunbar 1975, Mori 1979e, Dunbar 1984). Some harem units contained several followers, and their spatial association with the harem leader was rather variable and related to their kinship relationship. In general, maternal brothers had a greater tendency than any other dyad class to form cohesive and stable leader-follower associations that eventually would lead to clans.

The analysis of the parental units from which the first cohorts of males from Madrid hamadryas bands 1 and 2 recruited females when they established their own harem units indicates that they tended to take over females from nonnatal units ($z = 3.08$, $n = 15$, $P = 0.002$; Wilcoxon test). Although these preliminary results suggest that the study males recruited females from nonnatal units, one should be cautious in the evaluation of its implications. I have considered that a male's natal unit was that of his father, regardless of the unit where his mother stayed while he was immature or the unit where he socialized most anyway. Whatever the processes at work, our findings indicate that the selected sample of band 1 and band 2 males did not generally associate and mate with paternal sisters.

Table 11.4. Most Frequent Relation Between Frequency of Observed Versus Expected Occurrence of Each Dyad Class of Females in the Harem Units from Madrid Band 1, During Eleven Time Blocks, from January 1983 to March 1985

Class of Dyad	Observed > Expected	Observed < Expected	$P*$
Mother-daughter	11	0	.0009
Full sisters	3	6	Not significant
Paternal sisters	11	0	.0009
Maternal sisters	0	11	.0009
Distantly related	3	8	Not significant
Unrelated	0	11	.0009

*Sign test.

Gelada females also transfer across harem units within their natal bands, but they do not leave their natal bands (Dunbar & Dunbar 1975, Ohsawa 1979, Dunbar 1984). It is important to emphasize that although the gelada baboon is considered to be a male transfer species, Dunbar (1984: 174–175) reported that about 70% of gelada males established their harems within their natal bands (see also Dunbar 1980b). It has been suggested that at least some males from all-male units might be (matrilineally) related (Dunbar & Dunbar 1975, Dunbar 1984) and that in some cases harem holders and their followers are likely to be relatives (Dunbar personal communication). The cohesion of gelada teams and bands is believed to be based on female kin bonding and alliances that males do not seem to be capable of breaking to any significant extent. As in typical *Macaca* and savanna *Papio* troops (Melnick & Pearl 1987, Dunbar 1988a), unit and band fissioning in the gelada probably takes place along genealogical lines (Dunbar & Dunbar 1975; Dunbar 1984, 1986).

In sum, in both species a significant proportion of males and females breed in their natal bands. In the hamadryas, in addition, some females may breed outside their natal bands because they are forced to transfer by their current owners.

Social Relationships

The main characteristics of the social relationships that have been described for hamadryas and gelada baboons are summarized in table 11.5 and are examined next. It must be emphasized that in these species' multilevel societies, the majority of affinitive interactions among adults take place within the harem units, although occasionally some interactions also occur between individuals belonging to the same clan/team and band (hamadryas: Kummer 1968a, 1995; see, however, Sigg 1980: 272; Colmenares unpublished data; gelada: Dunbar & Dunbar 1975, Mori 1979a, Dunbar 1984).

Relationships Between Females

Hamadryas females have been reported to exhibit very weak affinitive relationships with one another (Kummer 1968a, 1990, 1995). However, a close examination of the surprisingly few sociograms of within-unit grooming interactions that have been published for this species indicates that this generalization may be unwarranted. The published sociograms of the hamadryas units that Kummer (1968a) studied in Ethiopia contain information only about distances between the adult members, and Kummer himself reported that in one of the units (i.e., Circum's), grooming was relatively frequent in one dyad that he suspected could comprise a mother and her daughter. Sigg's (1980) study of three wild and three captive two-female harem units showed that between-unit variation in the distribution of grooming and other social behaviors within the harems was fairly marked, which was especially remarkable since the units he studied were matched for demographic composition (see also Kummer 1968a: 74). Swedell (2002) has reported that wild hamadryas females may form affinitive bonds with one another. Unfortunately, kinship relationships between the adult females in all of these studies were unknown. In the Madrid colony of hamadryas baboons, grooming occurs significantly more often in mother-daughter dyads than in dyads of full or maternal sisters, of paternal sisters, and of unrelated females (table 11.5). In the gelada, female

Table 11.5. Summary of Social Relationships in Hamadryas and Gelada Baboons

Relationships	Hamadryas[a]	Gelada[b]	Kinship Effect
Between females			
Grooming	Most frequent in mother/daughter pairs[c]	Most frequent in mother/daughter pairs	In both species
Agonism	Not intense	Not intense	In both species
Cooperation	Weak	Strong	In geladas
Dominance system	Individualistic and egalitarian	Nepotistic and despotic	In geladas
Reconciliation	Low conciliatory tendency[d]: 13%	Unknown[e]	
Consolation	Lacking[d]	Unknown	
Between males			
Spatial association	Close among kin in clans and in units with followers[f]	Close among males from all-male units and in units with followers	In hamadryas[f] and probably in geladas[g]
Agonism	Mostly during takeover attempts	Mostly during takeover attempts	
Cooperation	Mostly passive among clan males and among leaders and followers	Mostly passive among leaders and followers. Active cooperation also reported among leaders and followers[h]	In hamadryas[f] and probably in geladas[g]
Dominance system	Age-graded; dependent rank between males from different clans and harems	Age graded	
Reconciliation	High conciliatory tendency: 36–60%[e]	Unknown[e]	Unknown
Consolation	Lacking[i]	Unknown	Unknown

Between the sexes		
Grooming	More intense in heterosexual dyads than in female dyads. In heterosexual dyads, the female is the most active groomer[j]	Less intense in heterosexual dyads than in female dyads. At least in some heterosexual dyads, the male may be more active than the female.
Coercion	Intense, especially by young and old leaders[k]	Not very intense. It can trigger counterchases by coalitions of females.
Cooperation	Males very active in the protection of their females and their offspring. Their policing interventions can be agonistic and peaceful.	Males active in the protection of their females, especially during interunit conflicts, and their offspring. Followers may join females' coalitions against the unit leader.
Reconciliation	High conciliatory tendency[d]: 50%	Unknown[e]
Consolation	The unit leaders look for their females for consolation. Consolatory tendency[d]: 21%	Unknown

[a]Unless otherwise indicated, the hamadryas data come from the Madrid colony (Colmenares unpublished data).

[b]Dunbar 1980b, 1983a, b, 1984; Dunbar & Dunbar 1975; Kawai 1979.

[c]In the analysis of the Madrid hamadryas data (years 1995, 1998, and 2001, $n = 15$ harem units), grooming was more intense in mother-daughter pairs than in any of the other classes of kin dyads (Kruskal-Wallis test, $\chi^2 = 8.58$, $df = 3$, $p = .035$): mother/daughter pairs ($n = 5$, mean = 7.88), full and maternal sisters ($n = 12$, mean = 2.47), paternal sisters ($n = 12$, mean = 1.26) and female nonkin ($n = 107$, mean = 1.35).

[d]Zaragoza & Colmenares in preparation.

[e]Swedell 1997.

[f]Colmenares 1992, Colmenares et al. in preparation.

[g]Some males from all-male units or from leader-follower associations are probably relatives (Dunbar & Dunbar 1975, Dunbar 1984, see text).

[h]Dunbar & Dunbar 1975, Dunbar 1984, Mori 1979b.

[j]Zaragoza & Colmenares in preparation, Silveira & Colmenares in preparation.

[j]Madrid hamadryas data (1995, 1998, 2001): heterosexual dyads > female dyads (7.80 vs. 1.69, $U = 1,173.5$, $n_1 = 64$, $n_2 = 136$, $P < .001$); male grooms the female < female grooms the male (1.79 vs. 6, $U = 826$, $n_1 = n_2 = 128$, $P < .001$).

[k]Colmenares & Anaya-Huertas 2001.

grooming is frequent and is presumed to occur mainly between close relatives, that is, mothers, daughters, and (maternal) sisters (Dunbar 1980a, 1983a, 1984, 1986; see also Dunbar 1982). Again, the number of published sociograms of grooming in gelada harem units is actually very small, although the available evidence suggests that variation in the patterns of grooming among gelada females is fairly large as well (Mori 1979b).

Although agonistic interactions among females within the units of hamadryas and geladas are not frequent (Kummer 1968a, Dunbar & Dunbar 1975), the analysis of its asymmetric distribution within dyads and of the contexts in which they take place suggests that in both species female dominance relations are well developed. Wild and captive hamadryas and gelada females do compete for access to the unit leader and other limited resources, and this competition-related antagonism is especially intense when females are rearranging their social relationships following a takeover by a new male or when new females join the units (hamadryas: Kummer 1968a, Sigg 1980, Abegglen 1984, Colmenares & Gomendio 1988, Colmenares unpublished data; gelada: Dunbar & Dunbar 1975; Kummer 1975; Mori 1979b; Dunbar 1980a, 1983a, 1984, 1986; Mori & Dunbar 1985). Sigg (1980) found that the dominant female in the two-female units that he studied tended to be the younger one and speculated that in general fully grown hamadryas females should be likely to outrank older females. This age-graded dominance system seems to be also applicable to gelada females. In this species, however, as in the multimale/multifemale social systems of *Macaca* and savanna *Papio* (review: Chapais 1995), the existence of a matriline rank among adult females has been proposed, although here daughters would end up outranking their mothers when they passed their prime (Dunbar 1980a, 1984; see also Datta 1992).

We know very little about the dominance system among female hamadryas baboons. Although the size of the harem unit does influence the intensity of contest and scramble competition among the females, it appears that at least in small harems (two to three females) the female dominance system tends to be individualistic and egalitarian (Colmenares unpublished data, Swedell 2002). Two factors might be responsible for this: the absence of female relatives within the unit and, when they are available, the unit leader's effective interference in females' conflicts, which often end up undecided (see below: Relationships Between the Sexes). We have found that during social conflicts among Madrid hamadryas females, they supported each other rather infrequently (19%: Colmenares & Zaragoza in preparation), with kinship accounting for the majority of those few instances of female cooperation. By contrast, Dunbar (1980a, 1983a, 1984) has reported that, in the gelada, females often give coalitionary support to their grooming partners, who are probably their relatives, especially when they are involved in aggressive encounters with extraunit females (Dunbar 1980a, 1989).

One measure of the intensity of bonding and of the value that individuals attach to other group members is the tendency to reconcile their conflicts and to provide consolation to their partners when they are distressed, for example, after a conflict (de Waal & Aureli 1997, Cords & Aureli 2000, van Hooff 2001). In a study of postconflict interactions in Madrid hamadryas band 2, Zaragoza & Colmenares (in preparation) found that the females reconciled with former female antagonists only if they belonged to the same unit and that even so the conciliatory tendency was rather low (13%). Moreover, neither as aggressors nor as victims did antagonist females seek or receive consolation from female bystanders, either in female-female conflicts or in female-male conflicts.

Relationships Between Males

As adults, especially if they have already established bonds with one or more adult females, both hamadryas and gelada males rarely groom with one another (hamadryas: Kummer 1968a, Abegglen 1984, Colmenares & Zaragoza in preparation; gelada: Dunbar & Dunbar 1975, Kummer 1975, Mori 1979d, Dunbar 1984). In both species, then, the main trace of affiliation among adult males within clans (hamadryas) and within harem units with followers comes in the way of tolerance on the part of the leaders of the proximity of other clan males and of the followers to their unit females and of grooming interactions between some of their females and the followers.

It has been argued that hamadryas clans and leader-follower associations may have evolved by kin selection (Abegglen 1984; Kummer 1990, 1995), assuming that males of such multimale units are relatives, that they actually cooperate with one another, and that by doing so they ultimately increase their inclusive fitness. Data from the Madrid colony of hamadryas baboons indicate that the first assumption may well be correct (Colmenares 1992, Colmenares et al. in preparation). As regards the second assumption, the data from the wild are equivocal at best (Kummer 1995). Our observations from the Madrid colony suggest, however, that by being member of a clan, males may gain benefits that males from single-male units do not obtain. Thus, clan males were less likely to be successfully challenged by outside males and, when they were eventually engaged in a fight, their females were less likely to run away (both because they were effectively guarded by other males from the clan and because they were probably less willing to desert). Finally, with regard to the third assumption, Colmenares et al. (in preparation) have found that clan males from Madrid band 2 supplanted single-harem males at food resources. On the other hand, followers have been observed to inherit their leaders' former units (Colmenares 1992, unpublished data). According to our observations, such nepotistic succession is never passively handed on from leaders to followers but is settled by fights between them over the ownership of the former's females (see also Kummer et al. 1978, Abegglen 1984, Kummer 1995).

In the gelada, Dunbar (1984) has suggested that by allowing an extra male to join the harem as a follower, the current leader may reduce the chances of his unit being the target of a takeover attempt by a bachelor male and may thus prolong his tenure as a breeding male. Nevertheless, Dunbar (1984: 177) made clear that the benefits that unit leaders derive from taking on a follower have nothing to do with the latter's playing any active role in supporting the unit leader during takeover attempts by rival males. It seems to work, however, because harems with followers reduce the effective size of the units (i.e., the number of unit females actually bonded with the harem male), thus increasing the females' loyalty to the leader and reducing the probability of being taken over by other males (Dunbar 1984, 1986).

In hamadryas, greetings (termed "notifying" by Kummer 1968a) are frequently exchanged by adult males in the context of rivalry over females when they are uncertain about each other's tendencies to attack or withdraw (Kummer et al. 1974, 1978; Abegglen 1984; Colmenares 1991; Colmenares et al. 2000). In two different large, captive multiharem colonies of hamadryas baboons (Lecoq Zoo, Montevideo, and Madrid Zoo), we have found that greeting is used frequently by males to reconcile with each other after a conflict (conciliatory tendencies were 60% and 36%, respectively: Silveira & Colmenares in preparation, Zaragoza & Colmenares in preparation). This relatively high proportion of reconciled conflicts among males might reflect the ambivalent nature of the relationships existing between the hamadryas males, many of whom may be relatives. On the one hand, their relationships are highly competitive, as they seem to be permanently willing to try to enlarge their harem units. On the other, however, they strongly depend on each other's tolerance, "respect" (respect of ownership, cf. Kummer et al. 1974) and cooperation to maximize their reproductive success and ultimately their inclusive fitness (see also Kummer 1995). In the gelada, a similar albeit less ritualized form of greeting, used in the same context of rivalry for the ownership of females, has been described by Kummer (1975) in an enclosure study, and by Dunbar and Dunbar (1975) and Mori (1979d) in their field studies.

Relationships Between the Sexes

It has been stated that the sociogram reflecting the distribution of grooming interactions within the harem units of hamadryas baboons produces a star-shaped pattern (Kummer 1968a), where the arrows go mainly from the harem females toward the unit leader. Although this pattern is clear in some of the few published sociograms available in the literature, it is also fair to acknowledge that there is large variation in the sociograms of grooming interactions in the units of this species (see for example Sigg 1980, Swedell in press). In the Madrid studies (Colmenares et al. 1994, Colmenares et al. 2002, Colmenares & Zaragoza in preparation), it has been found that grooming tends to be significantly more frequent in male-female pairs than in dyads of harem females and tends to be one-sided, from the female to the male (table 11.5). However, not all the females would choose the harem male as their main grooming partner, and this might not only be a consequence of the harem females competing for access to the unit leader (Colmenares et al. 1994) but might also reflect the operation of biomarket forces in harems of varying in size, which determine the value of the unit leader as a protector and therefore their attractiveness as a partner to invest in (Colmenares et al. 2002).

In the gelada, by contrast, the females' main grooming partners are other harem females, unless they do not have a close female relative available in the unit. Thus, as already noted, females tend to form two-female grooming cliques, which are believed to consist of mother-daughter pairs (Dunbar 1980a, 1984, 1986). Like the other harem members, the unit leader forms a strong grooming relationship with only one harem female (Dunbar & Dunbar 1975, Kummer 1975, Dunbar 1984), who does not have a female relative in the unit to groom (Dunbar 1980a, 1983b, 1986). Dunbar (1982) also claimed that the effect of the females' dominance rank on the relationships among the females and relationships between the females and the unit leader should be expected to be more marked when the harem females were unrelated, as they were in Kummer's enclosure study (Kummer 1975).

In the hamadryas, the unit leaders play a very active role in the control of and interference in agonistic interactions between the females of the unit and in the protection and support of their females when they are engaged in agonistic encounters with females from other units and when they are harassed by other males (Kummer 1967, 1968a, 1995). In a detailed study of intervention strategies in Madrid hamadryas band 2 (Colmenares & Zaragoza in preparation), we found that harem males protected their females in 46% of their conflicts. This figure went up to 66% if the females were attacked by other males, and it was set at 37% if the two antagonists were females. Not only did hamadryas unit leaders in our study intervene in the conflicts where their females were involved, providing protection and peacefully neutralizing its potentially centripetal effects, but they also frequently reconciled with their females (conciliatory tendency of 50%) and sought them for consolation when they themselves were involved in conflicts with other males (triadic affiliation tendency of 21%).

Mori (1979b) claimed that gelada unit leaders protected their females when they were engaged in interunit conflicts, but that their role in policing intraunit agonistic encounters was unimportant. This impression is also supported by the observations from Dunbar's studies (Dunbar & Dunbar 1975; Dunbar 1983a, 1984). On the other hand, Dunbar (1983c) reported that both the harem male and the alpha female did play an active role in the defense and cohesion of the unit.

As mentioned, hamadryas unit leaders have been reported to continuously monitor their females' spatial and social interactions within and outside the unit's social space and to actively control them by means of herding behaviors (Kummer 1968a, b, 1995). In the gelada, in contrast, unit leaders use this form of aggressive coercion on the females more rarely and less effectively. In fact, gelada females may form coalitions, sometimes joined by the unit's follower, against the unit leader if he herds them too harshly. It has been argued that the strong cohesion of hamadryas one-male units is aggressively enforced by the unit leader and that this male behavior is species typical and genetically based. Thus, the looser cohesion of one-male units led by anubis-hamadryas hybrid males has been taken to reflect the males' inability to use this technique efficiently enough due to their hybrid genotype (Nagel 1971, Müller 1980, Sugawara 1982). It must be emphasized that there is large variation in the frequency with which hamadryas males use this form of aggressive coercion (Kummer 1968a, 1995; Abegglen 1984; Kummer et al. 1985) and that both demographic factors and a male's age (and status class) account for a large proportion of such variation (Colmenares & Anaya-Huertas 2001).

Hamadryas and gelada unit leaders that have been defeated and have lost their females tend to remain spatially and socially attached to their former units as old followers where they play a very active role in the defense and protection of their offspring and, in some cases, cooperate with the current leader in the defense of their former females (Dunbar & Dunbar 1975; Mori 1979b, d; Abegglen 1984; Dunbar 1984, 1986; Colmenares 1992, unpublished data; Kummer 1995). In both hamadryas and geladas, infanticide has been reported (or suspected) to occur in captivity and in the wild, when new leaders take over females who carry unweaned infants, especially if the ensuing social situation is unstable (hamadryas: Sigg 1980, Rijksen 1981, Gomendio & Colmenares 1989, Colmenares unpublished data, Swedell 2000; gelada: Moos et al. 1985, Mori et al. 1997).

Discussion and Concluding Remarks

The first conclusion that should be drawn from this review concerns the important limitations imposed by the amount and quality of the data available on hamadryas and gelada baboons from which general principles are to be formulated (table 11.1). Although simulation analyses are useful and can help us in building up descriptive and explanatory models, they cannot replace the hard data that are needed to feed and test such models. In this respect, we need to increase both the number of populations studied and the length of the observations so that the social and reproductive lifespans of individually identifiable individuals can be properly traced. We also need to study groups in which the matrilineal, and ideally the patrilineal, genealogical relations between the group members are known rather than inferred and in which experimental manipulations of relevant ecological, demographic, and social variables can be conducted. Of course, studies in which paternity and the genetic structure of the population can be determined should provide definitive tests of key hypotheses relating kinship to social behavior and group structure (e.g., Kummer 1992, Hapke et al. 2001).

Given these limitations, one should be very cautious when interpreting the similarities and differences that have been reviewed here between the social systems of the hamadryas and the gelada baboons. At this stage, we simply do not know the range of "natural" variation potentially shown by the social systems of these two species and whether what is known about them can be considered to represent the species' typical or modal pattern. We should keep in mind that a given group's social organization is likely to be just the expression of general organizing principles within a particular ecological and sociodemographic context (Altmann & Altmann 1979, Dunbar 1979b, Rowell 1979). The investigators that have studied the social systems of the hamadryas and the gelada baboons and that have speculated about their probable evolution have tended to take the surface structure (Hinde 1976; see also Dunbar 1988a) of their study groups as representing the species' archetype of social structure (Kummer 1968b, 1990; Dunbar 1983d, 1986, 1988a). On the other hand, many of these very same authors have repeatedly underscored the existence of large variation even within their study populations and have highlighted its implications for a proper understanding of the proximate and ultimate causes of primate social structure (Abegglen 1984, Dunbar 1988a, Kummer 1995). Nowadays, we are also well aware of the fact that within the savanna species of the genus *Papio* there seems to be a larger potential than was previously thought for individuals to develop patterns of cross-sex bonding under certain ecological and demographic conditions, which resemble those observed in hamadryas baboons (reviews: Barton et al. 1996, Barton 2000; see also Henzi et al. 2000).

The societies of hamadryas and geladas share two distinctive characteristics: the mating system, which is polygynous and sequentially polygynandrous, and the overall social structure, which is multilevel and fusion-fission (table 11.2). However, it has been argued that their social structure is the outcome of different underlying social processes, which are linked to differences in dispersal patterns and to the resulting kinship structure of their nested groupings (e.g., Dunbar 1983d, 1988a; Kummer 1990). The data reviewed here suggest that, contrary to the conventional view but in accordance with the empirical information available, hamadryas females, like gelada females, do transfer across harem units and clans, but they do so mostly within their natal bands. Although genetic evidence now supports the view that at least some hamadryas females do breed outside their natal bands (Hapke et al. 2001), it must be emphasized, however, that female dispersal in hamadryas is not comparable to female dispersal in other primate species in which individuals of either sex disperse (e.g., Thomas's langurs [*Presbytis thomasi*]: Sterck 1998; mountain gorillas [*Gorilla gorilla beringei*]: Watts 1996, Sicotte 2001). In effect, hamadryas females do not leave their natal bands; they are forced to transfer by their leader males' herding behavior.

As regards male dispersal in hamadryas baboons, I think that the long-term data from Phillips-Conroy et al.'s (1991, 1992) and from Kummer and his group's (Sigg et al. 1982, Abegglen 1984) studies are compatible with the view that, at least under certain demographic conditions, some hamadryas males leave their natal bands and breed somewhere else. Nevertheless, compared to their savanna counterparts, hamadryas males appear to be more inclined to remain in their natal band or even clan (Abegglen 1984). However, whether this is a species-typical trait selected for in the hamadryas baboon or an epiphenomenon of the particular ecological and demographic conditions of the bands where this has been found remains to be elucidated. On the other hand, although geladas are categorized as a male transfer species, the evidence available indicates that a large proportion of males probably return and breed in their natal bands (Dunbar 1984).

The cohesion of the harems and higher order unit groupings (i.e., clans/teams and bands) and the nature of the social relationships within and between the sexes have also been reported to vary across the two species (tables 11.2 and 11.5). The cohesion of hamadryas harems is mainly the outcome of the relationship between the harem male and his females, especially in small units. In larger units, however, females may form strong bonds and alliances with kin, and the intense interfemale competition that is likely to arise under these demographic circumstances may contribute to fragment the unit. In this respect, large hamadryas units resemble the typically larger gelada units in that strong cooperative relationships between female relatives, especially between mother-daughter dyads, may play an important role in the unit's general cohesion and social dynamics. Differences in the size of the harems may also explain the reported species differences in the female dominance relationships, that is, individualistic and egalitarian in hamadryas versus nepotistic and despotic in geladas.

Both hamadryas and gelada males can be tolerant and even actively cooperate with one another. In the former species, such cooperation is based on male kin bonding, whereas in the latter species the role of male kinship remains to be investigated, although it should not be discarded. Hamadryas males may derive important fitness-related benefits from being members of multimale units, that is, clans and harems with followers. They may prolong their tenure as breeding males (also in gelada units with followers: Dunbar 1984), have improved access to limited resources, and increase their lifetime reproductive success. On the other hand, hamadryas and gelada young followers may also gain from delaying the beginning of their reproductive careers, directly if they end up inheriting the unit's females and enjoying a longer tenure, and indirectly, at least in hamadryas, by contributing to a related unit leader's reproductive success. The role of young and old followers can be seen as analogous to that of helper, more so if they are kin, as tends to be the case in hamadryas (see also Kummer 1990) and to some extent in geladas as well (Dunbar 1984). Although male kinship is not a necessary condition for primate males to exhibit cooperative behavior (e.g., van Hooff 2000), the high tolerance and intense conciliatory tendencies that characterize their relationships suggest that kinship might have a facilitating influence, at least among hamadryas males.

Cross-sex bonding seems to be more intensely exploited by hamadryas than by geladas. Again, this might reflect differences in partner choice options available in each species arising from differences in the ecological and demographic conditions where they have to play out their strategies and in the resulting life history patterns (see also Datta 1992 for an early application of this principle to the analysis of species differences in female dominance relationships). Thus, under the harsh ecological conditions where some populations of hamadryas baboons live, females' breeding rates are lower and interbirth intervals are longer than those shown by gelada females (Dunbar 1988a).

As a consequence, the size of both one-male units and matrilines tends to be smaller in hamadryas than in geladas, and this might have two effects, which would not be necessarily mutually exclusive. First, hamadryas females might not bond to female kin simply because they would not have any around. Second, even if they had the chance to do so because female kin were available, hamadryas females would nevertheless actively choose the unit leader male because he would represent a more profitable partner to invest in. An additional or alternative factor that might favor the stronger cross-sex bondedness of hamadryas relative to gelada has to do with the possibility that it reflects underlying genetic differences between both species in the males' disposition (cf. Mason 1978) to actively interfere with bonding tendencies between their harem females, whether or not they were kin, and/or to actively invest in bonding with their females. Among the species (or subspecies) of the genus *Papio* (Jolly 1993), hamadryas males stand out for showing the most intense forms of herding behavior (Barton et al. 1996, Barton 2000). Differences in the use of this behavior, which is believed to contribute to the differences observed between *P. hamadryas* and the other savanna *Papio* species in the cohesion of their social units, are considered to reflect underlying genetic differences, as hybrid males between hamadryas and olive baboons do differ in the effective use of herding and in the cohesion of the units they own (Kummer 1968b, Nagel 1971, Müller 1980, Sugawara 1982).

The stronger male-female bonding of hamadryas baboons compared to that of geladas is associated not only with the more frequent and intense use of herding by hamadryas males but also with their more active involvement in the policing of their females' conflicts. The case for a possible genetic contribution to the behavioral differences observed between hamadryas and gelada males is also supported by recent preliminary evidence indicating that olive and hamadryas baboons might differ from one another in neurobiological factors related to between-species differences in male dispersal patterns (Kaplan et al. 1999).

In his highly influential conceptual framework, Hinde (1976, 1983) singled out kinship as one of the most important principles that could account for the variable nature of social relationships reported to occur within primate groups. Although kinship has turned out to have predicting power in many primate species, the link between dispersal, kinship, and cooperation is not as clear-cut as was initially thought (Gouzoules & Gouzoules 1987, Moore 1992, Di Fiore & Rendall 1994, Strier 1994, Chapais 1995). In the multilevel societies of hamadryas and gelada baboons examined here, kinship also emerges as an important contributing variable explaining the shape of their social systems (see tables 11.2 and 11.5). The availability of male kin makes a difference for hamadryas male individuals in terms of their options to form long-term associations with other males (i.e., leader-follower and clans) and thereby increase their ability to gain access to food, their reproductive success, and their inclusive fitness (Colmenares et al. in preparation). The availability of female kin also makes a big difference for gelada female individuals in terms of their options to attain and maintain high dominance rank and thereby increase their lifetime reproductive success (Dunbar 1980a, 1984).

The hamadryas baboon's social system resembles that of the mountain gorilla and of other primate and nonprimate species with one-male/multifemale social structure (see Watts 2001). The strength and nature of the male-female bonds and the role of kinship in the structuring of the relationships within the sexes in hamadryas and gorillas are cases in point (for recent reviews see Watts 2000, 2001; Watts et al. 2000; Robbins 2001; Sicotte 2001). Infanticide avoidance has been proposed as the main pressure selecting for strong cross-sex bonding in gorillas in particular (see also Harcourt & Greenberg 2001) and primates in general (e.g., van Hooff & van Schaik 1992, van Schaik 1996, Palombit 2000). In one-male/multifemale social systems paternity certainty is increased and is associated with increased male investment in their offspring (Anderson 1992, Palombit 2000, Paul et al. 2000). I speculate that the male kinship system underlying leader-follower associations and clan units in hamadryas baboons may provide a further buffer against the occurrence of infanticide, as it does in the mountain gorilla's multimale units (Watts 2000, Robbins 2001, Sicotte 2001). This might also apply to the gelada.

Social systems are the outcome of the network and nature of the partnerships of their constituent individuals. And these are a reflection of three basic processes: the individuals' partner choices, the constraints on the expression of their preferences, and the compromises that individuals are finally bound to reach. According to the biological markets theory (Noë & Hammerstein 1995, Noë 2001), partner choice decisions are based on assessments of the value of the services or resources that alternative partners can provide and trade for, which is related to the levels of supply and demand and on the ratio of trading classes of partners, which determine the intensity of outbidding and contest competition within trading classes and the level of conflict between trading classes. High-ranking female nonkin in some macaque species (review: Chapais 1995), female kin in geladas (Dunbar 1980a, 1984), and harem leader males in hamadryas and gorillas (Watts 2000, 2001) seem to be in each case the most valuable partners to form strong bonds with. The common underlying principle in all cases is that individuals are designed to cultivate relationships which are profitable, that market forces determine the partnerships that will yield the highest payoffs in each particular case, and that time, social, and relationship constraints will ultimately influence the partnerships that are actually found. Sometimes bonding with nonkin can be more economic than bonding with kin. I believe that an approach to the analysis of social systems in general and of cross-sex bonding in particular based on the study of the impact of market forces and the effects of constraints on the expression of the partners' preferences (Colmenares et al. 2002) may help elucidate the organizational principles underlying the social decisions of individuals and the effect of a partner's qualities, including relatedness.

Acknowledgments I thank all the students, too numerous to be named individually, who have contributed long-term records over the past 30 years, and the Directorate of the Madrid Zoo for their facilities to carry out the long-term baboon research project. I am also very grateful to Hans Kummer for his support over the years, Robin Dunbar for kindly clarifying some issues on gelada patterns of dispersal and for comments, to Larissa Swedell for kindly allowing me access to unpublished material, and to the editors for inviting my participation and for their comments. This research has been partly supported by project grants PR180/91–3379 from the Universidad Complutense de Madrid, and PB92–0194, PB95–0377 and PB98–0773 from the MECyD and the MCyT to the author.

References

Abegglen, J. J. 1984. *On Socialization in Hamadryas Baboons.* Cranbury, NJ: Associated University Presses.

Altmann, J. 1979. Age cohorts as paternal sibships. *Behav. Ecol. Sociobiol.,* 6, 161–164.

Altmann, J., Alberts, S. C., Haines, S. A., Dubach, J., Muruthi, P., Coote, T., Geffen, E., Cheesman, D. J., Mututua, R. S., Saiyalel, S. N., Wayne, R. K., Lacy, R. C., & Bruford, M. W. 1996. Behavior predicts genetic structure in a wild primate group. *Proc. Natl. Acad. Sci. USA,* 93, 5797–5801.

Altmann, S. A. & Altmann, J. 1979. Demographic constraints on behavior and social organization. In: *Primate Ecology and Human Origins: Ecological Influences on Social Organization* (Ed. by I. S. Bernstein & E. O. Smith), pp. 47–63. New York: Garland STPM Press.

Anderson, C. M. 1992. Male investment under changing conditions among chacma baboons at Suikerbosrand. *Am. J. Phys. Anthropol.,* 87, 479–496.

Barton, R. A. 2000. Socioecology of baboons: the interaction of male and female strategies. In: *Primate Males: Causes and Consequences of Variation in Group Composition* (Ed. by P. M. Kappeler), pp. 97–107. Cambridge: Cambridge University Press.

Barton, R., Byrne, R. W., & Whiten, A. 1996. Ecology, feeding competition and social structure in baboons. *Behav. Ecol. Sociobiol.,* 38, 321–329.

Chapais, B. 1995. Alliances as a means of competition in primates: evolutionary, developmental, and cognitive aspects. *Ybk. Phys. Anthropol.,* 38, 115–136.

Clutton-Brock, T. H. 1989. Female transfer, male tenure and inbreeding avoidance in social mammals. *Nature,* 337, 70–72.

Clutton-Brock, T. H. & Parker, G. A. 1995. Sexual coercion in animal societies. *Anim. Behav.,* 49, 1345–1365.

Colmenares, F. 1991. Greeting behaviour between male baboons: oestrous females, rivalry and negotiation. *Anim. Behav.,* 41, 49–60.

Colmenares, F. 1992. Clans and harems in a colony of hamadryas and hybrid baboons: male kinship, familiarity and the formation of brother-teams. *Behaviour,* 121, 61–94.

Colmenares, F. & Anaya-Huertas, C. 2001. Male coercion in hamadryas baboons (*Papio hamadryas*): male competition and female choice. *Primate Rep.,* 60–1, 18–19.

Colmenares, F. & Gomendio, M. 1988. Changes in female reproductive condition following male takeovers in a colony of hamadryas and hybrid baboons. *Folia Primatol.,* 50: 157–174.

Colmenares, F., Hofer, H., & East, M. L. 2000. Greeting ceremonies in baboons and hyenas. In: *Natural Conflict Resolution* (Ed. by F. Aureli & F. B. M. de Waal), pp. 94–96. Berkeley, CA: University of California Press.

Colmenares, F., Lozano, M. G., & Torres, P. 1994. Harem social structure in a multiharem colony of baboons (*Papio* spp.): a test of the hypothesis of the "star shaped" sociogram. In: *Current Primatology, Vol. II: Social Development, Learning and Behaviour* (Ed. by J. J. Roeder, B. Thierry, J. R. Anderson, & N. Herrenschmidt), pp. 93–101. Strasbourg: Université Louis Pasteur.

Colmenares, F., Zaragoza, F., & Hernández Lloreda, M. V. 2002. Grooming and coercion in one-male units of hamadryas baboons: market forces or relationship constraints? *Behaviour,* 139, 1525–1553.

Cords, M. 1997. Friendships, alliances, reciprocity and repair. In: *Machiavellian Intelligence II: Extensions and Evaluations* (Ed. by A. Whiten & R. W. Byrne), pp. 24–49. Cambridge: Cambridge University Press.

Cords, M. & Aureli, F. 2000. Reconciliation and relationship qualities. In: *Natural Conflict Resolution* (Ed. by F. Aureli & F. B. M. de Waal), pp. 177–198. Berkeley, CA: University of California Press.

Crook, J. H. 1970. The socio-ecology of primates. In: *Social Behaviour in Birds and Mammals* (Ed. by J. H. Crook), pp. 103–166. London: Academic Press.

Datta, S. 1992. Effects of availability of allies on female dominance structure. In: *Coalitions and Alliances in Humans and Other Animals* (Ed. by A. H. Harcourt & F. B. M. de Waal), pp. 61–82. Oxford: Oxford University Press.

de Waal, F. B. M. & Aureli, F. 1997. Conflict resolution and distress alleviation in monkeys and apes. *Ann. NY Acad. Sci.,* 807, 317–328.

Di Fiore, A. & Rendall, D. 1994. Evolution of social organization: a reappraisal for primates by using phylogenetic methods. *Proc. Natl. Acad. Sci. USA,* 91, 9941–9945.

Dunbar, R. I. M. 1979a. Structure of gelada baboon reproductive units. I. Stability of social relationships. *Behaviour,* 69, 72–87.

Dunbar, R. I. M. 1979b. Population demography, social organization, and mating strategies. In: *Primate Ecology and Human Origins* (Ed. by I. S. Bernstein & E. O. Smith), pp. 65–88. New York: Garland STPM Press.

Dunbar, R. I. M. 1980a. Determinants and evolutionary consequences of dominance among female gelada baboons. *Behav. Ecol. Sociobiol.,* 7, 253–265.

Dunbar, R. I. M. 1980b. Demographic and life history variables of a population of gelada baboons (*Theropithecus gelada*). *J. Anim. Ecol.,* 49, 485–506.

Dunbar, R. I. M. 1982. Structure of social relationships in a captive gelada group: a test of some hypotheses derived from studies of a wild population. *Primates,* 23, 89–94.

Dunbar, R. I. M. 1983a. Structure of gelada baboon reproductive units. II. Social relationships between reproductive females. *Anim. Behav.,* 31, 556–564.

Dunbar, R. I. M. 1983b. Structure of gelada baboon reproductive units. III. The male's relationship with his females. *Anim. Behav.,* 31, 565–575.

Dunbar, R. I. M. 1983c. Structure of gelada baboon reproductive units. IV. Integration at group level. *Z. Tierpsychol.,* 63, 265–282.

Dunbar, R. I. M. 1983d. Relationships and social structure in gelada and hamadryas baboons. In: *Primate Social Relationships: An Integrated Approach* (Ed. by R. A. Hinde), pp. 299–307. Oxford: Blackwell.

Dunbar, R. I. M. 1984. *Reproductive Decisions: An Economic Analysis of Gelada Baboon Social Strategies.* Princeton, NJ: Princeton University Press.

Dunbar, R. I. M. 1985. Population consequences of social structure. In: *Behavioral Ecology: Ecological Consequences of Adaptive Behaviour* (Ed. by R. M. Sibly & R. H. Smith), pp. 507–519. Oxford: Blackwell.

Dunbar, R. I. M. 1986. The social ecology of gelada baboons. In: *Ecological Aspects of Social Evolution* (Ed. by D. Rubenstein & R. Wrangham), pp. 332–351. Princeton, NJ: Princeton University Press.

Dunbar, R. I. M. 1988a. *Primate Social Systems.* London: Croom Helm.

Dunbar, R. I. M. 1988b. The evolutionary implications of social behavior. In: *The Role of Behavior in Evolution* (Ed. by H. C. Plotkin), pp. 165–188. Cambridge, MA:MIT Press.

Dunbar, R. I. M. 1989. Reproductive strategies of female gelada baboons. In: *Sociobiology of Sexual and Reproductive Strategies* (Ed. by A. Rasa & E. Voland), pp. 74–92. London: Chapman & Hall.

Dunbar, R. I. M. & Dunbar, E. P. 1975. *Social Dynamics of Gelada Baboons.* Basel: S. Karger.

Dunbar, R. I. M. & Dunbar, E. P. 1977. Dominance and reproductive success among gelada baboons. *Nature,* 266, 351–352.

Dunbar, R. I. M. & Sharman, M. 1983. Female competition for access to males affects birth rate in baboons. *Behav. Ecol. Sociobiol.,* 13, 157–159.

Foley, R. A. & Lee, P. C. 1989. Finite social space, evolutionary pathways, and reconstructing hominid social behaviour. *Science,* 243, 901–906.

Gomendio, M. & Colmenares, F. 1989. Infant killing and infant adoption following the introduction of new males to an all-female colony of baboons. *Ethology,* 80, 223–244.

Gouzoules, S. & Gouzoules, H. 1987. Kinship. In: *Primate Societies* (Ed. by B. B. Smuts, D. L. Cheney, R. M. Seyfarth, R. W. Wrangham, & T. T. Struhsaker), pp. 299–305. Chicago: University of Chicago Press.

Hamilton, W. D. 1964. The genetical evolution of social behavior, I and II. *J. Theor. Biol.,* 7, 1–52.

Hapke, A., Zinner, D., & Zischler, H. 2001. Mitochondrial DNA variation in Eritrean hamadryas baboons (*Papio hamadryas*): life history influences population genetic structure. *Behav. Ecol. Sociobiol.,* 50, 483–492.

Harcourt, A. H. & de Waal, F. B. M. (eds.) 1992. *Coalitions and Alliances in Humans and Other Animals.* Oxford: Oxford University Press.

Harcourt, A. H. & Greenberg, J. 2001. Do gorilla females join males to avoid infanticide? A quantitative model. *Anim. Behav.,* 62, 905–915.

Henzi, P., Lycett, J. E., Weingrill, A., & Piper, S. E. 2000. Social bonds and the coherence of mountain baboon troops. *Behaviour,* 137, 663–680.

Hinde, R. A. 1976. Interactions, relationships and social structure. *Man,* 11, 1–17.

Hinde, R. A. 1983. A conceptual framework. In: *Primate Social Relationships* (Ed. by R. A. Hinde), pp. 1–7. Oxford: Blackwell.

Isbell, L. A. 1991. Contest and scramble competition: patterns of female aggression and ranging behavior among primates. *Behav. Ecol.,* 2, 143–155.

Jolly, C. J. 1993. Species, subspecies, and baboon systematics. In: *Species, Species Concepts, and Primate Evolution* (Ed. by W. H. Kimbel & L. B. Martin), pp. 67–107. New York: Plenum.

Kaplan, J. R., Phillips-Conroy, J., Babette Fontenot, M., Jolly, C. J., Fairbanks, L.A., & Mann, J. 1999. Cerebrospinal fluid monoaminergic metabolites differ in wild anubis and hybrid (anubis-hamadryas) baboons: possible relationships to life history and behavior. *Neuropsychopharmacology,* 20, 517–524.

Kawai, M. (ed.) 1979. *Ecological and Sociological Studies of Gelada Baboons.* Basel: Karger.

Kawai, M., Dunbar, R. I. M., Ohsawa, H., & Mori, U. 1983. Social organization of gelada baboons: social units and definitions. *Primates,* 24, 13–24.

Kummer, H. 1967. Tripartite relations in hamadryas baboons. In: *Social Communication Among Primates* (Ed. by S. A. Altmann), pp. 63–71. Chicago: University of Chicago Press.

Kummer, H. 1968a. *Social Organization of Hamadryas Baboons.* Basel: Karger.

Kummer, H. 1968b. Two variations in the social organization of baboons. In: *Primates: Studies in Adaptation and Variability* (Ed. by P. C. Jay), pp. 293–312. New York: Holt, Rinehart & Winston.

Kummer, H. 1975. Rules of dyad and group formation among captive gelada baboons (*Theropithecus gelada*). In: *Proceedings from the Symposia of the Fifth Congress of the International Primatological Society* (Ed. by S. Kondo, M. Kawai, A. Ehara, & S. Kawamura), pp. 129–159. Tokyo: Japan Science Press.

Kummer, H. 1978. On the value of social relationships to nonhuman primates: a heuristic scheme. *Soc. Sci. Info.,* 17, 687–705.

Kummer, H. 1984. From laboratory to desert and back: a social system of hamadryas baboons. *Anim. Behav.,* 32, 965–971.

Kummer, H. 1990. The social system of hamadryas baboons and its presumable evolution. In: *Baboons: behaviour and ecology, use and care.* Selected proceedings of the 11th Congress of the International Primatological Society. (Ed. by T. de Mello, A. Whiten, & R. W. Byrne), pp. 43–60. Brazil: IPS.

Kummer, H. 1992. Some impacts of paternity studies on primate ethology. In: *Paternity in

Primates: Genetic Tests and Theories (Ed. by R. D. Martin, A. F. Dixson, & E. J. Wickings), pp. 1–2. Basel: Karger.

Kummer, H. 1995. *In Quest of the Sacred Baboon.* Princeton, NJ: Princeton University Press.

Kummer, H., Abegglen, J. J., Bachmann, C. H., Falett, J., & Sigg, H. 1978. Grooming relationship and object competition among hamadryas baboons. In: *Recent Advances in Primatology, Vol. 1: Behaviour* (Ed. by D. J. Chivers & J. Herbert), pp. 31–38. London: Academic Press.

Kummer, H., Banaja, A. A., Abo-Kathwa, A. N., & Grandour, A. M. 1985. Differences in social behavior between Ethiopian and Arabian hamadryas baboons. *Folia Primatol.,* 45, 1–8.

Kummer, H., Götz, W., & Angst, W. 1974. Triadic differentiation: an inhibitory process protecting pair bonds in baboons. *Behaviour,* 49, 62–87.

Kummer, H. & Kurt, F. 1965. A comparison of social behavior in captive and wild hamadryas baboons. In: *The Baboon in Medical Research* (Ed. by H. Vagtborg), pp. 1–15. Austin, Texas: University of Texas Press.

Mason, W. A. 1978. Ontogeny of primate social systems. In: *Recent Advances in Primatology, Vol. 1: Behaviour* (Ed. by D. J. Chivers & J. Herbert), pp. 5–14. London: Academic Press.

Melnick, D. J. 1987. The genetic consequences of primate social organization: a review of macaques, baboons and vervet monkeys. *Genetica,* 73, 117–135.

Melnick, D. J. & Pearl, M. C. 1987. Cercopithecines in multimale groups: genetic diversity and population structure. In: *Primate Societies* (Ed. by B. B. Smuts, D. L. Cheney, R. M. Seyfarth, R. W. Wrangham, & T. T. Struhsaker), pp. 121–134. Chicago: University of Chicago Press.

Moore, J. 1984. Female transfer in primates. *Int. J. Primatol.,* 5, 537–589.

Moore, J. 1992. Dispersal, nepotism, and primate social behavior. *Int. J. Primatol.,* 13, 361–378.

Moos, R., Rock, J., & Salzert, W. 1985. Infanticide in gelada baboons (*Theropithecus gelada*). *Primates,* 26, 497–500.

Mori, A., Iwamoto, T., & Bekele, A. 1997. A case of infanticide in a recently found gelada population in Arsi, Ethiopia. *Primates,* 38, 79–88.

Mori, U. 1979a. Inter-unit relationships. In: *Ecological and Sociological Studies of Gelada Baboons* (Ed. by M. Kawai), pp. 83–92. Basel: Karger.

Mori, U. 1979b. Individual relationships within a unit. In: *Ecological and Sociological Studies of Gelada Baboons* (Ed. by M. Kawai), pp. 93–124. Basel: Karger.

Mori, U. 1979c. Development of sociability and social status. In: *Ecological and Sociological Studies of Gelada Baboons* (Ed. by M. Kawai), pp. 125–154. Basel: Karger.

Mori, U. 1979d. Unit formation and the emergence of a new leader. In: *Ecological and Sociological Studies of Gelada Baboons* (Ed. by M. Kawai), pp. 155–181. Basel: Karger.

Mori, U. 1979e. Social structure of gelada baboons. In: *Ecological and Sociological Studies of Gelada Baboons* (Ed. by M. Kawai), pp. 243–247. Basel: Karger.

Mori, U. & Dunbar, R. I. M. 1985. Changes in the reproductive condition of female gelada baboons following the takeover of one-male units. *Z. Tierpsychol.,* 67, 215–224.

Müller, H. 1980. Variations of social behaviour in a baboon hybrid zone (*Papio anubis* x *Papio hamadryas*) in Ethiopia. PhD dissertation, University of Zurich, Zurich.

Nagel, U. 1971. Social organization in a baboon hybrid zone. In: *Proceedings of the 3rd International Congress of Primatology* (Ed. by H. Kummer), pp. 48–57. Zurich: Karger.

Noë, R. 2001. Biological markets: partner choice as the driving force behind the evolution of mutualisms. In: *Economics in Nature: Social Dilemmas, Mate Choice and Biological Markets* (Ed. by R. Noë, J. A. R. A. M. van Hooff, & P. Hammerstein), pp. 93–118. Cambridge: Cambridge University Press.

Noë, R. & Hammerstein, P. 1995. Biological markets. *Trends Ecol. Evol.*, 10, 336–339.

Ohsawa, H. 1979. The local gelada population and environment of the Gich area. In: *Ecological and Sociological Studies of Gelada Baboons* (Ed. by M. Kawai), pp. 3–80. Basel: Karger.

Palombit, R. A. 2000. Infanticide and the evolution of male-female bonds in animals. In: *Infanticide by Males and Its Implications* (Ed. by C. P. van Schaik & C. H. Janson), pp. 239–268. Cambridge: Cambridge University Press.

Paul, A., Preuschoft, S., & van Schaik, C. P. 2000. The other side of the coin: infanticide and the evolution of affiliative male-infant interactions in Old World primates. In: *Infanticide by Males and Its Implications* (Ed. by C. P. van Schaik & C. H. Janson), pp. 269–292. Cambridge: Cambridge University Press.

Phillips-Conroy, J. E., Jolly, C. J., & Brett, F. L. 1991. The characteristics of hamadryas-like males living in anubis baboons troops in the Awash National Park, Ethiopia. *Am. J. Phys. Anthropol.*, 86, 353–368.

Phillips-Conroy, J. E., Jolly, C. J., Nystrom, P., & Hemmalin, H. A. 1992. Migration of male hamadryas baboons into anubis groups in the Awash National Park, Ethiopia. *Int. J. Primatol.*, 13, 455–476.

Pusey, A. E. & Packer, C. 1987. Dispersal and philopatry. In: *Primate Societies* (Ed. by B. B. Smuts, D. L. Cheney, R. M. Seyfarth, R. W. Wrangham, & T. T. Struhsaker), pp. 250–266. Chicago: University of Chicago Press.

Rasmussen, D. R. 1981. Evolutionary, proximate, and functional primate social ecology. In: *Perspectives in Ethology, Vol. 4: Advantages of Diversity* (Ed. by P. P. G. Bateson & P. H. Klopfer), pp. 75–103. New York: Plenum.

Rijksen, H. D. 1981. Infant killing: a possible consequence of a disputed leader role. *Behaviour*, 78, 138–168.

Robbins, M. M. 2001. Variation in the social system of mountain gorillas: the male perspective. In *Mountain Gorillas: Three Decades of Research at Karisoke* (Ed. by M. M. Robbins, P. Sicotte, & K. J. Stewart), pp. 29–58. Cambridge: Cambridge University Press.

Rodseth, L., Wrangham, R. W., Harrigan, A. M., & Smuts, B. B. 1991. The human community as a primate society. *Curr. Anthropol.*, 32, 221–254.

Rowell, T. E. 1979. How would we know if social organization were *not* adaptive? In: *Primate Ecology and Human Origins* (Ed. by I. S. Bernstein & E. O. Smith), pp. 1–22. New York: Garland STPM Press.

Rubenstein, D. I. & Wrangham, R. W. (eds.) 1986. *Ecological Aspects of Social Evolution*. Princeton, NJ: Princeton University Press.

Seyfarth, R. M. 1983. Grooming and social competition in primates. In: *Primate Social Relationships* (Ed. by R. A. Hinde), pp. 182–190. Oxford: Blackwell.

Sicotte, P. 2001. Female choice in mountain gorillas. In: *Mountain Gorillas* (Ed. by M. M. Robbins, P. Sicotte, & K. J. Stewart), pp. 59–87. Cambridge: Cambridge University Press.

Sigg, H. 1980. Differentiation of female positions in hamadryas one-male-units. *Z. Tierpsychol.*, 53, 265–302.

Sigg, H., Stolba, A., Abegglen, J. J., & Dasser, V. 1982. Life history of hamadryas baboons: physical development, infant mortality, reproductive parameters and family relationships. *Primates*, 23, 473–487.

Small, M. F. 1993. *Female Choices: Sexual Behavior of Female Primates.* Ithaca, NY: Cornell University Press.

Smuts, B. B. 1987. Sexual competition and mate choice. In: *Primate Societies* (Ed. by B. B. Smuts, D. L. Cheney, R. M. Seyfarth, R. W. Wrangham, & T. T. Struhsaker), pp. 385–399. Chicago: University of Chicago Press.

Smuts, B. B. & Smuts, R. W. 1993. Male aggression and sexual coercion of females in nonhuman primates and other mammals: evidence and theoretical implications. *Adv. Study Anim. Behav.*, 22, 1–63.

Stammbach, E. 1987. Desert, forest, and montane baboons: multilevel societies. In: *Primate Societies* (Ed. by B. B. Smuts, D. L. Cheney, R. M. Seyfarth, R. W. Wrangham, & T. T. Struhsaker), pp. 112–120. Chicago: University of Chicago Press.

Sterck, E. H. M. 1998. Female dispersal, social organization, and infanticide in langurs: are they linked to human disturbance? *Am. J. Primatol.*, 44, 235–254.

Sterck, L., Watts, D. P., & van Schaik, C. P. 1997. The evolution of female social relationships in nonhuman primates. *Behav. Ecol. Sociobiol.*, 41, 291–309.

Strier, K. B. 1994. Myth of the typical primate. *Ybk. Phys. Anthropol.*, 37, 233–271.

Sugawara, K. 1982. Sociological comparison between two wild groups of anubis-hamadryas hybrid baboons. *Afr. Study Mono.*, 2: 73–131.

Swedell, L. 2000. Two takeovers in wild hamadryas baboons. *Folia Primatol.*, 71, 169–172.

Swedell, L. 2002. Affiliation among females in wild hamadryas baboons (*Papio hamadryas hamadryas*). *Int. J. Primatol.*, 23, 1205–1226.

Swedell, L. & Woolley-Barker, T. 2001. Dispersal and philopatry in hamadryas baboons: a re-evaluation based on behavioral and genetic evidence. *Am. J. Phys. Anthropol. Sup.*, 32, 146.

Trivers, R. L. 1971. The evolution of reciprocal altruism. *Q. Rev. Biol.*, 46, 35–57.

Trivers, R. L. 1972. Parental investment and sexual selection. In: *Sexual Selection and the Descent of Man* (Ed. by B. Campbell), pp. 136–179. Chicago: Aldine.

van Hooff, J. A. R. A. M. 2000. Relationships among nonhuman primate males: a deductive framework. In: *Primate Males* (Ed. by P. M. Kappeler), pp. 183–191. Cambridge: Cambridge University Press.

van Hooff, J. A. R. A. M. 2001. Conflict, reconciliation and negotiation in non-human primates: the value of long term relationships. In: *Economics in Nature: Social Dilemmas, Mate Choice and Biological Markets* (Ed. by R. Noë, J. A. R. A. M. van Hooff, & P. Hammerstein), pp. 67–90. Cambridge: Cambridge University Press.

van Hooff, J. A. R. A. M. & van Schaik, C. P. 1992. Cooperation in competition: the ecology of primate bonds. In: *Coalitions and Alliances in Humans and Other Animals* (Ed. by A. H. Harcourt & F. B. M. de Waal), pp. 357–389. Oxford: Oxford University Press.

van Schaik, C. P. 1983. Why are diurnal primates living in groups? *Behaviour*, 87, 120–143.

van Schaik, C. P. 1989. The ecology of social relationships amongst female primates. In: *Comparative Socioecology: The Behavioural Ecology of Human and Other Mammals* (Ed. by V. Standen & R. A. Foley), pp. 195–218. Oxford: Blackwell.

van Schaik, C. P. 1996. Social evolution in primates: the role of ecological factors and male behaviour. *Proc. Br. Acad.*, 88, 9–31.

Watts, D. P. 1996. Comparative socio-ecology of gorillas. In: *Great Ape Societies* (Ed. by W. C. McGrew, L. F. Marchant, & T. Nishida), pp. 16–28. Cambridge: Cambridge University Press.

Watts, D. P. 2000. Causes and consequences of variation in male mountain gorilla life histories and group membership. In: *Primate Males* (Ed. by P. M. Kappeler), pp. 169–179. Cambridge: Cambridge University Press.

Watts, D. P. 2001. Social relationships of female mountain gorillas. In: *Mountain Gorillas* (Ed. by M. M. Robbins, P. Sicotte, & K. J. Stewart), pp. 215–240. Cambridge: Cambridge University Press.

Watts, D., Colmenares, F., & Arnold, K. 2000. Redirection, consolation, and male policing. In: *Natural Conflict Resolution* (Ed. by F. Aureli & F. B. M. de Waal), pp. 281–301. Berkeley, CA: University of California Press.

Whitten, P. L. 1987. Infants and adult males. In: *Primate Societies* (Ed. by B. B. Smuts, D. L. Cheney, R. M. Seyfarth, R. W. Wrangham, & T. T. Struhsaker), pp. 343–357. Chicago: University of Chicago Press.

Wrangham, R. W. 1980. An ecological model of female-bonded primate groups. *Behaviour,* 75, 262–300.

Wrangham, R. W. 1982. Mutualism, kinship, and social evolution. In: *Current Problems in Sociobiology* (Ed. by King's College Sociobiology Group), pp. 269–290. Cambridge: Cambridge University Press.

Wrangham, R. W. 1987. Evolution of social structure. In: *Primate Societies* (Ed. by B. B. Smuts, D. L. Cheney, R. M. Seyfarth, R. W. Wrangham, & T. T. Struhsaker), pp. 282–296. Chicago: University of Chicago Press.

12

The Impact of Kinship on Mating and Reproduction

Andreas Paul
Jutta Kuester

> Generally speaking, there is a remarkable absence of erotic feelings between persons living very closely together from childhood.
> —Edward Westermarck, *The History of Human Marriage* (1922)

Early Theories of Sex and Kinship

Taboos and the Origins of Culture

Theories of sex and kinship have long been among the favorites of the humanities, and the study of incest and incest avoidance, in particular, has often been regarded as being at the very heart of anthropological theory (see e.g., Shepher 1983, Thornhill 1991, Wolf 1995). The reason for this was succinctly stated by Claude Lévi-Strauss (1969: 24, 25): "The prohibition of incest . . . is the fundamental step because of which, by which, but above all in which, the transition from nature to culture is accomplished. . . . Before it, culture is still nonexistent; with it, nature's sovereignty over man is ended. The prohibition of incest is where nature transcends itself. . . . It brings about and is in itself the advent of a new order." The "new order" Lévi-Strauss had in mind was a system of reciprocity and exchange: "Incest would prevent people from widening their circle of friends," as Thomas Aquinas had already argued in his *Summa Theologica* 700 years ago, but "when a man takes a wife from another family he is joined in a special friendship with her relations" (quoted by Wolf 1995: 150). According to this view, which was championed by Lévi-Strauss and several other cultural anthropologists, incest regulations serve to create alliances by enforcing exogamous marriages—which are in "primitive" societies more often political and economic contracts than erotic relationships. Almost all other cultural anthropologists were also convinced that the prohibition of incest was the event "where nature transcends itself," that it was the origin of human culture, society, and morality, but the "alliance theory" was not the only conse-

271

quence of the prohibition of incest. Bronislaw Malinowski and Brenda Seligman, for example, argued that early humans invented the incest taboo in order to prevent the family from "the sexual impulse, which is in general a very upsetting and socially disrupting force. . . . A society which allowed incest could not develop a stable family; it would therefore be deprived of the strongest foundations for kinship, and this in a primitive community would mean absence of social order" (Malinowski 1931: 629–630).

"Instincts": Westermarck's and Freudian Views

Obviously, all these views were built not only on the premise that incest is a "natural phenomenon found commonly among animals" (Lévi-Strauss 1969: 18), but also on Freud's claim that incestuous desires are universal and deeply rooted in the human psyche (see below). But are these premises true? Naturalists were never so sure. Like Buffon, who argued in 1753 in the fourth volume of his *Histoire Naturelle* that without cross-breeding, "all grain, flowers and animals degenerate" (quoted by Wolf 1995: 3), Darwin reasoned at length in *The Variation of Animals and Plants Under Domestication* about the deleterious effects of close inbreeding in domesticated animals and plants, and he added with his characteristic caution: "Although there seems to be no strong inherited feeling in mankind against incest, it seems possible that men during primeval times may have been more excited by strange females than by those with whom they have habitually lived" (Darwin 1868: 104). About 20 years later, Edward Westermarck picked up this idea and elaborated it in *The History of Human Marriage* ([1891] 1922) and *The Origin and Development of Moral Ideas* (1906–1908). Westermarck not only noted that claims that incest in animals was the norm were completely unsupported by empirical data, but building on Darwin's theory of natural selection and his accounts of the deleterious effects of close inbreeding, he asserted:

> It is impossible to believe that a law which holds good for the rest of the animal kingdom, as well as for plants, does not apply to humans also. . . . Thus an instinct would be developed which would be powerful enough, as a rule, to prevent injurious unions. Of course it would display itself simply as an aversion on the part of individuals to union with others with whom they lived; but these, as a matter of fact, would be blood-relations, so that the result would be the survival of the fittest. (Westermarck [1891] 1922: 339, 352)

The initial reaction to Westermarck and his hypothesis that early childhood association inhibits later sexual attraction and that this inhibition is the result of natural selection was quite positive (see Wolf 1995, for an elegant, and much more comprehensive, treatment of the history of the Westermarck hypothesis). But at the turn of the century the scientific climate changed. While Darwin's "transmutation" theory (i.e., his theory of evolution) had been widely accepted, his (and Alfred Wallace's) theory of natural selection suffered a dramatic decline (Mayr 1982). It was in this climate that Frazer (1910) accused Westermarck of being "too much under the influence of Darwin." Moreover, if Westermarck were right, Frazer argued, there would be no incest taboo, and that criticism was picked up by Freud: "It is not easy to see why any deep human instinct should need to be reinforced by law. . . . The law only forbids men to do what their instincts incline them to do; what nature itself prohibits and punishes, it would be superfluous for the law to prohibit and punish. . . . Instead of assuming, therefore, from the legal prohibition of incest, we ought rather to assume that there is a natural instinct in favour of it" (Freud [1913] 1950: 97–98).

Westermarck responded that this view "implies a curious misconception of the origin of legal prohibitions" (Westermarck [1891] 1922, 2: 203). "Aversions which are generally felt," he noted, "readily lead to moral disapproval and prohibitory customs or laws" (Westermarck 1932: 249). Moreover, "even if social prohibitions might prevent unions between the nearest relatives, they could not prevent the desire for such unions. The sexual instinct can hardly be changed by prescriptions; I doubt whether all laws against homosexual intercourse, even the most draconic, have ever been able to extinguish the peculiar desire of anybody born with homosexual tendencies" (Westermarck [1891] 1922, 2: 192). That there is a "natural instinct" in favor of incestuous relationships, was, however, Freud's view, who maintained that "the findings of psychoanalysis make the hypothesis of an innate aversion to incestuous intercourse totally untenable" (Freud [1913] 1950: 123–124). According to Freud, "psychoanalytic investigations have shown beyond the possibility of doubt that an incestuous love-choice is in fact the first and the regular one, and it is only later that any opposition is manifested towards it, the causes of which are not to be sought in the psychology of the individual" (Freud [1920] 1953: 221). Curiously, not only Westermarck, but also Freud ([1913] 1950) referred to Darwin when he offered his version of the origin of the incest taboo. But although he was aware of the deleterious "racial" effects of close inbreeding, he did not invoke the then unpopular theory of natural selection. Instead, he presented a crude story of sons killing their primeval despotic father and then inventing the barrier against incest in the name of "deferred obedience" and social solidarity. Moreover, although strong arguments against the theory of the inheritance of acquired traits had been made before (Weismann 1883), Freud was convinced that this "barrier against incest . . . , like other moral taboos, had no doubt already become established . . . by organic inheritance" (quoted by Wolf 1995: 487).

Nevertheless, despite Freud's "absurd biology" (Wolf 1995), for a long time the view prevailed that Frazer and Freud were right, and Westermarck was wrong (e.g., Aberle et al. 1963, Rose et al. 1984; see also Dickeman 1992; but see Bischof 1972, 1985), mainly because many cultural anthropologists and social scientists felt uncomfortable with the idea that genes could be responsible for human behavioral inclinations. But Westermarck's idea was not so simple. Although he used the (now widely abandoned) term "instinct" (which Westermarck himself later abandoned in favor of the term "innate aversion"), his hypothesis was, in Wolf's words, "like all good evolutionary hypotheses, . . . an ecological hypothesis" (Wolf 1995: 5). In Westermarck's view, the ontogenetic environment of the individual played at least as crucial a role in the development of a sexual aversion as his or her genes. Consequently, Westermarck's hypothesis not only predicts that incest should be rare, but also when and why it occurs (Wolf 1995).

In the remainder of this chapter, we review some current ideas about inbreeding and its biological consequences, and about the impact of kinship on mating preferences in nonhuman primates. Although nonhuman animals played a marginal role in the historical debate on the origins of the incest taboo, the behavior of nonhuman primates is clearly particularly interesting because, according to the Freudian view, any avoidance of incest should be an exclusively human invention. Earlier reviews already reached the conclusion that this is not the case (see especially Pusey 1990, Wolf 1995), but during the last few years many more data have accumulated not only on the impact of maternal kinship, but also on the impact of paternal kinship on mating and reproduction. These data have raised new questions about the limits of kin discrimination among primates and its possible mechanisms.

Biological Theories of Kinship and Reproduction

Costs of Inbreeding

Evolutionary theory predicts that all living beings are designed to maximize their genetic fitness. If mating with close genetic relatives confers fitness costs on one or both parties, one should expect, therefore, that any inclination to mate with close relatives should be selected against; individuals with an innate aversion to incest should leave more descendants than individuals who lack such an aversion. Data from a wide range of taxa indeed indicate that inbreeding is costly (Pusey & Wolf 1996, Crnokrak & Roff 1999, chapters in Thornhill 1993), even to the point that the risk of extinction is significantly elevated (e.g., Nieminen et al. 2001). The reduction of fitness after close inbreeding can be caused by a number of genetic factors such as the unmasking of deleterious recessive alleles due to increased homozygosity, decreased heterozygosity, and/or reduced allozyme variability (e.g., Mitton 1993, Crnokrak & Roff 1999). Moreover, although levels of inbreeding depression vary across species, local conditions, and breeding history, costs of inbreeding tend to increase with increasing coefficients of relatedness between spouses (May 1979). Thus, from this perspective, at least, one should expect a strong negative impact of kinship on mating and reproduction—not only in humans, but also in other animals.

But there are at least three arguments against this view. The first is that the genetic load (i.e., the number of deleterious recessive genes in a population) may actually be purged by recurrent inbreeding, resulting in significantly lower levels of inbreeding depression after a few generations of inbreeding (Wright 1977). This could imply that, in small, isolated populations, negative effects of inbreeding might be largely absent and inbreeding is not selected against. The second argument is related; if inbreeding were really so harmful, why is inbreeding habitual in so many species (Hamilton 1967, Shields 1982, Werren 1993)? Finally, any proper analysis of the impact of kinship on mating and reproduction has not only to consider the potential costs of inbreeding, but also its potential benefits. The costs of dispersal, for example, may outweigh those of inbreeding (e.g., Moore & Ali 1984); furthermore, even if close inbreeding or incest (parent-offspring and sibling-sibling mating) is costly, moderate inbreeding (mating between individuals related at approximately the level of first cousins) might be beneficial because of genetic costs associated with outbreeding that can arise from the breakup of coadapted gene complexes and hence promote the loss of local adaptedness (e.g., Bateson 1983, Shields 1993). Indeed, not only in human populations is some level of assortative mating commonly observed, and models of "optimal outbreeding" have suggested that the "optimal balance between inbreeding and outbreeding" might be achieved by a sexual preference for fairly close relatives such as first cousins (Bateson 1982).

While these arguments could imply that the impact of kinship on mating and reproduction is not as negative as suggested above, conclusive evidence for the effects of purging and outbreeding appears to be limited (Pusey & Wolf 1996, Crnokrak & Roff 1999; for a more recent example see Visscher et al. 2001). Moreover, since chronic inbreeders must not only be able to cope with a high mortality rate for some generations, but also face the problem of accumulating genes with mild deleterious effects within lineages, except under very special circumstances, close inbreeding is unlikely to continue indefinitely (Werren 1993, O'Riain & Braude 2001).

The Red Queen

Fitness loss due to inbreeding clearly poses a threat to many taxa, including primates, and when environmental stress interacts with problems caused by inbreeding, the costs associated with inbreeding appear to be substantially higher than previously thought (Pusey & Wolf 1996, Crnokrak & Roff 1999, Meagher et al. 2000). But the accumulation of deleterious genes may be neither the only nor even the most important ultimate reason why inbreeding should be avoided. As a matter of fact, inbreeding undermines the main effect of sexual reproduction—the creation of genetic variation (Bischof 1985, Hamilton 1993). Inbreeding increases genetic variation between populations, but it invariably leads to a decrease in genetic variation among one's own offspring and within populations. In this case, inbreeding has an effect similar to that of asexual reproduction, but unlike asexually breeding species, inbreeders still suffer from the twofold costs of sexuality: they neither forgo "wasteful" males, nor do they transmit 100% of their genes to the next generation. Inbreeders, therefore, have to bear the costs of sex, but they do not enjoy its benefits.

What is the benefit of sex? According to the "Red Queen" hypothesis proposed by Hamilton and others, sexually reproducing species have an advantage in the never-ending coevolutionary race between hosts and their parasites; sex is maintained because the continuous production of new gene combinations provides an effective barrier against omnipresent, fast-breeding and fast-evolving parasites (e.g., Hamilton et al. 1990). If the idea is right, then the implications for the inbreeding-outbreeding debate are obvious; "eventually inbred lines should suffer from the same inflexibility in the face of parasites as parthenogens do and this will make them die out or else force them back to outbreeding" (Hamilton 1993: 441). Moreover, since large K-selected species with slow reproductive turnover are more vulnerable to parasites than small r-strategists, it is not surprising to see that both asexual reproduction and habitual inbreeding are far more common in small, short-lived species than in large, long-lived ones (Hamilton 1993). The argument received further support by the finding that endotherms, which are—as a rule—larger than ectotherms, generally suffer from higher fitness costs due to inbreeding than ectotherms (Crnokrak & Roff 1999). "Large, long-lived beings," Hamilton (1993: 444) concluded, "whether trees, people, or [hymenopteran] supercolonies, usually retain sex, and for the act itself prefer relative strangers—strangers who should come from as far away, ideally, as their parasites."

Obviously, from this point of view, the Freudian view that incestuous desires are universal and that the avoidance of incest is an entirely cultural phenomenon that has been invented by our remote ancestors to keep them from ruining themselves, their families, and their society, has serious limitations. Given the potential costs of inbreeding, it would be surprising if humans, as well as other primates, had not an evolved psychological mechanism designed to solve the problem.

Sex Differences

Modern evolutionary reasoning adds a further complication. While Darwin (1868), based on an observation of one of his informants, suggested that males may be more "excited by strange females" than females are by strange males, Havelock Ellis noted that "this instinct is probably even more marked in the female" (quoted by Wolf 1995: 229). Indeed, due to their limited reproductive potential, the opportunity costs of casual sex usually are far greater for females than for males, suggesting that females are not only much more choosy (Darwin 1871, Trivers 1972), but also much more averse to close or moderate inbreeding than are males—perhaps even to the point that any evolved psychological mechanism that enables humans and other animals to avoid incest might be restricted to females (Alberts 1999, Haig 1999).

Mechanisms of Kin Discrimination

Whenever the opportunity for incest or moderate inbreeding is present, any avoidance relies on the ability to recognize those individuals likely to be kin. The Westermarck effect is one likely candidate, but not the only one. Phenotype matching, which involves the matching of phenotypic traits of individuals to a standard (Grafen 1990) or a template (Waldman et al. 1988), is the most often invoked alternative. Studies on fish, mice, and humans, for example, provide evidence that odor cues mediated by genes of the major histocompatibility complex play a role in kin recognition and inbreeding avoidance (reviewed by Penn 2002). Visual cues may also be important since at least chimpanzees (*Pan troglodytes*) perceive facial similarities of related but unfamiliar individuals, although this ability seems to be limited to the recognition of mothers and sons (Parr & de Waal 1999). Whether nonhuman primates are able to perceive similarities between themselves and related but unfamiliar individuals is much less clear, however. Experiments with pigtailed macaques (*Macaca nemestrina*) suggested that these primates might be able to recognize unfamiliar paternal half siblings by phenotypic cues (Wu et al. 1980), but subsequent studies have failed to replicate the original findings (Fredrickson & Sackett 1984, Sackett & Fredrickson 1987, Welker et al. 1987, Erhart et al. 1997; but see Small & Smith 1981, MacKenzie et al. 1985). However, some other studies provide data suggesting that primates might be able to recognize paternal kin in the absence of familiarity (Smith 1995, Alberts 1999, Widdig et al. 2001).

Predictions

The predictions to be derived from these considerations are relatively straightforward:

1. Contrary to Freud and his followers, incest avoidance should not be restricted to humans; whenever inbreeding is more costly than outbreeding, it should be expected that other animals also avoid inbreeding.
2. If the Westermarck effect is real, it should be expected that early childhood association inhibits later sexual attraction—regardless of kinship. Because there is no a priori reason why the Westermarck effect should be restricted to humans, it can be expected that early childhood association inhibits later sexual attraction at least in other primates and, most likely, also in other relatively large-brained animals. Moreover, whenever there are no close relationships between fathers and their offspring or between paternal siblings (as is usually the case in promiscuous mating systems), no avoidance of close inbreeding should be detected later in life. If, in contrast, phenotypic cues are used, paternal kin discrimination should be expected.
3. If models of optimal outbreeding are correct, moderate inbreeding, that is, a preference for individuals that are not too distantly related, such as first cousins, should be expected. If, in contrast, the Red Queen version of inbreeding avoidance is correct, then moderate inbreeding should not be expected. Limits of kin discrimi-

nation (Chapais et al. 1997, 2001) may set an upper limit for inbreeding avoidance, however.

4. If predictions derived from sexual selection theory are correct, then females should be much more averse to close inbreeding than are males.

Patterns of Incest and Incest Avoidance in Primates

"Whatever the uncertainties regarding the sexual habits of great apes," Lévi-Strauss (1969: 31) wrote in *The Elementary Structures of Kinship*, " . . . it is certain these great anthropoids practise no sexual discrimination against their near relatives." While one may wonder how a prominent anthropologist could make such a strong claim without any supportive evidence (the first edition of *Les Structures Élémentaire de la Parenté* appeared in 1949, when almost nothing was known about the behavior of nonhuman primates in natural settings), it was common for social scientists to think that animals regularly bred with their parents, siblings, and other genetic relatives. Indeed, as Pusey and Packer (1987: 250) noted, "early field workers were so struck by the apparent permanence of group membership in social primates that it was widely believed that primate groups were essentially closed genetic units." Although this view was perpetuated until the early 1990s (Leavitt 1990), evidence to the contrary was available by the end of the 1960s; the first field studies on nonhuman primates that lasted more than a few weeks or months, begun by Japanese researchers in the late 1940s, not only demonstrated that male Japanese macaques regularly leave their natal group and breed elsewhere, but also that, given the opportunity, sexual interactions between mothers and sons were surprisingly rare, if not nonexistent (Itani & Tokuda 1958, cited in Wolf 1995, Imanishi [1961] 1965). Unfortunately, since these first reports on incest avoidance in a nonhuman primate species were published in Japanese (but see Tokuda 1961–1962), they went unnoticed in the Western Hemisphere. But there, too, evidence began to accumulate slowly that among free-ranging animals, mechanisms leading to the avoidance of mating with close kin might be more widespread than generally believed (Aberle et al. 1963, Kaufman 1965, Sade 1968).

Kinship, Sex, and Reproduction Among Maternal Relatives

Primates resemble most other vertebrates in showing high rates of natal dispersal by one or both sexes (Pusey & Packer 1987). Sex-biased dispersal greatly reduces the opportunity for close inbreeding. However, under some conditions members of the dispersing sex may stay in their natal group, so that close relatives sometimes encounter each other as potential mates (e.g., Ménard & Vallet 1996 for Barbary macaques [*M. sylvanus*], Alberts & Altmann 1995 for savanna baboons [*Papio cynocephalus*], Constable et al. 2001 for chimpanzees). Although under these conditions mating and reproduction among close maternal relatives has been observed (see, e.g., Constable et al. 2001 for a recent case of mother-son incest in chimpanzees), both observational and experimental studies provide abundant evidence that such cases are the exception rather than the norm. Avoidance of mating with close maternal relatives has been most thoroughly documented among Old World monkeys and apes (reviewed by Pusey 1990, Wolf 1995), but has also been observed in gregarious prosimians (Taylor & Sussman 1985, Pereira & Weiss 1991, Sauther 1991) and New World monkeys

(Rothe 1975; see also Abbott 1993, Savage et al. 1996, Baker et al. 1999). Only gibbon (*Hylobates* spp.) groups have been suggested to consist commonly of "incestuous pairs" because of some apparent instances of mother-son incest following the death of the father (Leighton 1987, Pusey 1990). However, recent observations indicate that gibbon social structure is more variable than previously supposed and that many cases of "incest" may actually have been matings between unrelated individuals (Fuentes 2000). In fact, genetic paternity analyses revealed that subadult males living in gibbon groups are not always related to the breeding pair (Oka & Takenaka 2001). Moreover, although nonhuman primate offspring living in nuclear family groups sometimes mate with their mother or father shortly after the death of the same-sex parent, such cases appear to be rare, and breeding vacancies are more often filled by unrelated individuals (Leighton 1987, König et al. 1988, Baker et al. 1999).

For two reasons, skepticism about the existence of evolved psychological inclinations against incest has prevailed, however (e.g., Bernstein 1988). First, many reports of incest avoidance in nonhuman primates have been anecdotal, and few studies have attempted to match observed frequencies of matings between relatives against expected numbers of matings to determine accurately whether animals are actively avoiding inbreeding or not. Several studies have provided such data, however (Loy 1971 for rhesus monkeys [*M. mulatta*]; Takahata 1982 for Japanese macaques [*M. fuscata*]; Murray & Smith 1983 for stump-tailed macaques [*M. arctoides*]; Paul & Kuester 1985 for Barbary macaques [*M. sylvanus*]; Alberts & Altmann 1995 for savanna baboons). Moreover, experiments explicitly designed to promote mother-son incest also strongly suggest that psychological inclinations against incest exist (Itoigawa et al. 1981). Second, it has been argued that observational data on mating activities do not accurately predict reproductive outcomes. Genetic studies have shown, however, that matrilineal kin not only tend to avoid mating, but also that they almost never produce offspring with each other (Inoue et al. 1990 for Japanese macaques; Pereira & Weiss 1991 for ring-tailed lemurs [*Lemur catta*]; Bauers & Hearn 1994 for stump-tailed macaques; Kuester et al. 1994 for Barbary macaques; Smith 1995 for rhesus monkeys; Constable et al. 2001 for chimpanzees).

When rare incestuous matings have been observed in a number of species, they tend to differ from other sexual interactions in a number of ways. Missakian (1973), for example, observed "sexual activity" among 8 out of 26 rhesus monkey mother-son pairs, but she noted that "typical consort behavior, consisting of following, extensive reciprocal grooming, repeated mount series, and copulations was not observed in any case of mother-son mating" (Missakian 1973: 237; see also Sade et al. 1984). Similarly, sexual interactions between maternally related Japanese macaques were found to be characterized by irregular mount series, little (if any) sexual interest by one or both individuals, and rare ejaculation (Enomoto 1978, Baxter & Fedigan 1979, Takahata 1982). Atypical sexual behavior between maternal relatives was also observed among savanna baboons, where natal males occasionally mount female maternal relatives, but almost never establish consortship relations with them. Moreover, sexual interactions during periods in which conception is likely to occur appear to be extremely uncommon (e.g., Alberts & Altmann 1995). Second, most sexual interactions between maternal relatives are initiated by young males, often ones that had just reached sexual maturity, but sometimes also during infancy (Starin 2001, Takahata et al. 2002; see also Pusey 1990). Adult males, in contrast, almost never mount their mothers or other maternal relatives (Missakian 1973, Packer 1979, Scott 1984, Alberts & Altmann 1995 for Old

World monkeys; Tutin 1979, Pusey 1980, Goodall 1986 for chimpanzees; Kano 1992 for bonobos [*Pan paniscus*]). Thus, it appears that a few young males sometimes "experiment" with their mothers or other maternal relatives during infancy or adolescence, but that even these somewhat exceptional individuals prefer unrelated females as sexual partners once they have achieved adulthood (Wolf 1995). Finally, once males attempt to copulate with their mothers or maternal sisters, their sexual advances are commonly rejected by the female (Taylor & Sussman 1985, Pereira & Weiss 1991, Sauther 1991 for ring-tailed lemurs; Rothe 1975 for common marmosets [*Callithrix jacchus*]; Starin 2001 for red colobus monkeys [*Colobus badius*]; Enomoto 1978 for Japanese macaques; Packer 1979 for savanna baboons; Goodall 1986, Nishida 1990, Constable et al. 2001 for chimpanzees).

The available evidence, therefore, indicates that incestuous matings between maternal relatives are not only rare but also atypical and often singular events. Moreover, when they occur, they are typically initiated by males but avoided by females. Systematic analyses incorporating measures to quantify responsibility for proximity maintenance within dyads also reveal that females are more responsible than males for avoiding mating with matrilineal kin (Manson & Perry 1993 for rhesus monkeys; Soltis et al. 1999 for Japanese macaques). Finally, among Japanese macaques, females were found to perform homosexual acts with almost all available nonkin partners, but unlike heterosexual dyads they consistently avoided their mother, daughters, or sisters, suggesting that under conditions where female mating strategies are unconstrained by male behavior, incest avoidance is even more strongly developed (Chapais & Mignault 1991). This is not to say that only females are psychologically inclined to avoid incest, however. Several studies have found that males also showed little, if any, interest in their mothers or other female relatives when the latter were in estrus (Enomoto 1978, Pusey 1980, Itoigawa et al. 1981, Paul & Kuester 1985, Smuts 1985, see also Di Bitetti & Janson 2001). When no other options are available, however, the sexual inhibition against incest appears to be more fragile in males than in females.

Kinship, Sex, and Reproduction Among Paternal Relatives

In most species of gregarious mammals, average male tenure in a group is shorter than the time it takes females to reach sexual maturity. Among those species where average male tenure exceeds this period, females disperse (Clutton-Brock 1989). Dispersal patterns, therefore, also greatly reduce the probability that fathers encounter their daughters as potential mates (see also Alberts & Altmann 1995). Moreover, in many species of nonhuman primates, males appear to be more attracted to older, experienced females and tend to ignore young, nulliparous females, probably because older females have a higher chance to conceive and to successfully rear their offspring (reviewed by Anderson 1986). At the same time, however, such a mechanism would also reduce the probability of inbreeding via the paternal line. Some observations indeed suggest that males (and females) tend to avoid mating with paternal relatives (Pusey 1990). Copulations between female gorillas and their presumed fathers, for example, appear to be very rare (Stewart & Harcourt 1987, Watts 1990), and female tufted capuchins (*Cebus apella*) and their presumed fathers strongly avoid each other as sexual partners (Di Bitetti & Janson 2001). But whether paternal relatives actively avoid incest as strictly as maternal relatives remains unclear in most cases, since only genetic methods of paternity determination can solve the problem.

Long-term genetic research on a captive colony of rhesus monkeys showed that paternal relatives sired offspring with each other as often as expected by chance, while the incidence of inbreeding between maternal relatives was much lower (Smith 1995). Similar results have been obtained for captive ring-tailed lemurs (Pereira & Weiss 1991), captive Japanese macaques (Inoue et al. 1990), and semifree-ranging Barbary macaques (Kuester et al. 1994). Moreover, observational studies incorporating genetic paternity data also revealed that among free-ranging savanna baboons (Alberts 1999) and semifree-ranging Barbary macaques (Kuester et al. 1994) paternal relatives did not avoid each other as mating partners. It seems therefore that, unlike maternal relatives, at least among species with a promiscuous mating system, paternal relatives do not regularly avoid close inbreeding, perhaps because they do not recognize each other as kin. Some findings do not appear to match this explanation, however. First, although in Smith's (1995) study the observed number of patrilineally inbred offspring was not lower than expected by chance, the number of offspring resulting from father/daughter incest was, which led him to suggest that "rhesus macaques avoid father/daughter matings" (p. 37). Second, although Alberts (1999) noted that paternal half siblings (the only kinship category analyzed in this study) "did not show significant avoidance of each other as consort partners," these same "half-sibs exhibited significantly lower cohesiveness within consortships than nonrelatives, consistent with a discrimination against paternal half-siblings" (pp. 1503, 1504). Third, among both gorillas (Stewart & Harcourt 1987) and tufted capuchins (Di Bitetti & Janson 2001), likely fathers and their presumed daughters consistently avoid each other as sexual partners. These observations indicate that, at least under some conditions, paternal relatives may recognize each other. We return to this point below (see Mechanisms of Inbreeding Avoidance).

Limits of Kin Discrimination

Lack of incest avoidance among paternal relatives indicates that primates are limited in their ability to discriminate kin, which may not come as a surprise. Similarly, lack of incest avoidance among maternal kin past a certain degree of kinship could reflect limitations in the ability to discriminate kin. Indeed, several observations indicate that the distribution of primate behavior varies according to the degree and type of kinship (for a review, see Chapais 2001). Among Japanese macaques, for example, frequencies of agonistic support (Chapais et al. 1997, 2001) as well as tolerance of cofeeding at a defensible food source (Bélisle & Chapais 2001) drop markedly beyond a degree of relatedness of $r = .125$ among direct kin (great-grandparent–great-grandoffspring), and among collateral kin even beyond $r = .25$ (siblings). Among aunt-niece/nephew dyads ($r_c = .125$), preferential treatment occurred rather inconsistently, suggesting that this class corresponds, at best, to a gray area for nepotism among macaques (Bélisle & Chapais 2001, Chapais et al. 2001).

But altruism does not appear to be the only behavioral syndrome that drops markedly beyond a relatedness threshold of $r = .25$. Among female Japanese macaques, homosexual interactions were found to be completely absent between mothers and daughters ($r = .5$) and between sisters and grandmothers and granddaughters ($r = .25$). In contrast, one-third of all aunt-niece dyads ($r = .125$) in the study group engaged in homosexual activity, which was not significantly different from the proportion of homosexual nonkin dyads, suggesting that beyond a relatedness threshold of $r = .25$ these individuals regarded each other as nonkin (Chapais & Mignault 1991; Chapais et al. 1997).

Evidence that the frequency of sexual interactions between maternal relatives increases beyond a certain degree of relatedness (usually $r = 0.125$) is not restricted to homosexual pairs, but also exists for heterosexual pairs of macaques (Takahata 1982, Kuester et al. 1994). In addition, longitudinal data from Japanese macaques suggest that the probability of inbreeding among remote kin is strongly affected by demographic factors (Takahata et al. 2002).

Thus, the combined evidence of the distribution of different categories of behaviors suggests that the observed kinship threshold might correspond to a general limit of kin discrimination among macaques (Chapais et al. 1997, 2001; Bélisle & Chapais 2001; Chapais 2001). Not all data appear to support this interpretation, however.

While it seems clear that mating inhibitions between maternal relatives decrease with decreasing degrees of relatedness, there is also evidence supporting the hypothesis that the limit of kin discrimination goes well beyond $r = .25$ (or .125). For example, research on Barbary macaques suggests that the threshold for mating inhibition goes well beyond a degree of relatedness of $r = .25$ (Kuester et al. 1994; table 12.1). In this study, not only closely related maternal kin ($r = .5-.25$) but also more remote maternal kin ($r = .125-.063$) showed significantly less frequent sexual interactions with each other than nonkin dyads (close kin versus nonkin: $\chi_1^2 = 82.51$, $P = .0000$; distant kin versus nonkin: $\chi_1^2 = 59.07$, $P = .0000$). Only dyads related by $r < .063$ (female/mothers cousin) did not significantly differ from unrelated dyads ($\chi_1^2 = 1.22$, $P = .270$). Moreover, the proportion of distantly related mating maternal dyads was also significantly lower than the proportion of mating dyads among all classes of paternal relatives ($\chi_1^2 = 80.39$, $P = .0000$).

Table 12.1. Sexual Interactions Between Familiar Maternal Relatives and Between Paternal Relatives Among Barbary Macaques[a]

Type of Relative	Dyads (N)	Dyads with Sex (%)	Dyad Years with Sex (%)
Maternal relatives			
Mother/son	66	4.5	1.8
Sister/brother	123	2.4	1.6
Grandmother/grandson	13	0.0	0.0
Niece/uncle	50	6.0	3.8
Aunt/nephew	75	1.3	0.8
Cousins	33	9.1	6.7
Female/mother's cousin	11	18.1	11.8
Unrelated[b]	2,161	34.0	n.a.
Paternal relatives			
Father/daughter	38	52.6	31.2
Brother/sister	85	51.8	32.8
Grandfather/granddaughter	2	0.0	0.0
Uncle/niece	4	50.0	22.2
Nephew/aunt	1	0.0	0.0
Cousins	3	33.3	20.0
Total	133	50.4	31.2

[a]Data from Kuester et al. 1994.
[b]From Paul and Kuester 1985.

Assuming that the distribution of incest avoidance is a reliable marker of kin discrimination, these data might suggest that the limit of kin discrimination in Barbary macaques differs from that in Japanese macaques. Although we will return to this question below (see Conclusion), data from other groups of Japanese macaques raise some skepticism. Enomoto (1978), for example, observed that among several heterosexual pairs of Japanese macaques, females either refused sexual advances by their uncles and nephews (three out of four dyads) or cousins and great-cousins (four out of eight dyads), or that both sexes were sexually disinterested in each other. Finally, data provided by Takahata et al. (2002) suggest that among Japanese macaques, not only close kin dyads ($r = .5–.125$) strongly avoid matrilineal inbreeding, but also that the proportion of distantly related kin mating dyads was still significantly lower than the proportion of nonkin mating dyads (see table I in Takahata et al. 2002). Thus, it appears that the limits of nepotism found in Japanese macaques do not correspond closely to a general limit of kin discrimination in this and other species. Limits of the profitability of altruism appear to be the more likely explanation (Chapais et al. 2001)

Mechanisms of Inbreeding Avoidance in Primates: Westermarck's Effect or Something Else?

Freud, Lévi-Strauss, and their associates were obviously wrong; there is overwhelming evidence that not only humans but also other primates practice sexual discrimination against their near relatives. Moreover, there is also overwhelming evidence that Westermarck was right; association with immatures inhibits later sexual attraction. The most intimate and long-lasting relationship a primate experiences is, of course, the mother-offspring relationship. One should expect, therefore, an especially strong mating inhibition between mothers and sons, and the available data clearly support this prediction. Next, one should expect a strong mating inhibition between maternal siblings, which also has been found in several studies. Violations do occur, however, but such violations are to be expected (Wolf 1995). Among female-bonded species, for example, juvenile males are typically much less kin oriented than their female peers. One should expect, therefore, that maternal sibling incest should occur more often among dyads where the brother is older (and therefore less familiar with his sibling) than when the sister is older. Exactly this pattern has been found in Barbary macaques (Kuester et al. 1994).

The lack of mating inhibitions among unfamiliar but related individuals provides further evidence in support of the Westermarck hypothesis. Paternal relatives are typically not characterized by close social bonds and early childhood association, and such conditions would clearly require other, more specialized kin discrimination mechanisms, such as phenotype matching. Lack of mating inhibitions among unfamiliar paternal relatives suggests that such mechanisms are absent among primates (but see below). Maternal relatives also do not hesitate to mate when they happen to be unfamiliar with each other (Kuester et al. 1994)—just as Oedipus did not hesitate to marry his own unfamiliar mother.

Mating inhibitions between familiar but unrelated individuals provide the counterproof for the Westermarck hypothesis, and indeed, among humans, Westermarck's idea received the strongest support from such cases (reviewed by Wolf 1995). But the evidence is not limited to humans. Cross-fostered rhesus monkeys, for example, practice avoidance of mating with members of their foster matriline (Smith 1995), just as "encultured" female chimpanzees (who sometimes appear to be attracted to strange human males) strongly avoid male members of their human foster families when in estrus (Fouts & Mills 1997; see also Coe et al. 1979 for unrelated chimpanzees raised together). Finally, among savanna baboons (Smuts 1985), Japanese macaques (Wolfe 1979), and Barbary macaques (Kuester et al. 1994), females who had had close relationships with particular but not necessarily related adult males during infancy showed no sexual behavior with them when the females reached sexual maturity.

All these data strongly suggest that familiarity (which at least among matrilineal relatives usually correlates with genetic relatedness), but not genetic relatedness per se, breeds sexual indifference. Nevertheless, some observations appear to suggest that other mechanisms may be at work. Smith (1995), for example, suggested that rhesus monkeys might be able to use phenotypic cues to avoid father-daughter mating, because offspring resulting from such incestuous dyads were underrepresented in his study. Given that in the closely related Barbary macaque the incidence of father-daughter matings quickly reached 100% after only two years of co-residence as sexually mature individuals (Kuester et al. 1994), this interpretation appears unlikely, however. Rather, Smith's finding may be more parsimoniously explained by his observation that in this colony natal females preferred to reproduce with young natal males that were rising in rank (Smith 1994).

Phenotypic matching has also been invoked as a possible explanation for weak consortship cohesiveness among savanna baboon paternal half siblings (Alberts 1999). Alberts' study, however, provides much stronger support for Westermarck's hypothesis than for any other hypothesis because most paternal half siblings were also members of the same age cohort, and members of the same age cohort avoided each other much more consistently (both in terms of consortship frequency and consortship cohesiveness) than individuals belonging to different age cohorts. In fact, while scores for consortship cohesiveness among paternal half siblings and unrelated pairs overlapped, this was not the case for members of the same and different age cohorts (see Alberts' figure 2).

Recent observations of rhesus macaques appear to support Alberts' hypothesis of paternal kin discrimination via phenotype matching, however (Widdig et al. 2001). According to this study, paternal half sisters scored significantly higher in all measures of affiliation than nonkin, and although this effect was most pronounced among peers, it was not restricted to them, suggesting that phenotype matching might also be involved in kin discrimination. Nevertheless, in light of the fact that neither rhesus macaques (Smith 1995) nor Barbary macaques (Kuester et al. 1994) appear to avoid paternal half siblings as mating partners, this result is somewhat surprising and clearly requires further investigation.

Like savanna baboons, however, Barbary macaques do not seem to be sexually interested in same-aged natal peers. Preliminary analyses in the Salem Barbary macaque colony revealed that age differences between natal fathers and mothers were significantly larger than between nonnatal fathers and natal mothers. Indeed, no female in this study ever produced an offspring with a same-aged natal peer (Paul & Kuester 1998). Thus, in order to avoid sibling incest, both baboons and Barbary macaques appear to follow the simple Westermarckian rule "discriminate against natal members of your own age cohort" (Alberts 1999). Age proximity, therefore, clearly has a regulating effect on social relationships (Widdig et al. 2001) and, as a result, on mating preferences (Alberts 1999).

Despite the overwhelming evidence in support of the Westermarck hypothesis (see also Pusey 1990, Wolf 1995), some authors still remain reluctant about its general application for mating inhibitions among primates. Starin (2001), for example, observed that among red colobus monkeys, immature males "on three occasions" attempted to mount their mothers who, in contrast to other females, immediately discouraged them from doing so. Moreover, since older males did not show any interest in their mothers, but "freely" mated with other females they had known since birth, Starin (2001: 463), reminiscent of Freud, argued "that individuals who live together . . . do develop and retain sexual interest in each other unless they are inhibited by countervailing social and demographic pressures," and "that the origin of, at least some forms of the incest taboo has a learned component." Apart from the fact that this hypothesis does not explain why mothers discouraged apparently rare advances of their immature sons, it would be surprising to see that male infant colobus monkeys had no special relationship with their mothers. Sheer acquaintance with other females does not suffice to breed sexual indifference.

Finally, Di Bitetti & Janson (2001) suggested that Westermarck's hypothesis might be an insufficient explanation for the strong mating inhibition between young female capuchins and their putative father (the alpha male), since these females did not hesitate to mate with the beta male, which they had also known since birth. But in this case as well, the Westermarck hypothesis appears to be the most parsimonious explanation; although capuchins are not known for showing intense male care for infants, infants are commonly left with the dominant male, who allows them to huddle or play near him (Robinson & Janson 1987). At food trees, the dominant male tolerates only those immatures he could have sired (Janson 1984). Thus, it appears that infant capuchin monkeys are much more familiar with the dominant male than with other males of their group. The same is true for gorillas, where close associations between dominant males and infants is the most likely explanation for later father-daughter incest avoidance (Stewart & Harcourt 1987, Watts 1990, Meder 1995).

Conclusion

Contrary to earlier assumptions, a wealth of evidence now indicates that kinship plays a profound role in regulating patterns of mating and reproduction among not only human but also nonhuman primates. Given that inbreeding is, under most circumstances, costly, the existence of evolved psychological mechanisms designed specifically to solve the problem of close inbreeding avoidance (as has been first proposed by Edward Westermarck) is not at all surprising. Moreover, there is no evidence that primates prefer opposite-sex individuals as mates that are at least as closely related as first cousins, as predicted by optimal outbreeding theories. Instead, at least females appear to follow the footsteps of the late Bill Hamilton (1993); as large, long-lived, and slow-breeding beings that are highly vulnerable to their much faster breeding pathogens, they prefer relative strangers—strangers who come from as far away, ideally, as possible.

"Ideally" remains an important restriction. While it is true that both male and female primates readily mate with strangers, they are clearly limited in their ability to discriminate between kin. Are there species differences in the degree of kin discrimination and inbreeding avoidance among primates? In an earlier article, Paul and Kuester (1985) hypothesized that behavioral mechanisms for the avoidance of close inbreeding might be more strongly developed among species such as Barbary macaques, where due to reduced natal dispersal, relatives encounter each other more frequently as potential mates than do, for example, rhesus or Japanese macaques, species in which almost all males leave their natal group at an early age (see also Walters 1987, Pusey 1990). Although comparative data necessary to test this hypothesis more rigorously are not yet available, based on current knowledge this hypothesis seems rather unlikely, however. Mating inhibitions among Barbary macaques, as well as among other primates, are best explained by the Westermarck hypothesis, and there is no indication that Barbary macaques are better able to use other cues to identify their kin than other primates are. If so, and if inbreeding avoidance among Barbary macaques was more strongly developed than among rhesus or Japanese macaques, this would imply that kinship ties among Barbary macaques were also more intimate than among those species. Indeed, this approach explains why in some species such as gorillas and capuchin monkeys, but not in others such as Barbary macaques, father-daughter matings are remarkably absent. At present, however, there is little evidence suggesting that bonds between maternal cousins or aunts and nephews are more strongly developed among Barbary macaques than among other macaque species. In fact, Barbary macaque infants spend much more time with unrelated individuals than do rhesus or Japanese macaque infants, suggesting that in this species early childhood association with maternal kin may be even weaker than in rhesus or Japanese macaques. Nevertheless, these bonds appear to be intimate enough to result in the Westermarck effect—the "remarkable absence of erotic feelings" between even rather distant maternal relatives.

Since the costs of inbreeding are much higher for females than for males, it is also not surprising to see that females are, in general, much more averse to mating with close relatives than males are. One might argue, then, that females should be more likely to disperse from their natal group than males (Waser et al. 1986; see also Moore 1993). However, among mammals in general, and among most primates, the opposite is true. Male dispersal is almost universal, and among many species male-biased dispersal is the norm (Pusey & Packer 1987). Such a pattern is to be expected, however, since in most cases females will have few difficulties finding an unrelated mate. If they have difficulties, they tend to disperse (see Pusey & Packer 1987, Clutton-Brock 1989, Moore 1993, chapter 4). For a male, whose reproductive success depends on the number of eggs he can fertilize, the situation is quite different, because if females avoid matings with closely related males (and most males, as well, do not show much interest in their female relatives), the number of potential mates in his natal group can be drastically reduced, which obviously constitutes a strong incentive for males to disperse (Perrin & Goudet 2001; see also Murray & Smith 1983, Pereira & Weiss 1991, Moore 1993, among others). Thus, although minimizing the risk of inbreeding may not be the only ultimate reason for dispersal, inbreeding avoidance, female mate choice, and male mating strategies appear to be important determinants of social structure.

Finally, to paraphrase Westermarck, it would be surprising if a law that holds good for the rest of the primate order does not also apply to humans. Indeed, the human data tell much the same story as the nonhuman primate data; early association inhibits later sexual attraction (see especially Wolf's fascinating 1995 volume). According to some researchers, however, Westermarck's original hypothesis may require a slight revision. Questionnaires on sexual experiences between siblings revealed that early childhood association has an inhibiting effect on genital intercourse, but not on other, nonprocreative forms of sexual behavior (Bevc & Silverman 1993, 2000; see also Nesse & Lloyd 1992, who maintain that children have intense conscious and unconscious sexual wishes toward their parents). Although it is not clear whether these interactions were an expression of "erotic feelings" or something else, Bevc and Silverman (2000) conclude that early childhood association does not deter sexual interest per se, but rather acts in a more precise manner by creating a specific barrier against intercourse. Reports on incestuous behavior among nonhuman primates commonly also consist of nonprocreative forms of sexual behavior, often performed by immature individuals. Among nonhuman primates, such interactions do not seem to be the norm, and they may not be among humans either. Bevc and Silverman state that individuals with or without such experiences were sought in their studies; hence their sampling procedures were not designed to generate a random sample. Therefore, whether the Westermarck hypothesis requires revision or not remains an open question. In addition, the precise way in which the Westermarck effect works and whether primates are able to recognize kin by other mechanisms remain important areas for further investigation.

Acknowledgments   We are grateful to Carol Berman and Bernard Chapais for inviting us to contribute to this volume. We thank the editors and an anonymous reviewer for constructive comments on an earlier version of the manuscript. Research on Barbary macaques by J. K. and A. P. was supported by grants from the Deutsche Forschungsgemeinschaft (An 131/1–6, Vo 124/15–1, 18–1, Pa 408/2–1). The support from W. Angst, E. Merz, G. de Turckheim, and the late C. Vogel is greatly acknowledged.

References

Abbott, D. H. 1993. Social conflict and reproductive suppression in marmoset and tamarin monkeys. In: *Primate Social Conflict* (Ed. by W. A. Mason & S. P. Mendoza), pp. 331–372. Albany, NY: SUNY Press.

Aberle, D. F., Bronfenbrenner, V., Hess, E. H., Miller, D. R., Schneider, D. M., & Spuhler, J. M. 1963. The incest taboo and the mating patterns of animals. *Am. Anthropol.*, 65, 253–265.

Alberts, S. C. 1999. Paternal kin discrimination in wild baboons. *Proc. Royal Soc. Lond., B,* 266, 1501–1506.

Alberts, S. C. & Altmann, J. 1995. Balancing costs and opportunities: dispersal in male baboons. *Am. Nat.*, 145, 279–306.

Anderson, C. 1986. Female age: male preference and reproductive success in primates. *Int. J. Primatol.*, 7, 305–326.

Baker, J. V., Abbott, D. H., & Saltzman, W. 1999. Social determinants of reproductive failure in male common marmosets housed with their natal family. *Anim. Behav.*, 58, 501–513.

Bateson, P. 1983. Optimal outbreeding. In: *Mate Choice* (Ed. by P. Bateson), pp. 257–277. Cambridge:Cambridge University Press.

Bateson, P. P. G. 1982. Preferences for cousins in Japanese quail. *Nature,* 295, 236–237.

Bauers, K. A. & Hearn, J. P. 1994. Patterns of paternity in relation to male social rank in the stumptailed macaque, *Macaca arctoides*. *Behaviour,* 129, 149–176.

Baxter, M. J. & Fedigan, L. M. 1979. Grooming and consort selection in a troop of Japanese monkeys (*Macaca fuscata*). *Arch. Sex. Behav.*, 8, 445–458.

Bélisle, P. & Chapais, B. 2001. Tolerated co-feeding in relation to degree of kinship in Japanese macaques. *Behaviour,* 138, 487–509.

Bernstein, I. S. 1988. Kinship and behavior in nonhuman primates. *Behav. Genetics,* 18, 511–524.

Bevc, I. & Silverman, I. 1993. Early proximity and intimacy between siblings and incestuous behavior: a test of the Westermarck hypothesis. *Ethol. Sociobiol.,* 14, 171–181.

Bevc, I. & Silverman, I. 2000. Early separation and sibling incest: a test of the revised Westermarck theory. *Evol. Hum. Behav.,* 21, 151–161.

Bischof, N. 1972. The biological foundations of the incest taboo. *Soc. Sci. Info.,* 11, 7–36.

Bischof, N. 1985. *Das Rätsel Ödipus: Die Biologischen Wurzeln des Urkonfliktes von Intimität und Autonomie.* Munich: Piper.

Chapais, B. 2001. Primate nepotism: what is the explanatory value of kin selection? *Int. J. Primatol.,* 22, 203–229.

Chapais, B. & Mignault, C. 1991. Homosexual incest avoidance among females in captive Japanese macaques. *Am. J. Primatol.,* 23, 171–183.

Chapais, B., Gauthier, C., Prud'homme, J., & Vasey, P. 1997. Relatedness threshold for nepotism in Japanese macaques. *Anim. Behav.,* 53, 1089–1101.

Chapais, B., Savard, L., & Gauthier, C. 2001. Kin selection and the distribution of altruism in relation to degree of kinship in Japanese macaques (*Macaca fuscata*). *Behav. Ecol. Sociobiol.,* 49, 493–502.

Clutton-Brock, T. H. 1989. Female transfer and inbreeding avoidance in social mammals. *Nature,* 337, 70–72.

Coe, C. L., Conolly, A. C., Kraemer, H. C., & Levine, S. 1979. Reproductive development and behavior of captive female chimpanzees. *Primates,* 20, 571–582.

Constable, J. L., Ashley, M. V., Goodall, J., & Pusey, A. E. 2001. Noninvasive paternity assignment in Gombe chimpanzees. *Mol. Ecol.,* 10, 1279–1300.

Crnokrak, P. & Roff, D. A. 1999. Inbreeding depression in the wild. *Heredity,* 83, 260–270.

Darwin, C. 1868. *The Variation of Animals and Plants Under Domestication.* London: John Murray.

Darwin, C. 1871. *The Descent of Man, and Selection in Relation to Sex.* London: John Murray.

Di Bitetti, M. S. & Janson, C. H. 2001. Reproductive socioecology of tufted capuchins (*Cebus apella nigritus*) in northeastern Argentina. *Int. J. Primatol.,* 22, 127–142.

Dickeman, M. 1992. Phylogenetic fallacies and sexual oppression: a review article on *Pedophilia: Biosocial Dimensions,* Jay R. Feierman, ed. New York: Springer Verlag, 1990. *Hum. Nat.,* 3, 71–87.

Enomoto, T. 1978. On social preference in sexual behavior of Japanese monkeys (*Macaca fuscata*). *J. Hum. Evol.,* 7, 283–293.

Erhart, E., Coelho, A., & Bramblett, C. 1997. Kin recognition by paternal half-siblings in captive *Papio cynocephalus. Am. J. Primatol.,* 43, 147–157.

Fouts, R. & Mills, S. T. 1997. *Next of Kin: What Chimpanzees Have Taught Me About Who We Are.* New York: William Morrow.

Frazer, J. G. 1910. *Totemism and Exogamy.* London: Macmillan.

Fredrickson, W. T. & Sackett, G. P. 1984. Kin preferences in primates (*Macaca nemestrina*): relatedness or familiarity? *J. Comp. Psychol.,* 98, 29–34.

Freud, S. [1913] 1950. *Totem and Taboo.* Reprint, London: Routledge.

Freud, S. [1920] 1953. *A General Introduction to Psychoanalysis.* Reprint, New York: Pocket Books.

Fuentes, A. 2000. Hylobatid communities: changing views on pair bonding and social organization in hominoids. *Ybk. Phys. Anthropol.,* 43, 33–60.

Goodall, J. 1986. *The Chimpanzees of Gombe.* Cambridge, MA: Harvard University Press.

Grafen, A. 1990. Do animals really recognize kin? *Anim. Behav.,* 39, 42–54.

Haig, D. 1999. Asymmetric relations: internal conflicts and the horror of incest. *Evol. Hum. Behav.,* 20, 83–98.

Hamilton, W. D. 1967. Extraordinary sex ratios. *Science,* 156, 477–488.

Hamilton, W. D. 1993. Inbreeding in Egypt and in this book: a childish perspective. In: *The*

Natural History of Inbreeding and Outbreeding: Theoretical and Empirical Perspectives (Ed. by N. Thornhill), pp. 429–450. Chicago: University of Chicago Press.

Hamilton, W. D., Axelrod, R., & Tanese, R. 1990. Sexual reproduction as an adaptation to resist parasites. *Proc. Natl. Acad. Sci. USA,* 87, 3566–3573.

Imanishi, K. 1961. The origin of the human family: a primatological approach. *Jpn. J. Ethnol.,* 25, 119–138 (in Japanese). Reprinted 1965 in: *Japanese Monkeys: A Collection of Translations* (Ed. by S. A. Altmann), pp. 113–140. Atlanta, GA: S. A. Altmann.

Inoue, M., Takenata, A., Tanaka, S., Kominami, R., & Tanaka, O. 1990. Paternity discrimination in a Japanese macaque group by DNA fingerprinting. *Primates,* 31, 563–570.

Itani, J. & Tokuda, K. 1958. Japanese monkeys of Koshima islet. In: *Nihon-Dobutsuki (Social Life of Animals in Japan), Vol. 3* (Ed. by K. Imanishi), pp. 1–233. Tokyo: Kobunsha (in Japanese).

Itoigawa, N., Negayama, K., & Kondo, K. 1981. Experimental study on sexual behavior between mother and son in Japanese monkeys (*Macaca fuscata*). *Primates,* 22, 494–502.

Janson, C. H. 1984. Female choice and mating system of the brown capuchin monkey *Cebus apella* (Primates: Cebidae). *Z. Tierpsychol.,* 65, 177–200.

Kano, T. 1992. *The Last Ape: Pygmy Chimpanzee Behavior and Ecology.* Stanford, CA: Stanford University Press.

Kaufman, J. H. 1965. A three year study of mating behavior in a free-ranging band of rhesus monkeys. *Ecology,* 46, 500–512.

König, A., Rothe, H., Siess, M., Darms, K., Gröger, D., Radespiel, U., & Rock, J. 1988. Reproductive reorganization in incomplete groups of the common marmoset (*Callithrix jacchus*) under laboratory conditions. *Z. Säugetierkunde,* 53, 1–6.

Kuester, J., Paul, A., & Arnemann, J. 1994. Kinship, familiarity and mating avoidance in Barbary macaques, *Macaca sylvanus. Anim. Behav.,* 48, 1183–1194.

Leavitt, G. C. 1990. Sociobiological explanations of incest avoidance: a critical review of evidential claims. *Am. Anthropol.,* 92, 971–993.

Leighton, D. R. 1987. Gibbons, territoriality and monogamy. In: *Primate Societies* (Ed. by B. B. Smuts, D. L. Cheney, R. M. Seyfarth, R. W. Wrangham, & T. T. Struhsaker), pp. 135–145. Chicago: University of Chicago Press.

Lévi-Strauss, C. 1969. *The Elementary Structure of Kinship.* Boston: Beacon.

Loy, J. 1971. Estrous behavior of free-ranging rhesus monkeys (*Macaca mulatta*). *Primates,* 12, 1–31.

MacKenzie, M. M., McGrew, W. C., & Chamove, A. S. 1985. Social preferences in stumptailed macaques (*Macaca arctoides*): effects of companionship, kinship, and rearing. *Dev. Psychobiol.,* 18, 115–123.

Malinowski, B. 1931. Culture. In: *Encyclopedia of the Social Sciences, Vol. 4* (Ed. by E. R. A. Seligman), pp. 621–646. London: Macmillan.

Manson, J. H. & Perry, S. E. 1993. Inbreeding avoidance in rhesus macaques: whose choice? *Am. J. Primatol.,* 90, 335–344.

May, R. M. 1979. When to be incestuous. *Nature,* 279, 192–194.

Mayr, E. 1982. *The Growth of Biological Thought: Diversity, Evolution and Inheritance.* Cambridge, MA: Harvard University Press.

Meagher, S., Penn, D. J., & Potts, W. K. 2000. Male-male competition magnifies inbreeding depression in wild house mice. *Proc. Natl. Acad. Sci. USA,* 97, 3324–3329.

Meder, A. 1995. Die Rolle von Vertrautheit, Alter, Dominanz und Aufzuchtsweise bei der Fortpflanzung von Gorillas in Zoos. *Zoologische Garten N.F.,* 65, 153–164.

Ménard, N. & Vallet, D. 1996. Demography and ecology of Barbary macaques (*Macaca sylvanus*) in two different habitats. In: *Evolution and Ecology of Macaque Societies*

(Ed. by J. E. Fa & D. G. Lindburg), pp. 106–131. Cambridge: Cambridge University Press.

Missakian, E. M. 1973. Genealogical mating activity in free-ranging groups of rhesus monkeys (*Macaca mulatta*) in Cayo Santiago. *Behaviour*, 45, 225–241.

Mitton, J. B. 1993. Theory and data pertinent to the relationship between heterozygosity and fitness. In: *The Natural History of Inbreeding and Outbreeding: Theoretical and Empirical Perspectives* (Ed. by N. Thornhill), pp. 17–41. Chicago: University of Chicago Press.

Moore, J. 1993. Inbreeding and outbreeding in primates: what's wrong with "the dispersing sex"? In: *The Natural History of Inbreeding and Outbreeding: Theoretical and Empirical Perspectives* (Ed. by N. Thornhill), pp. 392–426. Chicago: University of Chicago Press.

Moore, J. & Ali, R. 1984. Are dispersal and inbreeding avoidance related? *Anim. Behav.*, 32, 94–112.

Murray, R. D. & Smith, E. O. 1983. The role of dominance and intrafamilial bonding in the avoidance of close inbreeding. *J. Hum. Evol.*, 12, 481–486.

Nesse, R. M. & Lloyd, A. T. 1992. The evolution of psychodynamic mechanisms. In: *The Adapted Mind: Evolutionary Psychology and the Generation of Culture* (Ed. by J. H. Barkow, L. Cosmides, & J. Tooby), pp. 601–624. New York: Oxford University Press.

Nieminen, M., Singer, M. C., Fortelius, W., Schöps, K., & Hanski, I. 2001. Experimental confirmation that inbreeding depression increases extinction risk in butterfly populations. *Am. Nat.*, 157, 237–244.

Nishida, T. 1990. A quarter century of research in the Mahale Mountains: an overview. In: *The Chimpanzees of the Mahale Mountains* (Ed. by T. Nishida), pp. 3–35. Tokyo: University of Tokyo Press.

Oka, T. & Takenaka, O. 2001. Wild gibbons' parentage tested by non-invasive DNA sampling and PCR-amplified polymorphic microsatellites. *Primates*, 42, 67–73.

O'Riain, M. J. & Braude, S. 2001. Inbreeding versus outbreeding in captive and wild populations of naked mole-rats. In: *Dispersal* (Ed. by J. Clobert, E. Danching, A. A. Dohnt, & J. D. Nichols), pp. 143–155. New York: Oxford University Press.

Packer, C. 1979. Intertroop transfer and inbreeding avoidance in *Papio anubis*. *Anim. Behav.*, 27, 1–36.

Parr, L. A. & de Waal, F. B. M. 1999. Visual kin recognition in chimpanzees. *Nature*, 399, 647–648.

Paul, A. & Kuester, J. 1985. Intergroup transfer and incest avoidance in semifree-ranging Barbary macaques (*Macaca sylvanus*) at Salem (FRG). *Am. J. Primatol.*, 8, 317–322.

Paul, A. & Kuester, J. 1998. Mate choice in Barbary macaques: what do mothers and fathers have in common? *Folia Primatol.*, 69, 213–214.

Penn, D. J. 2002. The scent of genetic compatibility: sexual selection and the major histocompatibility complex. *Ethology*, 108, 1–21.

Pereira, M. E. & Weiss, M. L. 1991. Female mate choice, male migration, and the threat of infanticide in ringtailed lemurs. *Behav. Ecol. Sociobiol.*, 28, 141–152.

Perrin, N. & Goudet, J. 2001. Inbreeding, kinship and the evolution of natal dispersal. In: *Dispersal* (Ed. by J. Clobert, E. Danching, A. A. Dohnt, & J. D. Nichols), pp. 123–142. New York: Oxford University Press.

Pusey, A. 1990. Mechanisms of inbreeding avoidance in nonhuman primates. In: *Pedophilia: Biosocial Dimensions* (Ed. by J. R. Feiermann), pp. 201–220. New York: Springer.

Pusey, A. & Wolf, M. 1996. Inbreeding avoidance in animals. *Trends Ecol. Evol.*, 11, 201–206.

Pusey, A. E. 1980. Inbreeding avoidance in chimpanzees. *Anim. Behav.*, 28, 543–552.

Pusey, A. E. & Packer, C. 1987. Dispersal and philopatry. In: *Primate Societies* (Ed. by B. B. Smuts, D. L. Cheney, R. M. Seyfarth, R. W. Wrangham, & T. T. Struhsaker), pp. 250–266. Chicago: University of Chicago Press.

Robinson, J. G. & Janson, C. H. 1987. Capuchins, squirrel monkeys, and atelines: socioecological convergence with Old World primates. In: *Primate Societies* (Ed. by B. B. Smuts, D. L. Cheney, R. M. Seyfarth, R. W. Wrangham, & T. T. Struhsaker), pp. 69–82. Chicago: University of Chicago Press.

Rose, S., Kamin, L. J., & Lewontin, R. C. 1984. *Not in Our Genes*. London: Penguin.

Rothe, H. 1975. Some aspects of sexuality and reproduction in groups of captive marmosets (*Callithrix jacchus*). *Z. Tierpsychol.*, 37, 255–273.

Sackett, G. P. & Fredrickson, W. T. 1987. Social preferences by pigtail macaques: familiarity versus degree and type of kinship. *Anim. Behav.*, 35, 603–607

Sade, D. S. 1968. Inhibition of son-mother mating among free-ranging rhesus monkeys. *Sci. Psychoanal.*, 12, 18–38.

Sade, D. S., Rhodes, D. L., Loy, J., Hausfater, G., Breuggeman, J. A., Kaplan, J. R., Chepko-Sade, B. D., & Cushing-Kaplan, K. 1984. New findings on incest among free-ranging rhesus monkeys. *Am. J. Phys. Anthropol.*, 63, 212–213.

Sauther, M. L. 1991. Reproductive behavior of free-ranging *Lemur catta* at Beza Mahafali Special Reserve, Madagascar. *Am. J. Phys. Anthropol.*, 84, 463–477.

Savage, A., Giraldo, L. H., Soto, L. H., & Snowdon, C. T. 1996. Demography, group composition and dispersal in wild cotton-top tamarin (*Saguinus oedipus*) groups. *Am. J. Primatol.*, 38, 85–100.

Scott, L. M. 1984. Reproductive behavior of adolescent female baboons (*Papio anubis*) in Kenya. In: *Female Primates: Studies by Women Primatologists* (Ed. by M. Small), pp. 77–100. New York: Alan R. Liss.

Shepher, J. 1983. *Incest: A Biosocial View*. New York: Academic Press.

Shields, W. M. 1982. *Philopatry, Inbreeding Avoidance, and the Evolution of Sex*. Albany, NY: SUNY Press.

Shields, W. M. 1993. The natural and unnatural history of inbreeding and outbreeding. In: *The Natural History of Inbreeding and Outbreeding: Theoretical and Empirical Perspectives* (Ed. by N. Thornhill), pp. 143–169. Chicago: University of Chicago Press.

Small, M. F. & Smith, D. G. 1981. Interactions with infants by full siblings, paternal half-siblings, and nonrelatives in a captive group of rhesus macaques (*Macaca mulatta*). *Am. J. Primatol.*, 1, 91–94.

Smith, D. G. 1994. Male dominance and reproductive success in a captive group of rhesus macaques (*Macaca mulatta*). *Behaviour*, 129, 225–242.

Smith, D. G. 1995. Avoidance of close consanguineous inbreeding in captive groups of rhesus macaques. *Am. J. Primatol.*, 35, 31–40.

Smuts, B. B. 1985. *Sex and Friendship in Baboons*. New York: Aldine.

Soltis, J., Mitsunaga, F., Shimizu, K., Yanagihara, Y., & Nozaki, M. 1999. Female mating strategy in an enclosed group of Japanese macaques. *Am. J. Primatol.*, 47, 263–278.

Starin, E. D. 2001. Patterns of inbreeding avoidance in Temminck's red colobus. *Behaviour*, 138, 453–465.

Stewart, K. J. & Harcourt, A. H. 1987. Gorillas: variation in female relationships. In: *Primate Societies* (Ed. by B. B. Smuts, D. L. Cheney, R. M. Seyfarth, R. W. Wrangham, & T. T. Struhsaker), pp. 155–164. Chicago: University of Chicago Press.

Takahata, Y. 1982. The socio-sexual behavior of Japanese monkeys. *Z. Tierpsychol.*, 59, 89–108.

Takahata, Y., Huffman, M. A., & Bardi, M. 2002. Long-term trends in matrilineal inbreeding among the Japanese macaques of Arashiyama B troop. *Int. J. Primatol.*, 23, 399–410.

Taylor, L. L. & Sussman, R. W. 1985. A preliminary study of kinship and social organization in a semi-free-ranging group of *Lemur catta. Int. J. Primatol.*, 6, 601–614.

Thornhill, N. 1991. An evolutionary analysis of rules regulating human inbreeding and marriage. *Behav. Brain Sci.*, 14, 247–293.

Thornhill, N. (ed.) 1993. *The Natural History of Inbreeding and Outbreeding: Theoretical and Empirical Perspectives*. Chicago: University of Chicago Press.

Tokuda, K. 1961–1962. A study on sexual behavior in the Japanese monkey troop. *Primates*, 3, 1–40.

Trivers, R. L. 1972. Parental investment and sexual selection. In: *Sexual Selection and the Descent of Man, 1871–1971* (Ed. by B. Campbell), pp. 136–179. Chicago: Aldine.

Tutin, C. E. G. 1979. Mating patterns and reproductive strategies in a community of wild chimpanzees (*Pan troglodytes schweinfurtii*). *Behav. Ecol. Sociobiol.*, 6, 29–38.

Visscher, P. M., Smith, D., Hall, S. J. G., & Williams, J. A. 2001. A viable herd of genetically uniform cattle. *Nature*, 409, 303.

Waldman, B., Frumhoff, P. C., & Sherman, P. W. 1988. Problems of kin recognition. *Trends Ecol. Evol.*, 3, 8–13.

Walters, R. 1987. Kin recognition in nonhuman primates. In: *Kin Recognition in Animals* (Ed. by D. J. C. Fletcher & C. D. Michener), pp. 359–393. New York: Wiley.

Waser, M., Austad, S. N., & Keane, B. 1986. When should animals tolerate inbreeding? *Am. Nat.*, 128, 529–537.

Watts, D. P. 1990. Mountain gorilla life histories, reproductive competition, and socio-sexual behavior and some implications for captive husbandry. *Zoo Biol.*, 9, 185–200.

Weismann, A. 1883. *Über die Vererbung*. Jena: G. Fischer.

Welker, C., Schwibbe, M. H., Schaefer-Witt, C., & Visalberghi, E. 1987. Failure of kin recognition in *Macaca fascicularis. Folia Primatol.*, 49, 216–221.

Werren, J. H. 1993. The evolution of inbreeding in haplodiploid organisms. In: *The Natural History of Inbreeding and Outbreeding: Theoretical and Empirical Perspectives* (Ed. by N. Thornhill), pp. 42–59. Chicago: University of Chicago Press.

Westermarck, E. [1891] 1922. *The History of Human Marriage*. Reprint, London: Macmillan.

Westermarck, E. 1906–1908. *The Origin and Development of Moral Ideas*. London: Macmillan.

Westermarck, E. 1932. *Ethical Relativity*. London: Kegan Paul.

Widdig, A., Nürnberg, P., Krawczak, M., Streich, W. J., & Bercovitch, F. B. 2001. Paternal relatedness and age proximity regulate social relationships among adult female rhesus macaques. *Proc. Natl. Acad. Sci. USA*, 98, 13769–13773.

Wolf, A. P. 1995. *Sexual Attraction and Childhood Association: A Chinese Brief for Edward Westermarck*. Stanford, CA: Stanford University Press.

Wolfe, L. 1979. Behavioral patterns of estrous females of the Arashiyama West troop of Japanese macaques (*Macaca fuscata*). *Primates*, 20, 525–534.

Wright, S. 1977. *Evolution and the Genetics of Populations, Vol. 4*. Chicago: University of Chicago Press.

Wu, H. M. H., Holmes, W. G., Medina, S. R., & Sackett, G. P. 1980. Kin preference in infant *Macaca nemestrina. Nature*, 285, 225–227.

Part IV

Kin Bias: Proximate and Functional Processes

An adult rhesus (*Macaca mulatta*) daughter "admires" her new infant sister as it is held securely by their mother. Persistent bonds between older daughters and their mothers ensure that sisters become familiar partners. Photo by Carol Berman.

13

"Recognizing" Kin: Mechanisms, Media, Minds, Modules, and Muddles

Drew Rendall

> If he could learn to recognize those of his neighbours who really were close rela-
> tives and could devote his beneficial actions to them alone an advantage to inclu-
> sive fitness would at once appear.
>
> —W. D. Hamilton (1964: 21)

The explosion of research on kin-biased behaviors (the observable preferential treatment of genetic relatives) following Hamilton's theory of inclusive fitness has encompassed various issues, but one of special interest has been the proximate (unobservable internal) mechanisms by which kin are actually recognized. After all, the ability to recognize kin is an essential prerequisite to accruing the hypothesized advantage to inclusive fitness. Hence, in addition to comprehensive studies of the form and extent of kin-biased behaviors in diverse animal taxa, the decades following Hamilton's (1964) landmark articles have seen considerable focused research on the specific mechanisms by which organisms recognize their kin.

Research on primates is oddly out of step in this regard, for very little energy has been devoted to elucidating such mechanisms, despite the fact that kinship has been one of the most enduring foci of primate research over the course of several decades (predating inclusive fitness theory, in fact) such that it is now widely held to be one of the central organizing principles of primate social life. Of course, there may be good reasons for this mechanistic hole in our research on primate kinship. Many primate species are long-lived, socially complex organisms, which makes it practically (and ethically) challenging to conduct the sort of research typically required to test recognition mechanisms (see figure 13.1). It is also possible that we have grown complacent about the mechanisms by which kin are recognized, the decades of research documenting the ubiquity of kin-biased behaviors lulling us into assuming simply that they do.

Figure 13.1. A pair of chacma baboon *(Papio cynocephalus ursinus)* sisters from South Africa, relaxing together during a midday rest period. Biological kinship plays a central role in structuring primate social life, yet the mechanisms by which the animals recognize their kin remain poorly understood. Photo courtesy of Karen Rendall.

Understandable or not, the lacuna is not ideal. Increasingly, the general adaptive niche of primates is cast as a social one (Jolly 1966, Humphrey 1976). Challenges associated with managing a network of different social relationships—many of which stem from a web of overlapping kin relations—are being emphasized as the primary selective pressure shaping the behavioral and mental evolution of primates. Thus, it is not only supremely paradoxical that we know so little about how primates recognize their kin, but it is also (or at least should be) more than a little worrisome. We are in the vulnerable position of having the entire edifice of a now mature discipline constructed on a "black box" foundation. So, to protect and bolster our functional hypotheses of primate sociality and cognition, we are really obliged to probe inside this mechanistic box and (notwithstanding the practical challenges) do our best to unpack it a bit.

In what follows, I provide a brief review of the various mechanisms of kin recognition suggested from research on other taxa and consider some unresolved conceptual ambiguities in them, as a prelude to reviewing the evidence available for primates (excluding humans). I then summarize the research on primates, including work on the potential cues used to mediate recognition in different sensory modalities. Finally, I consider a few specific proximate and functional issues that flow naturally from this evidence and that are in some way also connected to questions about the basic architecture of primate "minds" as the probable locus of any potential recognition mechanisms.

Mechanisms

A great deal of work has been done on mechanisms of kin recognition in nonprimates. This active field has been subject to regular review and debate (e.g., Byers & Bekoff 1986, Grafen 1990, Blaustein et al. 1991; see reviews by Holmes & Sherman 1983, Hepper 1986, Sherman et al. 1997, and edited volumes by Fletcher & Michener 1987, Hepper 1991a). While research on a wide range of species has uncovered a host of different processes, the general mechanisms by which kin are recognized are typically categorized into four broad classes, organized here in order of the increasing directness (or specificity) with which they are thought to discriminate genetic relatedness:

1. *Spatial distribution*—One very indirect mechanism of kin recognition involves simply recognizing as kin those individuals encountered in a particular location. This means of recognition occurs where environments are extremely heterogeneous with respect to specific behaviors. For example, in some bird species, parents use spatial information associated with nest sites to recognize and invest in their offspring. That parents are using only spatial cues (at least in the early stages) is clear from the fact that they will ignore their own chicks if placed just outside the nest but accept and feed foreign chicks if placed inside the nest (Beecher et al. 1981). Brood parasitic species, which lay their eggs in the nest of a different species, are capitalizing on their host species' reliance on this form of recognition.

2. *Familiarity*—Another indirect mechanism involves recognizing as kin those individuals with whom one has become most familiar during development. The typical method used to affirm it involves some form of cross-fostering experiment in which nonkin reared together are found to prefer one another over kin reared apart (e.g., Holmes & Sherman 1982).

3. *Phenotype matching*—A more direct mechansim is thought to involve recognizing as kin those individuals whose phenotype best matches some kind of "template" of kin. It is inferred where organisms recognize kin in the absence of prior experience with them, or discriminate among equally familiar but differentially related kin. For example, among ground squirrels (*Spermophilus* spp.), related females reared apart (i.e., unfamiliar kin) are less tolerant than related females reared together (familiar kin), but they are more tolerant than unrelated and unfamiliar females. And within litters in which some offspring have been sired by different males, full sisters are more cooperative than half sisters despite being equally familiar (Holmes & Sherman 1982). Both results suggest that kinship can sometimes be assessed on the basis of phenotypic markers of underlying genetic relatedness.

Referents for the phenotypic template of kin may vary. In mammals, the extended period of mother-infant association makes one's mother a natural referent for phenotypic comparisons of unfamiliar maternal kin encountered later in life. Organisms might also use themselves as the referent (reviewed in Hauber & Sherman 2001). Mateo and Johnson (2000) report such a case of self-referent phenotype matching—also dubbed the "armpit effect" (Dawkins 1976)—in golden hamsters (*Mesocricetus auratus*). Characteristics of the phenotype used in matching may also vary and be genetic or environmental in origin, or some combination. For example, in honey bees (*Apis mellifera*), nest mates are recognized and permitted access to the hive based on olfactory cues associated with the comb wax that are present on all bees (resulting from their activity in the hive) and reflect a blend of salivary secretions and environmental materials combined during the manufacture of the wax (Breed 1998).

4. *Direct genetic detection*—An even more specific mechanism involves the ability to detect genetic relatedness directly. While theoretically possible, the reality of such a mechanism is often debated because its implementation seems to require so-called recognition alleles that simultaneously code for a reliable phenotypic cue, the recognition of the same cue in kin, and the tendency to behave differentially toward others on the basis of cue possession. This combination has seemed a heavy burden for a single gene (or complex) and one that is also ripe for exploitation by individuals who evolve only the first component. Plausible or not, such a mechanism is proposed to account for the avoidance of self-pollination in many plants and to influence behavior in at least one animal species (Grosberg & Quinn 1986). In the marine invertebrate *Botryllus schlosseri*, individuals settle on rocks and fuse with other individuals to form large colonies. Settlement and fusion depend on the genetic makeup of neighbors, specifically whether they share the same allele at a single locus of the major histocompatibility complex (MHC). Matching at this locus promotes fusion and growth of the colony, and only individuals who match at this locus will fuse. Hence, alleles at this locus appear to govern MHC phenotype, its recognition in others, and the tendency to fuse or not based on compatibility; in short, all of the components of so-called recognition alleles.

Mechanism Muddles

While widely recognized and used to organize research and theorizing on kin recognition, these four classes of mechanism are ambiguous in several respects. To begin with, although apparently representing a continuum of specificity with which genetic kinship is recognized, from quite indirect (1) to quite direct (4), the mechanisms are, in fact, all indirect. As a result, they can be (and have been) critiqued for the extent to which they really represent mechanisms for *recognizing* kinship. At one extreme, reliance on spatial distribution is clearly an indirect means of recognizing kin, and what is recognized is clearly not kinship per se but a reliable correlate. The same is obviously true for the mechanism of familiarity. However, at the other apparently direct extreme, where the recognition is clearly genetically mediated, it is just as clear (at least for the single proposed animal case) that the recognition of kinship is indirect and in fact not really about kinship either. Thus, experiments with *B. schlosseri* show that unrelated individuals who happen to share MHC alleles will also fuse, while related individuals who happen not to share MHC alleles will not fuse. What is recognized, then, is allele sharing at a single locus, not kinship per se. What allows MHC allele sharing to pick out relatives under natural conditions is the high allelic diversity at this locus combined with the organisms' limited dispersal distances, which means that the individuals within a local area matching at this locus are almost always going to be kin.

The upshot is that there is a straightforward (though often confused) distinction between purely mechanistic conceptions of kin recognition mechanisms and more functional ones. Mechanisms that actually identify and discriminate degrees of genetic relatedness (so-called true kin recognition mechanisms) appear rare. Far more common are mechanisms that simply function that way under natural conditions because they use as proxies variables that—owing to regularities in the environment—reliably correlate with kinship and sometimes degrees of it. If this point appears disappointingly simple or crude, given the centrality of kinship discriminations to inclusive fitness theory, it is worth remembering that the theoretical problems we face in trying to account for organismal behavior are not necessarily isomorphic with the practical problems facing the organisms themselves. In other words, although discriminating kinship may be central to our accounts of animal behavior, it may not be central to animal behavior. In fact, much of the elegance of natural selection lies in its ability to capitalize on regularities in the environment to obviate the need for special adaptive mechanisms in the organism.

Notwithstanding these ambiguities, it is also the case that the several classes of mechanism are not as distinct as they first appear (Blaustein 1983, Blaustein et al. 1987, Fletcher 1987, Blaustein & Porter 1990, Hepper 1991b). For example, it may often be difficult to distinguish between phenotype matching and direct genetic detection. If the template for matching in the former is "built in," the two are in fact the same. But even where the template in phenotype matching is not built in but acquired, the interpretation may be problematic, particularly if the referent for the acquired template is the organism itself. Here it is difficult to deny the organism access to its own phenotype and so practically difficult to distinguish an acquired self-referent form of phenotype matching from a mechanism of direct genetic detection. Further complicating matters, acquired self-referent phenotype matching may be indiscernible—both in practice and in principle—from familiarity via developmental association. That is, if an organism (ego) assesses the relatedness of others based on phenotypic cues they share with ego, cues that ego can in fact expose itself to during development (e.g., the smell of its own armpit), then it may be effectively discriminating unfamiliar others on the basis of their relative familiarity—that is, how similar to ego they smell, or more to the point, how little they *seem* to smell!

Given such ambiguities, it becomes difficult to label many forms of recognition. Consider the honeybees described earlier, where kin discriminations are based on an acquired phenotypic template resulting from a combination of organismal and environmental cues. Is this a form of phenotype matching or familiarity discrimination? In principle, the key distinction between phenotype matching and familiarity is that in the latter no specific aspect of the phenotype is used across the various individuals recognized. Rather, recognition is based on the idiosyncrasies of individual phenotypes. In contrast, phenotype matching (and recognition alleles) involves discriminating variation in the same phenotypic feature across all individuals. But even this distinction gets fuzzy. Once the idiosyncratic phenotypic cues of others are learned, they can surely be used to compare with (match) the phenotypes of unfamiliar individuals who share such features to a greater or lesser degree due to kinship. Casual human experience suggests that this is certainly possible, and Hepper (1991b) describes a case of exactly this phenomenon in rats (*Rattus norvegicus*). Cross-fostering experiments show that rats learn their kin based on developmental association (they prefer unrelated littermates over unfamiliar kin), but they also generalize to prefer the unfamiliar collateral relatives of their (unrelated) littermates over unfamiliar and unrelated animals, suggesting an ability to match individualistic phenotypes using what must amount to a gradient of familiarity (see also Holmes 1986, Sun & Muller-Schwarze 1997).

There is no easy fix here, but perhaps the ambiguities are ultimately moot. In the end, what matters is not what mechanism label we apply in any particular case but a complete description of the types of cues used to recognize kin, their origins, the processes by which they are perceived by others, and when and how such perceptual processes emerge in the lifetime of individuals.

Mechanisms in Primates

Notwithstanding the mechanism muddles just noted, the research on kin recognition in primates suggests surprisingly few different mechanisms. In fact, although scant, what evidence there is points fairly consistently to the importance of familiarity (see previous reviews by Gouzoules 1984, Walters 1987, Bernstein 1991).

One of the earliest systematic tests of kin recognition in primates suggested the possibility of a mechanism of phenotype matching (Wu et al. 1980). In this study, 16 pigtail macaques (*Macaca nemestrina*) were reared from birth without access to kin to disentangle the normally confounded effects of familiarity and relatedness. In paired tests, subjects later preferred unfamiliar paternal half siblings over unfamiliar nonkin. However, the actual effects were relatively weak, and the study was subsequently criticized on methodological grounds. Moreover, follow-up experiments using the same design but a much larger sample of subjects (total $N = 155$) revealed no preference for kin (either maternal or paternal) over nonkin when both were equally unfamiliar, but instead showed a consistent preference for familiar over unfamiliar animals regardless of their relatedness (Frederickson & Sackett 1984, Sackett & Frederickson 1987).

Five other studies have suggested some mechanism of phenotype matching. Using paternity exclusion methods to identify sires in captive rhesus monkeys and wild baboons, respectively, Berenstein et al. (1981) and Buchan et al. (2003) found that adult males either associated at higher rates with their genetic offspring than with unrelated infants or preferentially supported the former in agonistic disputes as juveniles. However, in the former study, additional analyses revealed that the infants were responsible for the bias—approaching fathers more than unrelated males—and that this was explained by their mothers' proximity to, and hence the infants' familiarity with, these males. A similar mechanism of mother-mediated familiarity has been proposed to account for other preferential father-offspring associations in baboons (Stein 1984; Smuts 1985). (See Gust et al. 1998 for a paternity exclusion study with no evidence for preferential father-offspring association.)

Paternity exclusion analyses were also used by Small and Smith (1981) to examine mothers' responses to infant handling as a function of the relatedness of handlers in captive rhesus monkeys. Across six mother-offspring pairs, full siblings "grabbed" infants more than did paternal half siblings or nonkin, while mothers resisted fewer grabs by paternal half siblings than nonkin though not full siblings. The apparent preferential tolerance of infant grabs made by paternal half siblings as compared to nonkin might be taken to indicate that mothers recognized paternal relatedness. However, the mothers' similar high resistance to the grabs of full siblings and nonkin complicates this interpretation, and suggests instead that some meaningful variation in the quality of what were operationalized simply as grabs might account for the pattern of maternal resistance.

More recently, Widdig et al. (2001) evaluated patterns of social affiliation (spatial proximity, grooming, and social approach) among 34 adult female rhesus monkeys whose maternal relatedness was known and whose paternal relatedness was estimated using genetic data (see Smith et al. 2003 for a similar study of baboons). By far the highest levels of affiliation occurred between maternal half sisters, while the lowest occurred between unrelated females. Interestingly, levels of affiliation for paternal half sisters fell between these two extremes, implying that females who were not maternally related nevertheless recognized their paternal kinship through some form of phenotype matching. Two findings tend to muddy this interpretation, however. First, there was a clear effect of familiarity on affiliation—regardless of kinship, individuals from the same birth cohort (peers) affiliated at higher rates than individuals from different cohorts (nonpeers). Second, while technically intermediate, levels of affiliation for paternal half sisters were, in fact, far more similar to those of nonkin. Considering nonpeers' levels of proximity (i.e., controlling for the effects of cohort familiarity), grooming and approach for paternal half sisters exceeded those for nonkin by 10% to 32%, while those for maternal half sisters exceeded those for paternal half sisters by 500% to 3,100%! Hence, by comparison to maternal half sisters, paternal half sisters affiliated in ways that were only marginally different from nonkin. Furthermore, this marginal difference could have arisen via one of two familiarity effects: (1) through indirect cohort familiarity, if females' associations with older (or younger) maternal siblings brought them into more frequent contact with those siblings' paternal half sister peers; or (2) through mother-mediated familiarity, if their mothers' attachment to a common male during their infancy and early development increased their contact with paternal half sisters of all ages.

Finally, Parr and de Waal (1999) recently reported that five chimpanzees (*Pan troglodytes*) matched facial photographs of mothers and their male (but not female) offspring at levels above chance when the mothers and offspring were both unfamiliar and unrelated to the subjects. Although strictly a test of recognizing others' kin, the phenomenon would presumably generalize to recognizing one's own kin, and indeed was interpreted as an example of kin recognition by phenotype matching. Although intriguing, the results are difficult to interpret because they are subject to the above-noted difficulty distinguishing familiarity from phenotype matching. In this study, the latter was invoked because subjects did not "know" the animals pictured—that is, they were unfamiliar. However, the match-to-sample design (and an extensive pretraining period with the photographs) effectively familiarized subjects with the mothers to whom the offspring photographs were then to be matched. As a result, subjects could simply have been selecting offspring photographs based on their similarity to the learned idiosyncrasies (i.e., familiarity) of mothers' faces. Of course, this leaves unexplained the apparent asymmetry in the facial resemblance between mothers and their sons versus daughters (but see below, Visual Recognition).

Apart from this study, most others consistently indicate that familiarity is the mechanism by which kin are recognized. For example, studying a heterogeneous group of 26 stumptailed macaques (*M. arctoides*), MacKenzie et al. (1984) found that individuals' early rearing environment and subsequent companionship/familiarity were the principal variables affecting patterns of social behavior, while kinship (either maternal or paternal) had only a minor and inconsistent statistical influence. Similarly, Martin (1997) stressed the primacy of familiarity over kinship per se in structuring social behavior in pigtail macaques. Of special interest here was the fact that sisters whose mother had died interacted significantly less than sisters whose mother was still alive, and that sisters born more than two years apart interacted significantly less than sisters born within two years of one another. Both results suggest that relative familiarity is the mechanism underlying kin-biased behavior.

Complementary results have been reported recently for baboons by Erhart et al. (1997). In this study, 23 male and female infants were reared singly with their mothers from 24 hours after birth through 90 days, after which they were combined in a peer group of unfamiliar (maternal and paternal) kin and nonkin. Between 43 and 83 months of age, reliable sex differences in behavior were apparent, but there were no significant differences between kin (either maternal or paternal) and nonkin in the frequency of affiliative, sociosexual, or agonistic behaviors. Hence, even with mothers to serve as a potential referent for maternal kin early in life, there was no evidence for an ability to recognize unfamiliar kin. Welker et al. (1987) reported similar results for 20 crab-eating macaques (*M. fascicularis*) separated from their mothers at birth and reared in a social group composed of maternal and paternal kin as well as nonkin. Individuals showed no special attraction to relatives. Instead, familiarity alone was the best predictor of social affiliation.

The primacy of familiarity in recognizing kin receives indirect support from other sources. For example, in captive primate colonies, cross-fostered infants are usually readily accepted if transferred early, before the infants and their birth mothers have had substantial contact. Studies of inbreeding avoidance are also consistent with the familiarity mechanism. Despite the apparently strong selective pressure to avoid mating with close relatives, the proximate mechanism mediating the avoidance of inbreeding appears again to be familiarity. While familiarity is typically sufficient to prevent significant mating between maternal kin (Chapais & Mignault 1991, Manson & Perry 1993, Kuester et al. 1994), it is not always so between paternal kin (e.g., Smith 1982, Alberts 1999). For example, in a recent study of baboons, paternal half siblings were not generally any less likely to consort with each other than were nonkin (Alberts 1999). However, by comparison, natal males and females of the same age cohort showed reduced levels of consorting, suggesting that familiarity might account for the asymmetry. Because natal animals of the same cohort have a higher than average probability of being paternally related in polygynous species (Altmann et al. 1996), cohort familiarity (through infant peer interactions) might function to reduce mating between similarly aged paternal relatives, while its absence among paternal kin of different cohorts might reduce the avoidance (see also Erhardt 1988).

In sum, although not extensive, and still admitting the possibility of other mechanisms, the available evidence for primates paints a fairly consistent picture—namely that kin are "recognized" via the familiarity accruing to them during development.

Media

Recognizing kin—by whatever mechanism—requires sufficient cue variation in some medium that kin can be discriminated from nonkin (at the least). Although again not extensive, there is some research for primates on the extent to which appropriate cue variation exists in various sensory media.

Multimedia Recognition

A few studies have tested simultaneously for recognition in multiple sensory domains. Perhaps contrary to expectations, the results have not always been straightforward. For example, Klopfer (1970) found that six female bushbabies (*Galago* spp.) reliably discriminated their own from an unrelated infant when allowed all but physical access as well as when limited to only vocal and olfactory contact. The abilities emerged gradually over the infants' first week, suggesting that females required some familiarity with their infants. Strangely, mothers' discrimination waned after a few weeks, apparently not simply due to habituation to the experimental paradigm. Jensen and Tolman (1962) examined mothers' acceptance of their own versus a strange infant after a period of separation in two pigtail macaques when their infants were five to seven months of age. One mother accepted only her own infant on the first reunion, while the other initially accepted the strange infant, discriminating only on the second separation-reunion trial. In contrast, Nakamichi and Yoshida (1986) found that, of four female Japanese monkeys (*Macaca fuscata*) separated from their infants and then provided all but tactile access to one, the actual mother showed the highest levels of infant interest. The effect was present from the time of first testing when infants were only three days old.

Reversing the direction of discrimination, Kaplan and Russell (1975) found that 12 infant squirrel monkeys (*Saimiri sciureus*), less than 12 weeks old, recognized the cloth surrogate on which they had been reared based on olfactory but not visual cues. This pattern was later replicated with discriminations of real mothers from nonmothers by 14 infant squirrel monkeys (Kaplan et al. 1977) in which olfactory cues were sufficient for discrimination and visual cues added nothing to discriminability for infants aged 8 to 24 weeks. The more gradual emergence of visual recognition is also suggested by a study of five infant Japanese monkeys in which a preference for mothers (over three other females) based on only visual cues did not emerge until the infants' 12th week of life, whereas it began to appear by about the 8th week when all but tactile cues were available (Nakamichi & Yoshida 1986). In contrast, with both olfactory and tactile (but not visual) cues available, only three of five infant crab-eating macaques (aged 7–16 months) showed a preference for their mother over another female (Berkson & Becker 1975).

Visual Recognition

There has been only one systematic study of primates' ability to recognize kin visually. In this study, Parr and de Waal (1999) found that chimpanzees matched the faces of mothers and sons at levels above chance (65–70% correct where chance was 50%) but could not do the same for mothers and daughters. This recognition asymmetry was functionally attributed to the male philopatric structure of chimpanzee societies. However, it is not clear if it reflects greater expression of, or greater perceptual sensitivity to, maternal features in the faces of sons versus daughters. To address this issue, Vokey et al. (in press) used both an auto-associative neural network and 23 human subjects to evaluate similarity in the original photographs. Using the original procedures, both the network and the human subjects matched photographs at levels comparable to the chimpanzees and also replicated the asymmetry. Thus, with no special sensitivity to chimpanzee faces, both the network and human subjects preferentially matched mothers and sons, suggesting that their faces might indeed be more similar. However, in a follow-up analysis, in which each photograph was cropped to remove borders and extraneous background material, the network and human subjects correctly matched both sons and daughters to their mothers, suggesting that some artifact of border or background might have accounted for the original matching asymmetry.

Vocal Recognition

More work has been done on recognition abilities in the vocal domain, probably because vocal signals are simply far easier to collect, manipulate, and present to animals experimentally than are either visual or olfactory signals. Despite greater emphasis, however, there is no ready consensus on the capacity for vocal recognition of kin.

An early study by Jensen (1965) indicated that for five pigtail macaque mothers, the agitation associated with being separated from their infant was reduced when auditory contact was restored with their own but not an unfamiliar infant, suggesting that mothers recognized their own infants' calls. Mothers' ability to discriminate appeared to develop with exposure to their infant over its first week of life. Jovanic et al. (2000) also reported improved discriminability of infant distress calls by 15 female rhesus monkeys as their infants got to be more than a week old. However, mothers' responses were weak overall and the effect did not characterize all females equally.

Several other studies have reported maternal recognition of infant contact vocalizations in macaques (Pereira 1986, Hammerschmidt & Fischer 1998), squirrel monkeys (Symmes & Biben 1985), and ring-tailed lemurs (*Lemur catta*) (Nunn 2000). However, half as many studies have reported no such ability (Simons et al. 1968, Simons & Bielert 1973). Reconciling these differences is problematic due to the variety of methodologies and the small samples used (*N* < 6 subjects). An exception to the latter problem is a study of wild baboons (*Papio cynocephalus ursinus*), in which 19 females discriminated the contact barks of their own infants from those of unrelated but familiar infants in the group (Rendall et al. 2000). Reversing the direction of discrimination, Hansen (1976) and Fischer et al. (2000) reported that four juvenile rhesus monkeys and 10 infant baboons, respectively, discriminated their mothers' contact calls from those of other females. Extending tests beyond the mother-infant dyad, Rendall et al. (1996) found that 27 adult female rhesus monkeys responded preferentially to the contact (coo) vocalizations of their adult female kin as compared to those of unrelated but familiar group females.

Studying the screams typically produced in primates by victims of aggressive attack, Cheney and Seyfarth (1980) found that four vervet monkey mothers (*Cercopithecus aethiops*) could discriminate their own juvenile's calls from those of an unrelated juvenile. Gouzoules et al. (1986) reported a similar result for 10 female rhesus monkeys. They also found that five females discriminated between the screams produced by closely related as opposed to distantly related juvenile kin. However, in both cases, the evidence for discrimination held for only one of two types of scream vocalization tested, and was manifest in only one of two possible response variables measured. Testing a much larger sample of 30 female squirrel monkeys, Kaplan et al. (1978) found evidence for an ability of mothers to discriminate the screams and other distress calls of their own infant from those of an unrelated infant in two of five response variables measured. However, it was not possible to rule out the contribution of olfactory cues in this study. In contrast, studying a similarly large sample of 30 adult rhesus monkeys (from the same population as that studied by Gouzoules et al. 1986) and extending research beyond the mother-infant dyad, Rendall et al. (1998) found no evidence that females could discriminate the screams of their adult female kin from those of unrelated females in the group. This finding was buttressed by the results of acoustic analyses, which revealed no reliable differences in the structure of screams produced by nine females. A recent study of wild baboons has produced similar results, with 17 adult females failing to discriminate between the screams of their own juvenile offspring and those of another juvenile in the group (Rendall n.d.). One possible explanation for these discrepant findings for scream vocalizations is that Rendall presented only one or two screams in each playback trial, whereas the other studies presented much longer bouts of calling. As a result, additional cues to identity may have been available through the temporal patterning of entire scream sequences. This possibility has yet to be tested systematically, however.

While a variety of studies have tested the capacity for kin recognition vocally, only a few have tackled the specific acoustic features used to mediate such recognition, or their origins. Studying the coo vocalizations of seven adult female rhesus monkeys, Hauser (1991) found that, although individually distinctive, the calls of kin were sometimes confused with one another in a discriminant analysis classification, suggesting that they might be structurally similar. Hauser (1992) later reported that several members of one matriline produced coos with a distinctive spectral structure seemingly traceable to the filtering effects of the nasal cavities. He speculated that this phenomenon might reflect convergence among relatives on a common, learned mode of call production. Studying a larger sample ($N = 17$) of the same population of rhesus monkeys, Rendall (1996, Rendall et al. 1998) confirmed that coo calls were individually distinctive, that this was also due primarily to differences in spectral structure, and that individually distinctive call structures were indeed more similar among related females. However, Rendall argued that the distinctive spectral structures were due to idiosyncratic differences in vocal tract morphology (including but not limited to the nasal cavities) that generated individually distinctive patterns of vocal tract resonance, and that similar call structures among kin were thus more likely to reflect their similar vocal tract morphologies arising naturally from similarity in their underlying genetic and environmental determinants (see also Owren & Rendall 1997, 2001).

Media Summary

Research in various sensory media suggests that cue variation may be sufficient to permit kin to be recognized in each of them. However, any very firm conclusions here would be premature because research has sampled only a few different kin relationships (mostly mother-offspring), has sometimes produced conflicting results suggesting variable recognition abilities between and possibly also within sensory domains, and has seldom isolated the actual cues used or their origins. Furthermore, a wider range of cue dimensions than traditionally considered might be salient, including such things as body or hair type, posture and gait patterns, and even more abstract behavioral qualities related to social disposition and personality (Widdig et al. 2001).

Minds, Modules, and More Muddles

In this final section, I want to consider a few additional mechanistic and functional issues focusing specifically on the possibility of specialized processes for signaling, detecting, and responding to kinship. Because the available evidence to go on is limited, the discussion is necessarily preliminary and speculative—perhaps adding to, more than resolving, existing muddles. However, I hope it will spur others to additional, far more productive, black box probing.

Kinship Cues and Kin Detection Modules

The evidence reviewed suggests an ability to recognize kin (at least some categories) in several sensory media. It is natural to ask then whether such abilities reflect the effects of selection specifically on cues for signaling kinship and/or on mental processes for detecting it—what might be conceived of in contemporary evolutionary psychological theorizing as a cognitive module. Certainly kin may look, smell, and sound alike, and others may detect such similarity, but these facts by themselves do not necessarily imply selection on either phenomenon specifically for purposes of kin recognition. As Grafen (1990) and others have noted, appropriate cue variation and its detection at the level of kin groups may be epiphenomenal of selection at other levels. Grafen specifically stressed the possibility that kin recognition might be epiphenomenal of recognition at higher organizational levels—e.g., recognition of group or species membership. In the case of primates, however, any potential epiphenomenal origins are more likely to come not from above but from below—from individualistic cue variation and recognition. Individual recognition is clearly central to the differentiated social relationships observed in many primate species, making it at least functionally plausible that this is the most salient level of cue variation and detection. There is also considerable evidence of salient individual variation in the olfactory and vocal signals of primates (e.g., Porter & Moore 1981; Hauser 1991; Rendall et al. 1996, 1998), in addition to obvious individual differences in appearance.

However, even at this level, cue variation and recognition may be largely epiphenomenal. That is, individualistic cue variation in the major dimensions we think of (faces, voices, odors), while heritable, may be idiosyncratic in origin, and sensitivity to the variation may be an unselected by-product of more general discrimination processes. Consider faces, which are extremely salient to human and nonhuman primates alike. Faces are highly individualistic, but it is not clear that this variability has been specifically selected—that is, that variation in faces is greater than variation in other body parts, for example. Human fingerprints are equally (perhaps more) individualistic, and yet the variation is unselected, presumably resulting from idiosyncratic epigenetic processes. Nor is it entirely clear that, despite the obvious salience of faces, we are more sensitive to variation in them than equivalent variation in other visual stimuli. While often offered as an evolutionarily specialized cognitive module (e.g., Farah 1996), face processing may reflect more general routines associated with the processing of visually complex stimuli, or stimuli of low spatial frequency, whatever their actual content (Diamond & Carey 1986, Phelps & Roberts 1994, Wright & Roberts 1996, Gauthier & Logothetis 2000).

Similar points may apply to olfactory and vocal recognition. For example, olfactory cues, particularly if derived from the MHC, are likely to be individualistic but arguably largely epiphenomenal in the context of both individual and kin recognition. MHC diversity is extreme, evolved in the service of immune system function for purposes of discriminating "self" from "other" so as to identify, attack, and destroy foreign materials entering the body. This diversity may incidentally produce detectable, individually distinctive body odors that in turn may be more similar among kin owing to their shared genomes (Porter & Moore 1981; Porter et al. 1985, 1986). However, there need never have been selection for recognition at either level to produce these outcomes. Instead, they emerge naturally—epiphenomenally—as a result of selection for immune system self-recognition.

In sum, there is some reason to think that kin recognition abilities in primates are founded on ubiquitous, highly functional—but perhaps largely epiphenomenal—cues to individual identity combined perhaps with fairly general processes of discrimination and generalization. Of course, this possibility then begs the question of whether, for the animals, kinship is a real psychological construct at all. That is, does the ability to recognize others individually obviate the need for any higher order categorization based on kinship, or is it nevertheless possible that the animals operate with something akin to a concept of kin?

Kin Concepts

For many the question of kin concepts may seem anathema, or at least irrelevant. What matters are functional, social-behavioral outcomes, not intangible mental constructs. For others, dedicated to elucidating the structure of animal minds, the question may be absolutely central. Whatever your slant, the issue is tricky. Categories and concepts are notoriously fuzzy constructs, difficult to define and even more difficult to demonstrate concretely. In fact, the reality of a kin concept may seem unlikely on strictly theoretical grounds. For example, the once firmly established perceptual categories (so-called prototypes) assumed to underlie discrimination and classification processes in humans and nonhumans alike now appear illusory. Instead, discrimination and classification appear to involve comparison of present stimuli to recollections of previously encountered individual exemplars using basic processes of stimulus generalization (reviewed in Pearce 1994, Wasserman 1995, Roberts 1998). As a result, any kind of *perceptual* category of kin is unlikely.

Somewhat better substantiated in animals—though not entirely uncontroversial—are more abstract, relational, or associative concepts, in which perceptually dissimilar stimuli are linked either by some common relational property (e.g., larger/smaller, same/different) or by association with some common functional outcome. This kind of concept is probably more germane to the subject of kinship, anyway, as it is arguably a relational or associative sort of construct (at least to humans). Still, it may seem implausible, requiring a sense of functional equivalence between all of the variably related members of the kin class. While not inconceivable, it is not easy to imagine a (nonlinguistic) functional rationale for the treatment of a broad range of biological relatives categorically as kin given the variable costs and benefits associated with different acts directed toward individuals of varying degrees of relatedness. It may be no easier to imagine a mechanism of developmental association that would create the required sort of perceived functional equivalence among many differently related individuals. Nevertheless, the possibility is intriguing, and there are a few relevant data.

Rendall et al. (1996) reported weak evidence for a categorical construction of kinship in free-ranging rhesus monkeys revealed through qualitative differences in the animals' responses to the contact calls of variably related group members. Gouzoules et al. (1984) reported that different classes of scream vocalization in rhesus monkeys were used roughly categorically during agonistic encounters with either kin or nonkin, suggesting some form of discrete labeling of group members with respect to kinship. This possibility is also suggested by patterns of reconciliation and redirected aggression, in which individuals interact preferentially with the kin of former opponents following a fight (e.g., Cheney & Seyfarth 1986, 1990), and by systematic research on various forms of social behavior in Japanese monkeys (Chapais & Mignault 1991; Chapais et al. 1997, 2001; Bélisle & Chapais 2001). In the latter, individuals clearly differentiate among close matrilineal relatives (i.e., between $r = .5$, .25, and .125). However, they also show consistent limits to their performance of different behaviors (agonistic aiding, tolerated cofeeding, and homosexual incest avoidance) vis-à-vis matrilineal relatedness, suggesting a stable nepotism boundary (relatedness threshold) that might reflect a higher order categorization of kin.

At the same time, however, several comprehensive analyses of kin-biased social behaviors show a more continuous (rather than discrete) pattern (e.g., Kaplan 1977, Kurland 1977, Kapsalis & Berman 1996). In those studies suggesting some kind of discontinuity, the boundary defining kin has been variable and loose. For example, in the work by Rendall et al. 1996, the boundary (which was equivocal) fell between matrilineal relatedness coefficients of $r = .063$ and $r = .016$, while the far more extensive work of Chapais and colleagues puts the threshold at $r = .25$. Even here, however, the boundary seems to vary (between $r = .25$ and $r = .125$) depending on the type of behavior involved and whether the kin concerned are vertical or collateral relatives. That is, individuals who are equally related genetically (great-grandmothers and their great-granddaughters versus aunts and nieces) fall on opposite sides of the kin boundary for certain kinds of behavior. Even within the same kin category (e.g., aunt-neice), behavior sometimes varies from one dyad to another, suggesting that the boundaries of kinship are loose rather than absolute.

One way to interpret this limited and variable evidence is that it supports the notion of abstract kin categories which, as in humans, vary according to the criteria emphasized in their construction. Another, more parsimonious, interpretation is that abstract kin concepts are not well supported, but rather that some more basic dimension of familiarity contributes to elastic boundaries for the performance of specific behaviors depending on the particular social regime (cf. Berman et al. 1997). Additional systematic work on a wider range of species, like that done by Chapais and his colleagues on Japanese monkeys, might help to discriminate these possibilities. Techniques similar to those used by Parr and de Waal (1999) could also be applied to the issue. For instance, subjects could be required to indicate for photographs of pairs of kin related by varying degrees whether they are the "same" or "different."

Why Familiarity?

Throughout this review, familiarity has featured prominently in the mechanistic processes underlying kin recognition in primates. Why? Given the critical implications for inclusive fitness, why do primates rely on the seemingly clumsy and indirect process of familiarity for discriminating kinship rather than some more specific mechanism?

Functional Hypotheses

Primate Life Histories

Perhaps the simplest functional answer is that "familiarity works!" Many species are comparatively long-lived with an extended period of mother-infant dependency and overlapping generations. This sort of life history incidentally creates a nested hierarchy of kin relationships that a mechanism of developmental familiarity could (and evidently does) readily discriminate (Chapais 2001). In fact, a bit ironically perhaps, familiarity could be a more specific mechanism of recognition than even those that might detect degrees of genetic relatedness directly, because a gradient of familiarity can effectively discriminate far more categories than are provided by degrees of relatedness alone. And this fact could make it even more functional. After all, discriminating kinship in the service of nepotistic relationships offers only one way to negotiate the social world to individual advantage. There are other ways (Trivers 1971), other functional social relationships; hence, other discriminations to be made. Viewed in this more general light, some dimension of familiarity might well serve as the recognition/discrimination mechanism common to all manner of social relationships, whether or not they involve kin.

Still, this explanation may not seem entirely satisfactory because the mechanism of familiarity appears patently inadequate in certain contexts, such as in the avoidance of inbreeding among paternal kin, and in the avoidance of cuckoldry in males.

Adaptive Anonymity

Here again, there may be good functional reasons why more "specific" mechanisms are not used that hinge on the potential asymmetry in the benefits of being recognized to the different parties to an interaction. Although it is natural to assume that kin always want to be recognized, there may be cases where this is not true, where it actually pays to be anonymous (Beecher 1991). The context of male cuckoldry may be just such a case. For example, while selection might favor quite specific mechanisms in fathers for recognizing their own offspring so as to decide how or if to invest, it is not so clear that offspring should uniformly want to be recognized by their fathers. Certainly infants conceived through extrapair copulations, who are facing an unrelated potential male caretaker, do not stand to benefit from being recognized. Hence, it may pay infants to conceal rather than reveal paternity. This possibility makes especially good functional sense in humans, for instance, where several features of mothers' behavior and physiology (e.g., continuous sexual receptivity and concealed ovulation) appear to have evolved specifically to conceal paternity. Under these circumstances, vulnerable newborns would not be predicted to undermine their mother's strategy and abruptly give away the game at birth. Instead, where the rate of extrapair paternity is at all appreciable (Pagel 1997), an infant would be predicted to conceal its paternity and shelter under the umbrella of the familiarity mechanisms that bond the male to its mother and that will, in time, bond the male to it.

Of course, this functional argument may seem biased in favoring infants over fathers in the battle to conceal versus detect paternity. However, the bias may be justified. By definition, the selection cost for errors is (at least potentially) far higher for infants (life versus death) than it is for fathers (wasted resources). More fundamentally, it may simply be easier for infants to be anonymous than it is for fathers to detect parentage. For example, the phenomenon of rapid fat deposition on the faces of newborns appears to be a simple but effective way to hide landmark facial features (cheeks, chin, noses, eyes) that otherwise might be used to assess paternity. Although there are surprisingly few empirical studies of actual facial resemblance between parents (either fathers or mothers) and their offspring, what evidence there is lends some support to these ideas (Porter et al. 1984, Bredart & French 1999; see also Treves 1997, 2000). How generally this logic of adaptive anonymity should apply is unclear, but it could conceivably also encompass the case of inbreeding avoidance among paternal kin, where there is again an asymmetry between males and females in the selective benefits associated with being recognized as kin in the mating context.

Mechanistic Hypotheses

Primates' reliance on familiarity for recognizing kin may also stem from basic mechanistic processes.

Constraints on Cue Production and Perception

The flip side of many primates' long developmental period—that *allows* for functional use of familiarity—is that this same life history may also *preclude* more direct processes for signaling and detecting kinship. The extended maturational period means that cues to identity that are tied to organismal structures subject to growth and development (as many are) will either unfold gradually or be ever changing. Given this life history, it might be difficult to signal kinship very accurately (especially early in development), even where it might be functional to do so. Conversely, detecting kinship using some kind of crystallized template also would be problematic given that the identity cues of those to be discriminated would be constantly changing. Because the benefits from nepotistic behavior can begin very early in life and extend throughout it, familiarity may be the only mechanism flexible enough to accommodate these constraints on cue production and perception. Certainly, familiarity begins to accrue from birth and is sensitive to experiential change throughout life.

Central Processing "Constraints"

Even more fundamentally, the reliance on familiarity may stem from basic mechanisms of central processing. In one view of the nervous system, cognitive operations—like recognition (discrimination, categorization, etc.)—are embodied in the pattern of activation of neuronal units linked in distributed networks (e.g., Clark 1993). One property of such neural networks is that they, like the immune system (von Boehmer & Kisielow 1991), require experience for their operation—they do not come preconfigured with the "knowledge" they contain but rather "learn" it through repeated exposure. In fact, in an important way, the contents of neural networks are really just the cellular instantiation of familiarity. What the network "knows" is simply what it is most familiar with (what it has had the most exposure to). Hence, if the brain processes associated with kin recognition operate anything like neural networks (and there is good reason to think that they do), then the use of familiarity to recognize kin may simply be a result of the way the central nervous system works. Of course, the flip side of this experiential constraint of neural networks (if it really is a constraint), is that the above-noted problems inherent in the production and perceptual tracking of changing identity cues evaporate because another property of neural networks is their experiential plasticity, which, in this case, would allow them to continuously update their "vision" of relatives.

Summary and Conclusion

The ability to recognize kin is foundational to a variety of evolutionarily functional kin-biased social behaviors, and several different proximate mechanisms have been proposed that might support such an ability. Although heuristically productive, these mechanisms are not necessarily as distinct as they appear to be, and future work might profit from continued efforts to resolve their remaining conceptual ambiguities. Such ambiguities notwithstanding, available evidence suggests that primates rely primarily on just one mechanism for recognizing kin, namely their familiarity with specific individuals during early development. Given the centrality of refined kinship discriminations to theoretical explanations of the evolution of social behavior, it seems puzzling that so obviously nepotistic primates should rely on a seemingly so indirect, crude, and error-prone mechanism for recognizing their kin. However, there are several possible explanations for this emphasis, including these: environmental regularities and life history characteristics obviate the need for any more direct mechanism; the variety of social discriminations made by many species (including but not limited to ones based on kinship) actively favor familiarity as a more versatile functional mechanism for recognition and discrimination generally; more direct or specific mechanisms would (contrary to intuition) not actually be favored in all contexts or by all parties to an interaction; and basic mechanistic processes of growth and development and central nervous system operation constrain other possibilities.

Although it might be tempting, then, to conclude that familiarity is the mechanism by which primates recognize their kin, full stop, this conclusion would be premature and might tend to seal the matter before it has been properly opened. Clearly, the species about which we know anything are few, and even for those about which we know something, we know little. Furthermore, even if data from additional species were to confirm that familiarity is the proximate basis for kin-biased behavior in primates generally, a great deal would remain to be explained. What are the salient dimensions of familiarity, and how do they contribute to the observed subtleties in kin-biased behavior? For example, what aspects of familiarity might contribute to variation in the nepotistic behavior of equally related vertical versus collateral relatives (Chapais et al. 2001)? Is cohort familiarity (Altmann et al. 1996) a phenomenon that produces mating avoidance generally, or does it so function only in polygynous species where age mates are likely to be paternal kin? If true only for polygynous species, what dimension of familiarity in nonpolygynous species might account for the avoidance of mating between maternal relatives of different cohorts (arguably less familiar) but allow mating among unrelated age mates (arguably more familiar)? Are familiarity effects specific to particular periods of development, and, if so, when do they occur and are they variable for different types of behavior or social relationship?

Obviously, a comprehensive account of the proximate mental mechanisms underlying kin-biased behaviors awaits answers to these and many other questions. Given the obvious centrality of kinship to primate social life, the importance of the latter in contemporary accounts of primate mental evolution, and the prevailing cognitive focus of the behavioral sciences generally, perhaps it is not too much to hope that the wait be short.

Acknowledgments   Many thanks to Bernard Chapais and Carol Berman for their invitation to contribute this chapter and their considerable help and feedback during its writing, to Warren Holmes for his many productive comments on a first version, and to John Vokey for innumerable spontaneous and enlightening discussions of topics related to kin recognition. The continuing support of the Natural Sciences and Engineering Research Council of Canada is gratefully acknowledged.

References

Alberts, S. C. 1999. Paternal kin discrimination in wild baboons. *Proc. Royal Soc. Lond., B, Biol. Sci.*, 266(1427), 1501–1506.

Altmann, J., Alberts, S. C., Haines, S. A., Dubach, J., Muruthi, P., Coote, T., Geffen, E., Cheesman, D. J., Mututua, R. S., Saiyalel, S. N., Wayne, R. K., Lacy, R. C., & Bruford, M. W. 1996. Behavior predicts genetic structure in a wild primate group. *Proc. Natl. Acad. Sci. USA*, 93, 5797–5801.

Beecher, M. D. 1991. Successes and failures of parent-offspring recognition in animals. In: *Kin Recognition* (Ed. by P. G. Hepper), pp. 94–127. Cambridge: Cambridge University Press.

Beecher, M. D., Beecher, I. M., & Hahn, S. 1981. Parent-offspring recognition in bank swallows (*Riparia riparia*). II. Development and acoustic bias. *Anim. Behav.*, 29, 95–101.

Bélisle, P. & Chapais, B. 2001. Tolerated co-feeding in relation to degree of kinship in Japanese macaques. *Behaviour*, 138, 487–509.

Berenstein, L., Rodman, P. S., & Smith, D. G. 1981. Social relations between fathers and offspring in a captive group of rhesus monkeys (*Macaca mulatta*). *Anim. Behav.*, 29, 1057–1063.

Berkson, R. & Becker, J. D. 1975. Facial expressions and social responsiveness of blind monkeys. *J. Abnorm. Psychol.*, 84, 519–523.

Berman, C. M., Rasmussen, K. L. R., & Suomi, S. J. 1997. Group size, infant development and social networks in free-ranging rhesus monkeys. *Anim. Behav.*, 53, 405–421.

Bernstein, I. 1991. The correlation between kinship and behaviour in non-human primates. In: *Kin Recognition* (Ed. by P. G. Hepper), pp. 6–29. Cambridge: Cambridge University Press.

Blaustein, A. R. 1983. Kin recognition mechanisms: phenotypic matching or recognition alleles? *Am. Nat.*, 121, 749–754.

Blaustein, A. R., Bekoff, M., Byers, J. A., & Daniels, T. 1991. Kin recognition in vertebrates: what do we really know about adaptive value? *Anim. Behav.*, 41, 1079–1083.

Blaustein, A. R., Bekoff, M., & Daniels, T. 1987. Kin recognition in vertebrates (excluding primates): mechanisms, functions and future research. In: *Kin Recognition in Animals* (Ed. by D. J. C. Fletcher & C. D. Michener), pp. 333–358. New York: Wiley & Sons.

Blaustein, A. R. & Porter, R. H. 1990. The ubiquitous concept of recognition with special reference to kin. In: *Interpretation and Explanation in the Study of Animal Behavior, Vol. 1: Interpretation, Intentionality and Communication* (Ed by M. Bekoff & D. Jamieson), pp. 123–148. Boulder, CO: Westview Press.

Bredart, S. & French, R. M. 1999. Do babies resemble their fathers more than their mothers? A failure to replicate Christenfeld & Hill (1995). *Evol. Hum. Behav.*, 20, 129–135.

Breed, M. D. 1998. Recognition pheromones of the honey bee. *Bioscience*, 48, 463–470.

Buchan, J. C., Alberts, S. C., Silk, J. B., & Altmann, J. 2003. True paternal care in a multi-male primate society. *Nature*, 425, 179–181.

Byers, J. A. & Bekoff, M. 1986. What does "kin recognition" mean? *Ethology*, 72, 342–345.

Chapais, B. 2001. Primate nepotism: what is the explanatory value of kin selection? *Int. J. Primatol.*, 22, 207–208.

Chapais, B., Gauthier, C., Prud'homme, J., & Vasey, P. 1997. Relatedness threshold for nepotism in Japanese macaques. *Anim. Behav.*, 53, 1089–1101.

Chapais, B. & Mignault, C. 1991. Homosexual incest avoidance among females in captive Japanese macques. *Am. J. Primatol.*, 23, 171–183.

Chapais, B., Savard, L., & Gauthier, C. 2001. Kin selection and distribution of altruism in relation to degree of kinship in Japanese macaques. *Behav. Ecol. Sociobiol.,* 49, 493–502.

Cheney, D. L. & Seyfarth, R. M. 1980. Vocal recognition in free-ranging vervet monkeys. *Anim. Behav.,* 28, 362–367.

Cheney, D. L. & Seyfarth, R. M. 1986. The recognition of social alliances by vervet monkeys. *Anim. Behav.,* 34, 1722–1731.

Cheney, D. L. & Seyfarth, R. M. 1990. The representation of social relations by monkeys. *Cognition,* 37, 167–196.

Clark, A. 1993. *Associative Engines: Connectionism, Concepts and Representational Change.* Cambridge, MA: MIT Press.

Dawkins, R. 1976. *The Selfish Gene.* Oxford: Oxford University Press.

Diamond, R. & Carey, S. 1986. Why faces are not so special: an effect of expertise. *J. Exp. Psychol.: Gen.,* 115, 107–117.

Erhardt, C. L. 1988. Absence of strongly kin-preferential behavior by adult female sooty mangabeys (*Cercocebus atys*). *Am. J. Phys. Anthropol.,* 76, 233–243.

Erhart, E. M., Coelho, A. M. Jr., & Bramblett, C. A. 1997. Kin recognition by paternal half-siblings in captive *Papio cynocephalus. Am. J. Primatol.,* 43, 147–157.

Farah, M. J. 1996. Is face recognition special? Evidence from neuropsychology. *Behav. Brain Res.,* 76, 181–189.

Fischer, J., Cheney, D. L., & Seyfarth, R. M. 2000. Development of infant baboons' responses to graded bark variants. *Proc. Royal Soc. Lond., B,* 267(1459), 2317–2321.

Fletcher, D. J. C. 1987. The behavioral analysis of kin recognition: perspectives on methodology and interpretations. In: *Kin Recognition in Animals* (Ed by D. J. C. Fletcher & C. D. Michener), pp. 19–54. New York: Wiley & Sons.

Fletcher, D. J. C. & Michener, C. D. 1987. *Kin Recognition in Animals.* New York: Wiley & Sons.

Frederickson, W. T. & Sackett, G. P. 1984. Kin preferences in primates (*Macaca nemestrina*): relatedness or familiarity? *J. Comp. Psychol.,* 98, 29–34.

Gauthier, I. & Logothetis, N. K. (2000) Is face recognition not so unique after all? *Cognit. Neuropsychol.,* 17, 125–142.

Gouzoules, H., Gouzoules, S., & Marler, P. 1984. Rhesus monkey (*Macaca mulatta*) screams: representational signaling in the recruitment of agonistic aid. *Anim. Behav.,* 32, 182–193.

Gouzoules, H., Gouzoules, S., & Marler, P. 1986. Vocal communication: a vehicle for the study of social relationships. In: *The Cayo Santiago Macaques: History, Behavior and Biology* (Ed by R. G. Rawlins & M. J. Kessler), pp. 111–129. Albany, NY: SUNY Press.

Gouzoules, S. 1984. Primate mating systems, kin associations, and cooperative behavior: evidence for kin recognition? *Ybk. Phys. Anthropol.,* 27, 99–134.

Grafen, A. 1990. Do animals really recognize kin? *Anim. Behav.,* 39, 42–54.

Grosberg, R. K. & Quinn, J. F. 1986. The genetic control and consequences of kin recognition by the larvae of a colonial marine invertebrate. *Nature,* 322, 456–459.

Gust, D. A., McCaster, T., Gordon, T. P., Gergits, W. F., Casna, N. J., & McClure, H. M. 1998. Paternity in sooty mangabeys. *Int. J. Primatol.,* 19, 83–93.

Hamilton, W. D. 1964. The genetical evolution of social behavior. I and II. *J. Theor. Biol.,* 7, 1–52.

Hammerschmidt, K. & Fisher, J. 1998. Maternal discrimination of offspring vocalizations in Barbary macaques (*Macaca sylvanus*). *Primates,* 39, 231–236.

Hansen, E. W. 1976. Selective responding by recently separated juvenile rhesus monkeys to the calls of their mothers. *Dev. Psychobiol.,* 9, 83–88.

Hauber, M. E. & Sherman, P. W. 2001. Self-referent phenotype matching: theoretical considerations and empirical evidence. *Trends Neurosci.*, 24, 609–616.

Hauser, M. D. 1991. Sources of acoustic variation in rhesus macaque (*Macaca mulatta*) vocalizations. *Ethology,* 89, 29–46.

Hauser, M. D. 1992. Articulatory and social factors influence the acoustic structure of rhesus monkey vocalizations: a learned mode of production? *J. Acoust. Soc. Am.*, 91, 2175–2179.

Hepper, P. G. 1986. Kin recognition: functions and mechanisms. *Biol. Rev.*, 61, 63–93.

Hepper, P. G. (ed.) 1991a. *Kin Recognition.* Cambridge: Cambridge University Press.

Hepper, P. G. 1991b. Recognizing kin: ontogeny and classification. In: *Kin Recognition* (Ed. by P. G. Hepper), pp. 259–288. Cambridge: Cambridge University Press.

Holmes, W. G. 1986. Kin recognition by phenotype matching in female Belding's ground squirrels. *Anim. Behav.*, 34, 38–47.

Holmes, W. G. & Sherman, P. W. 1982. The ontogeny of kin recognition in two species of ground squirrel. *Am. Zool.*, 22, 491–517.

Holmes, W. G. & Sherman, P. W. 1983. Kin recognition in animals. *Am. Sci.*, 71, 46–55.

Humphrey, N. K. 1976. The social function of intellect. In: *Growing Points in Ethology* (Ed. by P. P. G. Bateson & R. A. Hinde), pp. 303–317. Cambridge: Cambridge University Press.

Jensen, G. D. 1965. Mother-infant relationship in the monkey, *Macaca nemestrina. J. Comp. Physiol. Psychol.,* 59, 305–308.

Jensen, G. D. & Tolman, C. W. 1962. The effect of brief separation and mother-infant specificity. *J. Comp. Physiol. Psychol.*, 55, 131–136.

Jolly, A. 1966. Lemur social behavior and primate intelligence. *Science*, 153, 501–506.

Jovanic, T., Megna, N. L., & Maestripieri, D. 2000. Early maternal recognition of offspring vocals in rhesus. *Primates,* 41, 421–428.

Kaplan, J. 1977. Patterns of fight interference in free-ranging rhesus monkeys. *Am. J. Phys. Anthropol.,* 47, 279–287.

Kaplan, J., Cubiciotti III, D., & Redican, W. K. 1977. Olfactory discrimination of squirrel monkey mothers by their infants. *Dev. Psychobiol.*, 10, 447–453.

Kaplan, J. & Russell, M. 1975. Olfactory recognition in the infant squirrel monkey. *Dev. Psychobiol.*, 7, 15–19.

Kaplan, J. N., Wiship-Ball, A., & Sim, L. 1978. Maternal discrimination of infant vocalizations in squirrel monkeys. *Primates*, 19, 187–193.

Kapsalis, E. & Berman, C. M. 1996. Models of affiliative relationships among free-ranging rhesus monkeys (*Macaca mulatta*). I. Criteria for kinship. *Behaviour,* 133, 1209–1234.

Klopfer, P. H. 1970. Discrimination of young in galagos. *Folia Primatol.*, 13, 137–143.

Kuester, J., Paul., A., & Arnemann, J. 1994. Kinship familiarity and mating avoidance in Barbary macaques. *Anim. Behav.*, 48, 1183–1194.

Kurland, J. 1977. Kin selection in the Japanese monkey. In: *Contributions to Primatology, Vol. 12*, pp. 1–145. Basel: S. Karger.

MacKenzie, M. M., McGrew, W. C., & Chamove, A. S. 1985. Social preferences in stumptailed macaques (*Macaca arctoides*): effects of companionship, kinship and rearing. *Dev. Psychobiol.*, 18, 115–123.

Manson, J. H. & Perry, S. E. 1993. Inbreeding avoidance in rhesus macaques: whose choice? *Am. J. Phys. Anthropol.*, 90, 335–344.

Martin, D. A. 1997. Kinship bias: a function of familiarity in pig-tailed macaques (*Macaca nemestrina*). PhD dissertation, University of Georgia, Athens, GA.

Mateo, J. M. & Johnston, R. E. 2000. Kin recognition and the "armpit effect": Evidence for self-referent phenotype matching. *Proc. Royal Soc. Lond., B*, 267, 695–700.

Nakamishi, M. & Yoshida, A. 1986. Discrimination of mother by infant among Japanese macaques (*Macaca fuscata*). *Int. J. Primatol.*, 7, 481–489.

Nunn, C. L. 2000. Maternal recognition of infant calls in ringtailed lemurs. *Folia Primatol.*, 71, 142–146.

Owren, M. J. & Rendall, D. 1997. An affect-conditioning model of nonhuman primate vocalizations. In: *Perspectives in Ethology, Vol. 12: Communication* (Ed by D. W. Owings, M. D. Beecher, & N. S. Thompson), pp. 299–346. New York: Plenum Press.

Owren, M. J. & Rendall, D. 2001. Sound on the rebound: returning form and function to the forefront in understanding nonhuman primate vocal signaling. *Evol. Anthropol.*, 10, 58–71.

Pagel, M. 1997. Desperately concealing father: a theory of parent-infant resemblance. *Anim. Behav.*, 53, 973–981.

Parr, L. A. & de Waal, F. B. M. 1999. Visual kin recognition in chimpanzees. *Nature*, 399, 647–648.

Pearce, J. M. 1994. Discrimination and categorization. In: *Animal Learning and Cognition* (Ed. by N. J. Mackintosh), pp.109–134. San Diego, CA: Academic Press.

Pereira, M. E. 1986. Maternal recognition of juvenile offspring coo vocalizations in Japanese macaques. *Anim. Behav.*, 34, 935–937.

Phelps, M. T. & Roberts, W. A. 1994. Memory for pictures of upright and inverted faces in humans and squirrel monkey and pigeons. *J. Comp. Psychol.*, 108, 114–125.

Porter, R. H., Balogh, R. D., Cernoch, J. M., & Franchi, C. 1986. Recognition of kin through characteristic body odors. *Chem. Senses*, 11, 389–395.

Porter, R. H., Cernoch, J. M., & Balogh, R. D. 1984. Recognition of neonates by facial-visual characteristics. *Pediatrics*, 74, 501–504.

Porter, R. H., Cernoch, J. M., & Balogh, R. D. 1985. Odor signatures and kin recognition. *Phys. Behav.*, 34, 445–448.

Porter, R. H. & Moore, J. D. 1981. Human kin recognition by olfactory cues. *Phys. Behav.*, 27, 493–495.

Rendall, D. n.d. The function of agonistic scream vocalizations in primates: Recruiting aid or repelling attack, or both? Unpublished manuscript.

Rendall, D. 1996. Social communication and vocal recognition in free-ranging rhesus monkeys (*Macaca mulatta*). PhD dissertation, University of California at Davis.

Rendall, D. 2003. Acoustic correlates of caller identity and affect intensity in the vowel-like grunt vocalizations of baboons. *J. Acoust. Soc. Amer.*, 113, 3390–3402.

Rendall, D., Cheney, D. L., & Seyfarth, R. M. 2000. Proximate factors mediating "contact" calls in adult female baboons (*Papio cynocephalus ursinus*) and their infants. *J. Comp. Psychol.*, 114, 36–46.

Rendall, D., Owren, M. J., & Rodman, P. S. 1998. The role of vocal tract filtering in identity cueing in rhesus monkey (*Macaca mulatta*) vocalizations. *J. Acoust. Soc. Am.*, 103, 602–614.

Rendall, D., Rodman, P. S., & Emond, R. 1996. Vocal recognition of individuals and kin in free-ranging rhesus monkeys. *Anim. Behav.*, 51, 1007–1015.

Roberts, W. A. 1998. *Principles of Animal Cognition*. New York: McGraw Hill.

Sackett, G. P. & Frederickson, W. T. 1987. Social preferences by pigtailed macaques: familiarity versus degree and type of kinship. *Anim. Behav.*, 35, 603–606.

Sherman, P. W., Reeve, H. K., & Pfennig, D. W. 1997. Recognition systems. In: *Behavioural Ecology: An Evolutionary Approach*, 4th ed. (Ed by J. R. Krebs & N. B. Davies), pp. 69–96. Oxford: Blackwell Science.

Simons, R. C. & Bielert, C. F. 1973. An experimental study of vocal communication between mother and infant monkeys (*Macaca nemestrina*). *Am. J. Phys. Anthropol.*, 38, 455–462.

Simons, R. C., Bobbitt, R. A., & Jensen, G. D. 1968. Mother monkeys' (*Macaca nemestrina*) responses to infant vocalizations. *Percept. Motor Skills*, 27, 3–10.

Small, M. F. & Smith, D. G. 1981. Interactions with infants by full siblings, paternal half-siblings, and nonrelatives in a captive group of rhesus macaques (*Macaca mulatta*). *Am. J. Primatol.*, 1, 91.

Smith, D. G. 1982. Inbreeding in three captive groups of rhesus monkeys. *Am. J. Phys. Anthropol.*, 58, 447–451.

Smith, K., Alberts, S. C., & Altmann, J. 2003. Wild female baboons bias their social behaviour towards paternal half-sisters. *Proc. Royal Soc. Lond., B*, 270, 503–510.

Smuts, B. B. 1985. *Sex and Friendship in Baboons.* Hawthorne, NY: Aldine.

Stein, D. M. 1984. Ontogeny of infant-adult male relationships during the first year of life for yellow baboons (*Papio cyncocephalus*). In: *Primate Paternalism* (Ed by D. M. Taub), pp. 213–243. New York: Van Nostrand Reinhold.

Sun, L. & Muller-Schwarze, D. 1997. Sibling recognition in the beaver: a field test for phenotype matching. *Anim. Behav.*, 54, 493–502.

Symmes, D. & Biben, M. 1985. Maternal recognition of individual infant squirrel monkeys from isolation call playbacks. *Am. J. Primatol.*, 9, 39–46.

Treves, A. 1997. Primate natal coats: a preliminary analysis of distribution and function. *Am. J. Phys. Anthropol.*, 104, 47–70.

Treves, A. 2000. Prevention of infanticide: the perspective of infant primates. In: *Infanticide by Males and Its Implications* (Ed by C. P. van Schaik & C. H. Janson), pp. 223–238. Cambridge: Cambridge University Press.

Trivers, R. L. 1971. The evolution of reciprocal altruism. *Q. Rev. Biol.*, 46, 35–57.

Vokey, J. R., Rendall, D., Tangen, J. M., Parr, L., & de Waal, F. B. M. In press. On visual kin recognition and family resemblance in chimpanzees. *J. Comp. Psychol.*

Von Boehmer, H. & Kisielow, P. 1991. How the immune system learns about self. *Sci. Am.*, 265, 74–81.

Walters, J. R. 1987. Kin recognition in non-human primates. In: *Kin Recognition in Animals* (Ed. by D. J. C. Fletcher & C. D. Michener), pp. 359–393. New York: Wiley & Sons.

Wasserman, E. A. 1995. The conceptual abilities of pigeons. *Am. Sci.*, 83, 246–255.

Welker, C., Schwibbe, M. H., Schäfer-Witt, C., & Visalberghi, E. 1987. Failure of kin recognition in *Macaca fascicularis*. *Folia Primatol.*, 49, 216–221.

Widdig, A., Nurnberg, P., Krawczak, M., Jurgen Striech, W., & Bercovitch, F. 2001. Paternal relatedness and age proximity regulate social relationships among adult female rhesus macaques. *Proc. Natl. Acad. Sci. USA*, 98, 13769–13773.

Wright, A. A. & Roberts, W. A. 1996. Monkey and human face perception: inversion effects for human faces but not monkey faces or scenes. *J. Cognit. Neurosci.*, 8, 278–290.

Wu, H. M. H., Holmes, W. G., Medina, S. R., & Sackett, G. P. 1980. Kin preference in infant *Macaca nemestrina*. *Nature*, 285, 225–227.

14

Developmental Aspects of Kin Bias in Behavior

Carol M. Berman

Immatures in many nonhuman primate species both show varying degrees of kin bias in their social interaction with other members of their group and may be subject to the effects of kin bias on the part of group members from an early age. Although recent studies raise the possibility of a role for phenotypic matching in kin bias, current perspectives view preferences for kin among young primates primarily as outcomes of early association with the mother and her close associates (who tend to be kin) and with similarly-aged peers (chapter 13). However, this is usually where the discussion of the development of kin preference ends. Relatively little research has focused on the processes and precise mechanisms by which early associations translate into preferences for kin. Nor do we understand the processes by which preferences for kin remain stable or change over the course of development. In many cases, early kin-based preferences endure until adulthood and appear to form the basis for long-term affiliation and exchange of benefits. In other cases, the effects are more temporary, and hence potentially provide only shorter term benefits.

I define kin bias here as the tendency to affiliate disproportionately with kin within a group that contains individuals who vary in degrees of relatedness to one another. Thus kin bias refers to observable behavior favoring closer kin over more distantly related and unrelated group members. It is more specific than kin discrimination, a term that also includes avoidance of kin. Kin bias may be considered an outcome of unobservable kin recognition processes (see chapter 13), but kin recognition does not guarantee kin bias. This is because the display of kin bias may also depend on assessments of costs and benefits of favoring kin, and, especially in the case of youngsters, on constraints imposed by others (chapter 16).

When considering the development of kin bias, it is also useful to distinguish between different forms based on the extent to which kin-biased interaction represents an active preference on the part of the youngster to initiate interaction with kin or to tolerate interaction initiated by kin. This active form of kin bias contrasts with what may be termed passive kin bias, in which youngsters interact more with kin than nonkin but display no active preference or tolerance for them. This can occur when, for example, infants initiate contact disproportionately with kin because their mothers carry them preferentially to kin, or when youngsters are surrounded by kin and interact indiscriminately with whomever is near. Youngsters may also receive disproportionate amounts of positive interaction from kin as a result of either active or passive forms of kin bias on the part of the kin partner. Thus, a youngster may be the passive recipient of active forms of kin bias, if it does not display an active preference for kin in return. Preferential treatment of kin is, of course, a prerequisite for kin selection and kin-biased mutualism or reciprocity, although it is not sufficient by itself to demonstrate these evolutionary processes (e.g., Chapais 2001). While both passive and active kin bias can direct benefits toward kin, active kin bias would be expected to do so more efficiently and under a wider range of circumstances (Walters 1987).

This chapter is concerned with describing broad patterns of kin bias among immature nonhuman primates. Because excellent recent reviews of mother-offspring relationships (e.g., Fairbanks 1996, 2000) and father-offspring relationships (e.g., van Schaik & Paul 1996, Paul et al. 2000) are available, this chapter focuses on kin relationships more distant than parent-offspring relationships. Similarly, to avoid undue overlap with several more general reviews of kin-biased behavior (e.g., Gouzoules 1984, Gouzoules & Gouzoules 1987, Walters 1987, Bernstein 1991, Silk 2001, chapter 7), this chapter focuses on a few understudied areas of special relevance to the development of young relationships: transitions from passive to active forms of kin bias, hypothesized developmental processes leading to various degrees of kin bias, and the possible implications of developmental processes for social structure. What little work has been done on these issues primarily concerns infants; hence the treatment of juveniles and adolescents is particularly brief and speculative. Also by necessity, the chapter concentrates primarily on studies of kin bias toward maternal kin in matrilineally based societies. However, where possible, kin bias in patrilineally based groups and among patrilineally related kin is also described.

Methodological Considerations

Techniques for examining the extent to which individuals display kin bias are similar for adults and immatures and require consideration of many of the same issues (reviewed in chapter 7). Generally, to detect kin bias, researchers compare rates of affiliative, tolerant, and supportive interaction that youngsters engage in with kin and nonkin, taking into consideration the number of potential partners in each age-sex-kinship class within the group. To avoid obscuring distinctive characteristics of immature social networks, it is crucial to analyze immature subjects separately from adults. While this point may seem obvious, it is frequently not done, particularly in small social groups. Analyzing kin effects within age and sex classes is also important (but difficult in small groups) because youngsters' social networks are shaped by multiple factors in addition to kinship, especially their own and their partners' age and sex. Hence, age and sex effects can interact with kin effects among developing primates or mask them in some cases. Finally, as for adults, it is important to control for possible correlations between kinship and rank distance, particularly in many cercopithecine species where close kin (and/or their mothers) tend to acquire similar ranks within the group (de Waal 1991, Kapsalis & Berman 1996).

Where possible, it is more useful to operationally define kin relationships in terms of degrees of relatedness rather than as a dichotomous variable (kin vs. nonkin) (e.g., Kurland 1977, Kapsalis & Berman 1996, Chapais et al. 1997). In the case of youngsters, it is particularly important to analyze mother-offspring interaction ($r = .5$) separately from interaction with other kin categories, because rates of mother-offspring interaction are often high enough to obscure the presence or absence of kin effects in other kin categories (e.g., Ehardt 1988). In some cases, important distinctions have been made by further dividing kin categories into direct (lineal ascendants or descendents) and collateral (siblings, aunts/nieces, etc.) kin. For example, in an experimental study of Japanese macaques (*Macaca fuscata*), Chapais et al. (2001) found that adult females that were directly related to juveniles by as little as $r = .125$ (e.g., mothers, grandmothers, and great-grandmothers) aided the juveniles to outrank peers. However, collaterally related females consistently aided only those juveniles who were related by at least $r = .25$ (half siblings).

Many authors rely on measures of proximity to describe social bonds, emphasizing the tendency for proximity patterns to correlate with rates of more intimate affiliative interaction. However, others (Kurland 1977, Walters 1987, Bernstein 1991, Chapais 2001) point out that proximity patterns do not allow one to distinguish coincidental association between partners from active attraction. These authors emphasize the importance of directly examining affiliative and supportive interaction, and of taking into account variations in the spatial availability of different partner classes. This issue may be particularly important when attempting to identify passive versus active forms of kin bias in youngsters. Because youngsters tend to spend large amounts of time near their mothers, they may spend disproportionately large amounts of time with other kin coincidentally because both partners are attracted to the youngster's mother or because their mothers are close associates. Studies that use more intimate and directional forms of interaction, such as rates of approaches or grooming initiations, without taking proximity patterns into account may also risk this kind of confusion, if youngsters interact disproportionately with individuals who happen to be near them.

A final issue of concern for detecting active kin bias in youngsters, and particularly infants, is maternal control. Mothers in many species exert control over the interactions of their immature young with other members of the group and tend to be more tolerant of kin than nonkin (e.g., Spencer-Booth 1968, Hrdy 1976, Cheney 1977, Fairbanks 1990). Hence it may be necessary to distinguish association patterns that are likely and unlikely to be influenced by direct maternal control (de Waal 1996).

Types of Behavior

The most detailed studies of early kin bias have been on macaques and other female-bonded species, in which adult females show strong to moderate preferences for matrilineal kin (see reviews in Gouzoules 1984, Gouzoules & Gouzoules 1987, Walters 1987, Bernstein 1991, Silk 2001, chapter 7). In these species, infants and young juveniles generally manifest kin effects in proximity associations and in a wide range of affiliative behaviors directed both to and by other group members, for example, approaching, contact, embracing, grooming, mounting, huddling, and carrying (Berman 1982a; Glick et al. 1986a, b; Walters 1987; Janus 1989; de Waal 1996). Moreover, these effects generally persist when rank distance (Berman et al. 1997) or the spatial availability of partners (Berman 1982a, Glick et al. 1986b, Walters 1987) is controlled. Some types of interaction are also more likely to be reciprocal among kin than nonkin. For example, grooming among immature rhesus monkeys (*M. mulatta*) is more evenly distributed between siblings than between nonsibling pairs, and siblings are more likely than nonsiblings to be each other's most frequent (immature) grooming partner (Janus 1989). Behaviors that reflect tolerance (e.g., co-feeding and drinking) also tend to be more common among immatures and their matrilineal kin (Yamada 1963, de Waal & Luttrell 1989, de Waal 1993). Gentle handling, care, and adoption of infants have been described as highly kin biased in many species (Hrdy 1976, Johnson et al. 1980, Hamilton et al. 1982, Goodall 1986, Nicolson 1987, Pusey 1990, Silk 1999, Chism 2000; but see Hayaki 1988). Nevertheless there are exceptions, particularly among species in which mothers tolerate interaction between infants and a wide range of group members (e.g., Hrdy 1977, Dolhinow & DeMay 1982, Maestripieri 1994a, Ogawa 1995). Social play tends to be less kin correlated than other interactions in that infants and juveniles tend to prefer same-sex partners of similar size (age) when available. However, close kin may be frequent play partners when more suitable play partners are not available or when well-matched kin are available (reviews in Walters 1986, 1987).

The relationship between rates of aggression and kinship is complex and inconsistent (Walters 1987, Bernstein 1991). A number of studies report that older matrilineal kin are less likely than nonkin to harass, mishandle, or attack infants and young juveniles (e.g., Hrdy 1976, Berman 1980a, Silk et al. 1981, Cords & Aureli 1993, Silk 1999). However, others report higher rates of aggression involving immature kin than nonkin, particularly with other immatures (e.g., de Waal & Luttrell 1989). In some but not all cases (e.g., Glick et al. 1986b, Bernstein et al. 1993), the relationship disappears or reverses when variations in spatial availability are taken into consideration. Moreover, in general, the most intense and potentially serious forms of aggression are found less often among kin (Colvin & Tissier 1985, Janus 1991; but see Bernstein & Ehardt 1986). Bernstein and Ehardt (1986) suggested that much aggression between immatures and their kin serves to socialize youngsters through punishment. While this may account for high rates of aggression between youngsters and adult kin, Janus (1991) favored an explanation based on frequent conflicts over access to third parties, particularly mothers, to account for high rates of aggression between immature siblings. Immature siblings may also engage in frequent dominance interactions, particularly when one has targeted another for rank reversal (Walters 1980, Datta 1988).

When infants and juveniles are attacked, mothers and other close matrilineal kin are more likely to intervene than less closely related counterparts, especially when the action puts the aider at risk (Cheney 1977, Kurland 1977, Massey 1977, Kaplan 1978, Berman 1980a, Janus 1992, Bernstein et al. 1993, Hardy 1997). However, the extent to which close kin aid immatures in risky situations is also influenced by rank; low-ranking mothers and their female kin intervene on behalf of infants and juveniles less consistently in a less confident manner than their high-ranking counterparts (Cheney 1977, Berman 1980a, Horrocks and Hunte 1983, Chapais & Gauthier 1993). Both differential patterns of intervention by high- and low-ranking kin, and differential rates of aggression toward infants by older high-born (and often unrelated) and low-born group members have been proposed as the primary mechanisms by which infants initially learn to challenge only those infant peers whose mothers rank lower than their own (Horrocks & Hunte 1983, Chapais & Gauthier 1993), although the experimental studies necessary to support the relative importance of these and other hypotheses have not yet been completed. In species where females eventually acquire ranks in the adult female hierarchy similar to those of their mothers, the same mechanisms have been implicated in learning the identities of older individuals to target eventually for rank reversal. Experimental studies indicate that challenges to these targets, supported by mothers and other close female kin, are the primary means by which young females acquire independent ranks in the adult female hierarchy just below their own mothers, and hence by which kin-correlated hierarchies are generated (Chapais 1988a). Females are able to maintain their ranks after reversal through support by mothers, close kin, and other dominant females (Chapais 1988b, Chapais et al. 1991).

Although a few studies have described postconflict affiliation (reconciliation) in infants (Weaver & de Waal 2000), none has examined the extent to which infants reconcile following conflicts with kin and nonkin. No kin bias in reconciliation was seen among juvenile long-tailed macaques (*M. fascicularis*) when they were attacked by adults, and negative kin bias (i.e., more reconciliation with nonkin than kin) was seen when they were attacked by peers (Cords 1988, Cords & Aureli 1993). This represents an apparent developmental discontinuity in that adult female long-tailed macaques, like those in several other species, reconcile with kin more than nonkin (Aureli et al. 1997). Cords and Aureli (1993) tentatively suggest that negative kin bias may indicate more stability or security in the relationships of juvenile kin than nonkin and hence less need for relationship repair.

Early Developmental Processes

Initial Mirroring

Nonhuman primate infants typically begin by showing affiliative patterns toward kin and nonkin that reflect the relationships of their mothers (see figure 14.1). Hence infants in species where adult females show strong preferences for close matrilineal kin (e.g., rhesus: Sade 1965; Japanese macaques: Yamada 1963, Kurland 1977; pigtail macaques [*M. nemestrina*]: Rosenblum et al. 1975), also show strong tendencies to associate with close matrilineal kin (Rosenblum et al. 1975, Berman 1982a, Thierry 1985, Glick et al. 1986b; see also Nakamichi 1996). Conversely, infants in species where adult females display moderate or no preferences for close maternal kin (e.g., bonnet macaques [*M. radiata*]: Rosenblum et al. 1975; stump-tails [*M. arctoides*]: de Waal & Luttrell 1989, Butovskaya 1993; Tonkean macaques [*M. tonkeana*]: Thierry et al. 1990; moor macaques [*M. maura*]: Matsumura & Okamoto 1997; sooty mangabeys [*Cercocebus atys*]: Ehardt 1988, Gust & Gordon 1994), tend to do the same (e.g., Rosenblum et al. 1975, Thierry 1985). This mirroring of kinship networks is undoubtedly due at first to two tendencies: (1) mothers and infants maintain close proximity to one another, and hence infants are exposed disproportionately to the mothers' close associates and their offspring; and (2) in species where mothers exert a large degree of control over their infants' interactions with other group members, mothers typically display more tolerance for their close associates (and their associates' offspring) than for others. Hence, initial forms of kin bias most likely represent passive kin bias on the part of infants, although older kin may be actively and differentially attracted to very young infants (Ber-

Figure 14.1. A Tibetan macaque *(Macaca thibetana)* mother and her infant huddle with another adult female, one of the mother's closest associates. Early social networks of mothers and infants tend to resemble one another, whether or not the mothers display high degrees of kin bias. Photo by Carol Berman.

man 1982b). De Waal (1996) uses the term "dependent affiliation" to describe early relationships, highlighting parallels with the notion of dependent rank (Kawai 1958) and explicitly linking the phenomenon to maternal influences.

Transition to Active Kin Bias

Maternal Transmission

The same factors that serve to establish early passive kin bias on the part of the infant (i.e., close proximity with mothers and maternal control over infant interaction) are hypothesized to lead eventually to the development of active forms of kin bias typical of juveniles and adults (Walters 1987, Bernstein et al. 1993). As an infant becomes progressively more independent from its mother, it not only spends less time near her, its mother also exerts less control over its interactions with others. Nevertheless, infants generally continue to pursue relationships with the same individuals with whom they interacted previously (Berman 1982a, Nakamichi 1996). In this sense, the development of active kin bias can be seen as a process of social transmission through the mother (Galef 1988, de Waal 1996). The notion of maternal transmission is supported by evidence of moderate to high correlations between rates of peer-peer affiliation among immatures and rates of affiliation between the peer's mothers (captive rhesus: de Waal 1996; free-ranging Japanese macaques: Nakamichi 1996). Similarly, measures of degree of kin bias among free-ranging infant rhesus monkeys correlate moderately with those of their mothers and are positively related to amounts of time spent near mothers (Berman et al. 1997, Berman & Kapsalis 1999).

Timing of the Transition

As yet, no data definitively pinpoint the ages at which youngsters first develop active forms of kin bias. To do so would require controlling simultaneously for the effects of the mother's control over her infants and the spatial availability of kin and nonkin. Nevertheless, two sets of data illustrate the gradual nature of the development of kin bias when infants are not under the full control of the mother. De Waal (1996) examined the extent to which 24 captive immature female rhesus monkeys distributed their approaches disproportionately to related peers while controlling in part for maternal influences. Specifically, he compared approaches that took place within a meter of the mothers with those that took place beyond a meter from the mothers. De Waal found that immature females gave or received significantly more approaches from kin than nonkin when near the mother from the earliest weeks of life. Significant evidence of kin bias when more than a meter from the mother first appeared when the youngsters were 7 to 12 months of age, although a nonsignificant trend appeared at 4 to 6 months of age. De Waal (1996: 149) suggested that "rhesus infants develop a preference to approach related peers independently of their mothers somewhere between 3–6 months of age."

Similar preliminary data on 30 free-ranging rhesus infants (Berman unpublished data) examined approaches both among infant peers and between infants and adult females. Unlike de Waal's study, approaches by and to infants were examined separately, and a longer distance from the mother was considered (beyond 5 m) in an attempt to further reduce the potential for direct maternal influence on approaches. As in de Waal's study, in all cases, approaches made near the mother (within 5 m) showed evidence of kin bias from the earliest weeks of life. The age at which infants first approached related partners significantly more than unrelated partners when they were at a distance from the mother varied with the type of partner: 17 to 22 weeks for peer partners and 25 to 30 weeks for adult female partners. Interestingly, approaches by adult females to infants when at a distance from the mother showed no evidence of kin bias even by the time infants were 25 to 30 weeks of age (the last age period analyzed). Thus, at 25 to 30 weeks, infants displayed kin-biased approaches to adult females even in situations in which it was not reciprocated.

Hypothetical Mechanisms of Maternal Transmission

Theoretically, maternal transmission of kin bias may be proactive on the part of the mother or the infant. In the former, the mother transmits kin bias through active proximity seeking and control of her infant's interactions with kin and nonkin, responding primarily to perceived risks and opportunities for her infant (Maestripieri 1995). In the latter, the mother transmits kin bias by serving as a secure base (Bowlby 1969) for her infant, wherein the infant actively seeks proximity with the mother in response to perceived risk to itself or other sources of separation anxiety. In either case, staying near the mother brings the infant in contact with her close associates, limits its interactions with others, and provides it with opportunities to learn the mother's social preferences. Following this, hypothesized learning mechanisms for the maternal transmission of kin bias include familiarity based on differential amounts of exposure and interaction with kin and nonkin (Bekoff 1981), observational learning with the mother as model (Altmann 1980, Evans & Tomasello 1986), and active and selective maternal intervention (Spencer-Booth 1968, Timme 1995). At this point, our understanding of the relative contributions of each hypothesized mechanism is still fragmentary, and it is possible that all play a role in the development of kin bias.

Familiarity

Association with kin-biased mothers not only gives rise to passive kin bias, as described above, it also provides the conditions under which active preferences for kin are likely to develop. The tendency for individuals to form active preferences for familiar conspecifics, frequent interactants, and even familiar inanimate objects is well documented in a variety of taxa (review in Hinde 1974). This has been studied intensively, primarily in the context of parent-infant/chick relationships, e.g., imprinting in precocial birds (e.g., Bateson 1973) and parental attachment in humans (Bowlby 1969), nonhuman primates (Suomi 1995), and other mammals (Rajecki et al. 1978). However, the importance of prior association in other social relationships, including those of older youngsters and adults, has also been documented (MacKenzie et al. 1985, Welker et al. 1987, Martin 1997). Attachment studies have long noted the formation of strong bonds between infants and major caretakers even in cases where the quality of interaction is poor, although the extent to which this applies to other frequent interactants is less clear. Finally, these studies describe initial periods of broad social openness in young infants in which moderately novel stimuli are attractive, attachments are formed easily, and there is little fear of strangers, followed by a narrowing of social receptivity when unfamiliar conspecifics are more likely to be avoided, particularly when the infant is not near the mother. The advent of "stranger anxiety" tends to coincide with a stage of increased mobility on the part of the infant, and may serve to translate early patterns of exposure while under the direct control of the mother into longer term social preferences. It has also been suggested that extensive early social experience before the onset of stranger anxiety may delay its onset (Schaffer 1966) in humans and result in less stranger anxiety later in life.

Observational Learning

True imitation is probably not involved in the transmission of social preferences from mother to offspring, not only because it is believed to be rare in nonhuman primates (Tomasello & Call 1997), but also because, as currently conceived, it is concerned with the learning of novel acts or skills rather than the direction of common, species-typical behaviors toward particular partners. Nevertheless, infants may learn their mothers' preferences for kin through other forms of observational learning. Altmann (1980) suggests that infant baboons may be specially prepared to learn about their mothers' relationships with other females through her differential display of fear or distress in their presence, and that such emotional responses may be transmitted to their infants through a form of observational conditioning (e.g., Galef 1988) much the same way that a fear of snakes may be transmitted between conspecifics. Alternatively, mothers could serve as foci for directing infants' attention to preferred conspecifics, in a process of stimulus enhancement similar to the process by which they learn the mother's preferred foods (Hikami et al. 1990).

Maternal Intervention

Finally, mothers may influence their infants' development of social preferences both by disrupting interaction with distant kin or unrelated individuals or by actively encouraging infants to interact with kin. Selective disruption of interactions with nonkin has been widely reported in species that are generally intolerant of interaction between infants and other group members (Hrdy 1976), whereas active encouragement is not well documented and may be more likely to occur in relatively more tolerant species. In harmony with this suggestion are observations of a form of active encouragement among Barbary macaques (*M. sylvanus*), a relatively tolerant species. Mothers with daughters preferentially engage in ritualistic triadic interactions (in which adults simultaneously handle one of their infants) with related mothers, but mothers with sons do so with unrelated mothers (Timme 1995). Timme hypothesizes that this may be a mechanism by which daughters are encouraged to form more highly kin-biased social networks than sons. However, more work is needed to validate this function and to document the distribution of disruptive versus encouraging intervention among species.

Independent Learning

While maternal transmission undoubtedly plays an important role in the development of kin bias, infants may also learn directly through interaction with kin and nonkin. Indeed, recent evidence suggests that maternal transmission may not be sufficient to account for observed patterns of kin bias among free-ranging infant rhesus monkeys. First, by 25 to 30 weeks, individual infants display degrees of kin bias that are only moderately correlated with those of their mothers. Thus by this age, infants tend to modify rather than duplicate their mothers' patterns. Second, the degree to which individual infants display patterns similar to those of their mothers is not related to the amount of time they spend together (Berman & Kapsalis 1999).

The tendency to modify rather than duplicate the mother's patterns suggests more complicated processes than maternal transmission alone. One possibility is that older infants may be better able to express preferences for similarly-aged peers regardless of their maternal relatedness (e.g., Berman 1982a). In groups with high male reproductive skew, such attractions would not only dilute the mother's network, but also increase probabilities of forming relationships with paternal kin (Altmann 1979, Altmann et al. 1996).

Another possibility is that infant kin networks are products of socially biased independent learning (sensu Galef 1995). Galef defines socially biased independent learning as socially transmitted behavior that is modified by the subsequent consequences of its performance by the learner. Under this hypothesis, infants learn to prefer kin partly through the quality (as opposed to quantity) of direct experiences with kin and nonkin. Through this direct experience, infants could either reinforce or moderate patterns of affiliation acquired initially through the mother. Certainly, older kin and nonkin provide the raw experience from which such independent learning could derive. As discussed above, kin are more likely than nonkin to initiate friendly and supportive interaction (see figure 14.2) and less likely to threaten or attack infants. Two indications that free-ranging rhesus infants may respond directly to their own experiences with others are that (1) individual infant degrees of kin bias are associated with aspects of social risk in their immediate surroundings independently of mother-infant interaction, and (2) the infants that develop degrees of kin bias most like those of their mothers are those whose exposures to social risk most resemble those of the mother (Berman & Kapsalis 1999).

Note that even the independent learning component of the socially biased independent learning hypothesis does not deny a role for the mother in the development of kin bias, but rather builds upon associations between the mother and her close kin. The ways in which kin and nonkin treat a given infant are expected to depend largely on their relationships with the infant's mother (and on their propensities to associate the infant with its mother—see chapter 15). What this hypothesis suggests is that infants learn about social risk and opportunity both directly from the quality of their interactions with kin and nonkin and indirectly through their mothers. It does not involve social transmission through kin and nonkin (sensu Galef 1988), because infants do not necessarily acquire the same behavior patterns as their interactants.

Figure 14.2. Two immature rhesus monkey *(Macaca mulatta)* siblings groom together. While maternal transmission undoubtedly plays an important role in the development of kin bias, infants may reinforce or moderate their preferences for kin through direct interaction with kin and nonkin. Photo by Carol Berman.

Maternal Transmission versus Independent Learning

The balance between maternally mediated and more direct influences on kin bias most likely changes as infants mature. Maternal factors predominate in the early weeks (de Waal 1996), whereas direct influences may take on increasing importance over time. Such a changing balance of influences would allow for the maternal transmission of kin bias per se over generations, but also permit maturing individuals to modify their mother's patterns to some extent to conform more closely to their own unique attributes and perceptions of social circumstances. Infants' perceptions are likely to be based on a variety of factors, including their relative size and vigor, individual temperament (Suomi 1991), cues about social risk and opportunity provided by mothers and others (Altmann 1980), and expectations of aid from mothers and other kin (Berman 1980a, Chapais & Gauthier 1993, Chapais et al. 1997).

Sources of Variation

Sex and Lineage Rank

Many factors, including attributes of mother and infant, the social environment, and the external environment may potentially influence patterns of kin bias within species. However, current evidence indicates that the immature's sex is not a major source of variation in degree of kin bias until after infancy and the early juvenile period (Berman 1978; Ehardt-Seward & Bramblett 1980; Glick et al. 1986a, b; Berman & Kapsalis 1999). A few studies suggest that cercopithecine infants and juveniles from higher ranking matrilines spend more time with and play more with their own kin than do their counterparts from lower ranking lineages (rhesus: Berman 1982a; vervets [*Cercopithecus aethiops*]: Ehardt-Seward & Bramblett 1980, Fairbanks 1988b; Japanese macaques: Imakawa 1990). The same tendency among adults has been explained as the result of attraction to high rank, a tendency that reinforces association rates among high-ranking kin but not among low-ranking kin (Seyfarth 1980), but this hypothesis has not been tested specifically for youngsters.

Maternal Protectiveness/Restrictiveness

Most research on individual differences in kin bias has focused on variation in maternal behavior in response to the social milieu, and has given rise to the hypothesis that the development of kin bias is an outgrowth of a functional system by which mothers protect infants from injury and harassment. Briefly, this body of literature, reviewed below, suggests that infants develop higher degrees of kin bias when reared by protective or restrictive mothers than when reared by relaxed or tolerant mothers. Mothers adopt protective or restrictive behavior to the extent that they perceive their infants to be vulnerable to injury or harassment, and this depends in turn on the social milieu in which mothers must raise their infants.

Of particular interest has been the degree to which mothers and infants remain near one another and the extent to which mothers restrict their infants' interaction with other group members (Rosenblum et al. 1975, Berman 1980b, Berman & Kapsalis 1999). In classic research, Rosenblum and his colleagues compared mother-infant relationships and infant social networks in identically housed groups of pigtail and bonnet macaques. Pigtail macaque mothers, who show strong preferences for kin, are relatively intolerant of interaction between their infants and other group members, other than close kin. Bonnet macaque mothers, in contrast, show less kin bias in their own social preferences and are tolerant of interaction between their infants and a wide range of group members (see also Mason et al. 1993, but see Silk 1980). As a result, bonnet macaque infants have greater opportunities for positive social interaction with a wide range of group members virtually from birth. As they mature and develop their own social networks, bonnet macaques display social networks that show little or no kin bias, whereas pigtail macaques display marked preferences for kin. Similar links between maternal protectiveness and infant kin bias have also been seen within groups of free-ranging infant rhesus monkeys. The infants whose mothers played relatively large roles in maintaining proximity to their infants developed relatively high degrees of kin bias toward other group members (Berman & Kapsalis 1999).

Infant Vulnerability

Maternal proximity seeking and restrictiveness appear to be associated, both between and within species, with a number of factors related to the vulnerability of infants to injury, particularly attack or harassment by conspecifics. In a comparative study of rhesus, pigtail, and stump-tailed macaques, Maestripieri (1994a, b) found a relationship between relatively low levels of maternal tolerance for infant handling by other group members and relatively high ratios of abusive to affiliative infant handling by other adult females. In another comparative study of rhesus, long-tailed, and Tonkean macaques, Thierry (1985) found that Tonkean mothers, whose relationships with other females are marked by considerably less intense agonism than those of the other species, also were less restrictive with their infants, allowing them to interact more with a wide range of group members, including nonkin. In both studies, no differences were found between species in levels of interest in infants. Hence, differences in levels of interaction with unrelated group members were hypothesized to be results of differences in maternal styles. While the infants in these studies were at most only 10 to 12 weeks old, differences in infant initiative could be a factor in older, more mobile infants (see also Dolhinow & DeMay 1982, Fairbanks 1988b, Chism 2000).

Some of the many factors that mothers respond to with increased maternal proximity seeking and protectiveness (reviews in Nicolson 1987, Fairbanks 1996) include arboreality (e.g., Johnson & Southwick 1987; see also Chalmers 1972, Sussman 1977), cold temperatures (Hiraiwa 1981, Nicolson 1982), human disturbance (Johnson & Southwick 1987, Berman 1989), previous infant mortality (Fairbanks 1988a), the presence of potentially infanticidal males (Fairbanks & McGuire 1987), primiparity (Seay 1966, Mitchell & Stevens 1968, Hooley 1983, Berman 1984), low rank (Altmann 1980, Nicolson 1982), the absence of supportive kin (Fairbanks 1988b, Berman 1992), and provisioning (Hauser & Fairbanks 1988), particularly in combination with large group size and/or high social density (Berman et al. 1997, Maestripieri 2001). Although all these factors might be expected to boost levels of kin bias in infants, only group size/social density has been examined in detail.

Group Size/Social Density

In a natural experiment among free-ranging infants on Cayo Santiago, Puerto Rico, Berman et al. (1997) examined changes in maternal behavior, infants' social milieux, and infant kin bias in a single social group during 17 years of rapid expansion and fission and in two daughter groups following fission. As the groups expanded in size, they became increasingly dense; infants found themselves moderately near (<5 m) larger numbers of group members at any one time, and a larger proportion of the individuals found near infants were only distantly related or unrelated to them. In addition, distantly related and unrelated individuals in larger groups spent less time near the infant on a per capita basis than those in small groups; hence not only were there were more such individuals near infants, each of them was less familiar to the infant and presumably the mother. Mothers appeared to respond to these changes with increased protection of infants—those in larger groups spent more time in proximity to their infants and took larger roles in maintaining proximity to them. By 25 to 30 weeks of age, infants born into the larger groups developed social networks that were more highly focused on kin than those born into smaller groups. When the original group fissioned, all these trends reversed, giving rise to the hypothesis that changes in group size/ density, infant social milieux, maternal behavior, and infant kin bias were causally linked.

Additional support for the hypothesis that high social density poses risks to infants comes from research on captive rhesus. Mull and Berman (unpublished data) indicate that both an infant's probability of being the target of threat or aggression and a mother's probability of displaying signs of anxiety increase significantly with numbers of group members near the infant. Similar links have been found between maternal signs of anxiety, probability of infant harassment, and social density in comparisons of two captive rhesus colonies (Maestripieri 2001). Hence, these studies suggest that both risk of actual injury and mothers' perceptions of risk to their infants increase with numbers of individuals near the infant. Given this observation, it is not surprising that captive rhesus mothers in densely housed colonies restrict their infants more than those in less densely housed colonies (Maestripieri 2001), and that individual free-ranging rhesus mothers adjust their proximity to their infants on a minute-to-minute basis as a function of the number of group members (particularly males) nearby (Berman 1988).

While maternal responses to group size/social density appear important, infants and other individuals may also contribute to the increased development of kin bias in large/dense groups. For example, infants and other kin may also respond to increased group size/density by seeking more proximity with the infant's mother. Alternatively, infants and other kin may respond by seeking more contact directly with one another. Clearly, more research is needed on this issue, but these possibilities are in harmony with findings that free-ranging rhesus infants in large groups show more signs of anxiety than infants in small groups (Warfield 2001) and that degrees of kin bias of individual infants are positively correlated with mean numbers of other group members near the infant independently of maternal proximity seeking (Berman et al. 1997).

Implications for Social Structure

Intergenerational Transmission of Kinship Structure

The tendency for infants to form preferences for kin as an outcome of their associations with their mothers gives rise to a simple hypothesis for the intergenerational transmission of kin-based affiliative structures among daughters in matrilocal/male dispersal societies. It requires the assumption that patterns established by females in early infancy endure through adulthood, a notion that is supported by longitudinal research on macaques under stable captive and free-ranging conditions (Rosenblum et al. 1975, de Waal 1996, Nakamichi 1996). However, to the extent that older infants or juveniles respond to changes their social environments, or respond differently than their mothers to specific aspects of their environment, kin-based affliative structures may be modified in subsequent generations (e.g., Altmann & Altmann 1979). For example, during a period of environmental hardship, a daughter whose mother associates primarily with a number of closely related adult females could find herself with few related peers. Such a daughter may benefit from forming strong relationships with unrelated peers and their families, and such changes could be transmitted eventually to her own offspring.

Group Cohesion

Low availability of kin would be the rule in groups with low birth rates and high mortality, resulting ultimately in groupwide kinship structures characterized by low degrees of kin bias and high cohesion among families (see chapter 16). Conversely, the data on group growth and infant kin bias on Cayo Santiago (Berman et al. 1997) suggest that under different demographic conditions, processes related to the development of kin bias may ultimately affect group integrity. On Cayo Santiago, high birth rates and low mortality resulted in rapid group growth (Rawlins & Kessler 1986) that led in turn to cohorts of young whose social networks were progressively more highly focused on close kin (and away from distant kin and unrelated group members) (Berman et al. 1997). Presumably this contributed to a gradual weakening of bonds between families and eventually to the series of fissionings that took place along family lines (and in the absence of resource scarcity) (Chepko-Sade & Sade 1979). Alternatively, bonds between families may have weakened because older offspring and adults responded immediately and directly to their social environments (e.g., Judge & de Waal 1997). Clearly, more longitudinal research is needed on responses of different age classes to changes in group size/social density.

Dominance Style

Recently, links between maternal protectiveness, development of kin bias, and group cohesion have been placed within a larger framework of dominance style that posits covariation between these features and several additional aspects of social structure, including intensity of aggression, rigidity of dominance relationships, tolerance around limited resources, conciliatory tendencies, and kin-based dominance hierarchies (Thierry 1985, 2000; de Waal & Luttrell 1989; Maestripieri 1994a, b, c). In species with more despotic dominance styles, individuals (particularly adult females) rely heavily on kin to avoid harassment and serious aggression from others and to gain access to monopolizable resources. Dominance relationships tend to be linear and rigid, and conciliatory tendencies are low. Both aggressive alliances and affiliative behavior tend to be kin biased. Individuals in more egalitarian societies show many of the opposite tendencies, including less serious aggression, more counteraggression, and less factionalism along kinship lines. Given such differences in social milieux, infants in more despotic societies are more vulnerable to harassment or injury from unrelated group members than those in more egalitarian societies. As a result, mothers are hypothesized to adopt more protective care patterns, culminating in both the development of higher degrees of kin bias among infants and in the perpetuation of species-typical patterns of kin bias and factionalism over generations. Ultimately, links between dominance styles and kin bias are hypothesized to be the outcome of different levels of within-group contest competition for limited food resources among females (Sterck et al. 1997), emergent properties based on epigenetic processes (Thierry 1990), demographic processes (Datta 1989), phylogeny (Matsumura 1999), or species-specific levels of aggression (Hemelrijk 1999).

The dominance style concept was developed originally to explain variation between species, and it has been used successfully to explain differences between colobine and cercopithecine patterns of maternal tolerance and infant handling (Maestripieri 1994a, b; see also Hrdy 1976, McKenna 1987). However, recently Paul (1999) pointed to several examples in which predicted relationships between infant handling and dominance style are not seen. For example, infant carrying is common among relatively despotic vervet monkeys (Fairbanks 1990), but may be rare among more egalitarian moor macaques (Matsumura 1997). To fully evaluate the extent to which the development of kin bias can be accommodated within the dominance style framework, more detailed data are needed on the quality of infant handling, degrees of maternal tolerance for kin and nonkin, and the extent to which infants who are not extensively handled nevertheless associate with related and unrelated group members.

Kin Bias Among Juveniles and Adolescents

Continuity and Discontinuity

The extent to which juveniles and adolescents continue to display patterns of affiliation like their mothers varies with a number of factors, most importantly sex and species dispersal patterns (see below). The fact that deviations from maternal patterns are common requires explanation in terms of both function and mechanism. However, so far work in this area is sparse and fragmentary. In general, the juvenile period is characterized as one of high vulnerability to conspecific aggression, predation, and malnutrition due to juveniles' small size, lack of experience, and reduced maternal involvement (Janson & van Schaik 1993). At the same time, juveniles must develop the skills and social strategies needed for later successful reproduction. Thus, it is reasonable to hypothesize that many aspects of juvenile behavior, including affiliative patterns, have been selected to address both current needs for safety and nutrition and the development of skills and social relationships needed later in life (e.g., Fairbanks & Pereira 1993). Within this framework, attraction to kin, particularly older kin, can be seen as an effective means to gain tolerance around limited resources, some degree of protection from predators, and support against conspecifics. In some species, mothers and other close kin continue to defend juveniles and adolescents when attacked (e.g., rhesus: Kaplan 1978, Bernstein & Ehardt 1985; Japanese macaques: Kurland 1977; bonobos [*Pan paniscus*]: Furuichi 1997), although the extent to which they do so may vary by sex in highly dimorphic species (e.g., Pereira 1992, 1995). Due to their tolerance, kin are also likely to provide reliable opportunities for learning skills, such as infant care, tool use, or foraging skills (e.g., Hikami et al. 1990, Boesch 1993, Chism 2000).

In many cases, variation in kin bias patterns by sex and species dispersal patterns can be explained in terms of the future availability of partners and their potential value as future alliance partners. For example, as detailed below, in both female philopatric and male philopatric species, affiliative relationships among related juveniles and adolescents tend to be stronger between same-sex members of the philopatric sex than between same-sex members of the dispersing sex. Following the same principle, strong affiliative relationships often develop between related juveniles that tend to disperse together or join new groups containing known kin (Pereira & Altmann 1985, chapters in Pereira & Fairbanks 1993, see below).

Female Philopatric/Male Dispersal

In many female philopatric/male dispersal species in which adult females show preferences for matrilineal kin, juvenile and adolescent daughters continue to do so as well into adulthood (e.g., Rosenblum et al. 1975, de Waal 1996). In contrast, juvenile sons may show decreasing tendencies to affiliate with maternal female kin as they mature in favor of older males and/or male peers that provide well-matched play partners (e.g., Hayaki 1983, Pereira 1988; review in Walters 1986). In some cases, they maintain affiliative relationships with maternally (and perhaps paternally) related male siblings and peers with whom they may eventually disperse or to whose groups they eventually transfer. For example, free-ranging male adolescent rhesus monkeys frequently make their initial transfers into groups that contain older maternal brothers. When reunited, the brothers both spend disproportionate amounts of time together and form alliances together more frequently than expected by chance (Meikle & Vessey 1981; see also Cheney & Seyfarth 1983, van Noordwijk & van Schaik 1985; but see Melnick et al. 1984, Colvin 1986). However, in most cases these associations dissolve when the brothers disperse again, limiting potential benefits to their initial period away from their natal groups. In another example, maternally and/or paternally related juvenile male langurs (*Presbytis entellus*) ousted from one-male groups after a takeover tend to disperse together and/or to join all-male groups containing older kin. Although there is evidence that fathers defend juvenile sons in their new groups, it is not clear whether related juveniles establish or maintain preferential relationships with one another (Rajpurohit & Sommer 1993, Rajpurohit et al. 1995). In a counterexample, no kin-biased dispersal has been observed among young male yellow baboons (*Papio cynocephalus*) (Alberts & Altmann 1995).

Male Philopatric/Female Dispersal

In species such as woolly spider monkeys (*Brachyteles arachnoides*), where males are philopatric and virtually all females disperse prior to reproducing, adult females residing in the same group are not generally related. Although they tend to spend more time with one another than expected by chance, neither they nor their juvenile or adolescent daughters tend to form strong affiliative bonds with other females. Adult males, on the other hand, who are likely to be related to one another matrilineally and/or patrilineally, form cooperative bonds with one another. As juveniles, young males pursue relationships with these males, who are also likely to be their kin, and achieve increasingly close relationships with them as they enter adolescence (Strier 1993, 1994). However, adult brothers do not necessarily prefer one another over other males (Strier et al. 2002), and the extent to which either immature or adult males differentiate between other closely and distantly related males is not known. Patterns among primarily patrilocal chimpanzees (*Pan troglodytes*) are similar.

Although maternally related sisters are sometimes found in the same group, there is little evidence that they maintain strong bonds as adults either with each other or with unrelated females (Nishida 1989). Bonds between philopatric adult males, while strong, are not necessarily biased toward maternal kin (Goldberg & Wrangham 1997; Mitani et al. 2000, 2002). Immatures of both sexes interact primarily with their mothers, although males become increasingly attracted to older males over time. Juveniles of both sexes also display prominent grooming, allocare, and play relationships with young maternal siblings before adolescence, and are likely to adopt younger siblings if their mother dies (Goodall 1986; Pusey 1983, 1990; but see Nishida 1983, Hayaki 1988). More typically, however, relationships with younger siblings end abruptly when females disperse and males join male subgroups on a more permanent basis (Watts & Pusey 1993). Although some maternal brothers form strong affiliative and supportive relationships within male subgroups, many do not (Goodall 1986, Pusey 1990). Indeed, Mitani et al. (2002) found that adolescent and adult males show strong tendencies to affiliate and support males of similar age and rank, but do not gravitate toward maternal kin even when available. Mitani and colleagues explain this finding in terms of demographic constraints on what is functionally possible; if kin of similar age and status are rarely available within groups, preferences for them are unlikely to evolve (Sherman 1981; but see chapter 16).

Both Sexes Disperse

In species such as mountain gorillas (*Gorilla gorilla*), where most members of both sexes disperse, about 80% of adult females nevertheless spend part of their adult lives with other related adult females, because they either remain in their natal groups or disperse into groups containing female kin. When that occurs, related adult females tend to display preferential affiliative and supportive bonds with one another, but are more likely to do so when they are related maternally than paternally (e.g., Stewart & Harcourt 1987, Watts 1994). Similarly, juvenile males and females who co-reside in groups with related adult females other than their mothers also form preferential affiliative relationships with them. These relationships continue into adolescence for females, although older adolescent sons tend to switch their associations away from related adult females to unrelated estrous females (Watts & Pusey 1993). In contrast, maternal sibling relationships among immature gorillas are not consistently strong. Same-sex age peers (who may be paternally related) are preferred over maternal siblings as play partners, and although females show a lot of interest in infants, they do not interact more with maternal infant siblings than with maternally unrelated infants (Watts & Pusey 1993). In red howler monkeys (*Alouatta seniculus*), female survival and reproductive opportunities are greatly diminished by dispersal, and juvenile females vigorously pursue supportive relationships with both mothers and older sisters in an apparent effort to avoid eviction. As adults, females living in groups with kin have significantly higher reproductive success than those living with nonkin (Pope 2000). Male juveniles begin to pursue alliances with male peers, and choose to ally with maternal kin, fathers, and/or other natal peers (who are related paternally) when available. Small groups of adolescent and adult males later disperse together and cooperate to invade or defend another bisexual group. Coalitions of related males tend to endure longer and their tenure in groups is longer when the males are related than when they are not related (Pope 1990, Agoramorthy & Rudran 1993, Crockett & Pope 1993).

Paternal Kinship Relationships

While many examples of affiliation and alliances between paternal kin are available, evidence that primates distinguish patrilineally related group members from unrelated natal counterparts is only beginning to appear (Alberts 1999, Buchan et al. 2003). Studies of free-ranging rhesus (Widdig et al. 2001) and yellow baboons (Smith et al. 2003) suggest that adult females affiliate more with paternal sisters than with unrelated peers. However, we are still unclear about the cues that may be used. Since the rhesus mothers did not associate preferentially with the mothers of their infant's paternal siblings, Widdig and colleagues rule out maternal transmission as an explanation (but see chapter 13). Preferences for same-aged peers were able to explain most, but not all of the results, both supporting Altmann's suggestion that innate preferences for age mates could lead to kin bias among paternal siblings in groups in which one or a few males sire most of the offspring in a given year (Altmann 1979, Altmann et al. 1996), and raising the possibility of some form of phenotypic matching. However, when these relationships develop and whether they yield reproductive benefits remain unclear. Although such associations between paternal siblings could theoretically have important effects on inclusive fitness, they do not always endure or appear to yield such benefits. For example, Rajpurohit et al. (1995) suggest that paternal brothers in all-male langur groups are unlikely to contribute to each other's reproductive success, because few kin survive long enough to aid their brothers in taking over a bisexual group. Although young hamadryas (*P. hamadryas*) males form affiliative relationships with paternally related peers belonging to the same clan, Colmenares (1992) found that strong cooperative relationships between adult males occurred primarily between fathers and sons and between males that shared a common mother (maternal or full brothers), but not between other paternal half brothers. Finally, where early affiliative relationships between paternal brothers have been seen in gorillas, they have not endured through adolescence (Watts & Pusey 1993; but see Robbins 1996).

A possible exception may be found in Peruvian squirrel monkeys (*Saimiri sciureus*), a female philopatric species. Like red howler males, juvenile males tend to affiliate with age mates and subsequently migrate to new groups as a cooperating unit. However, unlike red howlers, these migration alliances endure over multiple migrations (Mitchell 1994). Some age mates are likely to be paternal brothers, but until definitive data on relatedness are available, it will remain unclear whether related males preferentially form such alliances or whether they form more effective alliances than nonkin.

Developmental Processes for Juveniles and Adolescents

Very little attention has been devoted to the processes by which juveniles and adolescents maintain and modify affiliative patterns related to kinship. Mothers in many species continue to be a focus of interest, particularly for daughters, bringing them in contact with their own close associates. Hence, it is theoretically possible that kinship patterns continue to be transmitted through the mother to some extent. Evans and Tomasello (1986) suggested a process akin to social referencing in humans (e.g., Feinman 1982) to explain social preferences among juvenile chimpanzees whose interactions with social partners took place at a distance from the mother. Although it was not possible to rule out earlier familiarity as an explanation in this study, it is possible that mothers continue to reinforce preferences that offspring formed earlier in life, or to modify them as the mother modifies her own. On the other hand, maternal transmission alone cannot account for qualitative deviations from maternal patterns such as those shown by juvenile and adolescent male woolly spider monkeys (Strier 1993, 1994) or male chimpanzees (e.g., Pusey 1990, Mitani et al. 2002).

Given the limited potential of maternal transmission and the apparent emergence of socially biased independent learning mechanisms by late infancy, it seems reasonable to propose that independent learning plays a major role in maintaining and modifying social preferences among juveniles and adolescents. However, precisely how such learning takes place remains to be discovered. Like infants, juveniles and adolescents are likely to respond to differences in the quality of interaction that others direct toward them. However, as they mature, youngsters are also likely to take increasingly active roles in establishing and maintaining affiliative relationships. Indeed, most studies characterize juveniles as active, selective, and persistent pursuers of affiliative relationships, guided ultimately by potential costs and benefits of present and future relationships with particular classes of individual (see Fairbanks 1993, Fairbanks & Pereira 1993). For example, both male and female juvenile vervet monkeys play primary roles in initiating interaction with older related males and females, but differ in their choices of social partner. Juvenile females favor adult females, particularly if they are high ranking, but show no preferences for brothers over unrelated natal males. In contrast, juvenile males favor adult males, particularly older brothers. In so doing, each sex chooses kin who represent for them potentially powerful protectors or future allies.

On a proximate level, selectivity may be linked to species- and sex-specific temperamental characteristics in youngsters that lead to differential attraction to individuals with particular attributes related to sex, age, rank, familiarity, or personality (Mendoza & Mason 1989, Clarke & Boinski 1995). Given that an individual's style of approach to unfamiliar social partners and situations constitutes one of the most consistent behavioral markers of temperament, it is not unreasonable to hypothesize that temperamental factors play an important role in the extent to which youngsters select social partners on the basis of familiarity/kinship and/or other factors. Another related hypothesis is that paternal siblings may be attracted to one another based on similar inherited personality profiles (Widdig et al. 2001), but at this point it is unclear how reliable such cues may be for paternity in a complex social group.

Temperamentally based attractions may be detectable at birth, although they may interact with maternal care patterns and other kinds of experience during infancy (Clarke & Boinski 1995). Nevertheless, their effect on affiliative patterns may not be fully expressed until later in life. For example, Watts and Pusey (1993: 152) describe how juvenile male chimpanzees "have great interest in joining parties that contain adult males, but often have difficulty in persuading their mothers to accompany them." As they mature, attraction to males, regardless of kinship, eventually overrides attraction to kin. In other cases, such as Fairbanks' vervet monkeys, maturing males and females continue to pursue affiliative relationships with kin, but do so to differing extents with different age and sex classes of partner.

Summary and Future Research

Nearly four decades have passed since Yamada (1963) and Sade (1965) first wrote about kin bias as a central organizing principle of affiliative relationships among macaques and described it as an outgrowth of offspring's enduring relationships with their mothers. Since then a number of hypotheses have been proposed about the course of the development of kin bias and the precise mechanisms involved, but the literature still offers few definitive answers. Nevertheless the available hypotheses, evidence, and speculations offer a rough overview about what may be going on, and highlight a number of issues that need further investigation.

The infant's earliest social networks are necessarily patterned after those of the mother as a result of tendencies to remain in close proximity and for mothers of some species to show more tolerance for interaction with her own close associates, who are usually kin. In social settings in which infants are vulnerable to harassment or injury from conspecifics, this form of passive kin bias probably functions primarily to protect infants by limiting their exposure to potentially harmful individuals and favoring exposure to those who are least likely to harm them and most likely to tolerate and protect them. This hypothesis is supported by links between infant vulnerability, maternal protectiveness, and infant kin bias within and between species. However, detailed data on these issues are lacking for many species (e.g., Paul 1999). The adaptive value of early kin relationships is also suggested by evidence of lower mortality rates for captive vervet and free-ranging Japanese monkeys who possess grandmothers (Fairbanks 1988b, Pavelka et al. 2002), but clearly this is just a beginning.

The transition to active forms of kin bias on the part of infants has not been pinpointed in any species, but evidence suggests that rhesus infants that are at a distance from their mothers display kin bias toward adult females and peers only after several months of life. We also have yet to sort out the relative contributions of several mechanisms by which infants are hypothesized to develop active preferences for kin. Maternally mediated mechanisms probably predominate at first, while socially biased independent learning, based on differences in the quality of interaction with kin and nonkin, may take on increasing importance over time. Sorting out these issues is likely to require both more refined analyses of longitudinal data and careful experimentation, such as that done by Chapais and colleagues on mechanisms of dominance acquisition. Additional longitudinal data are also crucial for further understanding the extent to which early experiences with kin influence the perpetuation and modification of kin-based social structures.

The extent to which juveniles and adolescents maintain affiliative patterns like those of their mothers varies widely by sex- and species-dispersal pattern. Our appreciation of this fact comes largely from recent data on a variety of taxa and is likely to increase as more species are studied. As infants mature into juveniles and adolescents, independent learning and self-initiative probably play increasingly large roles in maintaining or modifying social preferences. Who youngsters choose to pursue and how they respond to the overtures of others appears to be guided ultimately by current needs for protection, nutrition, and opportunities to learn skills as well as by the future availability of allies. Further studies on male philopatric species and on species in which both sexes disperse should be especially useful in distinguishing the extent to which kin bias in youngsters is based on present needs versus future availability of allies. In addition, more work is needed on the benefits those relationships actually provide. Questions about the role of kin bias in dispersal are proving particularly relevant in this regard: how does the availability of kin influence the choice to disperse, the choice of groups to join, and survival and reproduction after dispersal? Finally, more work is needed on possible links between individual, species-, and sex-specific temperamental characteristics and attraction to partners based on kinship/familiarity, sex, age, rank, and personality.

Acknowledgments I am very grateful for the helpful comments of Bernard Chapais, Barbara DeVinney, Frans de Waal, Melissa Gerald, and Bernard Thierry. My thanks also go to Meredith Dorner for assisting me in my search for the literature and to Monique Fortunato for clerical assistance. Finally, I am most grateful to Bernard Chapais for twisting my arm in the first place and for offering support and friendship throughout the process.

References

Agoramoorthy, G. & Rudran, R. 1993. Male dispersal among free-ranging red howler monkeys (*Alouatta seniculus*) in Venezuela. *Folia Primatol.,* 61, 92–96.

Alberts, S. C. 1999. Paternal kin discrimination in wild baboons. *Proc. Royal Soc. Lond., B*, 266, 1501–1506.

Alberts, S. C. & Altmann, J. 1995. Balancing costs and opportunities: dispersal in male baboons. *Am. Nat.*, 145, 279–306.

Altmann, J. 1979. Age cohorts as paternal sibships. *Behav. Ecol. Sociobiol.,* 6, 161–164.

Altmann, J. 1980. *Baboon Mothers and Infants.* Cambridge, MA: Harvard University Press.

Altmann, J., Alberts, S. C., Haines, S. A., Dubach, J., Muruthi, P., Coote, T., Geffen, E., Cheesman, D. J., Mututua, R. S., Saiyalele, S. N., Wayne, R. K., Lacy, R. C., & Bruford, M. W. 1996. Behavior predicts genetic structure in a wild primate group. *Proc. Natl. Acad. Sci. USA,* 93, 5797–5801.

Altmann, S. & Altmann, J. 1979. Demographic constraints on behavior and social organization. In: *Primate Ecology and Human Origins* (Ed. by I. Bernstein & E. O. Smith), pp. 47–63. New York: Garland STPM Press.

Aureli, F., Das, M., & Veenema, H. C. 1997. Differential kinship effect on reconciliation in three species of macaques (*Macaca fascicularis, M. fuscata, and M. sylvanus*). *J. Comp. Psychol.,* 111, 91–99.

Bateson, P. P. G. 1973. Internal influences on early learning in birds. In: *Constraints on Learning: Limitations and Predispositions* (Ed. by R. A. Hinde & J. G. Stevenson-Hinde). New York: Academic Press.

Bekoff, M. 1981. Mammalian sibling interactions: genes, facilitation environments, and the coefficient of familiarity. In: *Parental Care in Mammals* (Ed. by D. J. Gubernick & P. H. Klopfer), pp. 307–346. New York: Plenum Press.

Berman, C. M. 1978. *Social Relationships Among Free-Ranging Infant Rhesus Monkeys.* PhD dissertation, University of Cambridge, Cambridge, UK.

Berman, C. M. 1980a. Early agonistic experience and peer rank acquisition among free-ranging infant rhesus monkeys. *Int. J. Primatol.*, 1, 153–170.

Berman, C. M. 1980b. Mother-infant relationships among free-ranging rhesus monkeys on Cayo Santiago: a comparison with captive pairs. *Anim. Behav.*, 28, 860–873.

Berman, C. M. 1982a. The ontogeny of social relationships with group companions among free-ranging infant rhesus monkeys. I. Social networks and differentiation. *Anim. Behav.,* 30, 149–162.

Berman, C. M. 1982b. The ontogeny of social relationships with group companions among free-ranging infant rhesus monkeys. II. Differentiation and attractiveness. *Anim. Behav.,* 30, 63–170.

Berman, C. M. 1984. Variation in mother-infant relationships: traditional and nontraditional factors. In: *Female Primates: Studies by Women Primatologists* (Ed. by M. F. Small), pp. 17–36. New York: Alan R. Liss.

Berman, C. M. 1988. Demography and mother-infant relationships: implications for group structure. In: *Ecology and Behavior of Food-Enhanced Primate Groups* (Ed. by J. Fa & C. Southwick), pp. 269–296. New York: Alan R. Liss.

Berman, C. M. 1989. Trapping activities and mother-infant relationships on Cayo Santiago: a cautionary tale. *P. R. Health Sci. J.,* 8, 73–78.

Berman, C. M. 1992. Immature siblings and mother-infant relationships among free-ranging rhesus monkeys on Cayo Santiago. *Anim. Behav.,* 44, 247–258.

Berman, C. M. & Kapsalis, E. 1999. Development of kin bias among rhesus monkeys: maternal transmission or individual learning? *Anim. Behav.,* 58, 883–894.

Berman, C. M., Rasmussen, K. L. R., & Suomi, S. J. 1997. Group size, infant development and social networks in free-ranging rhesus monkeys. *Anim. Behav.,* 53, 405–421.

Bernstein, I. S. 1991. The correlation between kinship and behaviour in non-human primates. In: *Kin Recognition* (Ed. by P. G. Hepper), pp. 7–29. Cambridge: Cambridge University Press.

Bernstein, I. S. & Ehardt, C. L. 1985. Agonistic aiding: kinship, rank, age, and sex influences. *Am. J. Primatol.,* 8, 37–52.

Bernstein, I. S. & Ehardt, C. L. 1986. The influence of kinship and socialization on aggressive behaviour in rhesus monkeys (*Macaca mulatta*). *Anim. Behav.,* 34, 739–747.

Bernstein, I. S., Judge, P. G., & Ruehlmann, T. E. 1993. Kinship, association, and social relationships in rhesus monkeys (*Macaca mulatta*). *Am. J. Primatol.,* 31, 41–53.

Boesch, C. 1993. Aspects of transmission of tool-use in wild chimpanzees. In: *Tools, Language and Cognition in Human Evolution* (Ed. by K. R. Gibson & T. Ingold), pp. 171–183. New York: Cambridge University Press.

Bowlby, J. 1969. *Attachment and Loss. Vol. 1: Attachment.* New York: Basic Books.

Buchan, J., Alberts, S. C., Silk, J. B., & Altmann, J. 2003. True paternal care in a multi-male primate society. *Nature,* 425, 179–181.

Butovskaya, M. 1993. Kinship and different dominance styles in groups of three species of the genus *Macaca* (*M. arctoides, M. mulatta, M. fascicularis*). *Folia Primatol.,* 60, 210–224.

Chalmers, N. R. 1972. Comparative aspects of early infant development in some captive cercopithecines. In: *Primate Socialization* (Ed. by F. E. Poirier), pp. 63–82. New York: Random House.

Chapais, B. 1988a. Experimental matrilineal inheritance of rank in female Japanese macaques. *Anim. Behav.,* 36, 1025–1037.

Chapais, B. 1988b. Rank maintenance in female Japanese macaques: experimental evidence for social dependency. *Behaviour,* 104, 41–59.

Chapais, B. 2001. Primate nepotism: what is the explanatory value of kin selection? *Int. J. Primatol.,* 22, 203–229.

Chapais, B. & Gauthier, C. 1993. Early agonistic experience and the onset of matrilineal rank acquisition in Japanese macaques. In: *Juvenile Primates: Life History, Development, and Behavior* (Ed. by M. E. Pereira & L. A. Fairbanks), pp. 245–258. New York: Oxford University Press.

Chapais, B., Gauthier, C., Prud'homme, J., & Vasey, P. 1997. Relatedness threshold for nepotism in Japanese macaques. *Anim. Behav.,* 53, 1089–1101.

Chapais, B., Girard, M., & Primi, G. 1991. Non-kin alliances, and the stability of matrilineal dominance relations in Japanese macaques. *Anim. Behav.,* 41, 481–491.

Chapais B., Savard, L., & Gauthier, C. 2001. Kin selection and the distribution of altruism in relation to degree of kinship in Japanese macaques. *Behav. Ecol. Sociobiol.,* 49, 493–502.

Cheney, D. L. 1977. The acquisition of rank and the development of reciprocal alliances among free-ranging immature baboons. *Behav. Ecol. Sociobiol.,* 2, 303–318.

Cheney, D. L. & Seyfarth, R. M. 1983. Nonrandom dispersal in free-ranging vervet monkeys: social and genetic consequences. *Am. Nat.,* 122, 392–412.

Chepko-Sade, B. D. & Sade, D. S. 1979. Patterns of group splitting within matrilineal kinship groups: a study of social group structure in *Macaca mulatta* (Cercopithecidae: Primates). *Behav. Ecol. Sociobiol.,* 5: 67–87.

Chism, J. 2000. Allocare patterns among cercopithecines. *Folia Primatol.,* 71, 55–66.

Clarke, A. S. & Boinski, S. 1995. Temperament in nonhuman primates. *Am. J. Primatol.,* 37, 103–125.

Colmenares, F. 1992. Clans and harems in a colony of hamadryas and hybrid baboons: male kinship, familiarity and the formation of brother-teams. *Behaviour,* 121, 61–94.

Colvin, J. D. 1986. Proximate causes of male emigration at puberty in rhesus monkeys. In: *The Cayo Santiago Macaques: History, Behavior and Biology* (Ed. by R. G. Rawlins & M. J. Kessler), pp. 131–157. Albany, NY: SUNY Press.

Colvin, J. D. & Tissier, G. 1985. Affiliation and reciprocity in sibling and peer relationships among free-ranging immature male rhesus monkeys. *Anim. Behav.,* 33, 959–977.

Cords, M. 1988. Resolution of aggressive conflicts by immature long-tailed macaques, *Macaca fascicularis. Anim. Behav.,* 36, 1124–1135.

Cords, M. & Aureli, F. 1993. Patterns of reconciliation among juvenile long-tailed macaques. In: *Juvenile Primates: Life History, Development, and Behavior* (Ed. by M. E. Pereira & L. A. Fairbanks), pp. 271–284. New York: Oxford University Press.

Crockett, C. M. & Pope, T. R. 1993. Consequences of sex differences in dispersal for juvenile red howler monkeys. In: *Juvenile Primates: Life History, Development, and Behavior* (Ed. by M. E. Pereira & L. A. Fairbanks), pp. 104–118. New York: Oxford University Press.

Datta, S. B. 1988. The acquisition of dominance among free-ranging rhesus monkey siblings. *Anim. Behav.,* 36, 754–772.

Datta, S. B. 1989. Demographic influences on dominance structure among female primates. In: *Comparative Socioecology: The Behavioural Ecology of Humans and Other Mammals* (Ed. by V. Standen & R. A. Foley), pp. 265–284. Oxford: Blackwell Scientific Publications.

de Waal, F. B. M. 1991. Rank distance as a central feature of rhesus monkey social organisation: a sociometric analysis. *Anim. Behav.,* 41, 383–395.

de Waal, F. B. M. 1993. Co-development of dominance relations and affiliative bonds in rhesus monkeys. In: *Juvenile Primates: Life History, Development and Behavior* (Ed. by M. E. Pereira & L. A. Fairbanks), pp. 259–270. New York: Oxford University Press.

de Waal, F. B. M. 1996. Macaque social culture: development and perpetuation of affiliative networks. *J. Comp. Psychol.,* 110, 147–154.

de Waal, F. B. M. & Luttrell, L. M. 1989. Toward a comparative socioecology of the genus *Macaca*: different dominance styles in rhesus and stumptail monkeys. *Am. J. Primatol.,* 19, 83–109.

Dolhinow, P. & DeMay, M. G. 1982. Adoption: the importance of infant choice. *J. Hum. Evol.,* 11, 391–420.

Ehardt, C. L. 1988. Absence of strongly kin-preferential behavior by adult female sooty mangabeys (*Cercocebus atys*). *Am. J. Phys. Anthropol.,* 76, 233–243.

Ehardt-Seward, C. & Bramblett, C. A. 1980. The structure of social space among a captive group of vervet monkeys. *Folia Primatol.,* 34, 214–238.

Evans, A. & Tomasello, M. 1986. Evidence for social referencing in young chimpanzees (*Pan troglodytes*). *Folia Primatol.,* 47, 49–54.

Fairbanks, L. A. 1988a. Mother-infant behavior in vervet monkeys: response to failure of last pregnancy. *Behav. Ecol. Sociobiol.,* 23, 157–165.

Fairbanks, L. A. 1988b. Vervet monkey grandmothers: interactions with infant grandoffspring. *Int. J. Primatol.,* 9, 425–441.

Fairbanks, L. A. 1990. Reciprocal benefits of allomothering for female vervet monkeys. *Anim. Behav.,* 40, 553–562.

Fairbanks, L. A. 1993. Juvenile vervet monkeys: establishing relationships and practicing skills for the future. In: *Juvenile Primates: Life History, Development, and Behavior* (Ed. by M. E. Pereira & L. A. Fairbanks), pp. 211–227. New York: Oxford University Press.

Fairbanks, L. A. 1996. Individual differences in maternal style: causes and consequences for mothers and offspring. *Adv. Study Behav.,* 25, 579–611.

Fairbanks, L. A. 2000. Maternal investment throughout the life span in Old World monkeys. In: *Old World Monkeys* (Ed. by P. F. Whitehead & C. F. Jolly), pp. 341–367. Cambridge: Cambridge University Press.

Fairbanks, L. A. & McGuire, M. T. 1987. Mother-infant relationships in vervet monkeys: Response to new adult males. *Int. J. Primatol.,* 8, 351–366.

Fairbanks, L. A. & Pereira, M. E. 1993. Juvenile primates: dimensions for future research. In: *Juvenile Primates: Life History, Development, and Behavior* (Ed. by M. E. Pereira & L. A. Fairbanks), pp. 359–366. New York: Oxford University Press.

Feinman, S. 1982. Social referencing in infancy. *Merrill-Palmer Q.,* 28, 445–470.

Furuichi, T. 1997. Agonistic interactions and matrifocal dominance rank of wild bonobos (*Pan paniscus*) at Wamba. *Int. J. Primatol.,* 18, 855–875.

Galef, B. G., Jr. 1988. Imitation in animals: history, definition, and interpretation of data from the psychological laboratory. In: *Social Learning: Psychological and Biological Perspectives* (Ed. by T. R. Zentall & B. G. Galef, Jr.), pp. 3–28. Hillsdale, NJ: Lawrence Erlbaum.

Galef, B. G., Jr. 1995. Why behavior patterns that animals learn socially are locally adaptive. *Anim. Behav.,* 49, 1325–1334.

Glick, B. B., Eaton, G. G., Johnson, D. F., & Worlein, J. 1986a. Development of partner preferences in Japanese macaques (*Macaca fuscata*): effects of gender and kinship during the second year of life. *Int. J. Primatol.* 7, 467–479.

Glick, B. B., Eaton, G. G., Johnson, D. F., & Worlein, J. 1986b. Social behavior of infant and mother Japanese macaques (*Macaca fuscata*): effects of kinship, partner sex, and infant sex. *Int. J. Primatol.* 7, 139–155.

Goldberg, T. L. & Wrangham, R. W. 1997. Genetic correlates of social behaviour in wild chimpanzees: evidence from mitochondrial DNA. *Anim. Behav.,* 54, 559–570.

Goodall, J. 1986. *The Chimpanzees of Gombe: Patterns of Behavior.* Cambridge, MA: Harvard University Press.

Gouzoules, S. 1984. Primate mating systems, kin associations and cooperative behavior: evidence for kin recognition? *Am. J. Phys. Anthropol.* 27, 99–134.

Gouzoules, S. & Gouzoules, H. 1987. Kinship. In: *Primate Societies* (Ed. by B. B. Smuts, D. L. Cheney, R. M. Seyfarth, R. W. Wrangham, & T. T. Struhsaker), pp. 299–305. Chicago: University of Chicago Press.

Gust, D. A. & Gordon, T. P. 1994. The absence of a matrilineally based dominance system in sooty mangabeys, *Cercocebus torquatus atys. Anim. Behav.,* 47, 589–594.

Hamilton, W. J., III, Busse, C., & Smith, K. S. 1982. Adoption of infant orphan chacma baboons. *Anim. Behav.,* 30, 29–34.

Hardy, K. M. 1997. Agonistic support, affiliation and kinship: a study of social relationships among adult female and juvenile rhesus macaques. PhD dissertation, University of Pennsylvania, Philadelphia, PA.

Hauser, M. D. & Fairbanks, L. A. 1988. Mother-offspring conflict in vervet monkeys: variation in response to ecological conditions. *Anim. Behav.,* 36, 802–813.

Hayaki, H. 1983. The social interactions of juvenile Japanese monkeys on Koshima Islet. *Primates,* 24, 139–153.

Hayaki, H. 1988. Association partners of young chimpanzees in the Mahale Mountains National Park, Tanzania. *Primates,* 29, 147–161.

Hemelrijk, C. K. 1999. An individual-orientated model of the emergence of despotic and egalitarian societies. *Proc. Royal Soc. Lond., B,* 266(1417), 361–369.

Hikami, K., Hasegawa,Y., & Matsuzawa, T. 1990. Social transmission of food preferences in Japanese monkeys (*Macaca fuscata*) after mere exposure or aversion training. *J. Comp. Psychol.,* 104, 233–237.

Hinde, R. A. 1974. *Biological Bases of Human Social Behaviour.* New York: McGraw-Hill.

Hiraiwa, M. 1981. Maternal and alloparental care in a troop of free-ranging Japanese monkeys. *Primates,* 22, 309–329.

Hooley, J. M. 1983. Primiparous and multiparous mothers and their infants. In: *Primate Social Relationships: An Integrated Approach* (Ed. by R. A. Hinde), pp. 142–145. Sunderland, MA: Sinauer Associates.

Horrocks, J. & Hunte, W. 1983. Maternal rank and offspring rank in vervet monkeys: an appraisal of the mechanisms of rank acquisition. *Anim. Behav.,* 31, 772–782.

Hrdy, S. B. 1976. Care and exploitation of nonhuman primate infants by conspecifics other than the mother. In: *Advances in the Study of Animal Behaviour, Vol. 6* (Ed. by J. S. Rosenblatt, R. A. Hinde, E. Shaw, & C. Beer), pp. 101–158. London: Academic Press.

Hrdy, S. B. 1977. *The Langurs of Abu: Female and Male Strategies of Reproduction.* Cambridge, MA: Harvard University Press.

Imakawa, S. 1990. Playmate relationships of immature free-ranging Japanese monkeys at Katsuyama. *Primates,* 31, 509–521.

Janson, C. H. & van Schaik, C. P. 1993. Ecological risk aversion in juvenile primates: slow and steady wins the race. In: *Juvenile Primates: Life History, Development, and Behavior* (Ed. by M. E. Pereira & L. A. Fairbanks), pp. 57–74. New York: Oxford University Press.

Janus, M. 1989. Reciprocity in play, grooming, and proximity in sibling and nonsibling young rhesus monkeys. *Int. J. Primatol.,* 10, 243–261.

Janus, M. 1991. Aggression in interactions of immature rhesus monkeys: components, context and relation to affiliation levels. *Anim. Behav.,* 41, 121–134.

Janus, M. 1992. Interplay between various aspects in social relationships of young rhesus monkeys: dominance, agonistic help, and affiliation. *Am. J. Primatol.,* 26, 291–308.

Johnson, C., Koerner, C., Estrin, M., & Duoos, D. 1980. Alloparental care and kinship in captive social groups of vervet monkeys (*Cercopithecus aethiops sabaeus*). *Primates,* 21, 406–415.

Johnson, R. L. & Southwick, C. H. 1987. Ecological constraints on the development of infant independence in rhesus. *Am. J. Primatol.,* 13, 103–118.

Judge, P. G. & de Waal, F. B. M. 1997. Rhesus monkey behaviour under diverse population densities: coping with long-term crowding. *Anim. Behav.,* 54, 643–662.

Kaplan, J. R. 1978. Fight interference and altruism in rhesus monkeys. *Am. J. Phys. Anthropol.,* 49, 241–249.

Kapsalis, E. & Berman, C. M. 1996. Models of affiliative relationships among free-ranging rhesus monkeys, *Macaca mulatta.* I. Criteria for kinship. *Behaviour,* 133, 1209–1234.

Kawai, M. 1958. On the system of social ranks in a natural troop of Japanese monkey. I. Basic rank and dependent rank. *Primates,* 1, 111–130.

Kurland, J. A. 1977. *Kin Selection in the Japanese Monkey.* Basel: Karger.

MacKenzie, M. M., McGrew, W. C., & Chamove, A. S. 1985. Social preferences in stump-tailed macaques (*Macaca arctoides*): Effects of companionship, kinship and rearing. *Dev. Psychobiol.,* 18, 115–123.

Maestripieri, D. 1994a. Mother-infant relationships in three species of macaques (*Macaca mulatta, M. nemestrina, M. arctoides*). I. Development of the mother-infant relationship in the first three months. *Behaviour,* 131, 75–96.

Maestripieri, D. 1994b. Mother-infant relationships in three species of macaques (*Macaca mulatta, M. nemestrina, M. arctoides*). II. The social environment. *Behaviour,* 131, 97–113.

Maestripieri, D. 1994c. Social structure, infant handling and mothering styles in group-living Old World monkeys. *Int. J. Primatol.,* 15, 531–554.

Maestripieri, D. 1995. Assessment of danger to themselves and their infants by rhesus macaque (*Macaca mulatta*) mothers. *J. Comp. Psychol.,* 109, 416–420.

Maestripieri, D. 2001. Intraspecific variability in parenting styles of rhesus macaques (*Macaca mulatta*): the role of the social environment. *Ethology,* 107, 237–248.

Martin, D. A. 1997. Kinship bias: a function of familiarity in pigtailed macaques (*Macaca nemestrina*). PhD dissertation, University of Georgia, Athens, GA.

Mason, W. A., Long, D. D., & Mendoza, S. P. 1993. Temperament and mother-infant conflict in macaques: a transactional analysis. In: *Primate Social Conflict* (Ed. by W. A. Mason & S. P. Mendoza), pp. 205–227. Albany, NY: SUNY Press.

Massey, A. 1977. Agonistic aids and kinship in a group of pigtail macaques. *Behav. Ecol. Sociobiol.,* 2, 31–40.

Matsumura, S. 1997. Mothers in a wild group of moor macaques (*Macaca maurus*) are more attractive to other group members when holding their infants. *Folia Primatol.,* 68, 77–85.

Matsumura, S. 1999. The evolution of "egalitarian" and "despotic" social systems among macaques. *Primates,* 40, 23–31.

Matsumura, S. & Okamoto, K. 1997. Factors affecting proximity among member of a wild group of moor macaques during feeding, moving, and resting. *Int. J. Primatol.,* 18, 929–940.

McKenna, J. J. 1987. Parental supplements and surrogates among primates: cross-species and cross-cultural comparisons. In: *Parenting Across the Life Span: Biosocial Dimensions* (Ed. by J. Lancaster, J. Altmann, & A. Rossi), pp. 143–184. New York: Aldine de Gruyter.

Meikle, D. B. & Vessey, S. H. 1981. Nepotism among rhesus monkey brothers. *Nature,* 294, 160–161.

Melnick, D. J., Pearl, M. C., & Richard, A. F. 1984. Male migration and inbreeding avoidance in wild rhesus monkeys. *Am. J. Primatol.,* 7, 229–243.

Mendoza, S. P. & Mason, W. A. 1989. Primate relationships: social dispositions and physiological responses. In: *Perspectives in Primate Biology, Vol. 2* (Ed. by P. K. Seth & S. Seth), pp. 129–143. New Delhi: Today and Tomorrow's Printers and Publishers.

Mitani, J. C., Merriwether, D. A., & Zhang, C. 2000. Male affiliation, cooperation and kinship in wild chimpanzees. *Anim. Behav.,* 59, 885–893.

Mitani, J. C., Watts, D. P., Pepper, J. P., & Merriwether, D. A. 2002. Demographic and social constraints on male chimpanzee behaviour. *Anim. Behav.,* 64, 727–737.

Mitchell, C. L. 1994. Migration alliances and coalitions among adult male South American squirrel monkeys (*Saimiri sciureus*). *Behaviour,* 130, 169–190.

Mitchell, G. D. & Stevens, C. W. 1968. Primiparous and multiparous monkey mothers in a mildly stressful social situation: first three months. *Dev. Psychobiol.,* 1, 280–286.

Nakamichi, M. 1996. Proximity relationships within a birth cohort of immature Japanese monkeys (*Macaca fuscata*) in a free-ranging group during the first four years of life. *Am. J. Primatol.*, 40, 315–325.

Nicolson, N. A. 1982. Weaning and the development of independence in olive baboons. PhD dissertation, Harvard University, Cambridge, MA.

Nicolson, N. A. 1987. Infants, mothers, and other females. In: *Primate Societies* (Ed. by B. B. Smuts, D. L. Cheney, R. M. Seyfarth, R. W. Wrangham, & T. T. Struhsaker), pp. 330–342. Chicago: University of Chicago Press.

Nishida, T. 1983. Alloparental behavior in wild chimpanzees of the Mahale mountains, Tanzania. *Folia Primatol.*, 41, 1–33.

Nishida, T. 1989. Social interactions between resident and immigrant female chimpanzees. In: *Understanding Chimpanzees* (Ed. by P. G. Heltne & L. A. Marquardt), pp. 68–89. Cambridge, MA: Harvard University Press.

Ogawa, H. 1995. Triadic male-female-infant relationships and bridging behaviour among Tibetan macaques (*Macaca thibetana*). *Folia Primatol.*, 64, 153–157.

Paul, A. 1999. The socioecology of infant handling in primates: is the current model convincing? *Primates,* 40, 33–46.

Paul, A., Preuschoft, S., & van Schaik, C. P. 2000. The other side of the coin: infanticide and the evolution of affiliative male-infant interactions in Old World primates. In: *Infanticide by Males and Its Implications* (Ed. by C. P. van Schaik & C. H. Janson), pp. 269–292. Cambridge: Cambridge University Press.

Pavelka, M. S. M., Fedigan, L. M., & Zohar, S. 2002. Availability and adaptive value of reproductive and post-reproductive Japanese macaque mothers and grandmothers. *Anim. Behav.*, 64, 407–414.

Pereira, M. E. 1988. Effects of age and sex on intra-group spacing behaviour in juvenile savannah baboons, *Papio cynocephalus cynocephalus. Anim. Behav.*, 36, 184–204.

Pereira, M. E. 1992. The development of dominance relations before puberty in cercopithecine societies. In: *Aggression and Peacefulness in Humans and Other Primates* (Ed. by J. Silverberg & J. P. Gray), pp. 117–149. New York: Oxford University Press.

Pereira, M. E. 1995. Development and social dominance among group-living primates. *Am. J. Primatatol.*, 37, 143–175.

Pereira, M. E. & Altmann, J. 1985. Development of social behavior in free-living nonhuman primates. In: *Nonhuman Primate Models for Human Growth and Development* (Ed. by E. S. Watts), pp. 217–309. New York: Alan R. Liss.

Pereira, M. E. & Fairbanks, L. A. 1993. *Juvenile Primates: Life History, Development, and Behavior.* New York: Oxford University Press.

Pope, T. R. 1990. The reproductive consequences of male cooperation in the red howler monkey: paternity exclusion in multi-male and single-male troops using genetic markers. *Behav. Ecol. Sociobiol.*, 27, 439–446.

Pope, T. R. 2000. Reproductive success increases with degree of kinship in cooperative coalitions of female red howler monkeys (*Alouatta seniculus*). *Behav. Ecol. Sociobiol.*, 48, 253–267.

Pusey, A. E. 1983. Mother-offspring relationships in chimpanzees after weaning. *Anim. Behav.*, 31, 363–277.

Pusey, A. E. 1990. Behavioural changes at adolescence in chimpanzees. *Behaviour*, 115, 203–246.

Rajecki, D. W., Lamb, M. E., & Obmascher, P. 1978. Toward a general theory of infantile attachment: a comparative review of aspects of the social bond. *Behav. Brain Sci.,* 1, 417–464.

Rajpurohit, L. S. & Sommer, V. 1993. Juvenile male emigration from natal one-male troops in hanuman langurs. In: *Juvenile Primates: Life History, Development, and Behavior* (Ed. by M. E. Pereira & L. A. Fairbanks), pp. 86–103. New York: Oxford University Press.

Rajpurohit, L. S., Sommer, V., & Mohnot, S. M. 1995. Wanderers between harems and bachelor bands: male hanuman langurs (*Presbytis entellus*) at Jodhpur in Rajasthan. *Behavior*, 132, 255–299.

Rawlins, R. G. & Kessler, M. J. 1986. Demography of free-ranging Cayo Santiago macaques, 1976–1983. In: *The Cayo Santiago Macaques: History, Biology and Behavior* (Ed. by R. G. Rawlins & M. J. Kessler), pp. 47–72. Albany, NY: SUNY Press.

Robbins, M. M. 1996. Male-male interactions in heterosexual and all-male wild mountain gorilla groups. *Ethology*, 102, 942–965.

Rosenblum, L. A., Coe, C. L., & Bromley, L. J. 1975. Peer relations in monkeys: the influence of social structure, gender, and familiarity. *Origins Behav.*, 4, 67–98.

Sade, D. S. 1965. Some aspects of parent-offspring and sibling relations in a group of rhesus monkeys, with a discussion of grooming. *Am. J. Phys. Anthropol.* 23, 1–17.

Schaffer, H. R. 1966. The onset of fear of strangers and the incongruity hypothesis. *J. Child Psychol. Psychiatry*, 7, 95–106.

Seay, B. 1966. Maternal behavior in primiparous and multiparous rhesus monkeys. *Folia Primatol.*, 4, 146–168.

Seyfarth, R. M. 1980. The distribution of grooming and related behaviors among adult female vervet monkeys. *Anim. Behav.*, 28, 798–813.

Sherman, P. 1981. Kinship, demography and Belding's ground squirrel nepotism. *Behav. Ecol. Sociobiol.*, 8, 251–259.

Silk, J. B. 1980. Kidnapping and female competition among captive bonnet macaques. *Primates*, 21, 100–110.

Silk, J. B. 1999. Why are infants so attractive to others? The form and function of infant handling in bonnet macaques. *Anim. Behav.*, 57, 1021–1032.

Silk, J. B. 2001. Ties that bond: the role of kinship in primate societies. In: *New Directions in Anthropological Kinship* (Ed. by L. Stone), pp. 71–92. New York: Rowman & Littlefield.

Silk, J. B., Samuels, A., & Rodman, P. S. 1981.The influence of kinship, rank, and sex on affiliation and aggression between adult female and immature bonnet macaques (*Macaca radiata*). *Behaviour*, 78, 111–137.

Smith, K., Alberts, S. C., & Altmann, J. 2003. Wild female baboons bias their social behaviour towards paternal half-sisters. *Proc. Royal Soc. Lond., B*, 270, 503–510.

Spencer-Booth, Y. 1968. The behaviour of group companions toward rhesus monkey infants. *Anim. Behav.*, 16, 541–557.

Sterck, E. M. K., Watts, D. P., & van Schaik, C. P. 1997. The evolution of female social relationships in nonhuman primates. *Behav. Ecol. Sociobiol.*, 41, 291–309.

Stewart, K. J. & Harcourt, A. H. 1987. Gorillas: variation in female relationships. In: *Primate Societies* (Ed. by B. B. Smuts, D. L. Cheney, R. M. Seyfarth, R. W. Wrangham, & T. T. Struhsaker), pp. 155–164. Chicago: University of Chicago Press.

Strier, K. B. 1993. Growing up in a patrifocal society: sex differences in the spatial relations of immature muriquis. In: *Juvenile Primates: Life History, Development, and Behavior* (Ed. by M. E. Pereira & L. A. Fairbanks), pp. 138–147. New York: Oxford University Press.

Strier, K. B. 1994. Brotherhoods among atelins: kinship, affiliation, and competition. *Behaviour*, 130, 151–167.

Strier, K. B., Dib, L. T., & Figueira, J. E. C. 2002. Social dynamics of male muriquis (*Brachyteles arachnoides hypoxanthus*). *Behaviour*, 139, 315–342.

Suomi, S. J. 1991. Uptight and laid-back monkeys: individual differences in the response to social challenges. In: *Plasticity of Development* (Ed. by S. E. Brauth, W. S. Hall, & R. J. Dooling), pp. 27–56. Cambridge, MA: MIT Press.

Suomi, S. J. 1995. Influence of attachment theory on ethological studies of biobehavioral development in nonhuman primates. In: *Attachment Theory: Social, Developmental, and Clinical Perspectives* (Ed. by S. Goldberg, R. Muir, & J. Kerr), pp. 185–201. (Hillsdale, NJ: Analytic Press).

Sussman, R. W. 1977. Socialization, social structure, and ecology of two sympatric species of lemur. In: *Primate Bio-Social Development: Biological, Social, and Ecological Determinants* (Ed. by S. Chevalier-Skolnikoff & F. E. Poirier), pp. 515–528. New York: Garland.

Thierry, B. 1985. Social development in three species of macaque (*Macaca mulatta, M. fascicularis, M. tonkeana*): a preliminary report on the first ten weeks of life. *Behav. Proc.,* 11, 89–95.

Thierry, B. 1990. Feedback loop between kinship and dominance: the macaque model. *J. Theor. Biol.,* 145, 511–522.

Thierry, B. 2000. Covariation of conflict management patterns across macaque species. In: *Natural Conflict Resolution* (Ed. by F. Aureli & F. B. M. de Waal), pp. 106–128. Berkeley, CA: University of California Press.

Thierry, B., Gauthier, C., & Peignot, P. 1990. Social grooming in Tonkean macaques (*Macaca tonkeana*). *Int. J. Primatol.,* 11, 357–375.

Timme, A. 1995. Sex differences in infant integration in a semifree-ranging group of Barbary macaques (*Macaca sylvanus*, L. 1758) at Salem, Germany. *Am. J. Primatol.,* 37, 221–231.

Tomasello, M. & Call, J. 1997. *Primate Cognition.* New York: Oxford University Press.

van Noordwijk, M. A. & van Schaik, C. P. 1985. Male migration and rank acquisition in wild long-tailed macaques (*Macaca fascicularis*). *Anim. Behav.,* 33, 849–861.

van Schaik, C. P. & Paul, A. 1996. Male care in primates: does it ever reflect paternity? *Evol. Anthropol.,* 5, 152–156.

Walters, J. 1980. Interventions and the development of dominance relationships in female baboons. *Folia Primatol.,* 34, 61–89.

Walters, J. R. 1986. Transition to adulthood. In: *Primate Societies* (Ed. by B. B. Smuts, D. L. Cheney, R. M. Seyfarth, R. W. Wrangham, & T. T. Struhsaker), pp. 358–369. Chicago: University of Chicago Press.

Walters, J. R. 1987. Kin recognition in nonhuman primates. In: *Kin Recognition in Animals* (Ed. by D. F. Fletcher & C. D. Michener), pp. 359–393. New York: John Wiley.

Warfield, J. 2001. Self-scratching in free-ranging infant rhesus macaque monkeys (*Macaca mulatta*). *Am. J. Primatol.,* 54 (Suppl 1), 91.

Watts, D. P. 1994. Social relationships of immigrant and resident female mountain gorillas: relatedness, residence and relationships between females. *Am. J. Primatol.,* 32, 13–30.

Watts, D. P. & Pusey, A. P. 1993. Behavior of juvenile and adolescent great apes. In: *Juvenile Primates: Life History, Development, and Behavior* (Ed. by M. E. Pereira & L. A. Fairbanks), pp. 148–167. New York: Oxford University Press.

Weaver, A. C. & de Waal, F. B. M. 2000. The development of reconciliation in brown capuchins. In: *Natural Conflict Resolution* (Ed. by F. Aureli & F. B. M. de Waal), pp. 216–218. Berkeley, CA: University of California Press.

Welker, C., Schwibbe, M. H., Schaefer-Witt, C., & Visalberghi, E. 1987. Failure of kin recognition in *Macaca fascicularis*. *Folia Primatol.,* 49, 216–221.

Widdig, A., Nurnberg, P., Krawczak, M., Streich, W. J., & Bercovitch, F. B. 2001. Paternal relatedness and age proximity regulate social relationships among adult females rhesus macaques. *Proc. Natl. Acad. Sci. USA,* 98, 13769–13773.

Yamada, M. 1963. A study of blood-relationship in the natural society of the Japanese macaque—an analysis of co-feeding, grooming, and playmate relationships in Minoo-B-troop. *Primates,* 4, 43–65.

15

The Recognition of Other Individuals' Kinship Relationships

Dorothy L. Cheney
Robert M. Seyfarth

> He knew all the ramifications of New York's cousinships; and could not only elucidate such complicated questions as that of the connection between the Mingotts (through the Thorleys) with the Dallases of South Carolina, and that of the relationship of the elder branch of Philadelphia Thorleys to the Albany Chiverses . . . , but could also enumerate the leading characteristics of each family; as, for instance, . . . the fatal tendency of Rushworths to make foolish matches . . .
>
> —Edith Wharton, *The Age of Innocence*

What does it mean to say that someone has knowledge of his or her social companions? At the most basic level it is essential to distinguish familiar from unfamiliar individuals. This relatively simple form of social knowledge has been documented in many species and appears to be controlled by specific genes. Recent research with congenic mice, for example, has demonstrated that a gene controlling the regulation of oxytocin is necessary for forming social memories (Ferguson et al. 2000). Mice lacking this gene are unable to recognize familiar individuals, although the gene has no apparent influence on spatial or other types of memory. Evidently, social memory has a neural basis distinct from other forms of memory.

The ability to recognize individuals whom we have met before is a necessary first step to classifying relationships. However, it is also many steps removed from the sort of social calculus demanded by human society, where individuals must not only assess the relationships that they themselves have with others but also determine the social networks that comprise the relationships of others.

Hamilton's (1964) theory of kin selection predicts that individuals capable of discriminating close kin from others gain a reproductive advantage over those who are unable to make such discriminations. By contrast, the ability to take a third party's perspective, and discriminate among the relationships that exist among others, is not directly related to kin selection theory. Instead, it is thought to have evolved in social groups where alliances are common and where individuals must be able to predict who will be allied with whom (Harcourt 1988, 1992; Cheney & Seyfarth 1990; see below). The ability of monkeys and apes to recognize third-party relationships may therefore represent an evolutionary precursor to our own ability to recognize not just the immediate kin of others but also whole kinship systems that involve many individuals allied both formally, through societal rules, and informally, through friendship.

Like the doyens of Wharton's New York, monkeys seem to need formidable social calculations in order to survive and reproduce. Baboons (*Papio* spp.), for example, typically live in groups of 80 or more individuals that contain several matrilineal families arranged in a stable, linear dominance rank order. Members of the same matriline maintain similar ranks and close social bonds. Cutting across these stable, long-term relationships based on rank and kinship are more transient bonds; for example, bonds formed between a high- and a low-ranking mother whose infants are of the same age, or bonds formed between males and female "friends" with young infants. Parallels between the members of a typical baboon group and the "50 families" that made up New York society in Edith Wharton's day are difficult to avoid.

What sort of intelligence is required to navigate this social landscape? How do monkeys and apes acquire information about their social companions, and how do they store it in memory? The recognition of third-party relationships demands a nonegocentric view of the social group. Individuals must attend to interactions in which they are not themselves involved and which they do not personally experience. Does a baboon who apparently knows the matrilineal kin relations of other group members have an abstract concept of matrilineal kinship (Dasser 1988a, b), or has she learned to distinguish other individuals' kin through associative processes (Thompson 1995, Schusterman & Kastak 1998)? As we discuss below, there is no doubt that associative learning plays a major role in the recognition of third-party relationships, but these processes are not necessarily simple.

The recognition of third-party relationships does not demand a theory of mind. An animal can learn to recognize that two other individuals have a close social bond simply by attending to their behavior; it is not necessary also to recognize the motives and emotions that underlie this bond. Because there is little evidence that any nonhuman animals attribute mental states different from their own to others (chimpanzees [*Pan troglodytes*] may be an exception; see Tomasello & Call 1997; Hare et al. 2000, 2001), studies of the extent to which animals recognize other individuals' social relationships may demonstrate the extent and limits of social cognition in the absence of a theory of mind.

In this chapter we begin with a brief review of the data and ask: What must a monkey know, and how must her knowledge be structured, to account for her behavior? We restrict our review primarily to monkeys because we still know almost nothing about the recognition of third-party relationships in apes or prosimians.

We then compare the evidence from monkeys to what is known about the recognition of third-party relationships in nonprimate species. If, as has been hypothesized, the recognition of third-party relationships confers a selective advantage because it allows individuals to remember who associates with whom, who outranks whom, and who is allied to whom, we should expect to find evidence for this ability not just in nonhuman primates but also in any animal species that lives in large, stable social groups. We would also predict that selection should have acted less strongly on this ability in species living in small, egalitarian groups that are composed primarily either of close kin or of unrelated individuals. Thus, the ability to recognize the close associates of others should be evident in nonprimate species such as hyenas and lacking or less evident in some ape species, including gorillas (*Gorilla gorilla*) and orangutans (*Pongo pygmaeus*). It is not yet clear whether either of these predictions can be supported. In fact, recent research on birds and fish suggests that the ability to monitor other individuals' social interactions and to make assessments about, for example, their dominance ranks may be widespread even in species that do not live in social groups.

Knowledge of Other Individuals' Kinship Relationships

Some of the first evidence that monkeys recognize other animals' social relationships emerged as part of a relatively simple playback experiment originally designed to document individual vocal recognition in vervet monkeys (*Cercopithecus aethiops*) (Cheney & Seyfarth 1980). To test the hypothesis that mothers recognize their offspring by voice, we played the screams of a juvenile to subgroups of three adult females, one of whom was the juvenile's mother. As predicted, mothers consistently oriented toward the screams for longer durations than did control females. More interestingly, however, we found that in a significant number of cases one or both of the two control females looked at the mother. In so doing, the females behaved as if they were able to associate the screams with a particular juvenile, and that juvenile with its mother. They appeared to recognize, in other words, the social bonds that existed between particular juveniles and particular adult females (Cheney & Seyfarth 1980, 1982).

In an attempt to replicate and extend these results, we later conducted a more complex set of experiments on free-ranging baboons (*Papio cynocephalus ursinus*) (Cheney & Seyfarth 1999). In these experiments, two unrelated adult females (e.g., individuals C and F) heard a sequence of calls that mimicked a fight between two of their close relatives (e.g., C_2 and F_2). In separate trials, the same two subjects also heard two control sequences of calls. The first sequence mimicked a fight involving the dominant subject's kin and an individual unrelated to either female (e.g., C_2 and E_2); the second mimicked a fight involving two individuals who were unrelated to either female (e.g., E_2 and G_2).

After hearing the test sequence, a significant number of subjects looked toward the other female, suggesting that they recognized not just the calls of unrelated individuals but also those individuals' kin (or close associates). Moreover, in the minutes following playback of the test sequence, the more dominant subject was significantly more likely to supplant the more subordinate, suggesting that the two females now regarded each other as opponents, if only temporarily. Females' responses following the test sequence differed significantly from their responses following control sequences. Following the first control sequence, when only the dominant subject's relative appeared to be involved in the fight, only the subordinate subject looked at her partner. Following the second control sequence, when neither of the subjects' relatives was involved, neither subject looked at the other. Finally, following both control sequences, the two subjects were significantly more likely to approach each other and interact in a friendly manner than following the test sequence. Note that the increased likelihood of aggression following playback of the test sequence of calls did not occur simply because the two subjects were in relatively close proximity to each other, because they were also in proximity following playback of the two control sequences.

Other studies provide additional evidence of monkeys' abilities to distinguish both their own and other individuals' close associates. For example, in a playback study using the contact calls of rhesus macaques (*Macaca mulatta*), Rendall et al. (1996) found that females not only recognized the identities of different signalers but also categorized signalers according to matrilineal kinship. Taken together, these experiments argue that monkeys recognize the members of their group as distinct individuals and classify these individuals into what we call matrilineal families based on their associations with each other.

Natural patterns of aggression also reflect the knowledge that monkeys have of their group's social network. In many species, an individual who has just threatened or been threatened by another animal will often "redirect" aggression by threatening a third, previously uninvolved, individual. Such aggression is not always randomly directed toward any more subordinate animal. Instead, animals often specifically target a close matrilineal relative of their recent opponent (pigtail macaques [*M. nemestrina*]: Judge 1982; Japanese macaques [*M. fuscata*]: Aureli et al. 1992; vervets: Cheney & Seyfarth 1986, 1989).

Kin-biased redirected aggression also appears in more complex forms. In vervet monkeys, a fight between two individuals significantly increases the probability that their relatives too will subsequently be aggressive (Cheney & Seyfarth 1986, 1989). Kin-biased redirected aggression does not occur simply because opponents' kin are more likely to be in close proximity to one another and hence more likely to be available as targets of aggression. Instead, monkeys often seem actively to seek out their opponents' close associates. Their behavior suggests that they are able to recognize that certain types of social relationships share similar characteristics (Dasser 1988a, Cheney & Seyfarth 1990). When a monkey threatens her relative's opponent's kin, she acts as if she recognizes that the relationship between her opponent and her opponent's relatives is in some way similar to her own relationship with her own relative.

Given the many animal learning experiments that have focused on the ability of captive animals to categorize relatively arbitrary stimuli (see below), it seems surprising that only one has ever asked monkeys to classify fellow group members. In this study, Dasser (1988a, b) trained a female long-tailed macaque (*M. fascicularis*) to choose between slides of a mother-offspring pair and a slide of two unrelated individuals. Having been trained to respond to one mother-offspring pair, the subject was then tested with 14 novel slides of different mothers and offspring paired with an equal number of novel pairs of unrelated animals. In all tests, she correctly selected the mother-offspring pair. In so doing, she appeared to use an abstract category to classify pairs of individuals that was analogous to our concept of "mother-child affiliation."

In another test with a second subject that used a match-to-sample procedure, a mother was represented as the sample in the center of the screen, while one of her offspring and another animal of the same age and sex were given as positive and negative alternatives, respectively (Dasser 1988a, b). Having learned to select the offspring during training, the subject was presented with 22 novel combinations of mother, offspring, and an unrelated individual. She chose correctly in 20 of 22 tests. Dasser was able to rule out the possibility that mothers and offspring were matched solely because of physical resemblance. Instead, subjects seemed to classify individuals according to their degree of association.

To date, only one study has suggested that primates can recognize other individuals' kin on the basis of physical resemblance. In this study, captive chimpanzees were asked to sort photographs of unfamiliar individuals according to kinship. Although subjects were able to match unfamiliar sons and mothers, they were unable to match daughters and mothers (Parr & de Waal 1999). As a result, the experiment is difficult to interpret. Monkeys clearly treat their own close kin differently from less closely related individuals (Gouzoules & Gouzoules 1987; Silk 1987; Chapais et al. 1997, 2001; many chapters in this volume). It is unclear, however, whether they do so because they are sensitive to differences in degrees of genetic relatedness or whether they simply attend to differences in rates of interaction. Because kin typically interact at high rates, monkeys may not recognize other individuals' kin except as close associates.

Under natural conditions, it is difficult to determine whether animals distinguish among different categories of social relationships. Do monkeys recognize, for example, that mother-offspring bonds are distinct from sibling bonds or friendships even when all are characterized by high rates of interaction? In perhaps the only test of monkeys' ability to recognize different categories of social affiliation, Dasser (1988b) trained a long-tailed macaque to identify a pair of siblings and then tested her ability to distinguish novel sibling pairs from mother-offspring pairs, pairs of less closely related matrilineal kin, and unrelated pairs. Although the subject did distinguish siblings from unrelated pairs and pairs of less closely related individuals, she was unable to discriminate between siblings and mothers and offspring. This failure may have occurred because she had previously been rewarded for picking the mother-offspring pair. It is also possible, however, that the subject did not distinguish between different kinship categories and simply chose the pair that was more closely affiliated.

In sum, a number of independent studies have suggested that monkeys recognize third-party relationships. Upon hearing a juvenile scream, monkeys look toward that juvenile's close kin. A dispute between two individuals increases the probability that either they themselves or their relatives will behave aggressively toward the kin of their opponents. These changes in behavior do not imply that females consciously seek revenge against families whose members have recently fought with their own relatives, merely that hearing a particular combination of calls in some way changes their attitude and behavior toward certain other group members. The behavior of monkeys is influenced not just by their own interactions and experiences but also by the interactions and experiences of their close kin. Monkeys appear to view their social groups not just in terms of the individuals that comprise them but also in terms of a network of social relationships in which certain individuals are linked with several others.

Knowledge of Other Individuals' Dominance Ranks

Along with matrilineal kinship, linear, transitive dominance relations are a pervasive feature of social behavior in groups of Old World monkeys. Linear, transitive rank orders might emerge because individuals recognize the dominance relations that exist among others: A is dominant to B and B is dominant to C; therefore A must be dominant to C. Alternatively, monkeys might simply recognize who is dominant or subordinate to themselves. In the latter case, a transitive, linear hierarchy would emerge as the incidental outcome of paired interactions. The hierarchy would be evident to a human observer, but not to the monkeys themselves.

It seems probable, however, that monkeys do recognize other individuals' relative ranks. For example, dominant female baboons often grunt to mothers with infants as they approach the mothers and attempt to handle their infants. The grunts seem to function to facilitate social interactions by appeasing anxious mothers, because an approach accompanied by a grunt is significantly more likely to lead to subsequent friendly interaction than is an approach without a grunt (Cheney et al. 1995b). Occasionally, however, a mother will utter a submissive call, or "fear bark," as a dominant female approaches. Fear barks are an unambiguous indication of subordination; they are never given to lower ranking females.

To test whether baboons recognize the rank relations that exist among others, we played sequences of calls to female subjects in which a lower ranking female apparently grunted to a higher ranking female and the higher ranking female apparently responded with fear barks (Cheney et al. 1995a). As a control, subjects heard the same sequence of grunts and fear barks made causally consistent by the inclusion of additional grunts from a third female who was dominant to both of the others. For example, if the inconsistent sequence was composed of female F's grunts followed by female B's fear barks, the corresponding consistent sequence might begin with female A's grunts, followed by female F's grunts, and ending with female B's fear barks. Subjects responded significantly more strongly to the inconsistent sequences, suggesting that they recognized other individuals' relative ranks.

Patterns of grooming competition among female vervet monkeys provide further evidence that monkeys recognize other individuals' ranks (Seyfarth 1980). Such competition occurs when one female approaches two that are grooming, supplants one of them, and then grooms with the female that remains. Interestingly, in those cases when a female approaches two groomers who are both subordinate to herself, the lower ranking of the two groomers typically moves away, while the higher ranking remains (Cheney & Seyfarth 1990). The departure of the lowest ranking female does not appear to be influenced by the two higher ranking females' behavior toward her, because these two females do not threaten her in any overt way. Similarly, her behavior is not affected by aggression received in other contexts, because aggression rates are not linearly correlated with dominance rank. Instead, the two lower ranking females appear to recognize that, though they are both subordinate to the approaching female, one is more subordinate than the other. In order to accomplish this ranking, it seems necessary that females recognize not only their own status relative to other individuals but also other individuals' status relative to each other (Cheney & Seyfarth 1990).

Similar supplant patterns are evident in baboons. During one yearlong study of 19 females, a female approached and supplanted one of two lower ranking individuals on 95 occasions. In 86% of these cases, the lower ranking female retreated (Cheney, Seyfarth, & Silk unpublished data). Interestingly, in 77% of those cases when the higher ranking female departed, the lower ranking female had a young infant, while the higher ranking female either had no infant or had an older infant. Female baboons with young infants are very attractive to other females, who often compete to interact with their infants (Seyfarth 1976, 1977, 1980; Altmann 1980; Silk et al. 1999). It therefore seems possible that the higher ranking female retreated because she recognized that the lower ranking female was, at least temporarily, more attractive to the approaching individual than she was.

Assessments of other females' relative ranks and attractiveness also appear to influence social interactions among bonnet macaques (*M. radiata*) (Sinha 1998). On most of the occasions when a female was observed to approach two individuals who were both lower ranking than she was, the more subordinate of the two females was supplanted. However, in those few cases when the more dominant member of the dyad retreated, the subordinate female was apparently the more socially attractive to other females in the group, as measured by the amount of grooming received. When deciding whether or not to retreat or stay, therefore, females appeared to take into account not only their own and other individuals' relative ranks but also their relative social attractiveness.

Knowledge of More Transient Social Relationships

All of the studies discussed thus far focused on interactions among females in groups where matrilineal kin typically retain close bonds and similar ranks throughout their lives. It might seem, therefore, that an individual could simply memorize the close associates and relative ranks of other females and thereafter navigate easily through a predictable network of social relationships. Not all social and rank relationships, however, are as stable as those among matrilineal kin. Some types of social bonds are relatively transient, and some rank relationships—particularly among adult males—change often. Such unstable relationships cannot be ignored if an individual is to predict the behavior of others.

For example, under natural conditions, male and female hamadryas baboons (*Papio hamadryas*) form close, long-term bonds that can last for a number of years. Potential rivals appear to recognize the "ownership" of specific females by other males and refrain from challenging those males for their females (Kummer et al. 1974). Experiments conducted in captivity have shown that rival males assess the strength of males' relationships with their females before attempting to challenge them. They do not attempt to take over a male's females if the pair appears to have a close social bond (Bachmann & Kummer 1980). Although similar experiments have not yet been conducted with savanna baboons, observational data suggest that these baboons, too, recognize the temporary bonds, or "friendships," that are formed between males and lactating females. For example, Smuts (1985) observed that males who had recently been threatened by another male often redirected aggression toward the female friends of their opponent (see Dunbar 1983 for similar observations on gelada baboons [*Theropithecus gelada*]).

Monkeys also seem to recognize the bonds that exist between males and particular infants. In Tibetan macaques (*M. thibetana*), males are often closely affiliated with a particular infant in the group. Competitive interactions between males are mediated by the carrying of infants, and a male will frequently carry an infant and present it to another male. In a study of such carrying (or "bridging") behavior, Ogawa (1995) observed that males more frequently provided other males with those males' affiliated infants than with other, nonaffiliated infants.

Finally, there is evidence that monkeys recognize even very transient dominance relations among others. Studying captive bonnet macaques, Silk (1993, 1999) found that males consistently solicited allies that outranked both themselves and their opponents. Silk's analysis ruled out simpler explanations based on the hypotheses that males chose allies that outranked themselves, or that males chose the highest ranking individual in the group. Instead, soliciting males seemed to recognize not only their own rank relative to that of a potential ally but also the rank relation between the ally and his opponent. In this respect, males' knowledge of others' relative ranks was similar to that of the females described earlier. What made learning potentially more challenging, however, was the fact that male ranks were extremely unstable and transient. Of the 16 males in Silk's study, an average of 8.2 males changed dominance rank each month (Silk 1993).

The Recognition of Third-Party Social Relationships by Nonprimates

If the cognitive abilities that underlie the recognition of third-party relationships are selectively advantageous for any species that lives in large social groups, these abilities should not be restricted to only one taxon. Instead, we should find evidence for the recognition of third-party relationships in birds like white-fronted bee-eaters (*Merops bullockoides*) that live in large colonies comprising a number of different family groups (Emlen et al. 1995). In fact, however, little is yet known about the recognition of third-party relationships in nonprimate animals.

Within the Primate order, species that live in large groups have relatively larger brains than those that are solitary or live in small groups. A similar relation is found in carnivores, supporting the hypothesis that sociality has favored the evolution of large brains (Barton & Dunbar 1997; see also Jolly 1966, Humphrey 1976, Cheney & Seyfarth 1990). At the same time, however, primates have larger brains for their body size than all other vertebrates. Dunbar (2000) argues that this arises because primate social groups are not only larger but also more complex than those of other taxa. The pressure to compute and manage social relationships within these groups has favored the evolution of larger brains. Supporting this argument, primates appear to form more complex alliances than nonprimates. While alliances in birds and nonprimate mammals tend to involve support of offspring or other kin, primates also choose allies based on those individuals' relative dominance ranks or fighting ability (Harcourt 1988, 1992).

To date, only one study of spotted hyenas (*Crocuta crocuta*) has explicitly attempted to replicate the experiments on social knowledge conducted with monkeys. Like many species of Old World monkeys, hyenas live in social groups comprising matrilines in which offspring inherit their mothers' dominance ranks (Smale et al. 1993, Engh et al. 2000). Holekamp et al. (1999) played recordings of cubs' "whoop" calls to mothers and other breeding females. As in the case of vervet monkeys and baboons, females responded more strongly to the calls of their offspring and close relatives than to the calls of unrelated cubs. In contrast to vervets and baboons, however, unrelated females did not look at the cubs' mothers. One explanation for these negative results is that hyenas are unable to recognize third-party relationships, despite living in social groups that are superficially similar to those of many primates. It also remains possible, however, that some aspect of the experimental protocol—for example, the type of call stimulus used—caused unrelated subjects to ignore the vocalization.

Indeed, hyenas' patterns of alliance formation suggest that they do attend to other individuals' interactions and that they extrapolate information about other individuals' relative ranks from their observations. During competitive interactions over meat, hyenas often solicit alliance support from other, uninvolved individuals. When choosing to join ongoing skirmishes, hyenas who are dominant to both of the contestants almost always support the more dominant of the two individuals (Engh et al. in press). Similarly, when the ally is intermediate in rank between the two opponents, it inevitably supports the dominant individual. These data provide some of the first evidence that species other than primates form alliances based on both their allies' and opponents' relative ranks.

Spotted hyenas are one of the few nonprimate species that live in large, stable social groups composed of multiple matrilines. Most animal species are either transiently social or live in monogamous pairs or small family groups. As a result, the social context for recognizing other individuals' kin or close associates rarely arises. It is therefore unclear whether the lack of evidence for the recognition of third-party social relationships in most animal species occurs because of the lack of cognitive capacity or because of the lack of opportunity to investigate such a capacity. Most hypotheses concerned with the selective advantages that might give rise to the recognition of third-party social relationships have posited that this ability should be manifested in relatively large-brained species that live in large social groups composed of individuals of varying degrees of genetic relatedness and fighting ability. For several reasons, these assumptions may not be completely valid.

First, there is evidence that the need to monitor and manage other individuals' social relationships may favor larger brains even in species that lack a cortex, including insects. For example, in paper wasps (*Polistes dominulus*) there is a significant increase in the size of the antennal lobes and collar (a substructure of the calyx of the mushroom body) in females that nest colonially—with other queens—as opposed to solitary breeders (Ehmer et al. 2001). This increase in neural volume may be favored because sociality places increased demand on the need to discriminate between familiar and unfamiliar individuals and to monitor other females' dominance and breeding status. Clearly, therefore, the neural correlates of sociality need not be restricted to higher mammals.

Second, there is evidence of at least a rudimentary form of third-party recognition even in species that do not live in complex social groups, or even any group at all. At least some fish, for example, are able to assess the relative dominance of two potential rivals by monitoring, or eavesdropping, on the competitive interactions of other individuals. In one experiment involving fighting fish (*Betta splendens*), males who had witnessed two other males involved in an aggressive interaction were subsequently more likely to approach the loser of that interaction than the winner (Oliveira et al. 1998). Their behavior was not influenced by any inherent differences between the two males, because they were equally likely to approach the winner and the loser in competitive interactions they had not observed.

Similarly, several studies have shown that male songbirds eavesdrop on the singing contests of territorial neighbors in order to assess other individuals' relative dominance. Peake et al. (2001) presented male great tits (*Parus major*) with the opportunity to monitor an apparent competitive interaction between two neighbors by simulating a singing contest using two loudspeakers. The relative timing of the singing bouts (for example, the degree of overlap between the two songs) provided information about each "contestant's" apparent dominance. Following the singing interaction, one of the "contestants"—a loudspeaker—was introduced into the male's territory. Males responded significantly less strongly to singers that had apparently just "lost" the interaction. Eavesdropping appears to be selectively advantageous, because it allows individuals to extract information about other males' relative fighting ability without having to engage in a match themselves (see also, e.g., McGregor & Dabelsteen 1996, Naguib et al. 1999).

In sum, it seems likely that the ability to monitor and assess other individuals' interactions is present in many species of animals, including those lacking large brains and those living in small or transient societies. Hyenas' ability to recognize other individuals' relative ranks may be similar to that of monkeys. Similarly, through their observations of dyadic interactions, fish and birds assess other individuals' dominance and fighting ability (see, e.g., White & Galef 1999, Doutrelant & McGregor 2000 for similar data on mate assessment by birds and fish).

Comparison with Primates

Although the ability to monitor other individuals' social interactions appears to be widespread among animals, it remains unclear whether any species other than monkeys are capable of identifying the social bonds that exist among multiple individuals in multiple family groups. This lack of evidence may be due in part to the fact that few nonprimate species live in large social groups composed of many breeding males and females of varying degrees of genetic relatedness, with the result that the opportunity to demonstrate this ability seldom arises. It is also possible, however, that there are fundamental differences between the cognitive capacities of primates and nonprimates, which are reflected both in brain size and degrees of social complexity. At least some learning experiments with captive animals indicate that nonprimates may be generally less adept than primates at recognizing relational categories. These experiments, however, are far from conclusive.

Under captive conditions, monkeys and apes can readily be taught to solve problems that require recognition of a relation between objects. In oddity tests, for example, a subject is presented with three objects, two of which are the same and one of which is different. Monkeys and apes achieve high levels of accuracy in such tests even when tested with novel stimuli (Harlow 1949, D'Amato et al. 1985; see also reviews by Tomasello & Call 1997, Shettleworth 1998). Baboons and chimpanzees can also learn to make abstract discriminations about relations between relations, matching patterns containing repeated samples of the same item with similar "same" patterns (Premack 1983, Oden et al. 1988, Fagot et al. 2001). In all cases, subjects' performance suggests the use of an abstract hypothesis, because concepts like "odd" specify a relation between objects independent of their physical features (Roitblat 1987). In a similar manner, the concept "closely bonded" can be applied to any two individuals and need not be restricted to specific pairs that look alike.

Judgments based on relations among items have been demonstrated more often in nonhuman primates than in other taxa, and primates seem to recognize abstract relations more readily than at least some other animals. Although it is possible, for example, to train pigeons (*Columba livia*) to recognize relations such as "same," the procedural details of the test appear to be more critical for pigeons than they are for monkeys, and relational distinctions can easily be disrupted (Herrnstein 1985, Wright et al. 1988, Wasserman et al. 1995). Rather than attending to the relations among stimuli, pigeons seem predisposed to focus on absolute stimulus properties and to form item-specific associations (reviewed by Shettleworth 1998).

Some tests of transitive inference have also suggested that there may be fundamental cognitive differences between primates and nonprimates. While monkeys and apes readily acquire some representation of series order, species like pigeons seem to attend primarily to the association between adjacent pairs (D'Amato & Colombo 1989, von Fersen et al. 1991, Treichler & van Tilberg 1996, Zentall et al. 1996, Brannon & Terrace 1998; see also reviews by Tomasello & Call 1997, Shettleworth 1998).

Other studies, however, suggest few differences between primates and other animals in the ability to make relational distinctions. For example, the African gray parrot Alex (*Psittacus erithacus*) is reported to make explicit same/different judgments about sets of objects (Pepperberg 1992). Similarly, sea lions (*Zalophus californianus*) (Schusterman & Krieger 1986, Schusterman & Gisener 1988) and dolphins (*Tursiops truncates*) (Herman et al. 1993, Mercado et al. 2000) have been taught to respond to terms such as "left" and "bright" that require the animals to assess relations among a variety of different objects. Finally, a number of species, including parrots (Pepperberg 1994) and rats (*Rattus norvegicus*) (Church & Meck 1984, Capaldi 1993), are able to assess quantities, suggesting that relatively abstract concepts of numerosity may be pervasive among animals (reviewed by Shettleworth 1998). Clearly, more data are needed from both natural and laboratory studies before we can make any definitive conclusions about any cognitive differences between primates and other animals.

Underlying Mechanisms

Although there now exists good evidence for the recognition of third-party relationships in monkeys, we still know relatively little about the mechanisms that might underlie this recognition. Because kin typically associate at high rates, it is difficult to determine whether monkeys have different concepts about different categories of kin or whether they simply take note of other individuals' close companions. Similarly, we do not yet know whether monkeys recognize any distinction between close kin and nonkin "friends" who interact at high rates.

Several authors (e.g., Heyes 1994, Thompson 1995, Schusterman & Kastak 1998, Schusterman et al. 2000) have argued that the social behavior of nonhuman primates can be explained by relatively simple processes of associative learning, in which kin are classified into the same "functional" or "equivalence" class purely by virtue of their high interaction rates.

In captivity, many animal species can be taught to place dissimilar stimuli in the same functional class. Objects that have been associated with a particular reward will come to be treated as functionally equivalent even if they share no obvious physical similarities (Dube et al. 1993, Fields 1993, Heyes 1994, Sidman 1994, Wasserman & Astley 1994, Thompson 1995). Reviewing work on equivalence class formation by a captive sea lion, Schusterman and Kastak (1993, 1998; see also Schusterman et al. 2000) suggest that equivalence judgments constitute a general learning process that underlies much of the social behavior of animals, including the recognition of social relationships: "both social and nonsocial features of the environment can become related through behavioral contingencies, becoming mutually substitutable even when sharing few or no perceptual similarities" (1998: 1088). Thus, for example, a baboon or vervet monkey learns to classify members of the same matriline together because they share a history of common association and functional relations. And when one monkey threatens the close relative of a recent opponent, she does so because members of the same matriline have effectively become interchangeable.

There is no doubt that associative processes provide a powerful and often accurate means for animals to assess the relations that exist among different stimuli, including members of their own species. Indeed, it seems unlikely that a monkey could form a concept such as "closely bonded" without attending to social interactions and forming associations between one individual and another. To a large extent, learning about other individuals' social relationships is by definition dependent on some form of associative conditioning. However, in several respects the "equivalence classes" that make up nonhuman primate groups may exhibit complexities not present in laboratory experiments.

First, no single behavioral measure underlies the associations between individuals. It is, of course, a truism that monkeys can learn which other individuals share a close social relationship by attending to patterns of association. Matrilineal kin, for example, almost always associate at higher rates than nonkin. But no single behavioral measure is either necessary or sufficient to recognize such associations. For example, aggression often occurs at equally high rates within and between families, and different family members may groom each other and associate with each other at different rates (Cheney & Seyfarth 1986). To recognize that two individuals are closely bonded despite relatively high rates of aggression or relatively low rates of grooming, a monkey must take note of a variety of different patterns of aggression, reconciliation, grooming, and proximity. There is no threshold or defining criterion for a "close" social bond. By contrast, equivalence classes in experiments with captive animals are established by repeatedly presenting arbitrary visual stimuli aligned in groups of two or three. Reward and spatial or temporal juxtaposition therefore suffice as a basis for the formation of association of stimuli within an equivalence class.

Second, class members are sometimes mutually substitutable and sometimes not. In discussing the experiment in which an adult female vervet hears a juvenile's scream and then looks at the juvenile's mother (Cheney & Seyfarth 1980), Schusterman and Kastak (1998: 1093) argue that the scream, the juvenile, and the juvenile's mother form a three-member equivalence class in which any one of the three stimuli can be substituted for another. But in fact the call, the juvenile, and the mother are not interchangeable in this manner. A female who has a close bond with the juvenile's mother, for example, may interact little or not at all with the juvenile himself.

Third, some social relationships are transitive, while others are not. In animal learning experiments, members of the same equivalence class are not only mutually substitutable but also exhibit transitivity: if $A_1 > B_1$, then $A_2 > B_2$. In a similar manner, if infant A_1 and juvenile A_2 both associate at high rates with a particular adult female, it is usually correct to infer that the juvenile and infant are also closely bonded and will support one another in an aggressive dispute (e.g., Altmann et al. 1996). Similarly, if A is dominant to B and B is dominant to C, it is usually true that A is dominant to C (Cheney & Seyfarth 1990).

In other cases, however, transitivity cannot be assumed. If infant baboon A_1 and juvenile baboon A_2 both associate at high rates with the same adult female and she associates with an adult male "friend," we can infer that the male is probably also closely allied to the infant. However, it would be incorrect to assume that he is equally closely allied to the juvenile, who may instead be more closely allied to another male who was previously the mother's friend (Seyfarth 1978, Smuts 1985, Palombit et al. 1997). Baboon females from the same matriline often form friendships with different males; conversely, the same male may form simultaneous friendships with females from two different matrilines. In the latter case, the existence of a close bond between a male and two females does not predict a close bond between the two females. In fact, their relationship is as likely to be competitive as it is friendly (Palombit et al. 2001).

Fourth, class membership can be relatively transient. While female dominance rank and membership in a kin group constitute relatively stable, predictable associations, other social relationships change often and unpredictably. For example, when a low-ranking female forms a close friendship with a dominant male, she gains access to feeding sites from which she might normally be excluded by higher ranking females. This preferential access disappears, however, when the friendship ends (Seyfarth 1978, Smuts 1985, Palombit et al. 2001).

Finally, the magnitude of the items classified under natural conditions is often considerably larger than it is in the laboratory. As group size increases, the number of possible combinations of dyadic and triadic social relationships increases exponentially. Although groups of baboons and macaques regularly include more than 50 individuals and five or more matrilines, most laboratory experiments require that subjects classify a limited number of items into just two distinct categories. Moreover, most laboratory studies do not require subjects to form equivalence classes solely on the basis of their observations, without explicit reward.

There is in principle no reason why the mechanisms adopted by animals in laboratory stimulus equivalence tests should be qualitatively different from those used by the same species when classifying conspecifics. Indeed, experiments with captive pigeons, rats, and sea lions suggest that animals actively seek rules by which to impose a structure on the items to be classified (Terrace 1987, Dallal & Meck 1990, Macuda & Roberts 1995, Reichmuth-Kastak et al. 2001), indicating that explicit reward and training are not essential components of class formation. It is to be hoped that future research will begin to identify the mechanisms spontaneously adopted by animals when classifying both conspecifics and arbitrary stimuli.

Third-Party Relationships and Theory of Mind

Finally, it seems possible that the ability to group other individuals' social interactions into categories such as "closely bonded" or "subordinate" may represent an evolutionary precursor to mental state attribution. An individual that possesses a theory of mind recognizes that other individuals' beliefs, knowledge, and emotions may be different from his own. We humans tend to view social relationships in terms of the emotions and motives that underlie them. Individuals who share a close social bond are assumed to like or love each other, whereas individuals whose interactions are antagonistic are assumed to dislike, envy, or even hate each other. We recognize that two individuals may like each other even when we ourselves feel no fondness for them. Similarly, our recognition of other individuals' friends or enemies is not based a single social metric or a particular interaction rate; even a single interaction may be sufficient for us to form a prediction about two individuals' feelings for each other. If a monkey is able to recognize that two other group members share a close social bond, what additional cognitive steps might allow her to sum her observations into a single emotional category like "affection"? Interestingly, in humans the recognition that another individual may have different likes and dislikes from oneself (e.g., for different types of food) begins to emerge around 18 months of age, long before children are able to attribute different beliefs to others (Repacholi & Gopnik 1997). In a similar manner, monkeys' recognition of third-party relationships may involve primitive inferences about other individuals' affective and motivational dispositions. Another goal of future research should be to determine whether a rudimentary understanding of other individuals' emotions might exist even in species that are unable to attribute more complex mental states like ignorance or false belief to others.

Summary

The ability to recognize third-party relationships is adaptive and may be widespread in animals. The degree to which primates may differ from nonprimate species in their ability to identify and monitor other individuals' social interactions is not yet clear. Similarly, we do not yet know whether monkeys discriminate among different types of social bond— whether they distinguish, for example, among the bonds formed by mothers and offspring, sisters, or friends. Moreover, the degree to which there is a quantitative or qualitative threshold for learning to recognize that two other individuals share a close bond is not known. Humans often use observations of other individuals' social interactions to make inferences about those individuals' emotions or motives for one another. Whether or not nonhuman primates are capable of similar emotional attributions is not known.

References

Altmann, J. 1980. *Baboon Mothers and Infants*. Cambridge, MA: Harvard University Press.

Altmann, J., Alberts, S. C., Haines, S. A., Dubach, J., Muruthi, P., Coote, T., Geffen, E., Cheesman, D. J., Mututua, R. S., Saiyalele, S. N., Wayne, R. K., Lacy, R. C., & Bruford, M. W. 1996. Behavior predicts genetic structure in a wild primate group. *Proc. Natl. Acad. Sci. USA*, 93, 5797–5801.

Aureli, F., Cozzolino, R., Cordischi, C., & Scucchi, S. 1992. Kin-oriented redirection among Japanese macaques: an expression of a revenge system? *Anim. Behav.*, 44, 283–291.

Bachmann, C. & Kummer, H. 1980. Male assessment of female choice in hamadryas baboons. *Behav. Ecol. Sociobiol.*, 6, 315–321.

Barton, R. A. & Dunbar, R. 1997. Evolution of the social brain. In: *Machiavellian Intelligence II: Extensions and Evaluations* (Ed. by A. Whiten & R. W. Byrne), pp. 240–263. Cambridge: Cambridge University Press.

Brannon, E. & Terrace, H. 1998. Ordering of the numerosities 1 to 9 by monkeys. *Science*, 282, 746–749.

Capaldi, E. J. 1993. Animal number abilities: implications for a hierarchical approach to instrumental learning. In: *The Development of Numerical Competence* (Ed. by S. T. Boysen & E. J. Capaldi), pp. 191–209. Hillsdale, NJ: Lawrence Erlbaum Associates.

Chapais, B., Gauthier, C., Prud'homme, J., & Vasey, P. 1997. Relatedness threshold for nepotism in Japanese macaques. *Anim. Behav.*, 53, 1089–1101.

Chapais, B., Savard, L., & Gauthier, C. 2001. Kin selection and the distribution of altruism in relation to degree of kinship in Japanese macaques (*Macaca fuscata*). *Behav. Ecol. Sociobiol.*, 49, 493–502.

Cheney, D. L. & Seyfarth, R. M. 1980. Vocal recognition in free-ranging vervet monkeys. *Anim. Behav.*, 28, 362–367.

Cheney, D. L. & Seyfarth, R. M. 1982. Recognition of individuals within and between groups of free-ranging vervet monkeys. *Am. Zool.*, 22, 519–529.

Cheney, D. L. & Seyfarth, R. M. 1986. The recognition of social alliances among vervet monkeys. *Anim. Behav.*, 34, 1722–1731.

Cheney, D. L. & Seyfarth, R. M. 1989. Reconciliation and redirected aggression in vervet monkeys. *Behaviour*, 110, 258–275.

Cheney, D. L. & Seyfarth, R. M. 1990. *How Monkeys See the World: Inside the Mind of Another Species.* Chicago: University of Chicago Press.

Cheney, D. L. & Seyfarth, R. M. 1999. Recognition of other individuals' social relationships by female baboons. *Anim. Behav.*, 58, 67–75.

Cheney, D. L., Seyfarth, R. M., & Silk, J. B. 1995a. The responses of female baboons (*Papio cynocephalus ursinus*) to anomalous social interactions: evidence for causal reasoning? *J. Comp. Psychol.*, 109, 134–141.

Cheney, D. L., Seyfarth, R. M., & Silk, J. B. 1995b. The role of grunts in reconciling opponents and facilitating interactions among adult female baboons. *Anim. Behav.*, 50, 249–257.

Church, R. M. & Meck, W. H. 1984. The numerical attribute of stimuli. In: *Animal Cognition* (Ed. by H. L. Roitblat, T. G. Bever, & H. S. Terrace), pp. 445–464. Hillsdale, NJ: Lawrence Erlbaum Associates.

Dallal, N. & Meck, W. H. 1990. Hierarchical structures: chunking by food type facilitates spatial memory. *J. Exp. Psychol. Anim. Behav. Processes*, 16, 69–84.

D'Amato, M. & Colombo, M. 1989. Serial learning with wild card items by monkeys (*Cebus apella*): implications for knowledge of ordinal rank. *J. Comp Psychol.*, 103, 252–261.

D'Amato, M., Salmon, P., & Colombo, M. 1985. Extent and limits of the matching concept in monkeys (*Cebus apella*). *J. Exp. Psychol. Anim. Behav. Processes*, 11, 35–51.

Dasser, V. 1988a. A social concept in Java monkeys. *Anim. Behav.*, 36, 225–230.

Dasser, V. 1988b. Mapping social concepts in monkeys. In: *Machiavellian Intelligence: Social Expertise and the Evolution of Intellect in Monkeys, Apes, and Humans* (Ed. by R. W. Byrne & A. Whiten), pp. 85–93. Oxford: Oxford University Press.

Doutrelant, C. & McGregor, P. K. 2000. Eavesdropping and mate choice in female fighting fish. *Behaviour*, 137, 1655–1669.

Dube, W. V., McIlvaine, W. J., Callahan, T. D., & Stoddard, L. T. 1993. The search for stimulus equivalence in nonverbal organisms. *Psychol. Rec.*, 43, 761–778.

Dunbar, R. 1983. Structure of gelada baboon reproductive units. III. The male's relationship with his females. *Anim. Behav.*, 31, 565–575.

Dunbar, R. 2000. Causal reasoning, mental rehearsal, and the evolution of primate cognition. In: *The Evolution of Cognition: Vienna Series in Theoretical Biology* (Ed. by C. Heyes & L. Huber), pp. 205–219. Cambridge, MA: MIT Press.

Ehmer, B., Reeve, H. K., & Hoy, R. R. 2001. Comparison of brain volumes between single and multiple foundresses in the paper wasp *Polistes dominulus*. *Brain Behav. Evol.*, 57, 161–168.

Emlen, S. T., Wrege, P. H., & Demong, N. J. 1995. Making decisions in the family: an evolutionary perspective. *Am. Sci.*, 83, 148–157.

Engh, A. L., Esch, K., Smale, L., & Holekamp, K. E. 2000. Mechanisms of maternal rank "inheritance" in the spotted hyaena, *Crocuta crocuta*. *Anim. Behav.*, 60, 323–332.

Engh, A. L., Siebert, E., Greenberg, D., & Holekamp, K. In press. Primate-like cognition in a social carnivore. *Anim. Behav.*

Fagot, J., Wasserman, E. A., & Young, M. 2001. Discriminating the relation between relations: the role of entropy in abstract conceptualization by baboons (*Papio papio*) and humans (*Homo sapiens*). *J. Exp. Psychol. Anim. Behav. Proc.*, 27, 316–328.

Ferguson, J. N., Young, L. J., Hearn, E. F., Matzuk, M. M., Insel, T. R., & Winslow, J. T. 2000. Social amnesia in mice lacking the oxytocin gene. *Nat. Genet.*, 25, 284–288.

Fields, L. 1993. Foreword: special issue on stimulus equivalence. *Psychol. Rec.*, 43, 543–546.

Gouzoules, S. & Gouzoules, H. 1987. Kinship. In: *Primate Societies* (Ed. by B. Smuts, D. Cheney, R. Seyfarth, R. Wrangham, & T. Struhsaker), pp. 299–305. Chicago: University of Chicago Press.

Hamilton, W. D. 1964. The genetical evolution of social behaviour. I and II. *J. Theor. Biol.*, 7, 1–51.

Harcourt, A. H. 1988. Alliances in contests and social intelligence. In: *Machiavellian Intelligence: Social Expertise and the Evolution of Intellect in Monkeys, Apes, and Humans* (Ed. by R. W. Byrne & A. Whiten), pp. 132–152. Oxford: Oxford University Press.

Harcourt, A. 1992. Coalitions and alliances: are primates more complex than non-primates? In: *Coalitions and Alliances in Humans and Other Animals* (Ed. by A. Harcourt & F. de Waal), pp. 445–471. New York: Oxford University Press.

Hare, B., Call, J., Agnetta, B., & Tomasello, M. 2000. Chimpanzees know what conspecifics do and do not see. *Anim. Behav.*, 59, 771–785.

Hare, B., Call, J., & Tomasello, M. 2001. Do chimpanzees know what conspecifics know? *Anim. Behav.*, 61, 139–151.

Harlow, H. F. 1949. The formation of learning sets. *Psychol. Rev.*, 56, 51–65.

Herman, L. M., Pack, A. A., & Morrel-Samuels, P. 1993. Representational and conceptual skills of dolphins. In: *Comparative Cognition and Neuroscience* (Ed. by H. L. Roitblat, L. M. Herman, & P. E. Nachtigall), pp. 403–442. Hillsdale, NJ: Lawrence Erlbaum Associates.

Herrnstein, R. J. 1985. Riddles of natural categorization. *Phil. Trans. Royal Soc. Lond., B*, 308, 129–144.

Heyes, C. M. 1994. Social cognition in primates. In: *Animal Learning and Cognition* (Ed. by N. J. Mackintosh), pp. 281–305. New York: Academic Press.

Holekamp, K. E., Boydston, E. E., Szykman, M., Graham, I., Nutt, K. J., Birch, S., Piskiel, A., & Singh, M. 1999. Vocal recognition in the spotted hyaena and its possible implications regarding the evolution of intelligence. *Anim. Behav.*, 58, 383–395.

Humphrey, N. K. 1976. The social function of intellect. In: *Growing Points in Ethology* (Ed. by P. Bateson & R. A. Hinde), pp. 303–318. Cambridge: Cambridge University Press.

Jolly, A. 1966. Lemur social behavior and primate intelligence. *Science*, 153, 501–506.

Judge, P. 1982. Redirection of aggression based on kinship in a captive group of pigtail macaques. *Int. J. Primatol.*, 3, 301.

Kummer, H., Goetz, W., & Angst, W. 1974. Triadic differentiation: an inhibitory process protecting pair bonds in baboons. *Behaviour*, 49, 62–87.

Macuda, T. & Roberts, W. A. 1995. Further evidence for hierarchical chunking in rat spatial memory. *J. Exp. Psychol. Anim. Behav. Proc.*, 21, 20–32.

McGregor, P. K. & Dabelsteen, T. 1996. Communication networks. In: *Ecology and Evolution of Acoustic Communication in Birds* (Ed. by D. E. Kroodsma & E. H. Miller), pp. 409–425. Ithaca, NY: Cornell University Press.

Mercado, E., Killebrew, D. A., Pack, A. A., Macha, I., & Herman, L. M. 2000. Generalization of "same-different" classification abilities in bottlenosed dolphins. *Behav. Proc.*, 50, 79–94.

Naguib, M., Fichtel, C., & Todt, D. 1999. Nightingales respond more strongly to vocal leaders of simulated dyadic interactions. *Proc. Royal Soc. Lond. B*, 266, 537–542.

Oden, D., Thompson, R., & Premack, D. 1988. Spontaneous transfer of matching by infant chimpanzees (*Pan troglodytes*). *J. Exp. Psychol. Anim. Behav. Proc.*, 14, 140–145.

Ogawa, H. 1995. Recognition of social relationships in bridging behavior among Tibetan macaques (*Macaca thibetana*). *Am. J. Primatol.*, 35, 305–310.

Oliveira, R. F., McGregor, P. K., & Latruffe, C. 1998. Know thine enemy: fighting fish gather information from observing conspecific interactions. *Proc. Royal Soc. Lond., B*, 265, 1045–1049.

Palombit, R. A., Cheney, D. L., & Seyfarth, R. M. 2001. Female-female competition for male "friends" in wild chacma baboons (*Papio cynocephalus ursinus*). *Anim. Behav.*, 61, 1159–1171.

Palombit, R. A., Seyfarth, R. M., & Cheney, D. L. 1997. The adaptive value of "friendships" to female baboons: experimental and observational evidence. *Anim. Behav.*, 54, 599–614.

Parr, L. A. & de Waal, F. B. M. 1999. Visual kin recognition in chimpanzees. *Nature*, 399, 647–648.

Peake, T. M., Terry, A. M. R., McGregor, P. K., & Dabelsteen, T. 2001. Male great tits eavesdrop on simulated male-male vocal interactions. *Proc. Royal Soc. Lond., B*, 268, 1183–1187.

Pepperberg, I. M. 1992. Proficient performance of a conjunctive, recursive task by an African gray parrot (*Psittacus erithacus*). *J. Comp. Psychol.*, 106, 295–305.

Pepperberg, I. M. 1994. Numerical competence in an African gray parrot (*Psittacus erithacus*). *J. Comp. Psychol.*, 108, 36–44.

Premack, D. 1983. The codes of man and beast. *Behav. Brain Sci.*, 6, 125–167.

Reichmuth-Kastak, C., Schusterman, R. J., & Kastak, D. 2001. Equivalence classification by California sea lions using class-specific reinforcers. *J. Exp. Anal. Behav.*, 76, 131–158.

Rendall, D., Rodman, P. S., & Emond, R. E. 1996. Vocal recognition of individuals and kin in free-ranging rhesus monkeys. *Anim. Behav.*, 51, 1007–1015.

Repacholi, B. & Gopnik, A. 1997. Early reasoning about desires: evidence from 14- and 18-month-olds. *Dev. Psychol.*, 33, 12–21.

Roitblat, H. 1987. *Introduction to Comparative Cognition*. New York: W. H. Freeman.

Schusterman, R. J. & Gisiner, R. 1988. Artificial language comprehension in dolphins and sea lions: the essential cognitive skills. *Psychol. Rec.*, 38, 311–348.

Schusterman, R. J. & Kastak, D. A. 1993. A California sea lion (*Zalophus californianus*) is capable of forming equivalence relations. *Psychol. Rec.*, 43, 823–839.

Schusterman, R. J. & Kastak, D. A. 1998. Functional equivalence in a California sea lion: relevance to animal social and communicative interactions. *Anim. Behav.*, 55, 1087–1095.

Schusterman, R. J. & Krieger, K. 1986. Artificial language comprehension and size transposition by a California sea lion (*Zalophus californianus*). *J. Comp. Psychol.*, 100, 348–355.

Schusterman, R. J., Reichmuth, C. J., & Kastak, D. 2000. How animals classify friends and foes. *Curr. Dir. Psychol. Sci.*, 9, 1–6.

Seyfarth, R. M. 1976. Social relationships among adult female baboons. *Anim. Behav.*, 24, 917–938.

Seyfarth, R. M. 1977. A model of social grooming among adult female monkeys. *J. Theor. Biol.*, 65, 671–698.

Seyfarth, R. M. 1978. Social relations among adult male and female baboons. II. Behavior throughout the female reproductive cycle. *Behaviour*, 64, 227–247.

Seyfarth, R. M. 1980. The distribution of grooming and related behaviors among adult female vervet monkeys. *Anim. Behav.*, 28, 798–813.

Shettleworth, S. 1998. *Cognition, Evolution, and Behaviour*. Oxford: Oxford University Press.

Sidman, M. 1994. *Equivalence Relations and Behavior: A Research Story*. Boston: Authors Cooperative.

Silk, J. B. 1987. Social behavior in evolutionary perspective. In: *Primate Societies* (Ed. by B. Smuts, D. Cheney, R. Seyfarth, R. Wrangham, & T. Struhsaker), pp. 318–329. Chicago: University of Chicago Press.

Silk, J. B. 1993. Does participation in coalitions influence dominance relationships among male bonnet macaques? *Behaviour*, 126, 171–189.

Silk, J. B. 1999. Male bonnet macaques use information about third party rank relationships to recruit allies. *Anim. Behav.*, 58, 45–51.

Silk, J. B., Seyfarth, R. M., & Cheney, D. L. 1999. The structure of social relationships among female savanna baboons. *Behaviour*, 136, 679–703.

Sinha, A. 1998. Knowledge acquired and decisions made: triadic interactions during allogrooming in wild bonnet macaques, *Macaca radiata*. *Phil. Trans. Royal Soc. Lond., B*, 353, 619–631.

Smale, L., Frank, L. G., & Holekamp, K. E. 1993. Ontogeny of dominance in free-living spotted hyaenas: juvenile rank relations with adult females and immigrant males. *Anim. Behav.*, 46, 467–477.

Smuts, B. 1985. *Sex and Friendship in Baboons*. Chicago: Aldine.

Terrace, H. S. 1987. Chunking by a pigeon in a serial learning task. *Nature*, 325, 149–151.

Thompson, R. K. R. 1995. Natural and relational concepts in animals. In: *Comparative Approaches to Cognitive Science* (Ed. by H. Roitblat & J. A. Meyer), pp. 175–224. Cambridge: MIT Press.

Tomasello, M. & Call, J. 1997. *Primate Cognition*. Oxford: Oxford University Press.

Treichler, F. & van Tilburg, D. 1996. Concurrent conditional discrimination tests of transitive inference by macaque monkeys: list linking. *J. Exp. Psychol. Anim. Behav. Proc.*, 22, 105–117.

von Fersen, L., Wynne, C., Delius, J., & Staddon, J. 1991. Transitive inference formation in pigeons. *J. Exp. Psychol. Anim. Behav. Proc.*, 17, 334–341.

Wasserman, E. A. & Astley, S. L. 1994. A behavioral analysis of concepts: application to pigeons and children. In: *Psychology of Learning and Motivation, Vol. 31* (Ed. by D. L. Medin), pp. 73–132. New York: Academic Press.

Wasserman, E. A., Hugart, J. A., & Kirkpatrick-Steger, K. 1995. Pigeons show same-different conceptualization after training with complex visual stimuli. *J. Exp. Psychol. Anim. Behav. Proc.*, 21, 248–252.

White, D. J. & Galef, B. G. 1999. Social effects on mate choices of male Japanese quail, *Coturnix japonica*. *Anim. Behav.*, 57, 1005–1012.

Wright, A., Cook, R., & Rivera, J. 1988. Concept learning by pigeons: matching to sample with trial-unique video picture stimuli. *Anim. Learning Behav.*, 16, 436–444

Zentall, T., Weaver, J., & Sherburne, L. 1996. Value transfer in concurrent-schedule discriminations by pigeons. *Anim. Learning Behav.*, 24, 401–409.

16

Constraints on Kin Selection in Primate Groups

Bernard Chapais
Patrick Bélisle

In sum, deployment of altruism beyond closest kin is entailed not by kin selection theory but by a constellation of factors. At present, the optimal allocation of altruism is unknown.

—Stuart Altmann, 1979: 958

The central argument developed here is that kin selection theory (Hamilton 1964) describes what unconstrained individuals could potentially do to increase their inclusive fitness, not what they can actually do. Kin selection does not take into account the various constraints on nepotism that operate among individuals living with several kin. For this reason, although kin selection is a central and necessary component of a theory of kin favoritism, it can hardly predict how individuals should allocate their aid among their relatives in social groups.

Hamilton's classic rule states that altruism is profitable to the donor when the benefits to the recipient (b) devalued by their degree of relatedness (r) are larger than the costs (c) to the donor (i.e., when $br > c$). Thus, for altruism to be profitable between any two kin, their degree of relatedness must be greater than the ratio c/b. Hence, Hamilton's equation determines the degree of relatedness beyond which a specified form of altruism ceases to be profitable from a donor's perspective. Otherwise stated, Hamilton's rule defines the maximal deployment of altruism among kin—that is, the subset of kin, among all those present, that an individual would gain by aiding. This simple rule creates an interesting decision-making problem to animals coresiding on a regular basis with several kin and able to discriminate their degree of kinship with the kin. Clearly, one cannot assume that individuals have an unlimited capacity, whether in terms of time, energy, or access, to aid their kin, nor that their aid toward a given recipient is always required; that recipient's needs might have already been satisfied by another relative. How, then, should individuals allocate their time and energy among their kin?

Primates are particularly useful to address this problem because, in several primate socie-ties, individuals co-reside with kin belonging to a variable number of categories (Pusey & Packer 1987, chapter 4) and are able to discriminate some of them. Consider, first, matrilocal (female philopatric) societies. Studies confirm that behavioral patterns of male dispersal and female philopatry accord well with genetic structure within groups: relatedness is higher among females than among males, and degree of kinship determined by known maternity and by paternity exclusion analysis correctly predicts degree of genetic relatedness (Altmann et al. 1996; de Ruiter & Geffen 1998). In these societies, kin discrimination is inferred from the preferential treatment of others based on kinship (nepotism)—that is, on certain corre-lates of kinship that are perceived by the animals. Although relatively little is known about the relevant correlates, a number of studies point to familiarity (prior association during development) as being a fundamental one (MacKenzie et al. 1985, Sackett and Fredrikson 1987, Welker et al. 1987, Kuester et al. 1994, Berman et al. 1997, Martin 1997, Bernstein 1999, chapters 12, 13). Similarly, the discrimination of different kin categories by animals is inferred from their differential treatment of these categories. Nepotism has been frequently reported in various behavioral areas among matrilineal relatives coresiding in matrilocal groups (Gouzoules 1984; Walters 1987; Bernstein 1991; Kapsalis & Berman 1996a, b; Silk 2001; Widdig et al. 2001; chapters 7, 14). Matrilineal kin recognition has been experimen-tally demonstrated between mother and offspring (e.g., Nakamishi & Yoshida 1986, Pereira 1986, Hammerschmidt & Fisher 1998) and between maternal siblings (Sackett & Fredrikson 1987). Although no experimental studies have been carried out on other categories of matri-lineal kin, their discrimination may be inferred from data on the distribution of behaviors according to degree of kinship. Such distribution curves are pictured in figure 16.1. As a first approximation, we may infer that relatives interacting at levels significantly higher than those of nonkin recognize each other as kin.

Thus, matrilocal societies satisfy the conditions required to analyze the distribution of nepotism among matrilineal kin. But matrilineal kinship is not restricted to matrilocal groups. Patrilocal (male philopatric) species such as chimpanzees (*Pan troglodytes*), bo-nobos (*Pan paniscus*), and species of the Atelinae (Moore 1992, Morin et al. 1994, Strier 1994, Hashimoto et al. 1996, Gerloff et al. 1999, Vigilant et al. 2001, chapter 4) also contain matrilineal relatives. In these species, maternal kin are mostly males because females leave their natal group. However, the number of matrilineal kin categories that can be discrimi-nated is probably much smaller in patrilocal groups than in matrilocal ones, being possibly limited to mother-offspring pairs and maternal brothers. This is largely because paternity in promiscuous primate groups is most often not recognized (Bernstein 1991; Paul et al. 1992, 1996; Gust et al. 1998; Ménard et al. 2001; but see Buchan et al. 2003). As a result, a male can hardly recognize, for example, the sons of his maternal brothers because even if he recognizes his brothers, the latter do not recognize their sons. Not surprisingly, therefore, male matrilineal kinship was found not to translate into preferential relationships in chim-panzees (Goldberg & Wrangham 1997; Mitani et al. 2000, 2002), although nepotism has been observed between mothers and offspring and between maternal brothers in chimpan-zees and bonobos (Goodall 1986, Furuichi 1997). Patrilocal primate groups also contain various classes of patrilineal kin (chapter 8). However, several studies found no evidence of discrimination among patrilineal kin (Fredrikson & Sackett 1984; Paul et al. 1992, 1996; Kuester et al. 1994; Erhardt et al. 1997; Gust et al. 1998; Ménard et al. 2001; chapters 12, 13) whereas others reported discrimination between fathers and offspring and between pater-

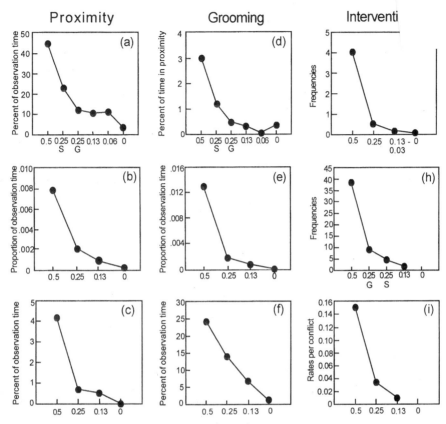

Figure 16.1. Distribution curves of three categories of behaviors (proximity, social grooming, and interventions in conflicts) in relation to degree of relatedness in Japanese macaques
(*Macaca fuscata*) (a, b, d, e, h), rhesus macaques (*M. mulatta*) (c, f, g) and pigtail macaques
(*M. nemestrina*) (i). Curves (a) and (d): modified from Kurland (1977). Curves (b) and (e):
modified from Glick et al. (1986). Curves (c) and (f): modified from Kapsalis and Berman
(1996a). Curve (g): from data in Kaplan (1977). Curve (h): from data in Massey (1977). In
all cases, data were corrected for the number of recipients per kin category. Proximity was
defined as being within 1 m in (a), within 0.6 m in (b), and between 0.6 and 5 m in (c). S:
dyads of sisters. G: grandmother-grandoffspring dyads.

nal siblings (Alberts 1999, Widdig et al. 2001, Buchan et al. 2003, Smith et al. 2003)
suggesting that patrilineal kin recognition may be species or group specific. For this reason
we limit our analysis of the distribution of nepotism to matrilineal kin in matrilocal societies.

What does a distribution curve of altruism look like? At present it is very difficult to
answer this question for at least three reasons. First, data on the distribution of behaviors
among matrilineal kin (figure 16.1) cannot be assumed to reflect the sole effect of kinship.
They might also reflect the influence of factors correlating with kinship, such as rank distance, in societies characterized by matrilineal dominance systems. In these societies, degree
of kinship correlates positively with proximity in the rank order; hence, what is identified

as a kin bias might in fact reflect a rank proximity bias. None of the curves pictured in figure 16.1 control for the effect of rank distance. However, some studies in which the effect of rank distance on rates of interactions was factored out (de Waal 1991; Kapsalis & Berman 1996a, b; Silk et al. 1999) found that kinship explained an important part of the variance in behavior, in other words, that nepotism was real.

Second, the importance of kinship as a factor regulating the social life of primates varies significantly between species. For example, species of the genus *Macaca* are characterized by different dominance styles, from egalitarian to despotic (de Waal & Luttrell 1989; Thierry 1990, 2000; Chaffin et al. 1995; Petit et al. 1997, Matsumura 1998). The role of kinship in patterning social relationships varies accordingly, being more important in despotic species (Thierry 1990, 2000; Butovskaya 1993; Aureli et al. 1997). On this basis alone, one expects the distribution of behavioral kin biases in relation to degree of kinship to vary substantially among species.

Third, the existing distribution curves of nepotism (figure 16.1) most often mix two functional categories of kin-biased interactions: altruistic and cooperative. In altruistic nepotism the donor incurs a net cost without being repaid later, whereas in cooperative nepotism each participant derives a net benefit. Therefore, the theoretical predictions concerning the distribution of altruistic and cooperative interactions differ markedly, and for this reason they should be analyzed separately, as they are here.

In this chapter, we begin by reviewing the various factors that affect the distribution of altruism among matrilineal kin in matrilocal groups, including time constraints, the number of kin present, the individuals' needs, and the differential availability of kin. Because these factors most often curtail the deployment of altruism, we argue that the observed limit of altruism in terms of degree of relatedness probably falls short of the limit of the profitability of altruism defined by Hamilton's equation. We then report experiments carried out on Japanese macaques that assess the deployment of altruism when it is unconstrained by the above factors. Last, we analyze the distribution of cooperative interactions among matrilineal kin. We do so by distinguishing the two assumed fitness components of kin cooperation, direct (personal fitness) and indirect (inclusive fitness), and their respective predictions on the distribution of nepotism.

The Deployment of Altruism Among Kin

As stated by Altmann (1979), kin selection alone is only one of a number of factors that affect the distribution of altruism among kin. We review here three such additional sets of factors: (1) time constraints limiting the donors' capacity to aid others, (2) the cumulative impact of kin altruism by several donors on the satisfaction of a recipient's needs, and (3) the correlation between degree of kinship and spatial availability. These three factors concur in producing declines in the distribution of altruism, in most cases sharp ones, and therefore limit substantially the deployment of altruism.

Some other factors that may affect the distribution of altruism among kin are not considered here because, although they may create variations in rates of nepotism across kin dyads, they are unlikely to affect specifically distant kin dyads and determine the maximal deployment of altruism. Among these are the differential value of recipients according to their reproductive value (e.g., Hamilton 1964, Chapais & Schulman 1980, Schulman & Rubenstein 1983, Combes & Altmann 2001) and variation in the level of competition within kin dyads, hence on the level of constraints on nepotism (e.g., Chapais et al. 1994; see also West et al. 2001).

Hamilton's Equation in the Context of Time Constraints

According to Hamilton's equation, for altruism to be profitable between any two kin, r must be greater than c/b. Thus the values of b and c determine the relatedness limit of profitable altruism, or relatedness threshold for altruism (Chapais 2001). In turn, the relatedness threshold for altruism defines the subset of potential beneficiaries of altruism among all kin present. However, because individuals are limited in the time and energy they can devote to the care of others, they may be unable to attend to the needs of all the kin that are within the relatedness threshold for altruism. How, then, should they allocate aid among them?

A number of researchers have assumed that individuals should aid their kin in proportion to their degree of relatedness (e.g., Kurland 1977), giving proportionately more to daughters than to granddaughters, more to granddaughters than to cousins, and so forth. However, Altmann (1979) argued that such proportional investment amounted to "commiting a common gambler's fallacy, of distributing the stakes (i.e., investment) in proportion to the odds (degree of kinship)," and that in order to maximize its gain in inclusive fitness, an animal should instead "bet only on that outcome with the highest expected value" (p. 958), in other words, favor only its closest relative. After the needs of the closest relative have been satisfied, the donor is free to attend to those of its next closest kin and so on, until it has exhausted all the time and energy at its disposal.

Consider, first, the case of behaviors that extend over time, such as infant carrying, infant monitoring, lactating, and grooming. These behaviors do not necessarily have a fixed and constant cost-benefit ratio. The benefits of being fed or groomed are likely to decrease over time. Taking grooming as an example, Altmann (1979) reasoned that the benefits of grooming to the groomee (e.g., its hygienic function) followed some sort of diminishing returns curve and that in this situation a donor would at some point benefit more by ceasing to groom its closest kin and beginning to groom its next closest one. This idea was later modeled mathematically and confirmed by Weigel (1981), Schulman and Rubenstein (1983), and Reiss (1984).

When should a female, for example, stop grooming her (single) daughter and start grooming her granddaughter? The payoff of grooming for the donor is given by $br - c$. Suppose that the benefits of grooming at the beginning of the interaction have a value of 20 and the costs a value of 2. The donor's initial payoff from grooming a daughter is $20 \times 0.5 - 2 = 8$. This value decreases over time as the benefits of grooming decrease. The donor's initial payoff from grooming a granddaughter is $20 \times 0.25 - 2 = 3$. Thus, the female should continue grooming her daughter for as long as her payoff is greater than that from grooming a granddaughter (i.e., >3). This point in time is determined by the rate at which the payoff of grooming diminishes over time. If the rate is slow (i.e., the recipient's needs are satisfied slowly), the donor might spend all the grooming time she has available on her daughter before she should switch to her granddaughter. In this case, the deployment of grooming among kin categories would be extremely limited.

In contrast, if the payoff of grooming diminishes rapidly over time, the female should switch rapidly to her granddaughter, then to her next closest kin, and so on until she has no time left to groom other kin. Thus, the rate of diminishing returns from grooming determines the optimal mean duration of grooming between kin, hence the number of kin a female can groom on any given day. Note that this number cannot be greater than the number of kin that are within the relatedness threshold for altruism. In the above example, the c/b ratio of grooming $(2 \div 20 = 0.1)$ sets the threshold at $r > .1$.

Suppose that for any given set of values (c/b ratio and rate of diminishing returns), the average female can attend to the needs of all members of her n closest kin categories, but to only a fraction of those of her next closest kin. This would produce a distribution curve with a sudden decline past the nth kin category, with an intermediate average value for the next kin class, followed by a drop to values comparable to those of nonkin for the next kin category. Thus, the integration of time constraints to Hamilton's equation produces curves characterized by an abrupt decline somewhere upstream in relation to the relatedness threshold for altruism.

Consider now the case of behaviors that are brief and well delineated in time, such as instances of protection against aggressors, and suppose that a donor has the opportunity of aiding the same beneficiary more than once every day. Because these interactions occur at different times and are relatively independent of each other, the benefits to the recipient and the costs to the donor (the c/b ratio) are similar every time the interaction takes place. In this situation, the number of kin a donor can aid among all those that are within the relatedness threshold for altruism should be determined by the total amount of time and energy the donor can devote to its kin in this context, and by its kin's needs, in this case the number of opportunities for protection (e.g., aggressions received or solicitations for help). Suppose the average female can attend to the needs of her n closest kin categories and to a fraction of the needs of her next closest kin. This produces a curve characterized by a sudden decline past the nth kin category, as was the case with grooming. However, brief and infrequent interactions such as interventions entail substantially lower time constraints. For this reason, the deployment of protracted behaviors should be markedly more limited than that of brief interactions. For example, individuals could lack time to groom distant kin but have enough time to intervene in the conflicts of these same kin.

However, looking at the curves pictured in figure 16.1, the deployment of interventions does not seem more extensive than that of grooming. This might be because other factors are also at play, namely the cost-benefit ratio of behaviors. For example, if certain types of interventions in conflicts are substantially more costly than grooming, their profitability could be limited to close kin while the profitability of grooming could include more distant kin. In such a situation, time constraints and the cost-benefit ratio of behaviors would be working in opposite directions, illustrating some of the difficulties of predicting the allocation of altruism. As we shall see, further difficulties relate to variations in mean lineage size.

The Effect of Mean Matriline Size

Another component of the time constraints affecting an individual's ability to aid its kin is the sheer number of close kin in that individual's group, more specifically the average number of close relatives per kin category. The larger that average number, the larger the number of potential close kin recipients of any individual, hence the less time left to help more distant kin.

Variation in the reproductive rates of groups may profoundly affect the size of matrilines. For example, Dunbar (1988) compared three hypothetical populations: one with a low reproductive rate characterized by an age at first birth at 6 years and an interbirth interval of 3 years, another with a medium reproductive rate (first birth at 4 years and interbirth interval of 2.5 years), and a third with a high reproductive rate (first birth at 4 years and interbirth interval of 1 year). Dunbar calculated that the mean size of matrilines (counting only the number of mature females) varied considerably, from 1.5 kin (range: 1–2) in the low reproductive rate group to 5.4 kin (range: 1–9) in the group with the highest reproductive rate. To these figures one should add the number of sexually immature females, which will be greater in expanding populations. This simple simulation clearly shows that the genealogical environment of females may vary considerably depending on their life history regime, which is in part affected by ecological variables (e.g., Dunbar 1988, Lee & Bowman 1995). In turn, the genealogical environment in which females are brought up may profoundly affect their social relationships (Berman 1988, Datta 1992, Hill & Okayasu 1996, Berman et al. 1997, chapter 6).

In larger matrilines (greater numbers of close kin), the deployment of altruism should be more limited and the curves should start declining sooner (at a higher r) than in smaller matrilines. Thus, if the deployment of any specific form of altruism were found to be the same regardless of matriline size, this would suggest that the limit of the profitability of altruism (its c/b ratio) determines the deployment. On the other hand, if the deployment of altruism were found to be more extensive in smaller matrilines, this would suggest that time constraints limit the deployment of altruism in large matrilines. To carry out this type of analysis, one needs data on the distribution of the same contextual category of altruistic behaviors in matrilines differing in size, or in groups varying in their mean matriline size. To our knowledge, such data are not available at present. Thus, an undetermined portion of the variation in the distribution curves of figure 16.1 may reflect differences in matriline size.

The Cumulative Satisfaction of the Recipients' Needs

As just seen, any donor may have several potential recipients and face time constraints as a result, but it is also true that any recipient may have several potential donors. For example, young females acquiring their rank in their group's matrilineal dominance order may be aided by kin belonging to various categories (Berman 1980; Datta 1983, 1988; Chapais & Gauthier 1993; Chapais et al. 1997). Thus, a different category of factors affecting the distribution of altruism has to do with the total effects of kin altruism from several kin on the satisfaction of a recipient's needs. To take an extreme example, if an infant's total needs for milk are satisfied daily by its mother, one does not expect other kin to nurse the infant. But if only 90% of the infant's needs, say for transportation, are satisfied by its mother, the infant will be dependent on others for the remaining 10%. As another example, if a female's total needs for grooming have been satisfied by her n closest kin, we do not expect the female to be groomed by more distant kin. In general, an individual's level of dependence on any given relative should depend on the number of closer kin that satisfy that individual's needs. The larger that number, the less dependent the individual on more distant kin. Thus, as degree of kinship decreases, so do dyadic levels of dependence between kin. In other words, rates of nepotism are expected to decrease with decreasing degree of kinship.

This principle holds true whether behaviors are time-consuming or brief and whether mean matriline size is large or small, although rates of nepotism should decrease more abruptly with decreasing relatedness in large matrilines than small ones. Thus, the fact that kin altruism is cumulative is expected to produce sharp declines in the distribution curves of altruism in most situations.

The Correlation Between Degree of Kinship and Spatial Availability

The factors previously considered have in common that they predict a sudden drop in the distribution of altruism past a certain degree of kinship. In contrast, another factor, the differential spatial availabilty of kin, is expected to generate a progressive diminution of behavior frequencies with decreasing degree of kinship, and therefore to soften the expected slopes. In matrilocal societies, maternal investment extends throughout the lifespan, mothers maintaining supportive and affiliative relationships with their adult daughters (Fairbanks 2000). In theory, the sole existence of lifelong bonds between mothers and daughters produces a positive correlation between degree of relatedness and level of proximity among the other categories of matrilineal kin. This is because, for instance, two sisters would spend time together if only because they were attracted to the same mother. Thus, one expects to find positive correlations between kinship and proximity among matrilocal females in the absence of any intrinsic attraction between matrilineal kin other than mothers and daughters (Chapais 2001). Various studies have reported positive correlations between degree of kinship and amount of time spent in proximity in primates (e.g., Kurland 1977, Kapsalis & Berman 1996a), and there is some indication that these correlations are mediated by proximity to the mother (Martin 1997, Berman & Kapsalis 1999). Some of these correlations are illustrated in figure 16.1(a–c).

As a result, from ego's perspective closer kin usually are more available than distant kin. For example, individuals would be expected to witness the conflicts or hear the solicitations for help of their closer kin more often than they witness or hear those of more distant kin. Thus, the less closely related two individuals are, the smaller the number of opportunities they have to interact. Hence, even in the absence of time constraints, one expects frequencies of nepotism to decrease with degree of kinship. The slope of this relation should vary according to the degree of social cohesiveness and spatial dispersion within groups, which depend on factors such as group size, feeding competition, and predation pressure (e.g., Sterck et al. 1997). Spatial dispersion is known to vary considerably both within and between species (e.g., Matsumura & Okamoto 1997).

One may control for the effect of the differential spatial availability of kin on the distribution of behaviors. By correcting behavioral frequencies for differences in time spent in proximity across kin categories, one should obtain behavior rates that are independent of the differential availability of kin, hence constant across categories. Some studies include this correction (e.g., curve d in figure 16.1), but the distributions are nevertheless declining, indicating that factors other than the differential availability of kin are at play. These include the profitability of altruism, time constraints, and the decreasing dependence levels on kin with decreasing r.

The covariation between degree of kinship and kin availability has another important consequence for the deployment of altruism. Even though it may be in a female's interest always to favor her closest kin, she will be unable to do so if the latter are not always around. Thus, the partial unavailability of a female's closest kin frees up time and energy for her to aid more distant kin that she would not aid if her closest kin were always at hand. On this basis, the partial unavailability of a female's favorite kin is expected to extend the deployment of altruism past the relatedness limit of nepotism determined by time constraints alone. This factor would tend to spread otherwise highly skewed distribution curves (figure 16.1).

Deployment of Altruism: Conclusions

We conclude that existing data on the deployment of altruism in primate groups cannot be taken to reflect the limit of the profitability of altruism defined by Hamilton's equation. In other words, our knowledge of the role of kin selection in the observed deployment of altruism in primate groups is approximate at best. Various factors converge to curtail the deployment of altruism well before the relatedness threshold for altruism: donors may simply lack time to aid distant kin, especially in groups composed of large lineages and for time-consuming activities, they may also be less often available to interact with them, and recipients may be cumulatively less dependent on more distant kin. To take an example, a niece could receive little from her aunt because (1) the aunt is busy taking care of her own daughters and grandaughters, (2) the aunt sees and hears her niece less often than her closer kin, and (3) the niece's needs are satisfied for the most part by her own mother, grandmother, and sisters.

Furthermore, altruism may evolve not only through kin selection but through reciprocal altruism (Trivers 1971) as well. Because reciprocal altruism is a form of cooperation, we examine its impact on the distribution of nepotism in the section on cooperation (see below). But at this stage, we note that the occurrence of reciprocal altruism would potentially extend the deployment of altruism well beyond the relatedness threshold for altruism, among distant kin and nonkin. This is one more reason not to assume that the observed relatedness limit of altruism in any group is determined by Hamilton's equation. Note, however, that reciprocal altruism cannot explain the occurrence of unilateral altruism between kin. Thus, one way of assessing the impact of kin selection is to focus on instances of altruism that are distributed unilaterally among kin (Chapais 2001).

Given that the exact role of the terms of Hamilton's equation (b, c, and r) in the distribution of altruism probably lies hidden under the above constraints, what would be the deployment of altruism if the constraints are removed? What kin would a female aid if she had plenty of time to do so, equal access to all of them, if she was the only donor available, and if she could not expect returns from the recipient? In other words, what is a female's potential for altruism? How large is the discrepancy between the empirical limit of altruism in natural groups and the theoretical limit determined by Hamilton's equation? In the following section, we report experiments that address these questions.

The Deployment of Altruism in the Absence of Group Constraints

The impact of Hamilton's equation on the distribution of altruism was experimentally studied in Japanese macaques (*Macaca fuscata*) and reported in a related set of articles (Chapais

1988; Chapais et al. 1997, 2001). Female Japanese monkeys establish matrilineal dominance orders by socially inheriting the rank of their mother (Kawamura 1965). In these experiments, an adult female was isolated with a group of juvenile females, the lowest ranking of which was a familiar relative of the adult female. Due to her larger size, the adult female was dominant to all juveniles from the outset; hence she had the opportunity of acting nepotistically by performing low-cost interventions in favor of her kin against dominant peers that were not related to her. Several categories of adult female kin were tested separately: mothers, grandmothers, sisters, great-grandmothers, aunts, grandaunts, and cousins. In each test, the adult female could intervene in the conflicts involving her young kin, which would most often outrank its peers as a result, or the adult female could act as if unconcerned by peer conflicts, in which case rank reversals rarely took place. The adult female's interventions were assumed to be altruistic because the costs were small, being limited to the expenditure of time and energy, and the adult female incurred no risk of retaliation from the juveniles. At the same time, interventions provided important benefits to the recipients in terms of rank increments (Chapais et al. 2001).

This experimental protocol satisfies the conditions required to assess the role of Hamilton's equation on the distribution of altruism. First, the two components of time constraints were eliminated because in each experiment the donor had a single kin to help, and to do so only one intervention or a brief series of interventions was needed. Second, the problem of the lower availability of more distant kin was also eliminated because all juvenile kin tested were equally available and had equal opportunities of being helped. Third, the possibility that the recipient might not need help due to the contribution of closer kin was nonexistent, the adult kin being the only one available. Fourth, the potential confounding effect of reciprocal altruism on kin selection was minimized because altruism was unilateral, from the adult female to the juvenile one.

In this context, the distribution of altruism can be expected to be determined mainly by the behavior's cost-benefit ratio. The lower the c/b ratio (i.e., the higher the benefits to the recipients and/or the smaller the costs to the donor), the more extensive the expected deployment of altruism among kin. In the above experiments a single form of altruism was tested, characterized by a particularly low c/b ratio (see above), hence by an expected extensive deployment.

Seven categories of kin relations were tested. Juveniles (males and females combined) could outrank dominant peers in the presence of either their mother ($r = .5$), grandmother ($r = .25$), older sister ($r = .25$), or great-grandmother ($r = .125$), but rarely in the presence of their aunt ($r = .125$), and never in the presence of great-aunts or cousins ($r = .063$). In other words, altruism toward younger kin extended up to $r = .125$ among lineal kin (great-grandmother/great-grandoffspring dyads), and $r = .25$ among collateral kin (siblings), inconsistently to $r = .125$ (avunculates). In theory, behavioral categories characterized by lower c/b ratios (higher benefits and/or lower costs) could extend the deployment of altruism even further. But it should be noted that situations such as the present experiments, in which the costs to the donor were low and the benefits to the recipient high, may be rare. Indeed, several categories of interactions might provide recipients with benefits higher than rank increments, for example, protection from predators or adult males. However, such interactions simultaneously entail substantial costs to adult females, so that their cost-benefit ratio is probably higher than in the above experiments, hence their expected deployment more limited. According to this reasoning, the relatedness limits of altruism found in the experiments reported above might approach the limit of the profitability of altru[...]cies, hence the deployment of altruism when females are minimally constra[...]

It would be interesting to compare the relatedness threshold for altruism [...] the limit of altruism observed in primate groups in which the constraints [...] cussed curtail its deployment. Altruism is well documented for mother-offs[...] various areas of behavior, including protection and food sharing. Among siblings and grand-mother-grandoffspring dyads ($r = .25$), the evidence is much less abundant and often equivocal (Kurland 1977; Silk 1982, 1987; Walters 1987; Schaub 1996) for various reasons, including the difficulty of assessing the altruistic nature of behaviors and of teasing apart altruistic and mutualistic subcategories of interactions (Chapais 2001). Beyond $r = .25$, clear evidence for kin-selected altruism is almost nonexistent. Because data on altruism in primate groups are often equivocal, assessing the discrepancy between the potential deployment of altruism and the observed one is difficult. However, considering the adult females' potential for altruism revealed by the above experiments, reported instances of altruism appear remarkably limited, both quantitatively and in terms of degree of kinship. This observation supports the view that the coresidence of several kin constrains the deployment of altruism below expectations based solely on the terms of Hamilton's equation.

We have argued that in general the observed deployment of altruism in primate groups probably underestimates the altruistic capacity of adult females. But it is also possible to overestimate the deployment of altruism by misappraising the cost-benefit ratio of behaviors, for example, by neglecting their possible self-serving aspects. We use co-feeding as an illustration.

If two individuals with a clear dominance relationship eat at the same food source, the dominant individual might tolerate the subordinate's presence because the costs of monopolizing the resource, in terms of energy spent and/or risk of injuries, are too high relative to the benefits of consuming it all. In this situation, tolerance would be imposed upon the dominant individual and might serve the purpose of buying peace while eating (Wrangham 1975) or of avoiding risky fights ("tolerated theft," Blurton-Jones 1984). Tolerance would be only apparently altruistic and the "donor" would in fact be serving its self-interest. Thus, for co-feeding to qualify as truly altruistic, food must be monopolizable by the dominant individual, who should yield food to the subordinate in the absence of any constraint. The failure to take into account the costs of food defense may lead to an overestimation of the deployment of altruism among kin. For example, Yamada (1963) presented data on co-feeding in relation to degree of kinship in free-ranging Japanese macaques. These data were later interpreted by Barash (1982: 80) as an illustration of food sharing, altruism, and kin selection operating presumably among several kin classes, including avunculates ($r = .125$). However, the extent to which food was monopolizable by the dominant kin in Yamada's (1963) study is unknown.

Again, the experimental approach is useful here. Bélisle and Chapais (2001) carried out experiments on Japanese macaques in which two adult females with a clear dominance relationship had the opportunity of eating at a translucent food box ($46 \times 20 \times 17$ cm) containing a small quantity of highly prized food. When co-feeding, the two females were facing each other at arm's length and fed from two holes (6 cm in diameter) located at each end of the box. Because the box was easily monopolizable by the dominant female, co-feeding was assumed to imply altruism. Five categories of dyads were tested: mother-daugh-

., grandmother-granddaughter, sister-sister, aunt-niece, and nonkin. Co-feeding decreased significantly with degree of kinship, and rates of co-feeding between aunts and nieces were similar to those of nonkin. These results suggested that the relatedness limit of co-feeding was $r = .25$ (sisters, grandmother-granddaughter dyads).

Although in this experiment (Experiment 1) the food box was monopolizable to a large extent, its defense might nevertheless have imposed some costs to the donor (e.g., chasing the subordinate). If this was indeed the case, the tolerance exhibited by the dominant individual might in part serve the purpose of avoiding these costs. To test this possibility, a pilot experiment was carried out (Experiment 2) in which the same food box was used but with the two holes located on the same side of the box, 20 cm apart (Bélisle 2002). When co-feeding, the two females sat side by side, and the dominant female could defend the food box more easily, using only visual or vocal threats, instead of more energetic forms of aggression such as moving around the box, lunging at the subordinate, or chasing her. To prevent the dominant female from feeding from the two holes simultaneously, a small partition was fixed perpendicularly to the side of the box between the two holes so that the two females were in contact at the shoulder level.

Experiment 2 focused on mother-daughter and grandmother-granddaughter dyads, the kin categories with the highest levels of co-feeding in Experiment 1. The dyads tested were the same. The costs of food defense to the dominant females were measured by rates of supplants and chases. As expected, these costs were significantly lower in Experiment 2 (mean rate per min. = 0.01 compared to 1.22 in Experiment 1). In theory, therefore, dominant females stood less to gain by tolerating the subordinate females in Experiment 2. Accordingly, their levels of tolerance decreased significantly. Co-feeding time was significantly lower, dominant females directed more aggression to subordinate females, from a larger mean distance, and their aggression was more intense on average (Bélisle 2002).

These preliminary results indicate that a significant portion of the time spent co-feeding in Experiment 1 was not altruistic on the part of the dominant female but rather selfish, serving to reduce the costs of food defense. Assuming that the levels of co-feeding obtained in Experiment 2 reflect "pure" altruism, these data suggest that the deployment of altruistic co-feeding is more limited than suggested by Experiment 1, being restricted to mother-daughter pairs.

The Deployment of Cooperation Among Kin

We have hitherto examined the contribution of kin selection theory to understanding the distribution of altruistic behaviors. Kin also engage in cooperative relationships, in the course of which both participants derive net benefits. One possibility is for the participants to gain immediate mutual benefits that are higher than what each individual would gain by not cooperating. This is called mutualism (Pusey & Packer 1997). Another possibility is for two cooperators to reciprocate altruistic acts so that interactions that are immediately costly provide mutual benefits in the long run (Trivers 1971). However, theoretical investigations indicate that the conditions under which reciprocity is expected to evolve are very limited (reviewed by Pusey & Packer 1997). Moreover, empirical instances of reciprocity have proven extremely difficult to identify because in most cases it is not possible to eliminate the possibility that the presumed donors are also gaining in the course of the interaction, in other words that the interaction is mutualistic (Bercovitch 1988; Noë 1990, 1992; Chapais et al. 1991; Hemelrijk 1996; Pusey & Packer 1997).

Whether cooperation in primates consists mostly, if not exclusively, of instances of mutualism, or whether it consists of a mixture of mutualism and reciprocal altruism, it is expected to be kin biased for at least two reasons. First, due to the long-term nature of bonds between mothers and daughters in matrilocal groups (Fairbanks 2000), members of the same matrilineage are generally more available and more familiar to each other compared to other individuals. As argued elsewhere, this factor alone may bias the selection of kin as social partners for cooperative activities, whether in the form of repeated mutualism or reciprocity (Chapais 2001).

Second, independently of this availability argument, Wrangham (1982) proposed that it pays more to engage in mutualism with kin than with nonkin because in the former case the interactants gain both directly through personal benefits and indirectly through the benefits accruing to their kin (inclusive fitness benefits). Note that the same reasoning applies to reciprocal altruism, which should also be more profitable between kin than between nonkin: in addition to deriving personal benefits upon being reciprocated, donors increase their inclusive fitness when behaving altruistically with their kin. That cooperation among kin provides fitness benefits, both direct and indirect, was empirically demonstrated in red howler monkeys (*Alouatta seniculus*). In this species, females form long-term cooperative coalitions to defend feeding territories. Females have higher reproductive success in coalitions of kin than in coalitions of nonkin, and their reproductive success increases with the average degree of relatedness among them (Pope 2000).

Assuming, then, that cooperation is kin biased and provides both direct and indirect fitness benefits, how should cooperation be distributed among kin categories? To answer this question, it is useful to analyze separately, as if they were independent, the direct and the indirect fitness components of kin cooperation. To maximize the indirect component (benefits to its kin), an individual should follow Altmann's (1979) principle developed in relation to altruism—that is, the individual should cooperate most often with its closest relative until they have satisfied their common needs, then with its next closest relative, and so on. As this was the case with altruism, the expected distribution curve of cooperation should be affected by time constraints acting on cooperators, decreasing needs for cooperation with decreasing r, and the decreased availability of kin with decreasing r. Thus, considering only the indirect fitness component of cooperation, one would expect rates of mutualism and reciprocal altruism to be highest among closest kin and then to decline rapidly. Note that one expects no relatedness limit in the distribution of cooperative nepotism—no equivalent of a relatedness threshold for altruism—because cooperation is always advantageous regardless of degree of kinship.

However, the direct fitness component of cooperation is probably more fundamental than the indirect one because personal benefits are not devalued by degree of relatedness. Moreover, cooperation is profitable and common between nonkin (Pusey & Packer 1997), suggesting that when cooperation takes place between kin, inclusive fitness benefits may be secondary in importance, additional, and perhaps incidental (Chapais 2001). Thus, considering only the personal benefits accruing from cooperation, the distribution of cooperative tasks among kin should be independent of degree of kinship and be determined, instead, by the adequacy of the partners' personal attributes (age, dominance rank, length of tenure, specific skills, etc.) in relation to the task involved (types of coalitions, hunting, huddling, grooming, etc.).

Taking into account simultaneously the two fitness components of cooperation between kin, cooperating with a competent nonkin might be more profitable than cooperating with a less qualified close kin—that is, obtaining important direct benefits with a nonkin might pay more than obtaining important inclusive fitness benefits with a close kin. This reasoning would apply particularly in the case of activities requiring specific characteristics and/or skills because such attributes are not expected to covary with degree of kinship. Thus, one would expect activities such as cooperative hunting or coalition formation, which require specific individual attributes, to show relatively weak kin biases. In contrast, activities such as huddling, which requires no specific attributes except body size, can be performed with any individual, kin or nonkin. Hence, concentrating one's huddling time on close kin would maximize inclusive fitness without reducing the personal benefits derived from huddling. In sum, one would expect attribute-independent forms of cooperation to be kin biased (obviously, if kin are available), but attribute-dependent forms of cooperation to be much less kin biased.

Some data provide preliminary tests of these predictions. Huddling and physical contact entail immediate mutual benefits in terms of thermoregulation (Anderson 1984, Takahashi 1997) and social comfort. These attribute-independent cooperative activities are strongly kin biased, as expected (Gouzoules 1984, Walters 1987, Bernstein 1991). In contrast, other forms of cooperation are more attribute dependent and therefore are expected to be less kin biased. For example, in species with matrilineal dominance systems, higher ranking females form conservative alliances against lower ranking females to maintain their rank above the latter. Such interventions are mutualistic because they provide immediate benefits to both the intervener and the beneficiary, at no risk of retaliation by the target (Chapais et al. 1991, Prud'homme & Chapais 1996). One necessary attribute of the female beneficiary is that she be higher ranking than the targeted female. Thus, conservative alliances are a form of attribute-dependent cooperation. Although females could exclusively support their close kin against lower ranking females, they commonly support unrelated dominant females against them (Chapais 1983, Hunte & Horrocks 1986, Netto & van Hooff 1986, Chapais et al. 1991), and the distribution of these interventions is not strongly kin biased.

As another example, Mitani et al. (2000) found that three forms of presumably attribute-dependent cooperation (participation in alliances, meat sharing, and patroling) were distributed independently of matrilineal kinship (assessed by mtDNA) in chimpanzees. Part of these results may reflect the fact that the discrimination of matrilineal kinship in patrilocal societies is limited to mother-son pairs and maternal brothers (see above). But even then, the absence of cooperation between maternal brothers is significant and suggests that inclusive fitness effects on cooperation are secondary in chimpanzees (Goldberg & Wrangham 1997, Mitani et al. 2000) and that partner selection is based on other individual attributes, such as age and dominance rank similarities (Mitani et al. 2002). This is consistent with the above predictions.

The case of social grooming is more complex. Grooming is a highly heteregenous, multicontextual and multifunctional activity. For this reason, it may be attribute dependent in some contexts and attribute independent in others. For example, individuals might exchange grooming primarily for its hygienic function (Hutchins & Barash 1976, Barton 1985, Tanaka & Takefushi 1993) or tension reduction effects (Schino et al. 1988, Boccia et al. 1989). The observation that grooming is often reciprocal (Janus 1989, Muroyama 1991, Barrett et al. 1999) lends support to this hypothesis. In this context, the benefits of grooming would be relatively independent of the participants' characteristics and skills, and its distribution should therefore be kin biased (if kin are available). However, available data do not make it possible to test this prediction adequately because although grooming is known to be strongly kin biased (e.g., figure 16.1 d–f, Kurland 1977, Kapsalis & Berman 1996a, Schino 2001), we do not know whether grooming that serves a hygienic or tension reduction function is more kin biased than grooming serving other functions.

On the other hand, individuals might perform grooming as a means of obtaining other types of benefits. Grooming may provide groomers with immediate benefits if it allows dominant individuals to remain near valuable subordinates or if it increases the tolerance levels of dominant individuals (Hemelrijk et al. 1992, Muroyama 1994, Henzi & Barrett 1999). It might also increase the probability of obtaining delayed benefits such as food (de Waal 1997) or coalitionary support (Seyfarth 1977, Seyfarth & Cheney 1984, Hemelrijk 1994). In these situations, grooming would be strongly attribute dependent and its distribution would not be expected to be particularly kin biased. Some of these functions of grooming are still debated (e.g., contrast Henzi & Barrett 1999 and Schino 2001) and one of the issues is whether grooming is altruistic and part of a system of reciprocal exchange, or mutualistic (Silk 1982, 1987; Dunbar & Sharman 1984; Walters 1987; Dunbar 1988; Hemelrijk et al. 1992; Maestripieri 1993; Barrett et al. 1999; Chapais 2001). Clearly, however, attributes such as oestrus state, dominance status, and motherhood play a fundamental role in the distribution of grooming, a role that often overshadows that of kinship.

Conclusion

Altmann's (1979) conclusion cited at the beginning of this chapter—"the optimal allocation of altruism is unknown"—is still largely intact 25 years later. Considering the sheer number of factors involved in the distribution of nepotism, the difficulties of measuring some of the most basic of these—especially the costs and benefits of behaviors—and given the paucity of studies on the subject, this should come as no surprise. Our goal in this chapter is mainly to pose the problem in all its complexity and to indicate the type of data needed to further our understanding of it.

Kin biases in behavior have been reported so often in the primate literature that one may easily assume that all categories of beneficent behaviors should be nepotistically distributed. This is not necessarily expected, however. Although gains in inclusive fitness may be maximized by cooperating with one's closest kin, gains in personal fitness are maximized by cooperating with the most competent partners regardless of degree of kinship. For this reason, one expects cooperative activities requiring specific individual characteristics or skills (attribute-dependent cooperation) not to be strongly kin biased. This may explain the absence of correlations between kinship and cooperation in many behavioral areas.

Two other categories of beneficent behaviors are expected to be strongly kin biased: attribute-independent cooperation and unilateral altruism. By selecting kin partners for cooperative activities that require no specific characteristics or skills, individuals are maximizing their inclusive fitness at no cost to their personal fitness. This kind of cooperation should be strongly kin biased, though not as much as altruism. This is because cooperation is profitable between nonkin as well as between kin, whereas altruism, especially unilateral altruism, is profitable only between certain kin. Thus, one can think of a gradient in the degree to which behaviors should be kin biased, with unilateral altruism at one extreme, attribute-dependent cooperation at the other, and attribute-independent cooperation somewhere between the two.

Hamilton's equation determines the maximal deployment of any given form of altruism among kin, depending on its specific cost-benefit ratio. However, the coresidency of several kin creates a number of constraints that are expected to curtail the deployment of altruism: donors may lack time and energy to aid distant kin, they may be less available to interact with them, and potential recipients may be less dependent on more distant kin because closer kin satisfy their needs. These factors are expected to produce highly skewed distributions of altruism and limit the deployment of altruism well within the relatedness limit for profitable altruism. Much research needs to be done to assess the impact of these factors on the distribution of altruism. Experiments that eliminated several of these constraints revealed that altruism extended to $r = .125$ among lineal descendants and to $r = .25$—inconsistently to .125—among collateral kin. Few observational studies have reported unequivocal cases of kin-selected altruism beyond the mother-offspring bond. These results support the contention that various constraints limit the operation of kin selection in primate groups. We provisionally conclude that the observed distributions of altruism most often underestimate the altruistic potential of individuals.

Acknowledgments   We thank Stuart Altmann, Carol Berman, Dario Maestripieri, Jean Prud-'homme, Joan Silk, and Shona Teijeiro for helpful comments on an earlier version of this manuscript.

References

Alberts, S. C. 1999. Paternal kin discrimination in wild baboons. *Proc. Royal Soc. Lond.,* 266, 1501–1506.

Altmann, J. 1979. Altruistic behaviour: the fallacy of kin deployment. *Anim. Behav.,* 27, 958–962.

Altmann, J., Alberts, S. C., Haines, S. A., Dubach, J., Muruthi, P., Coote, T., Geffen, E., Cheesman, D. J., Mututua, R. S., Saiyalele, S. N., Wayne, R. K., Lacy, R. C., & Bruford, M. W. 1996. Behavior predicts genetic structure in a wild primate group. *Proc. Natl. Acad. Sci. USA,* 93, 5797–5801.

Anderson, J. 1984. Ethology and ecology of sleep in monkeys and apes. *Adv. Study Behav., 14,* 165–229.

Aureli, F., Das, M., & Veenema, H. C. 1997. Differential kinship effect on reconciliation in three species of macaques (*Macaca fascicularis, M. fuscata,* and *M. sylvanus*). *J. Comp. Psychol.,* 111, 91–99.

Barash, D. P. 1982. *Sociobiology and Behavior,* 2nd ed. New York: Elsevier.

Barrett, L., Henzi, S. P., Weingrill, T., Lycett, J. E., & Hill, R. A. 1999. Market forces predict grooming reciprocity in female baboons. *Proc. Royal Soc. Lond.,* 266, 665–670.

Barton, R. A. 1985. Grooming site preferences in primates and their functional implications. *Int. J. Primatol.,* 6, 519–532.

Bélisle, P. 2002. Apparentement et co-alimentation chez le macaque Japonais (*Macaca fuscata*). PhD dissertation, University of Montreal.

Bélisle, P. & Chapais, B. 2001. Tolerated co-feeding in relation to degree of kinship in Japanese macaques. *Behaviour,* 138, 487–509.

Bercovitch, F. B. 1988. Coalitions, cooperation and reproductive tactics among adult male baboons. *Anim. Behav.,* 36, 1198–1209.

Berman, C. M. 1980. Early agonistic experience and rank acquisition among free-ranging infant rhesus monkeys. *Int. J. Primatol.*, 1, 153–170.

Berman, C. M. 1988. Demography and mother-infant relationships: implications for group structure. In: *Ecology and Behavior of Food-Enhanced Primate Groups* (Ed. by J. Fa & C. Southwick), pp. 269–296. New York: Alan R. Liss.

Berman, C. M. & Kapsalis, E. 1999. Development of kin bias among rhesus monkeys: maternal transmission or individual learning? *Anim. Behav.*, 58, 883–894.

Berman, C. M., Rasmussen, K. L. R., & Suomi, S. J. 1997. Group size, infant development and social networks in free-ranging rhesus monkeys. *Anim. Behav.*, 53, 405–421.

Bernstein, I. S. 1991. The correlation between kinship and behaviour in non-human primates. In: *Kin Recognition* (Ed. by P. G. Hepper), pp. 7–29. Cambridge: Cambridge University Press.

Bernstein, I. 1999. Kinship and the behavior of nonhuman primates. In: *The Non-Human Primates* (Ed. by P. Dolhinow & A. Fuentes), pp. 202–205. MountainView, CA: Mayfield.

Blurton-Jones, N. G. 1984. A selfish origin for human food sharing: tolerated theft. *Ethol. Sociobiol.*, 5, 1–3.

Boccia, M. L., Reite, M., & Laudenslager, M. 1989. On the physiology of grooming in a pigtail macaque. *Phys. Behav.*, 45, 667–670.

Buchan, J. C., Alberts, S. C., Silk, J. B., & Altmann, J. 2003. True paternal care in a multi-male primate society. *Nature*, 425, 179–181.

Butovskaya, M. 1993. Kinship and different dominance styles in groups of three species of the genus *Macaca* (*M. arctoides, M. mulatta, M. fascicularis*). *Folia Primatol.*, 60, 210–224.

Chaffin, C. L., Friedlen, K., & de Waal, F. B. M. 1995. Dominance style of Japanese macaques compared with rhesus and stumptail macaques. *Am. J. Primatol.*, 35, 103–116.

Chapais, B. 1983. Dominance, relatedness and the structure of female relationships in rhesus monkeys. In: *Primate Social Relationships: An Integrated Approach* (Ed. by R. A. Hinde), pp. 208–217. Oxford: Blackwell.

Chapais, B. 1988. Experimental matrilineal inheritance of rank in female Japanese macaques. *Anim. Behav.*, 36: 1025–1037.

Chapais, B. 2001. Primate nepotism: what is the explanatory value of kin selection? *Int. J. Primatol.*, 22, 203–229.

Chapais, B. & Gauthier, C. 1993. Early agonistic experience and the onset of matrilineal rank acquisition. In: *Juvenile Primates: Life-History, Development, and Behaviour* (Ed. by M. E. Pereira & L. A. Fairbanks), pp. 246–258. New York: Oxford University Press.

Chapais, B., Gauthier, C., Prud'homme, J., & Vasey, P. 1997. Relatedness threshold for nepotism in Japanese macaques. *Anim. Behav.*, 53, 1089–1101.

Chapais, B., Girard, M., & Primi, G. 1991. Non-kin alliances, and the stability of matrilineal dominance relations in Japanese macaques. *Anim. Behav.*, 41, 481–491.

Chapais, B., Prud'homme, J., & Teijeiro, S. 1994. Dominance competition among siblings in Japanese macaques: constraints on nepotism. *Anim. Behav.*, 48, 1335–1347.

Chapais, B., Savard, L., & Gauthier, C. 2001. Kin selection and the distribution of altruism in relation to degree of kinship in Japanese macaques. *Behav. Ecol. Sociobiol.*, 49, 493–502.

Chapais, B. & Schulman, S. 1980. An evolutionary model of female dominance relationships in primates. *J. Theor. Biol.*, 82, 47–89.

Combes, S. L. & Altmann, J. 2001. Status change during adulthood: life-history by-product or kin selection based on reproductive value? *Proc. Royal Soc. London.*, 268, 1367–1373.

Datta, S. B. 1983. Relative power and the acquisition of rank. In: *Primate Social Relationships: An Integrated Approach* (Ed. by R. A. Hinde), pp. 93–103. Oxford: Blackwell.

Datta, S. B. 1988. The acquisition of dominance among free-ranging rhesus monkey siblings. *Anim. Behav.*, 36, 754–772.

Datta, S. B. 1992. Effects of the availabilty of allies on female dominance structure. In: *Coalitions and Alliances in Humans and Other Animals* (Ed. by A. H. Harcourt & F. B. M. de Waal), pp. 61–82. New York: Oxford University Press.

de Ruiter, J. R. & Geffen, E. 1998. Relatedness of matrilines, dispersing males and social groups in long-tailed macaques (*Macaca fascicularis*). *Proc. Royal Soc. Lond.*, 265, 79–87.

de Waal, F. B. M. 1991. Rank distance as a central feature of rhesus monkey social organisation: a sociometric analysis. *Anim. Behav.*, 41, 383–395.

de Waal, F. B. M. 1997. The chimpanzee's service economy: food for grooming. *Evol. Hum. Behav.*, 18, 375–386.

de Waal, F. B. M. & Luttrell, L. M. 1989. Toward a comparative socioecology of the genus *Macaca*: different dominance styles in rhesus and stumptail monkeys. *Am. J. Primatol.*, 19, 83–109.

Dunbar, R. I. M. 1988. *Primate Social Systems*. Ithaca, NY: Cornell University Press.

Dunbar, R. I. M. & Sharman, M. 1984. Is social grooming altruistic? *Z. Tierpsychol.*, 64, 163–173.

Erhart, E. M., Coelho, A. M., & Bramblett, C. A. 1997. Kin recognition by paternal half-siblings in captive *Papio cynocephalus*. *Am. J. Primatol.*, 43, 147–157.

Fairbanks, L. A. 2000. Maternal investment throughout the life span in Old World monkeys. In: *Old World Monkeys* (Ed. by P. F. Whitehead & C. J. Jolly), pp. 341–367. Cambridge: Cambridge University Press.

Fredrikson, W. T. & Sackett, G. 1984. Kin preferences in primates, *Macaca nemestrina:* relatedness or familiarity? *J. Comp. Psychol.*, 98, 29–34.

Furuichi, T. 1997. Agonistic interactions and matrifocal dominance rank of wild bonobos (*Pan paniscus*) at Wamba. *Int. J. Primatol.*, 18, 855–875.

Gerloff, U., Hartung, B., Fruth, B., Hohmann, G., & Tautz, D. 1999. Intracommunity relationships, dispersal pattern and paternity success in a wild living community of bonobos (*Pan paniscus*) determined from DNA analysis of faecal samples. *Proc. Royal Soc. Lond.*, 266, 1189–1195.

Glick, B. B., Eaton, G. G., Johnson, D. F., & Worlein, J. 1986. Social behavior of infant and mother Japanese macaques, *Macaca fuscata*: effects of kinship, partner sex, and infant sex. *Int. J. Primatol.*, 7, 139–155.

Goldberg, T. L. & Wrangham, R. W. 1997. Genetic correlates of social behaviour in wild chimpanzees: evidence from mitochondrial DNA. *Anim. Behav.*, 54, 559–579.

Goodall, J. 1986. *The Chimpanzees of Gombe*. Cambridge, MA: Harvard University Press.

Gouzoules, S. 1984. Primate mating systems, kin association, and cooperative behavior: evidence for kin recognition? *Ybk. Phys. Anthropol.*, 27, 99–134.

Gust, D. A., McCaster, T., Gordon, T. P., Gergits, W. F., Casna, N. J., & McClure, H. M. 1998. Paternity in sooty mangabeys. *Int. J. Primatol.*, 19, 83–94.

Hamilton, W. D. 1964. The genetical theory of social behavior. I and II. *J. Theor. Biol.*, 7, 1–52.

Hammerschmidt, K. & Fisher, J. 1998. Maternal discrimination of offspring vocalizations in Barbary macaques (*Macaca sylvanus*). *Primates*, 39, 231–236.

Hashimoto, C., Furuichi, T., & Takenaka, O. 1996. Matrilineal kin relationships and social behavior of wild bonobos (*Pan paniscus*): sequencing the D-loop region of mitochondrial DNA. *Primates*, 37, 305–318.

Hemelrijk, C. K. 1994. Support for being groomed in long-tailed macaques, *Macaca fascicularis*. *Anim. Behav.*, 48, 479–481.

Hemelrijk, C. K. 1996. Reciprocation in apes: from complex cognition to self-structuring. In: *Great Ape Societies* (Ed. by W. C. McGrew, L. F. Marchant, & T. Nishida), pp. 185–195. Cambridge: Cambridge University Press.

Hemelrijk, C. K., van Laere, G. J., & van Hooff, J. A. R. A. M. 1992. Sexual exchange relationships in captive chimpanzees? *Behav. Ecol. Sociobiol.*, 30, 269–275.

Henzi, S. P. & Barrett, L. 1999. The value of grooming to female primates. *Primates*, 40, 47–59.

Hill, D. A. & Okayasu, N. 1996. Determinants of dominance among female macaques: nepotism, demography and danger. In: *Evolution and Ecology of Macaque Societies* (Ed. by J. E. Fa & D. G. Lindburg), pp. 459–472. New York: Cambridge University Press.

Hunte, W. & Horrocks, J. A. 1986. Kin and non-kin interventions in the aggressive disputes of vervet monkeys. *Behav. Ecol. Sociobiol.*, 20, 257–263.

Hutchins, M. & Barash, D. P. 1976. Grooming in primates: implications for its utilitarian function. *Primates*, 17, 145–150.

Janus, M. 1989. Reciprocity in play, grooming, and proximity in sibling and non-sibling young rhesus monkeys. *Int. J. Primatol.*, 10, 243–261.

Kaplan, J. R. 1977. Patterns of fight interference in free-ranging rhesus monkeys. *Am. J. Phys. Anthropol.*, 47: 279–288.

Kapsalis, E. & Berman, C. M. 1996a. Models of affiliative relationships among free-ranging rhesus monkeys, *Macaca mulatta*. I. Criteria for kinship. *Behaviour*, 133, 1209–1234.

Kapsalis, E. & Berman, C. M. 1996b. Models of affiliative relationships among free-ranging rhesus monkeys, *Macaca mulatta*. II. Testing predictions for three hypothesized organizing principles. *Behaviour*, 133, 1235–1263.

Kawamura, S. 1965. Matriarchal social ranks in the Minoo-B group: a study of Japanese monkeys. In: *Japanese Monkeys: A Collection of Translations* (Ed. by S. A. Altmann), pp. 105–112. Atlanta, GA: Emory University Press.

Kuester, J., Paul, A., & Arnemann, J. 1994. Kinship, familiarity and mating avoidance in Barbary macaques, *Macaca sylvanus*. *Anim. Behav.*, 48, 1183–1194.

Kurland, J. A. 1977. *Kin Selection in the Japanese Monkey*. Basel: Karger.

Lee, P. C. & Bowman, A.1995. Influence of ecology and energetics on primate mothers and infants. In: *Motherhood in Human and Nonhuman Primates: Biosocial Determinants* (Ed. by R. D. Martin & D. Skuse), pp. 47–58. Basel: Karger.

MacKenzie, M. M., McGrew, W. C. & Chamove, A. S. 1985. Social preferences in stump-tailed macaques (*Macaca arctoides*): effects of companionship, kinship and rearing. *Dev. Psychobiol.*, 18, 115–123.

Maestripieri, D. 1993. Vigilance costs of allogrooming in macaque mothers. *Am. Nat.*, 141, 744–753.

Martin, D. A. 1997. Kinship bias: a function of familiarity in pig-tailed macaques (*Macaca nemestrina*). PhD dissertation, University of Georgia, Athens, GA.

Massey, A. 1977. Agonistic aids and kinship in a group of pigtail macaques. *Behav. Ecol. Sociobiol.*, 2, 31–40.

Matsumara, S. 1998. Relaxed dominance relations among female moor macaques (*Macaca maurus*) in their natural habitat, South Sulawesi, Indonesia. *Folia Primatol.*, 69, 345–356.

Matsumura, S. & Okamoto, K. 1997. Factors affecting proximity among members of a wild troop of moor macaques during feeding, moving, and resting. *Int. J. Primatol*, 18, 929–940.

Ménard, M., von Segesser, F., Scheffrahn, W., Pastorini, J., Vallet, D., Belkacem, G., Martin, R. D., & Gauthier-Hion, A. 2001. Is male-infant caretaking related to paternity and/or mating activities in wild Barbary macaques (*Macaca sylvanus*)? *C. R. Acad. Sci., Paris*, 324, 601–610.

Mitani, J. C., Merriwether, D. A., & Zhang, C. 2000. Male affiliation, cooperation, and kinship in wild chimpanzees. *Anim. Behav.*, 59, 885–893.

Mitani, J. C., Watts, D. P., Pepper, J. W., & Merriwether, A. D. 2002. Demographic and social constraints on male chimpanzee behaviour. *Anim. Behav.*, 64, 727–737.

Moore, J. 1992. Dispersal, nepotism, and primate social behavior. *Int. J. Primatol.*, 13, 61–378.

Morin, P. A., Moore, J. J., Chakraborty, R., Jin, L., Goodall, J., & Woodruff, D. S. 1994. Kin selection, social structure, gene flow, and the evolution of chimpanzees. *Science*, 265, 1193–1201.

Muroyama, Y. 1991. Mutual reciprocity of grooming in female Japanese macaques (*M. fuscata*). *Behaviour*, 119, 161–170.

Muroyama, Y. 1994. Exchange of grooming for allomothering in female pata monkeys. *Behaviour*, 128, 103–119.

Nakamishi, M. & Yoshida, A. 1986. Discrimination of mother by infant among Japanese macaques (*Macaca fuscata*). *Int. J. Primatol.*, 7, 481–489.

Netto, J. W. & van Hooff, J. A. R. A. M. 1986. Conflict interference and the development of dominance relationships in immature *Macaca fascicularis*. In: *Primate Ontogeny, Cognition and Social Behaviour* (Ed. by J. G. Else & P. C. Lee), pp. 291–300. Cambridge: Cambridge University Press.

Noë, R. 1990. A veto game played by baboons: a challenge to the use of the prisoner's dilemma as a paradigm for reciprocity and cooperation. *Anim. Behav.*, 39, 78–90.

Noë, R. 1992. Alliance formation among male baboons: shopping for profitable partners. In: *Coalitions and Alliances in Humans and Other Animals* (Ed. by A. H. Harcourt & F. B. M. de Waal), pp. 285–322. Oxford: Oxford University Press.

Paul, A., Kuester, J., & Arnemann, J. 1992. DNA fingerprinting reveals that infant care by male Barbary macaques (*Macaca sylvanus*) is not parental investment. *Folia Primatol.*, 58, 93–98.

Paul, A., Kuester, J., & Arnemann, J. 1996. The sociobiology of male-infant interactions in Barbary macaques, *Macaca sylvanus*. *Anim. Behav.*, 51, 155–170.

Pereira, M. E. 1986. Maternal recognition of juvenile offspring coo vocalizations in Japanese macaques. *Anim. Behav.*, 34, 935–937.

Petit, O., Abegg, C., & Thierry, B. 1997. A comparative study of aggression and conciliation in three cercopithecine monkeys (*Macaca fuscata, Macaca nigra, Papio papio*). *Behaviour*, 434, 415–432.

Pope, T. 2000. Reproductive success increases with degree of kinship in cooperative coalitions of female red howler monkeys (*Alouatta seniculus*). *Behav. Ecol. Sociobiol.*, 48, 253–267.

Prud'homme, J. & Chapais, B. 1996. Development of intervention behavior in Japanese macaques: testing the targeting hypothesis. *Int. J. Primatol.*, 17, 429–443.

Pusey, A. E. & Packer, C. 1987. Dispersal and philopatry. In: *Primate Societies* (Ed. by B. B. Smuts, D. L. Cheney, R. M. Seyfarth, R. W. Wrangham, & T. T. Struhsaker), pp. 250–266. Chicago: University of Chicago Press.

Pusey, A. E. & Packer, C. 1997. The ecology of relationships. In: *Behavioral Ecology*, 4th ed. (Ed. by J. R. Krebs & N. B. Davies), pp. 254–283. Oxford: Blackwell.

Reiss, M. J. 1984. Kin selection, social grooming and removal of ectoparasites: a theoretical investigation. *Primates*, 25, 185–191.

Sackett, G. P. & Fredrikson, W. T. 1987. Social preferences by pigtailed macaques: familiarity versus degree and type of kinship. *Anim. Behav.*, 35, 603–607.

Schaub, H. 1996. Testing kin altruism in long-tailed macaques (*Macaca fascicularis*) in a food-sharing experiment. *Int. J. Primatol.*,17, 445–467.

Schino, G. 2001. Grooming, competition and social rank among female primates: a meta analysis. *Anim. Behav.*, 62, 265–271.

Schino, G., Scucchi, S., Maestripieri, D., & Turillazzi, P. G. 1988. Allogrooming as a tension- reduction mechanism: a behavioral approach. *Am. J. Primatol.*, 16, 43–50.

Schulman, S. R. & Rubenstein, D. I. 1983. Kinship, need, and the distribution of altruism. *Am. Nat.*, 121, 776–788.

Seyfarth, R. M. 1977. A model of social grooming among adult female monkeys. *J. Theor. Biol.*, 65, 671–698.

Seyfarth, R. M. & Cheney, D. L. 1984. Grooming, alliances, and reciprocal altruism in vervet monkeys. *Nature*, 308, 541–543.

Silk, J. B. 1982. Altruism among female *Macaca radiata*: explanations and analysis of patterns of grooming and coalition formation. *Behaviour*, 79, 162–187.

Silk, J. B. 1987. Social behavior in evolutionary perspective. In: *Primate Societies* (Ed. by B. B. Smuts, D. L. Cheney, R. M. Seyfarth, R. W. Wrangham, & T. T. Struhsaker), pp. 318–329. Chicago: University of Chicago Press.

Silk, J. B. 2001. Ties that bond: the role of kinship in primate societies. In: *New Directions in Anthropological Kinship* (Ed. by L. Stone), pp. 71–92. Lanham, MD: Rowman & Littlefield.

Silk, J. B., Seyfarth, R. M., & Cheney, D. L. 1999. The structure of social relationships among female savanna baboons in Moremi reserve, Botswana. *Behaviour*, 136, 679–703.

Smith, K., Alberts, S. C., & Altmann, J. 2003. Wild female baboons bias their social behaviour towards paternal half-sisters. *Proc. Royal Soc. Lond., B*, 270, 503–510.

Sterck, E. H. M., Watts, D. P., & van Schaik, C. P. 1997. The evolution of female social relationships in nonhuman primates. *Behav. Ecol. Sociobiol.*, 4, 291–309.

Strier, K. 1994. Brotherhoods among atelins: kinship, affiliation, and competition. *Behaviour*, 130, 151–167.

Takahashi, H. 1997. Huddling relationships in night sleeping groups among wild Japanese macaques in Kinkayan Island during winter. *Primates*, 38, 57–68.

Tanaka, I. & Takefushi, H. 1993. Elimination of external parasites (lice) is the primary function of grooming in free-ranging Japanese macaques. *Anthropol. Sci.*, 101, 187–193.

Thierry, B. 1990. Feedback loop between kinship and dominance: the macaque model. *J. Theor. Biol.*, 145, 511–521.

Thierry, B. 2000. Covariation of conflict management patterns across macaque species. In: *Natural Conflict Resolution* (Ed. by F. Aureli & F. B. M. de Waal), pp. 106–128. Berkeley, CA: University of California Press.

Trivers, R. L. 1971. The evolution of reciprocal altruism. *Q. Rev. Biol.*, 46, 35–57.

Vigilant, L., Hofreiter, M., Siedel, H., & Boesch, C. 2001. Paternity and relatedness in wild chimpanzee communities. *Proc. Natl. Acad. Sci. USA*, 98, 12890–12895.

Walters, J. R. 1987. Kin recognition in nonhuman primates. In: *Kin Recognition in Animals* (Ed. by D. F. Fletcher & C. D. Michener), pp. 359–393. New York: John Wiley.

Weigel, R. M. 1981. The distribution of altruism among kin: a mathematical model. *Am. Nat.*, 118, 191–201.

Welker, C., Schwibbe, M. H., Schäfer-Witt, C., & Visalberghi, E. 1987. Failure of kin recognition in *Macaca fascicularis*. *Folia Primatol.*, 49, 216–221.

West, S. A., Murray, M. G., Machado, C. A., Griffin, A. S., & Herre, E. A. 2001. Testing Hamilton's rule with competition between relatives. *Nature*, 409, 510–513.

Widdig, A., Nürnberg, P., Krawczak, M., Jürgen Streich, W., & Bercovitch, F. B. 2001. Paternal relatedness and age proximity regulate social relationships among adult female rhesus macaques. *Proc. Natl. Acad. Sci.*, 98, 13769–13773.

Wrangham, R. W. 1975. The behavioural ecology of chimpanzees in the Gombe National Park, Tanzania. PhD dissertation, Cambridge University.

Wrangham, R. W. 1982. Mutualism, kinship and social evolution. In: *Current Problems in Sociobiology* (Ed. by King's College Sociobiology Group), pp. 269–289. Cambridge: Cambridge University Press.

Yamada, M. 1963. A study of blood-relationship in the natural society of the Japanese macaques. *Primates*, 4, 45–65.

Part V

The Evolutionary Origins of Human Kinship

An "ima" feast among Efe hunter-gatherers and Lese horticultualists of the Ituri, northeastern Congo, January 1981. For several weeks, an Efe girl and Lese girl who reached menarche had been secluded together in an "ima" hut. The feast celebrating the girls' maturity brought kin and potential suitors from many miles around (Grinker, R. R. 1994. *Houses in the Rainforest: Ethnicity and Inequality Among Farmers and Foragers in Central Africa.* Berkeley, CA: University of California Press). Photo by Richard Wrangham and Elizabeth Ross.

17

Human Kinship: A Continuation of Politics by Other Means?

Lars Rodseth
Richard Wrangham

Our aim in this chapter is to describe and analyze the most common patterns of human kinship. In the anthropological literature, these patterns are often overshadowed by the enormous diversity of ethnographic cases. Human societies vary widely in the terms by which kin are classified, the rules that govern behavior toward kin in different categories, prohibitions on (or preferences for) marriage between kin, norms of residence, succession, and inheritance, and the treatment that kin are actually afforded. Our challenge is to capture the central tendencies of human kinship without forcing cultural variation into a single mold.

Here we focus on those relationships among adult kin that are most clearly responsible for creating human social groups. First we consider some of the special problems raised by human kinship, which is seen by many anthropologists as a domain of cultural meanings and social institutions, rather than a suite of behavioral tendencies or biological adaptations. Our contention is that there are uniquely human patterns of kin-oriented behavior comparable to such patterns in other species. Though the behavioral dimension does not exhaust what is interesting about human kinship, we argue for a careful examination of this dimension, if only to sharpen our understanding of how and why human social groups are different from those of other primates.

To characterize this difference, we turn to patterns of postmarital residence, which are often equated with patterns of sex-biased dispersal and philopatry in nonhuman animals. Male kinship alliances are next examined in view of their near universal importance in shaping the public politics of human groups. We then consider kinship among women, whose social bonds are often less conspicuous than men's but are also crucial to the organization of society. Relationships between male and female kin are examined in two distinct contexts. First, men and women often maintain lifelong bonds with their opposite-sex kin, and we discuss the evolutionary significance of this phenomenon. Second, as a result of marriage, people everywhere have "in-laws"—that is, affinal kin. Affinal relationships tend to create or reinforce alliances between kin groups, in a manner that is peculiarly human. We then discuss the strategic extension of kinship across time and space, which may give rise to descent groups, exchange networks, and other kinds of regional or national systems.

In conclusion, we return to the question of what, if anything, human kinship has to do with biological relatedness. Our own view is that such kinship does indeed involve much more than shared ancestry or relatedness by "blood." Even if natural selection has strongly favored genetic altruism in humans as in other animals (Hamilton 1964, Alexander 1979), concepts of relatedness have been put to a vast array of uses in human societies. These concepts and their uses are not explained by the theory of kin selection alone (Fox 1979, Jones 2000). In the end, we argue, the myriad ways in which people conceptualize and act toward "kin" involve not only their knowledge of who is descended from the same ancestor or who shares the same bodily substance, but also their sense of who has been and should remain an ally in political competition.

Problems of Human Kinship

For a century or more, the study of kinship was an abiding concern of social and cultural anthropology (Kuper 1982, 1988; Trautmann 1987). In the 1970s, however—just when the implications of sexual selection and kin selection were being explored by a handful of anthropologists (Fox 1972, 1975; Chagnon & Irons 1979)—the cultural study of kinship fell into disarray. As a result of criticisms leveled by Schneider (1972, 1980, 1984), many anthropologists began to suspect that the entire ethnographic record of kinship systems was ideologically tainted. Fieldworkers had tacitly assumed that for human beings everywhere "blood is thicker than water," but this was now argued to be "an integral part of the ideology of European culture" projected onto other societies (Schneider 1984: 174). At the same time, traditional accounts of "that lineage stuff" began to seem quaint and esoteric in comparison with such issues as gender, ethnicity, and globalization (Brettell 2001). By the 1990s, it seemed that kinship as a topic of anthropological investigation was "dead or moribund" (Peletz 1995: 345).

The study of kinship was never entirely abandoned, but on the basis of Schneider's critique, it did undergo a major transformation. Before the critique, many anthropologists held that procreation was universally recognized in human societies as the foundation of kinship and that "kinship systems" amounted to the varying cultural interpretations of certain natural facts: men impregnate women, women give birth to children, and so on (Fox 1967: 31). Today, this position still finds a few defenders—the most prominent of whom is Scheffler (1991, 2001)—but is rejected by most anthropologists who specialize in kinship. Now to invoke, a priori, the facts of human reproduction is widely viewed as ethnocentric, for the simple reason that not all human groups conceive of "kinship" or "relatedness" as a result of pregnancy and parturition. Having offspring, it is argued, is only one of several ways in which humans create kinship. Other processes, such as living together, sharing food, or undergoing rites of passage, are considered in many societies to be as important as sexual reproduction in determining who is related to whom (e.g., Kelly 1993: 521, Holy 1996: 171). In these societies, what Westerners would call blood relatives tend to form just a subset—often a rather small subset—of those defined as kin.

Considerations such as these have forced social and cultural anthropologists to rethink what they mean by kinship, and this in turn has sparked a major revival of interest in kinship studies (Strathern 1992a, b; Bouquet 1993; Holy 1996; Carsten 1997, 2000; Schweitzer 2000; Franklin & McKinnon 2001; Stone 2001). The consensus in this literature is that biological relatedness should enter into the definition of kinship only if the natives themselves conceive of it in terms of shared "blood" or other substances transmitted through sexual reproduction. Such an approach, which stresses varying local conceptions and sentiments rather than "objective" measures of genealogical distance, would seem at first to be antithetical to any form of evolutionary analysis. As we suggest below, however, people's beliefs and feelings about kinship are of great evolutionary interest in their own right. Such beliefs and feelings should count as phenomena to be analyzed and explained, right alongside the behavioral phenomena that evolutionists are more accustomed to study.

Cultural beliefs in general are clearly related to patterns of behavior, but the match is not perfect. Thus, in applying evolutionary theory to human kinship, we must carefully distinguish between what people believe (or honestly say) about their behavior and what they actually do. The danger of confusing these was pointed out more than forty years ago by Leach (1961: 30):

> The field worker has three distinct "levels" of behaviour pattern to consider. The first is the actual behaviour of individuals. The average of all such individual behaviour patterns constitutes the second, which may be fairly described as "the norm." But there is a third pattern, the native's own description of himself and his society, which constitutes "the ideal." Because the field worker's time is short and he must rely upon a limited number of informants, he is always tempted to identify the second of these patterns with the third. Clearly the norm is strongly influenced by the ideal, but I question whether the two are ever precisely coincident.

When ideals, norms, and actual behaviors are blurred or collapsed together, what emerges is the familiar ethnographic depiction of "the Nuer," "the Trobriander," or "the Balinese" as an integrated subject who acts and thinks in one customary way (Clifford 1988: 39–40, Rodseth 1998: 57). Despite the criticism of such integrated portraits, the fieldworker will always be "tempted," as Leach put it, to identify what the natives actually do with their own idealized concepts of what they do.

Students of animal behavior, by contrast, never have to face this temptation. However much primatologists might like to study the ideals of a baboon (*Papio* spp.) or chimpanzee (*Pan troglodytes*) population, they are obliged to study actual behavior instead. This has at least one fortunate result, insofar as primatologists are less likely than ethnographers to accept an idealized description of behavior as an adequate population-wide characterization. Even the "average" or most frequent behavior is likely to be recognized by a primatologist as precisely that: a modal pattern around which there is often significant variation.

At the same time, primatology is vitally interested in a question that sociocultural anthropology has long neglected: what accounts for central tendencies? Despite the enormous variation in human kinship behavior, there are general patterns that cry out for explanation. Ironically, primatologists were drawn to the analysis of human kinship in the 1970s, about the same time that most anthropologists were letting it go. Thus, in the face of anthropological objections, it has been largely up to primatologists and other behavioral biologists to characterize the fundamental patterns of human society. A preliminary analysis of these patterns was provided by Rodseth et al. (1991) and here we attempt to carry this analysis forward. The essential justification for doing so is the consolidated evidence showing that each major social pattern characteristic of our species is found in other primate species, but in different combinations. The complex of fission-fusion grouping, communal defense by male coalitions, lifelong bonds between males, lifelong bonds between females, and exclusive sexual alliances is uniquely human, but the distribution of these elements, especially among the apes, may well shed light on human social evolution (Wrangham 1987a, 1999, 2001; Wrangham & Peterson 1996). While the primatology of human kinship has thus far been largely descriptive, description paves the way for explanation.

The Transcendence of Residence

Nonhuman primates usually form relationships only with members of their own local group. Moreover, either males or females, but not both sexes, tend to remain in the same local group throughout their lives. As a result, sex-biased patterns of dispersal and philopatry largely determine whether males or females are likely to form enduring relationships with their natal kin (Wrangham 1980; Pusey & Packer 1987; van Schaik 1989, 1996). Among humans, patterns of postmarital residence have often been treated as if they were functionally equivalent to nonhuman patterns of dispersal and philopatry. Most societies in the ethnographic record do indeed have residence rules, according to which one sex is expected to marry out while the other is expected to remain with natal kin and receive a marriage partner from another household. If a woman, for example, moves in with her husband and his parents, the pattern is said to be patrilocal, while if a man moves in with his wife and her parents, it is said to be matrilocal. On the surface, patrilocality seems to resemble (or indeed to exemplify) what primatologists would call male philopatry, while matrilocality seems similarly to resemble or exemplify female philopatry.

Exploring such resemblances is a useful exercise, but its significance can be easily overstated. In particular, it seems to us that a human pattern of postmarital residence does not give decisive shape to the kinship system in the way that a pattern of dispersal and philopatry tends to do in nonhuman primates. For example, in most nonhuman primate groups, sex-biased dispersal severs the relationships between emigrants and their natal kin. Among humans, however, matrilocal residence does not normally cut the ties between a married man and his natal kin. Even patrilocal residence, which often requires the bride to move to a distant community, does not by itself prevent association between a married woman and her natal kin. In the case of humans, then, and in sharp contrast to the case of all other primates, "dispersal patterns predict patterns of kinship and cooperation only weakly, if at all" (Rodseth et al. 1991: 230).

Why is postmarital residence such an unreliable guide to patterns of kinship behavior? First, postmarital residence is often inferred from the reported rules of residence, which may vary from actual patterns of residence "on the ground" (chapter 18). In a patrilocal society, for example, a significant number of men may reside with their wives' families, but these "exceptional" cases tend to be ignored in the ethnographic characterization of the society as a whole. How to handle such variation is a problem, of course, in the collection of all behavioral data, but the problem is especially acute when the natives themselves contend that they do things in only one (idealized) way.

Second, postmarital residence is solely a matter of one's household affiliation. As a result, what is usually described as "locality" says nothing about wider group membership, propinquity to natal kin, how far people tend to move after marriage, or how much of a difference any such move makes in terms of their social relationships. In the Canadian Great Plains, for example, "Slightly under one third of the women married to ranchers came from outside the immediate region. In families where the wife was from outside, her parents were much less involved in the life of her family of procreation than the parents of her husband who lived in the region" (Kohl & Bennett 1965: 103–104). Yet under the same category of "patrilocal residence" would fall a very different kind of case, in which a wife had moved just down the road, or even just next door, to join her husband's household. Now the wife's parents might be involved on a daily or hourly basis with her family of procreation, offering valuable assistance to her and her children. Examples such as this could be multiplied, but the point is probably clear: "residence patterns do not provide a sufficiently fine grained indication of the proximity of kin" (Smuts 1992: 15).

The classification of residence would certainly be more informative if it referred not merely to household affiliation but to "the *zone* within which one resides" (Kopytoff 1977: 555). Yet remaining in the same zone with one's natal kin does not guarantee that one will maintain relationships with those kin. In fact, patterns of social relationships may be difficult or impossible to read from the residential arrangement on the ground (Rodseth & Novak 2000). To complicate matters further, the frequency of interaction between two individuals is not always a reliable indicator of the strength or importance of their relationship (Rodseth & Novak in press). The sexual division of labor, for example, often ensures that a husband and wife spend very little of their waking time together, even when they are in fact the closest of allies.

In the study of nonhuman primates, spatial proximity is a reasonably good index of social relationships, because relationships are maintained only through regular interaction. In humans, however, language makes it possible for relationships to be decoupled from interaction for months or years at a time. This "release from proximity" is one of the distinctive features of human sociality (Rodseth et al. 1991; Gamble 1998, 1999: 40–41), and clearly depends on uniquely human capacities to (1) monitor the behavior of absent others through gossip and other verbal reports, (2) reach precise (i.e., verbalized) understandings of rules and agreements, and (3) enforce such rules and agreements through criticism and the mobilization of public opinion (Boehm 1993, 1999: 187–191, Dunbar 1996: 171–173). Linguistic culture is indeed a web that humans themselves have spun (Geertz 1973: 5), a web that tends to hold us fast in "social space" even as it allows us greater freedom of movement through physical space.

What might be called the "transcendence of residence" is a special case of the release from proximity. Because humans form relationships not only on a local basis but across wide expanses of space and time, their residence patterns cannot be easily equated with patterns of dispersal and philopatry in nonhuman primates. Humans who marry out do, in a narrow sense, disperse from their natal households, but not necessarily from their natal bands, villages, or neighborhoods, and even if they move far away, their links with natal kin may well be preserved. As a result, dispersal tends to enhance and expand the kinship networks of human beings, while it usually diminishes such networks in nonhuman primates (Rodseth et al. 1991: 232).

In the analysis of human kinship, there is no alternative but to document what Radcliffe-Brown (1940) called the "actually existing network" of social relationships. This network is seldom if ever coterminous with residential units such as bands and villages, descent groups such as lineages and clans, or even ethnic and linguistic communities such as tribes and nations. Yet the study of residence, descent, and ethnicity has consistently taken precedence in the anthropological literature over the study of actual relationships. Most ethnographic accounts of "social organization" are strangely silent on a fundamental question: who are one's primary social allies? Instead of addressing this question directly by recording actual patterns of alliance—who provides support against enemies, remains loyal over the long term, and can be called upon in a crisis—anthropologists have tended to translate this question into one about propinquity, relatedness, or culture. As a result, we seldom catch a glimpse of even one person's actually existing network, let alone the network that links people of various communities into regional and transnational systems.

In fact, as Radcliffe-Brown pointed out more than 60 years ago, the human network spreads over the entire world. Most nonhuman primates, by comparison, live in rather neatly bounded local groups, not entirely isolated but with a strong tendency toward social and physical separation from their neighbors. There is little to be gained by treating human groups as if they were marked by anything like this kind of localism (Lesser 1961, Wolf 1982, Wilmsen 1989).

Consanguineal Kinship

A human society consists of an indefinite number of crosscutting (and thus interlocking) social networks, which can be artificially represented, for various analytical purposes, as separate systems. How such systems are to be delimited depends on one's objective, but in any case at least four dimensions of sociality must be carefully distinguished: (1) spatial proximity, (2) frequency of interaction, (3) stability of relationships, and (4) reckoning of kinship. By "relationships," it should be noted, we mean something quite specific. A relationship involves some degree of advantage to the participants, who gain more by coordinating their action than they would by acting alone (Wrangham 1982: 270; see also van Schaik 1989). In this sense, relationships are not simply a matter of friendly interaction and in many cases do not even require such interaction on a regular basis (Rodseth & Novak in press). Here we are interested in all four dimensions of human sociality, but especially in stable relationships that are founded on kinship.

Ever since Morgan (1871), anthropologists have distinguished between two kinds of kinship, consanguineal and affinal, or what Westerners in general call relatedness by birth and relatedness by marriage. For now we adopt this traditional terminology, postponing the question of whether consanguineal kinship should be reckoned in terms of genetic relatedness (the r coefficient) or in terms of indigenous concepts. This approach assumes that in most human societies there is enough overlap between genetic measures and indigenous notions to allow a comparison of biological relatedness in nonhuman primates with human patterns of consanguinity, however that notion is locally defined.

While bonds of kinship are usually maintained throughout life by both men and women, the sexes differ in how they organize and deploy these bonds. Men rely heavily on their male kin as allies in what Wrangham (1982) called interference mutualism—cooperative behavior that benefits the participants while imposing a cost on a third party. Interference mutualism in one form or another, including team sports, business rivalries, party politics, and tribal warfare, looms large in the social lives of men everywhere. Women also engage in interference mutualism, though seldom in the form of physical violence against members of their own sex. What is especially notable, particularly in contrast to the pattern of men, is the extent to which women rely on their female kin as partners in *non*interference mutualism—cooperative behavior that benefits the participants without imposing a cost on anyone. Here we elaborate on these sex differences before turning to the often neglected topic of kinship bonding between the sexes.

Fraternal Interest Groups

The public and political structures of most human societies have been dominated by coalitions of men (Brown 1991; Boehm 1992, 1999). Before the emergence of the modern state, such coalitions were usually based on kinship. The ethnographic record is teeming with examples of fraternal interest groups, including clans, clubs, cults, guilds, and military orders, which incorporate and reinforce the bonds among male kin (van Velzen & van Wetering 1960). Even in a matrilocal setting, a man often belongs to a "brotherhood" (or what is more likely a "cousinhood")—a network of related males who may have grown up in different households but have united as a local power group, often within a men's house or other exclusively masculine setting (Murphy 1957, Ember & Ember 1971). Under these circumstances, a man typically remains aloof from his conjugal home while investing most of his energy in warfare, ritual, and other public activities (Whiting & Whiting 1975, Draper & Harpending 1988, Rodseth & Novak 2000).

Though they may or may not be militaristic, fraternal interest groups are usually capable of serving a military function. Most populations in the ethnographic record are not (or were not at the time of study) protected by standing armies or other military specialists, but relied instead on "self-help"—the capacity of their hunters, herders, or farmers to turn themselves temporarily into warriors for the defense of the community. Unlike standing armies, furthermore, temporary power groups must draw upon existing ties of kinship and friendship. In general, they cannot be constituted of men who are strangers to each other, but must take advantage of whatever loyalties have grown up within the everyday context of family and community life.

The military potential of fraternal interest groups is reflected in their cross-cultural distribution. The strength of such groups, as demonstrated by Paige and Paige (1981), varies with the prevailing economic system. Societies that subsist by foraging, fishing, and swidden cultivation tend to have weak fraternal interest groups, while those practicing pastoralism, intensive horticulture, and agriculture tend to have strong ones. In short, alliances among men seem to be most fully elaborated when food and other resources are concentrated in small, dense patches (Rodseth & Novak 2000: 351–352). This trend has been explained by the fact that herds, crops, and other clumped resources are easily raided or seized by organized groups of men (Paige & Paige 1981: 72–78; see also Otterbein & Otterbein 1965, Otterbein 1968). Unilineal descent groups, which are generally tenuous or absent in foraging bands, tend to develop along with food production as a means of seizing, defending, and allocating wealth. In effect, according to Wolf (1982: 89):

... kinship works in basically different ways in two kinds of situations, those in which resources are widely available and open to anyone with the ability to obtain them, and those situations in which access to resources is restricted and available only to claimants with a "kinship license." In the first case, the ties of kinship grow out of the give-and-take of everyday life and link people who are in habitual interaction with one another. In the second case, the circle of kinship is drawn tightly around the resource base by means of stringent definitions of group membership.

Even in the absence of food production, there is an important distinction between those foraging societies that store food and those that do not (Testart 1982). The ethnographic record suggests that hunter-gatherers without storage live in unsegmented societies—small, politically independent bands with a minimal elaboration of subgroups (Kelly 2000: 44–45). With storage, however, foraging groups tend to develop an increasingly segmental organization, characterized by a hierarchy of social units (Kelly 2000: 71). Kelly considers this to be "a pivotal point in cultural evolution" and a direct response to the threat of organized attack: "When food is stockpiled, it inevitably becomes a key military objective of raiding parties" (Kelly 2000: 68). Thus, even among hunter-gatherers, the "military vulnerability" (Paige & Paige 1981: 56) or "alienability" (Manson & Wrangham 1991: 374) of food supplies seems to be an important determinant of fraternal interest group strength.

Societies without storage are also likely to be "warless," in Kelly's terms—not free of violence, but free of armed conflict that is carried out collectively (Kelly 2000: 3–7). In this respect, Kelly's reasoning converges with the arguments of Knauft (1987, 1991, 1994), who holds that "fraternal interest groups and the kind of violence associated with them are seldom of importance in simple human societies" (Knauft 1991: 406). Men in these societies do not consistently reside with their male relatives, but tend to live in bilateral kin groups with "shifting, open, and flexible" membership. According to Knauft (1991), then, any apparent similarities between the power groups of related males observed in chimpanzee communities and some human (especially "tribal") societies are "not homologies based on phylogenetic continuity through human evolution" (p. 407). Instead, humans have spent most their evolutionary history in social groups radically unlike those of the apes, including chimpanzees. As a result, it is argued, "Simple human societies constitute a major anomaly for models which propose an evolutionary similarity between great ape and human patterns of violence" (Knauft 1991: 391).

In response, it should first be noted that the differences between great ape and simple human societies are indeed fundamental and profound. By comparison with a chimpanzee community, a foraging band of !Kung or Mbuti is strikingly egalitarian, with much less competition among males over females, a uniquely human pattern of food sharing, and a remarkably shifting, open, and flexible membership (see also Boehm 1999). There is one difference of special interest in the present context: a chimpanzee community, unlike a foraging band, is a social isolate. Chimpanzees of different communities do not mingle peacefully, let alone exchange food or rely on each other for help in a crisis. Among foragers, by contrast, "Contiguous local groups are likely to aggregate during periods of seasonal abundance, with these gatherings providing an occasion for socializing, exchange, courtship, ritual performance, or some combination of these" (Kelly 2000: 45). Social relationships that have been formed at such gatherings may be maintained throughout the year and reinforced by ad hoc visits between bands. As a result, one's band members are just a small subset of the kin (whether real or fictive) on whom one may call for social and material support (Wiessner 1982, 1998). Even those societies that Knauft calls simple and Kelly calls unsegmented are thus embedded within a wider network and are never as autonomous as a chimpanzee community.

In this sense, the local band is not the appropriate social unit to compare to the chimpanzee community, though it has often been treated as such in earlier analyses (e.g., Manson & Wrangham 1991, Rodseth et al. 1991, Boehm 1999: 91). With an average size of perhaps 25 members (Kelly 1995: 210–211), a local band is somewhat smaller than a chimpanzee community, which averages more than 40 members (calculated from table 5.2 in Boesch & Boesch-Achermann 2000: 93). Yet the social network of a human forager is much larger than that of any ape. Among the !Kung, for example, an estimated 2,000 individuals were involved in the regional exchange network known as *hxaro* (Wiessner 1982). For any given participant, "The geographical spread of partners extended from camps within a 5 km radius to those up to 200 km away" (Wiessner 1998: 139). A number of anthropologists have drawn a distinction between the minimum band of a few nuclear families who camp together and the maximum band that assembles intermittently and may contain some 500 persons (e.g., Birdsell 1968, Tindale 1974, cf. Kelly 1995: 209–210). In complex or segmental hunter-gatherers, the tribe emerges as a unit of organization above the level of the maximum band (Jones 1984: 40):

> I define a tribe as an agglomeration of bands with contiguous estates, the members of which spoke a common language or dialect, shared the same cultural traits such as burial customs, art and body and decoration styles, usually intermarried, had a similar pattern of seasonal movement, and habitually met together for economic and other reasons. The pattern of peaceful relations between bands, such as marriage and trade, tended to be within such a tribal agglomeration and that of hostile ones or war outside it. Such a tribe occupied a territory that was the sum of the estates of its constituent bands.

A tribe in this sense is vast in comparison with any other primate society. In California, for example, 31 hunter-gatherer tribes from five linguistic groups averaged 2,200 members, with a range of 500 to 13,650 (calculated from data in Heiser & Elsasser 1980).

As these examples suggest, hunter-gatherers participate in various social networks, some tightly nucleated and others widely dispersed across the landscape. Even if local groups never unite for political or military purposes, they are woven together by social and economic relationships reinforced by periodic gatherings and celebrations. Foraging society, in this perspective, resembles a greatly expanded chimpanzee community with relatively stable parties (bands) that are physically isolated from each other for weeks or months at a time but remain loosely integrated at the regional level. Such regional integration, as argued above, depends heavily on language, which allows members of different local groups to reach precise understandings, coordinate their movements, and plan for gatherings at remote times and places.

Perhaps the greatest difference, then, between chimpanzee and human society is this: chimpanzees seem to lack even the potential to develop segmental organization, while humans, no matter how simple their existing organization, always carry this potential—and seem quick to realize it as circumstances arise. Ironically, then, what Knauft describes as the openness and flexibility of local bands is precisely what allows members of such bands to unite at higher levels of organization, and thus to wage war, if necessary, over stored food or other concentrated resources. When this happens, a boundary sharpens between one band or tribe and another, and their conflict may follow a pattern of raiding not unlike that between chimpanzee communities (Manson & Wrangham 1991, Boehm 1992, Wrangham 1999).

As long as this does not happen, however, the level of intergroup violence in simple human societies may appear to be much lower than in the case of chimpanzees. When the local band is treated as an organizational type, distinct from other kinds of human societies and representative of the way humans lived throughout most of their evolutionary history, we are left with the impression that phylogenetic continuity with chimp-like patterns of violence has been decisively broken. The real break, however, lies not in the patterns of violence per se but in the organizational level at which violence can be carried out. In chimpanzees, the community is the largest social unit that can be mobilized for purposes of aggression. In humans, there is no inherent limit on the size of the unit that can be so mobilized—only practical limits imposed by physical geography, population size, and the prevailing technology of transportation and communication. In this light, the phylogenetic question is not so much "Why did human foragers evolve such peaceful societies?" but "Why did humans in general evolve the capacity for segmental organization?"—a capacity that tends to create peace at lower levels of social organization even as it mobilizes people for potential conflict at higher levels.

Part of the answer to the latter question is clear enough, and by now familiar: segmental organization is made possible by linguistic communication and its social concomitant, the release from proximity. Beyond this, however, humans have a special technology, a body of cultural knowledge and skills, for creating and maintaining segmental organization. This special technology is kinship—not in the strictly biological sense of genetic relatedness, but in the cultural sense of who is conceived and said to be related to whom, for what purposes, and with what consequences. In this respect, our analysis converges with the approach taken by Jones (2000: 789), who argues that "classificatory and nongenealogical kinship . . . are important because they commonly govern not merely terminology for different types of relatives but the assignment of roles, rights, and duties." When such roles, rights, and duties are extended from close kin to distant kin and even to strangers, they provide the scaffolding of segmental organization on an ever larger scale. Without an extended concept of "brotherhood," for example, there would be no hierarchy of "tribal" units, no way to unite members of different local groups into lineages, clans, tribes, or nations. Human kinship, in this light, is a continuation of politics by other means.

These points are elaborated below (see Kinship Expanded and Exploded). Here we pursue the causes of male-male alliances, not within segmental societies already engaged in warfare, but within the societies described by Kelly (2000) as unsegmented and warless. What advantages could there be to male bonding, even in foraging bands with little or no wealth to be stolen or defended?

The Encounter-Rate Hypothesis

To account for the ubiquity of men's kinship bonds, we argue for an "encounter-rate hypothesis" derived from comparative analysis of intergroup violence in humans and African apes (Wrangham 1987b, 1999; Manson & Wrangham 1991). This hypothesis assumes that humans everywhere live in fission-fusion communities—social networks that break up into variable subgroups or parties throughout the day, with individuals sometimes traveling independently of each other. Fission-fusion sociality allows for extreme imbalances of power between parties from rival communities: large parties dominate small ones and may be able to inflict violence with impunity, especially on lone individuals. Thus, for members of a large party, the costs of interparty violence are drastically lowered, while the benefits may be considerable. Even if the victim has nothing to steal, the assailant has a valuable opportunity in this context to intimidate his rivals for territorial control and to demonstrate his power and prowess before an audience of peers. While the advantages of such showing off have been mainly discussed in the context of hunting (e.g., Hawkes 1990, 1991; Hawkes & Bliege Bird 2002), they are clearly present in a wide range of other human activities, including intergroup violence. The analogy, furthermore, between animal prey and human enemies seems to have been invoked to motivate or justify aggression in many cultural contexts (Bloch 1991).

Because small parties are vulnerable to attack by larger parties, those individuals most likely to confront groups of strangers or enemies have the greatest need of allies for personal protection. The encounter-rate hypothesis suggests that if one sex is constrained in its mobility (typically by parenting activities), the more widely ranging sex encounters strangers or enemies more often and consequently has a greater need for allies. Given that individuals prefer, ceteris paribus, to ally with kin over nonkin, those of the more mobile sex will tend to bond with their adult, same-sex kin.

Among nonhuman primates, the combination of male kinship bonds and fission-fusion communities occurs only in our two closest relatives (chimpanzees and bonobos [*Pan paniscus*]), and in spider monkeys (*Ateles* spp.) (and probably their close relatives, muriqui [*Brachyteles arachnoides*]). Explanations for the human pattern should take these parallels into account. In both chimpanzees and spider monkeys, males range farther than females and encounter outsiders more often. The prevalence of male bonding (and the lack of female bonding) in these species is thus consistent with the encounter-rate hypothesis. In bonobos, by contrast, the sexes are equally mobile and tend to encounter outsiders at an equivalent rate. Under these conditions, according to the hypothesis, there will be some degree of bonding in both males and females. Such is indeed the case in bonobos, though the alliances among females appear to be stronger than those among males (Parish 1996, Parish & de Waal 2000). Why this should be so remains unclear, especially given that females migrate from their natal groups and must therefore bond with nonkin. What is clear is that female bonobos are as likely as males to venture into border areas, usually in large parties of both sexes. Members of neighboring communities may chase each other or compete through vocal displays, but they have also been known to mingle peacefully (Idani 1991). The chimpanzee pattern of lethal intergroup aggression carried out by predominantly male "border patrols" has not been observed in bonobos (Wrangham & Peterson 1996: 215–216).

In humans, we observe daily patterns of grouping and ranging that are similar in some respects to those of chimpanzees. Everyday activities are usually performed by hunter-gatherers (and most other humans) in relatively small parties, perhaps comparable in size to those observed by Goodall (1986: 154): "chimpanzees typically travel, feed, and sleep in parties of five or less (excluding infants and juveniles)." A comparison with bonobos is instructive: although chimpanzees at Kibale and bonobos at Lomako were found to have a similar average party size (between five and six individuals), "chimpanzees have many small parties and some very large parties, while bonobos exhibit a less bimodal distribution" (Chapman et al. 1994: 53). Although quantitative data on human party sizes are extremely scarce, what can be culled from the ethnographic record suggests that humans are in this regard more like chimpanzees than they are like bonobos. In fact, the distribution of human party sizes is probably trimodal, with individuals tending to disperse during daily activities in parties of less than five, but congregating at night in parties of five or more and in still larger groups for purposes of ritual, politics, sports, or warfare.

Such variability in grouping patterns means that imbalances of power between human parties are likely to be extreme. The fact that people, unlike either species of *Pan*, tend to rendezvous each night in nucleated camps or settlements may well be linked to their need for protection from large parties of other human beings (Alexander 1979: 220–233). Yet the human pattern is also set apart by the extraordinary distances that foraging men, in particular, tend to walk in a day (averages of c. 10 km appear typical, twice the median distance for daily travel by male chimpanzees). In pursuit of prey or other resources, hunters may spend most of their waking hours far from home, following unpredictable routes that cross the pathways of other human beings. Foraging women, especially those with small children who cannot walk on their own, appear to range within a much more restricted area (Kelly 1995: 268–269; MacDonald & Hewlett 1999: 508–509, table 2). As a result, in the course of their daily activities, women are much less likely than men to encounter strangers or known enemies.

Who these enemies might be is not a simple matter in the case of humans. The identity of a human group can be so ambiguous that the notion of an encounter with "outsiders" becomes problematic. When relations between neighboring groups are openly hostile, any encounter with outsiders may be dangerous. In most ethnographic contexts, however, over-lapping social networks and incremental ethnic and linguistic differences make for a broad and shifting gray area between friend and foe. This human pattern contrasts sharply with the pattern of chimpanzees, in which members of different communities, as far as is known, are consistently antagonistic (Wrangham 1999).

In the absence of a state, however, which imposes peace within its borders and provides a legal framework for intergroup relations, human societies may have been more like chimpanzee communities, with tense or hostile relations between neighboring bands or tribes. Because the !Kung, the Mbuti, and most other foragers in the ethnographic record have long lived under the protection (and the domination) of a state, they probably have less need to guard against dangerous neighbors than they would have had in a stateless condition. At the same time, of course, an expanding state can introduce violence into an otherwise peaceful population of foragers or horticulturalists (Ferguson & Whitehead 1992). According to Knauft (1991: 403), in fact, such expansion is virtually the only cause of large-scale or protracted conflict in most bands and simple tribes: "Apart from the influence of state socie-ties, collective enmities in simple societies tend to be minimal and to occur between groups that lie outside the extensive networks of affiliation that link adjacent bands and territories. Travel to and armed conflict against persons in such areas tend to be infrequent."

The emphasis in this passage, as in the recent book by Kelly (2000), is on *collective* enmities between organized groups—in a word, war, as opposed to more random acts of violence between individuals or small parties from different local groups. According to Kelly's (2000: 4–5) definition, war is carried out collectively, with advance planning, and is grounded in a principle of "social substitution": any member of an enemy group may be considered a legitimate target, because he or she is substitutable for the specific individual who has attacked a member of one's own group. War in this sense is argued to originate under conditions of "restricted resource availability relative to population in environments rich in subsistence resources," especially "within a circumscribed environment" such as an island or a coastal region (Kelly 2000: 143). Like Knauft (1991: 402–403), Kelly (2000: 143) speculates that such conditions are unlikely to have existed in most parts of the world until quite recently:

> In open continental environments unsegmented foragers tend to move away from conflict so that an encounter between two groups of hunters seeking to exploit the same area is likely to engender wider spacing between their respective bands in the future. It is only when there is no opportunity to withdraw that such incidents become frequent enough to lead to adaptive modification on the part of the groups involved. Actual fighting replaces a display of strength that eventuates in withdrawal of the weaker party. Open confrontation gives way to ambush of trespassers. The identity of the specific individual responsible for a death is obscured, and retaliation thus necessarily entails social substitution. A form of collective violence predicated on targeting perpetrators of trespass, game theft, and homicide thus gives way to retaliatory violence against the compatriots of such malefactors.

Kelly's argument is designed to answer the question of why men war, not why men bond. Understandably, then, it pays little attention to the incidental and often random acts of intergroup violence that have probably peppered the histories even of warless societies. The implications of such violence are hardly trivial, however, for the hunters and other men whose daily activities take them far from home. Moving away from conflict is of course one solution for people who can afford to do so—but the victim of an attack must first survive the attack in order to move away. In the meantime, without recourse to an encompassing legal framework enforced by a state, men are thrown back on their own social devices, which usually means their male kin.

The encounter-rate hypothesis thus predicts that men will tend to form alliances with their same-sex kin, even in unsegmented and warless societies, because (1) they tend to face an unpredictable but persistent threat of attack by members of other groups and (2) they gain by expanding the territory within which they are safe from such attack. The hypothesis does not require that a state of war (in Kelly's sense of socially sanctioned, collective conflict) exists between the groups whose members encounter each other as strangers or known enemies. In fact, strangers can be viewed in many contexts as important sources of information or potential partners in exchange, rather than dangerous interlopers or murderous assailants—yet which of these they will turn out to be cannot be known in advance. The very unpredictability of their long-distance encounters may increase the pressure on men to maintain alliances with male kin who reside nearby and whose political loyalties can be periodically monitored, revitalized, and put to the test.

To some extent, this argument converges with the approach taken by Kristen Hawkes in this volume and in a series of recent articles (Hawkes 1990, 1991, 1993, 2000; Hawkes & Bliege Bird 2001; Hawkes et al. 2001). Like Hawkes, we argue that the social activities of men within the public sphere have cascading effects throughout the social organization of human groups (Rodseth & Novak 2000). Among these social activities is hunting. Long assumed to be a strategy for provisioning immediate family members, hunting is better understood, according to Hawkes, as the principal context within which men compete for status. As a "costly signal" of a man's abilities, hunting conveys reliable information even as it provisions the hunter's audience, who are thus especially motivated to heed the content of the signal. A successful hunter in this way attracts followers and allies, including both women and men, and reaps the long-term benefits of high social standing.

Hawkes's argument is designed to answer the question of why men hunt, not why men bond. Yet the logic by which she analyzes hunting is easily extended to other public activities through which men compete for social alliances. In particular, both intergroup conflict and trading, along with hunting, must be recognized as crucial to the formation of male bonds. What these activities have in common is clear enough: hunting, raiding, and trading can all function as costly signals, providing reliable information about a man's abilities while yielding benefits to the audience in the form of meat, plunder, or exotic goods. At the same time, as noted by MacDonald and Hewlett (1999: 508), such activities involve ranging over long distances and running the risk of encounters with strangers or enemies. The encounter-rate hypothesis thus applies in all three cases, and can explain why most human males—whether hunters, warriors, traders, or simply travelers—will attempt to form and maintain social bonds, preferably (though not exclusively) with their adult, same-sex kin.

Objections to this argument could be raised on the grounds that many foragers (and a majority of those living in "simple" societies) do not follow a pattern of patrilocal residence (e.g., Barnard 1983: 196–197, Knauft 1991: 405–406, chapter 18). Instead, group membership is usually characterized as flexible or fluid (Lee & DeVore 1968: 7–12), with young men often moving away from their natal kin to live temporarily or permanently in another band. The performance of brideservice, for example, requires a husband to provide game and labor to his in-laws during the early years of his marriage (Collier & Rosaldo 1981), and this practice "tends to separate brothers, since each is drawn into the orbit of his respective wife's family for a time" (Kelly 2000: 48). Yet the very openness and flexibility of band composition suggests that one's band of residence is not of crucial importance, and that a man's actual network of allies is dispersed across multiple local groups. In this light, the suggestion by Kelly (2000: 49) that brideservice tends to undercut the fraternal bond is open to question and remains an empirical issue that cannot be decided on the basis of residence patterns alone. If male kin are to remain political allies, especially in pedestrian societies without telecommunications, they must of course reside within a practical visiting distance, but this does not require them to live in the same household or even in the same local group.

Sisterhood and Its Discontents

Among hunter-gatherers, the sexual division of labor allows for two kinds of foraging strategy, with different consequences for grouping. Men's goal of acquiring high-quality resources, albeit distributed at low density, can mean that it pays men (purely from the perspective of foraging) to travel alone or in small parties. Women's goal of acquiring easily found and relatively low-quality foods, which are then improved by preparation, including cooking, means that they face much less pressure to disperse during foraging. As a result, all the women in a camp often forage together, return together relatively early to the sleeping camp, and spend most of their time within talking distance while in camp.

From the point of view of the encounter-rate hypothesis, it seems anomalous that hunting men spend so much time alone or with just a few others, while women tend to gather and prepare food in larger parties. Given that males are the more mobile sex in both humans and chimpanzees, men might be expected to follow the consistently gregarious pattern of chimpanzee males (Wrangham & Smuts 1980, Wrangham 2000). At the same time, the gregariousness of women would seem to pose a challenge to the encounter-rate hypothesis, which cannot account for bonding in the less mobile sex.

Here again the distinction between bonding and interacting is crucial: the encounter-rate hypothesis predicts patterns of bonding, but these may or may not be reflected in patterns of interaction. In fact, in the case of humans, the stability and importance of a bond are only weakly, and sometimes inversely, correlated with the frequency of interaction. This would seem to be the result of a curvilinear relationship between interaction rate and bond strength: interaction goes up as a bond is formed, but can then decline as the bond is solidified. The drop in interaction between human partners is often precipitous, resulting in a paradoxical pattern: those who seldom interact may be mutually avoidant, indifferent to each other, or solidly bonded, depending on the history of their relationship. This presents a striking contrast with the usual pattern of other social animals, in which bonds to a much greater extent can be maintained only through regular interaction.

In this light, solitary foraging does not indicate political isolation, any more than foraging together indicates close-knit bonds. Patterns of interaction within a human group may be largely determined by ecological or economic exigencies, with little or no effect on the network of established relationships. To see the "actually existing network," as Radcliffe-Brown (1940) called it, one would have to wait for a moment of crisis, when old relationships are activated, favors are called in, and alliances are put to the test. Such a crisis is sometimes ecological or economic (as in a case of natural disaster), but more commonly it is social and political—the result of potentially violent struggle within and between human groups (Alexander 1979, 1987, 1990). And indeed, as the threat of violence escalates, men would be expected to travel more consistently with companions, as they do, for example, among the Etoro of Papua New Guinea (Kelly 1993: 64).

To what extent are women as well as men vulnerable to attack by strangers or known enemies? Insofar as women are less mobile than men, their vulnerability would seem to be diminished. Yet the threat of sexual violence and other forms of male harassment must loom consistently for many women, however restricted in their ranging. In the case of men, it could be argued, encounters with dangerous others often occur far from home, but in the case of women such encounters are more likely to take place in the vicinity (or even within) the household. Given the costs of male harassment, a female's best strategy may be to bond with a particular male who can provide protection to her and her offspring (Wrangham 1979: 353–355). This "bodyguard hypothesis" has been used to account for the evolution of sexual alliances in a variety of mammals (Smuts & Smuts 1993, Mesnick 1997), including humans (Smuts 1992, 1995; Hawkes et al. 2001; chapter 19). A related explanation proposes that hominid females, with the advent of cooking, sought alliances with particular males who could protect accumulated food from thieves or freeloaders (Wrangham et al. 1999, Wrangham 2001).

Both the bodyguard hypothesis and the theft hypothesis construe the human family as a fundamentally political institution—a response to conflicts of interest and potentially violent interactions with conspecifics—as opposed to the many traditional accounts that portray it as a "little firm" based on the exchange of goods and services between husband and wife (e.g., Malinowski 1913, Murdock 1949, Washburn & Lancaster 1968, Sahlins 1972, Lovejoy 1981). This is not to deny the importance of the sexual division of labor in human social evolution (Lancaster & Lancaster 1987, Lancaster 1997). Humans everywhere expect males and females to specialize in different productive tasks, and the overall productivity of a conjugal family is no doubt increased through the pooling of gendered knowledge and skills (Kaplan et al. 2000). The question remains, however, why humans form relatively stable and exclusive sexual relationships. Men and women of the same band might well pool the products of their respective labors, yet remain free to mate without restriction and to spend each night with a different set of companions. Provisioning a household does not by itself require a man to sleep in that household or otherwise maintain a presence there; protecting a household, on the other hand, requires just such a presence, especially at night. From this perspective, the "pair bond" evolved not to promote or take advantage of the sexual division of labor, but to counter harassment, theft, and other social dangers.

From the female perspective, however, a bond with other females might be a workable (and perhaps preferable) alternative to a sexual alliance or pair bond. Although female bonding is precluded in some species by intense scramble competition (Wrangham 1980, Sterck et al. 1997), large patches of food would seem to allow females to band together in "sororal interest groups." Such a group might effectively deter harassment by mobbing a threatening male, as females do in savanna baboons and a number of other cercopithecines (Collins et al. 1984, Smuts 1985, Whitten 1987). In humans, scramble competition between females is relaxed, allowing them to forage together in large parties. The ecological reasons for this are a matter of debate (Rodseth & Novak in press), but the social consequences are clear: women are able to form important cooperative relationships with each other, even if they are unrelated by birth.

A striking parallel is observed in the case of bonobos (Kano 1992, Parish 1996, Parish & de Waal 2000). At sexual maturity, a female bonobo transfers to another community, where she develops close alliances with other, usually unrelated, females: "Establishing an artificial sisterhood, female bonobos may be said to be secondarily bonded" (de Waal 2001: 57). The same may be said for wives in a patrilocal household: having married into a network of distant relatives or strangers, they nevertheless develop and maintain close bonds with other women. Such bonds are often of crucial importance to their social and reproductive fortunes. Women help each other extensively in foraging or cultivation, food preparation, and child care; they also support each other in political struggles against both female and male opponents—though seldom in the form of physical conflict (the cases of women's violence reviewed by Kelly [2000: 31–38] seem to consist entirely of one-on-one fighting rather than coordinated aggression by female coalitions). The question, then, is why women in general do not use their "sisterhood" in the way that female bonobos do—why they do not, in other words, (1) consistently favor the company of females over that of males, (2) support each other in physical aggression against males who threaten or harass them, and thereby (3) gain a degree of social power equivalent to that of the males in the community (cf. Smuts 1992: 13–14).

The answer is almost certainly related to other differences between human and bonobo social organization. First, the sexual division of labor makes it difficult for either women or men to subsist entirely on their own, without the assistance of the opposite sex. As a result, women cannot associate exclusively with their female allies, though they may do so over long periods. Female bonobos, by contrast, subsist with little or no assistance and could, in principle, forage alone or with other females only. Second, alliances among male bonobos are poorly developed. This presents a stark contrast with the human case, in which males almost universally form important and enduring bonds with each other. Such bonds, according to the encounter-rate hypothesis, ultimately serve a military function, and this points to a third contrast with the bonobo pattern. Lethal aggression has never been observed in bonobos, and their intergroup encounters are not necessarily hostile. This suggests that bonobo males have little need to form alliances with each other in the way that both chimpanzee and human males do.

In any case, with the consolidation of male coalitions, women seem to be blocked from using alliances with other females to counter male harassment or domination (Smuts 1992, 1995). Keeping the company of her female allies surrounds a woman with sentinels and witnesses, perhaps, but not bodyguards. In fact, in the atmosphere of violence created by rival groups of men, women may find themselves increasingly dependent on male defenders. This in turn is likely to undermine existing bonds among women as they compete with each other to form alliances with men—whether these are sexual partners, on the one hand, or fathers, sons, and brothers, on the other. In general, the more isolated a woman is from her consanguineal kin, the more dependent she is upon her husband and in-laws for protection, and the more vulnerable she is to their exploitation or abuse (Smuts 1992: 14–15; see also Kelly 2000: 165–166). By contrast, if she moves only a short distance at marriage and can easily return to her natal home, she is often able to call upon her consanguineal ties to buffer her against abuse by her affines.

The Other Pair Bond

Relationships between adult, opposite-sex kin, including those between mother and son, father and daughter, and brother and sister, have received relatively little attention in the primatological literature. This is understandable, given that these relationships tend to be ephemeral in nonhuman primates. If either males or females but not both remain as adults with their natal kin, it follows that opposite-sex siblings will be separated at sexual maturity. Intergenerational relationships are sometimes more enduring, and mother-son relationships in particular are of considerable significance in the lives of chimpanzees, bonobos, and many monkeys. In general, however, parents do not survive long after their offspring have reached adulthood.

In humans, by contrast, relationships between male and female kin can be quite important, in part because overlapping generations allow filial relations to continue well into an offspring's adulthood. Thus a father can provide support and protection even to his mature daughters, while a mother often relies on her son as her principal protector and advocate in her old age. A comparison of mother-son relationships in humans, bonobos, and chimpanzees is especially revealing. Mother-son bonds are more important among bonobos than among chimpanzees. Humans perhaps fall in between these two species. Bonobo females are socially dominant, and males can spend much of their time in the company of their mothers, which affords them valuable alliances. Chimpanzees are forced by feeding competition to split up into smaller parties, and females are subordinate to males, so mothers have less power to aid their sons. Among humans, mothers have less social power than they do among bonobos, but through their own personalities and the voices of their male allies, they often have important influence. In patrilocal households in particular, the close, enduring bond between mother and son tends to enhance her influence and authority even as it "inhibits the formation of strong attachments between her son and his wife" (Michaelson & Goldschmidt 1971: 348).

Such a tradeoff between the mother-son bond and the husband-wife bond illustrates a more general tension in human societies between consanguineal and affinal relationships (Schneider 1961, Fox 1993). While most humans find it necessary to marry, the purposes served by a conjugal bond can often be served by a consanguineal one, as long as the consanguines are of opposite sex (Kelly 1993: 419–422). A brother and sister, in particular, can and sometimes do play the social and economic roles that we usually associate with husband and wife, as Kelly (1993: 397–398) documents in the case of the Etoro:

> A brother is a woman's protector and is responsible for accumulating half the compensation in the event that she is accused of witchcraft. A sister has a continuing obligation to assist in weeding her brother's garden throughout her life. Brother and sister often coreside and thus participate jointly in the same communal gardens and sago-processing operations. . . . It is noteworthy that it is this cross-sex sibling relationship, rather than the conjugal relation, that receives public social emphasis.

The Etoro case is no mere ethnographic oddity, but part of a much wider trend in human social organization. Although Western kinship systems stress the conjugal bond at the expense of bonds between siblings (Hsu 1965, Nuckolls 1993), in many societies the emphasis is more balanced or is entirely reversed (Marshall 1981, Kensinger 1985, Weisner 1989, Kolenda 1993). Especially when ties between husband and wife are tenuous or fragile, the brother-sister bond seems to serve as a functional equivalent—the other pair bond—across a wide range of cultural settings (Goodale 1981, Pollock 1985, de Munck 1993). In a critical review of the literature on the avunculate—the special role played in many societies by the mother's brother—Fox (1993: 211) argues that

> . . . brother and sister always constitute a "reserve" unit of quasi-parental behavior even if the mating unit is husband-wife. There will therefore be a constant tug-of-war between the two "bonds" (the "husband-wife" bond and the "brother-sister" bond) generalized into a "consanguine versus affine" conflict. . . . It is not so much . . . that the brother "relinquishes" the sister to the husband, as that the "mother" relinquishes, often with great reluctance, both the brother and the sister to their spouses.

Once she has borne children, a woman is often "pulled back into the consanguine fold," within which her brother is usually the "closest responsible consanguine male" (Fox 1993: 211)—a quasi-paternal figure who tends to invest especially in his sister's son (see also Schneider 1961, Hartung 1985, Bloch & Sperber 2002). While the avunculate takes many forms, it is described by Fox as "the first true cultural incursion into nature" (1993: 193) and even "the defining principle of humanity and culture" (1993: 227):

> It seems to be a peculiarly human thing to allow the asexual brother-sister tie to take over certain aspects of the parental role from the husband-wife tie. This gives rise to avuncular responsibilities that may flower into full blown matrilineal succession and inheritance, or to the classical indulgences of the patrilineal avunculate, or to the sacred duties towards mothers' brothers in bilateral systems, or even to the "love triangle" conflicts with them where the power of the maternal uncle threatens to rob the young males of their breeding preferences.

What makes the avunculate possible in humans and in no other species? As Fox has long argued (1972, 1975, 1980), humans are unusual (if not unique) in combining sexual alliances with bonds of consanguineal kinship. In other primates, sexual alliances are typical of one-male groups, while consanguineal bonds are typical of multimale groups. The two patterns are fully merged only in the social systems of humans and hamadryas baboons (Rodseth et al. 1991). In this light, it is perhaps unsurprising that affinal and consanguineal bonds should remain in tension, with each human society striking its own uneasy balance between the "one-male group" of the conjugal family and the "multimale group" of the lineage or clan.

Yet a combination of affinal and consanguineal bonds does not by itself suggest that a mother's brother would invest in his sister's offspring. The very possibility of a mother and her offspring being "pulled back into the consanguineal fold" depends on her ability to maintain bonds with her natal kin, even after she has married and perhaps moved to her husband's community. Again we are reminded of the release from proximity as a result of language, which enables a human being to maintain an invisible network of allies, beyond the local group, the immediate social setting, and the visible interactions of everyday life.

Kinship Expanded and Exploded

More than 20 years ago, in a discussion of stateless or "primitive" societies, Wolf (1982: 89) could still point to "kinship" as a relatively unexamined category in anthropology: "It is common to describe these populations as bound together by 'kinship,' but less common to inquire what kinship is." Writing on the eve of Schneider's (1984) critique, Wolf could not have foreseen the theoretical crisis that would rapidly ensue, a crisis that would swirl around the question of "what kinship is" and would cast doubt on one of the most basic of anthropological assumptions, the idea that kinship, in the sense of relatedness by blood, provides the foundation of social life in "primitive" populations.

What triggered this crisis? In *A Critique of the Study of Kinship*, Schneider (1984) argued that anthropologists had merely assumed the universality of certain "fundamental axioms," including what he formally designated as "Blood Is Thicker Than Water." Such an axiom was problematic in at least two ways, according to Schneider. First, it was supposed to be a natural law, or at least a statement of biological fact, when anthropology in Schneider's sense was fundamentally concerned not with nature or biology but with "the symbols and meanings and their configuration that a particular culture consists of" (p. 196). Second, there were good empirical reasons to doubt that Blood Is Thicker Than Water was sufficient to account for all the manifestations of "kinship" (p. 199):

> Even if this axiom were true as a biological fact, even if the most extensive scien-tifically acquired evidence showed it to be true—as true, for example, as the socio-biologists claim it to be, and that is a very strong position—the point remains that culture, even were it to do no more than recognize biological facts, still adds some-thing to those facts.... But the axiom that Blood Is Thicker Than Water does not hold water even for the sociobiologists. I need offer no more evidence for that state-ment than to call attention to the fact that even the sociobiologists do not claim to be able to account for the so-called extension of kinship. They only claim to account for *some* aspects of *some* of the relations between very close kin. This leaves a good deal to be accounted for.

Schneider's deep aversion to naturalistic explanation, along with his own axiomatic as-sumptions about what anthropology is or ought to be, have been scrutinized by Kuper (1999) and need not detain us here. For our purposes, what is interesting about this passage is the way it reveals "the so-called extension of kinship" as Schneider's trump card. Having rejected the fundamental assumption that kinship systems are all anchored in the same "facts of life," Schneider is still compelled to point out that such facts could not in any case account for the way kinship categories (and their associated obligations and sentiments) are extended from relations between "very close kin" to more distant relations and even to strangers (cf. Sahlins 1976, Alexander 1979: 152–155).

The notion that kinship can be extended in this way has a long history, reaching back to Morgan (1871), Malinowski (1913), and Radcliffe-Brown (1924). Much of what is described in the anthropological literature as kinship does indeed involve rather elastic concepts and practices, as succinctly summarized by Wolf (1982: 89):

> Some people have "a lot of kinship," others much less. Coresidence is often more significant than genealogy. . . . Thus, both Kroeber and Titiev have argued that coresidence underlay the formation of lineages. . . . Particular populations also vary greatly in how far they "extend" patterns of kinship found within familial entities to more distantly related families. . . . Patterns of kinship may be used to expand the scope of social and ideological linkages, and such linkages may become major operative factors in the jural and political realm.

When patterns of kinship are thus expanded, social practices begin to deviate from simple biological expectations. In northern India, for example, each rural village is regarded by local residents as an agnatic group—a network of patrilineal kin—even when its members are not in fact related by birth (Marriott 1955: 177–178): "People use agnatic terms for each other systematically, taking note of fictional generational standing throughout the village, ignoring all actual differences of caste and lineage. . . . In the countryside around [the study site of] Kishan Garhi, village is as important a fact for identifying a person as is clan in aboriginal Australia."

While the Australian comparison might seem to pit a case of "real" kinship against the "fictive" kinship of the Indian villagers, it is important to note that the membership of a clan may itself be founded on residence rather than ancestry (Kroeber 1952: 210, Titiev 1943). As Langness (1964: 174) observed in the context of the New Guinea Highlands, "the sheer fact of residence in a . . . group can and does determine kinship. People do not necessarily reside where they do because they are kinsmen; rather they become kinsmen because they reside there" (see also Strathern 1973).

By stretching the boundaries of kinship to include all who live on the same land, or share from the same hearth, or once nursed at the same breast, people are often seeking to create or enhance social bonds in ways that make political and economic sense. When men in a patrilineage term all their father's brother's sons their own brothers, they enhance the solidarity of their group. Since this naming system tends to occur in societies with valuable, clumped resources that must be defended from raiders, it seems appropriate to the social need. Similarly, when neighboring lineages claim descent from a common ancestor, they are more easily united against a common enemy. And when members of a foraging band use close kin terms to refer to distant kin in other bands, they are cultivating a regional network of allies who may grant them safe haven or access to resources as the need arises.

One of the most important ways in which humans expand their social networks is through intergroup affinity (Rodseth et al. 1991: 236). Reflecting the influence of Tylor (1889), many anthropologists have argued that rules of exogamy tend to promote alliances between members of different descent groups or local communities. While neighboring populations often live in tension between alliance and war, acknowledged kin relationships play an important role in peacemaking. Alliance by marriage has probably been as effective in sculpting intergroup relations in small-scale societies as it has been in monarchical states. The scarcity of data on unpacified hunter-gatherer groups makes the role of affinal alliances in their history difficult to assess, but there is every likelihood that ties between local bands have been promoted and maintained through spousal exchange.

Strikingly, affinal relationships are often treated by anthropologists and natives alike as a kind of kinship, even when they involve no relatedness by birth—no shared "blood" or other biogenetic substance. In some societies, of course, cousin marriage and other forms of endogamy tend to ensure that relatives by marriage are related by birth as well, while in all societies affines are likely to be biologically related to some of the same individuals in the next generation. Yet the very blurring of these categories demonstrates that ties of kinship can be remarkably elastic.

As Wiessner (1998: 134) argues, "the extension of kinship terms and relationships of mutual support to affinal relatives, distant kin, or even nonkin through cultural classification . . . is a critical adaptation of *Homo sapiens.*" In this light, to treat extended kinship as exceptional or epiphenomenal is to miss a fundamental point: this so-called fictive kinship provides the very sinews of most human societies. In a tribal or peasant context, membership in almost any group beyond one's immediate family is likely to involve extended kinship ties, and the more segmental the organization, the more important such ties become. To have a human society at all—not a social isolate like a chimpanzee community, but a network of networks, each of which is capable of elaborate segmentation—one must have extended kinship. Arguably, in fact, most of what we call kinship is generated by the same political conditions—persistent conflicts of interest, the danger of violent struggle—that lead to social segmentation in the first place. From this perspective, kinship is not so much *used* for political purposes as it is *constructed* for such purposes, and begins to seem more of an elaborate historical artifact or invention than a simple reflection of Hamilton's rule (Hamilton 1964, cf. Jones 2000).

In the eyes of most Westerners, however, kinship is a given, a set of natural facts, more or less accurately represented in systems of kinship reckoning. This is in keeping with a general Euro-American tendency, in Schneider's (1995: 222) view, to treat knowledge as "*discovered*, not invented"—something we accumulate "when the 'facts' of nature which are hidden from us mostly, are finally revealed." If another people's kinship system does not reflect these "facts," this is often interpreted from the Western point of view as an anomaly, a failure to understand the facts, or perhaps a convenient fiction to serve some special purpose. Because it contradicts what we know about sexual reproduction, extended kinship in particular has seemed an oddity or an atavism, diagnostic of a "primitive" mentality or a "backward" way of life.

More charitably, the extension of kinship might be attributed to the fact that people in primitive and peasant societies rely for their survival on the safety net of the local collectivity, rather than the formal institutions of the state or the market. In modern industrial societies, by contrast, the lives of young adults are marked by extreme social and spatial mobility and a prevailing ethic of individualism. As a result, people in these societies are often assumed to be less influenced by kinship—and especially extended kinship—than any others in the history of our species.

In at least one vital way, however, modern industrial societies seem to increase the use of extended kinship—increase it so much, in fact, that it no longer appears to be a matter of kinship at all. With the advent of the modern state in the eighteenth century, collective identities were supposed to be constructed on the basis of citizenship rather than ethnicity, "soil" rather than "blood" (Maine 1861)—yet the surging tides of nationalism in the nineteenth and twentieth centuries suggest otherwise (Verdery 1999: 41):

> In my view, the identities produced in nation-building processes do not displace those based in kinship but—as any inspection of national rhetorics will confirm—reinforce and are parasitic upon them. National ideologies are saturated with kinship metaphors: fatherland and motherland, sons of the nation and their brothers, mothers of these worthy sons, and occasionally daughters. Many national ideologies present their nations as large, mostly patrilineal kinship (descent) groups that celebrate founders, great politicians, and cultural figures as not just heroes but veritable "progenitors," forefathers—that is, as ancestors. . . . Nationalism is thus a kind of ancestor worship, a system of patrilineal kinship, in which national heroes occupy the place of clan elders in defining the nation as a noble lineage.

A link between nationalism and kinship, as Verdery notes, has been suggested in the work of several other anthropologists, including Schneider himself (1977). Yet the argument that national identities are anchored in "myths of ethnic descent" has been most fully elaborated by Smith (1984, 1991, 1999), whose distinction between "blood ties, real or alleged" and "spiritual kinship" is especially relevant in the present context (1999: 58):

> The biological link . . . ensures a high degree of communal solidarity, since the [ethnic] community is viewed as a network of interrelated kin groups claiming a common ancestor, and thereby marking them off from those unable to make such a claim. . . . Against such biological modes of tracing descent, we find another equally important set of generational linkages: those that rest on a cultural affinity and ideological "fit" with the presumed ancestors. . . . The aim is to recreate the heroic spirit (and the heroes) that animated "our ancestors" in some past golden age; and descent is traced, not through family pedigrees, but through the persistence of certain kinds of "virtue" or other distinctive cultural qualities. . . . What I shall argue is that within given ethnic communities since the French Revolution (or slightly earlier), both kinds of national myth-making emerge and persist in an often contrapuntal relationship.

By now we are able to appreciate the full implications of extended kinship. Not only is such kinship used, as Wolf (1982: 89) put it, "to expand the scope of social and ideological linkages," it provides modern people with their most important social identities, their basic ways of thinking about themselves and those around them, in the largest, most powerful societies in human history. At this point, kinship has not only expanded but exploded. Having reached the limits of plausibility, the idiom of kinship ruptures and gives way to a new rhetoric and a new way of thinking, nationalism, which seems to float free of kinship but continues to draw much of its vitality from metaphors of brotherhood, fatherland, and veneration of the ancestors (see also Jones 2000: 793–794).

How are we to explain this extraordinary human tendency to project the language, obligations, and sentiments of kinship onto larger and larger social groups, "right up to nations of hundreds of millions" (Alexander 1979: 221)? Extended kinship might be understood as a system that exploits the power of biological kinship, that is, the tendency of humans to favor and trust those they believe to be relatives by birth. This approach would explain the extension of kinship in much the way that Westermarck (1922: 192) explained the "remarkable lack of erotic feeling between persons who have been living closely together from childhood"—as an incidental effect of a set of innate dispositions that have tended to work in the usual environments of human evolution, but are easily set awry under modern conditions (see especially Wolf 1995). Such a comparison with the Westermarck effect would seem especially appropriate when kinship is extended to unrelated persons within the household or its immediate neighborhood. This is in fact the kind of case that recent definitions of kinship seem specifically designed to cover:

> Kinship has to be understood as a culturally specific notion of relatedness deriving from shared bodily and/or spiritual substance and its transmission. In the West as well as in many non-Western cultures this may be seen as resulting from processes of sexual reproduction; in yet other cultures this may be seen as resulting from sharing the same food, living on the same land, or whatever. (Holy 1996: 171)

> I have shown that it is possible . . . to incorporate a broader conception of kinship as shared substance, often ratified by other means. . . . Substances may include, but are not limited to, food, shelter, and body fluids. Biogenetic substances remain a part of this new model. (Galvin 2001: 122)

Such definitions leave little doubt that "kinship" can be based on a number of different processes, but what is especially striking is the clustering of these processes around notions of shared substance, usually based on nurturing and affiliation within the same house, compound, or other domestic setting.

What is equally striking, however, is the apparent inadequacy of such domestic processes to account for the extension of kinship to persons far beyond the household, to distant relatives or in-laws, even to strangers. In a sense, there has always been a problem for anthropologists in this regard (Fortes 1953: 30). Kinship in the domestic sphere and kinship in the public or political sphere seem to work in quite different ways, and might well be classified as quite different phenomena, were it not for the fact that humans so often use kinship terms to apply to persons in both domains. Yet extended kinship, as Wolf (1982: 92–93) describes it, is

> not the same as kinship on the level of filiation and marriage; it is concerned with jural allocation of rights and claims, and hence with political relations between people. On the level of filiation and marriage, kinship sets up individuated linkages among shareholders in social labor; extended kinship, in contrast, organizes social labor into labor pools and places controls over the transfer of labor from one pool to another. The persistence of the idiom of kinship in the jural-political realm, however, poses a problem. . . . Why should the language of kinship persist in this different setting?

Contrary to Wolf, we would suggest that the language of kinship has not in fact been extended from the level of filiation and marriage to the level of political relations. Kinship works differently in the domestic and public domains not because it has been artificially extended to serve political purposes, but because the conjugal bond and the bond of "brotherhood" among men involve different kinds of politics. To capture the practical advantages of kinship, Wolf invokes the language of labor, but we would opt instead for the language of war. Both kinds of language are metaphorical, and each captures an important dimension of what human kinship is all about. Yet the relationship between kinship and work has received far more attention than the relationship between kinship and violence.

The household, with its division of labor by sex, is one important setting in which human psychology has been sculpted by natural selection. A setting of nearly equal importance, however, lies in the public domain, where men in particular have always met to negotiate their alliances (Rodseth & Novak 2000). Here a different kind of Westermarck effect has perhaps been at work, "fooling" humans into thinking and feeling that their allies were not merely affines, cousins, or coresidents, but blood brothers (and sometimes sisters), descendants of the same heroic father (and sometimes mother). Ironically, then, precisely because it was a distortion of "real" genetic relationships, this kind of kinship would come to have a powerful selective advantage, not so much in the household or at the hearth, but in the plaza, the men's house, the barracks, and the church.

Conclusion

In spite of the wide variation in their behavior toward kin, humans exhibit a unique set of kinship tendencies, including lifelong bonds among relatives of both sexes and the use of sexual alliances to influence political relationships between families or other kin groups. The political dimension of human kinship is manifested in many ways, most obviously in the tendency of men to unite with male relatives (or those classified as such) in potentially violent competition against more distantly related males. Women, by comparison, seem more inclined to use kinship as a basis for mutualism that does not involve collective violence against other women. Yet humans of both sexes display an intense interest in naming and classifying kin, often assigning and adopting social roles on the basis of kinship categories. In this sense, men and women are always woven together in a kinship network that transcends local groupings and allows them to mobilize at higher or lower levels of organization, in response to the threat or the opportunity of the moment.

While some aspects of human kinship are explicable in terms immediately familiar to primatologists, a cultural framework is needed to understand the ways in which kinship is extended beyond genetic relationships. Although this extension appears to take human kinship beyond biology, we conclude that the cultural manipulation of kinship itself tends to serve biologically explicable goals.

References

Alexander, R. D. 1979. *Darwinism and Human Affairs*. Seattle: University of Washington Press.

Alexander, R. D. 1987. *The Biology of Moral Systems*. New York: Aldine.

Alexander, R. D. 1990. *How Did Humans Evolve? Reflections on the Uniquely Unique Species*. Ann Arbor, MI: University of Michigan. Museum of Zoology, Special Publications, No. 1.

Barnard, A. 1983. Contemporary hunter-gatherers: current issues in ecology and social organization. *Annu. Rev. Anthropol.*, 12, 193–214.

Birdsell, J. 1968. Some predictions for the Pleistocene based on equilibrium systems for recent hunter-gatherers. In: *Man the Hunter* (Ed. by R. B. Lee & I. DeVore), pp. 229–240. Chicago: Aldine.

Bloch, M. 1991. *Prey into Hunter: The Politics of Religious Experience*. Cambridge: Cambridge University Press.

Bloch, M. & Sperber, D. 2002. Kinship and evolved psychological dispositions: the mother's brother controversy reconsidered. *Curr. Anthropol.,* 43, 723–748.

Boehm, C. 1992. Segmentary "warfare" and the management of conflict: comparison of East African chimpanzees and patrilineal-patrilocal humans. In: *Coalitions and Alliances in Humans and Other Animals* (Ed. by A. H. Harcourt & F. B. M. de Waal), pp. 137–173. Oxford: Oxford University Press.

Boehm, C. 1993. Egalitarian behavior and reverse dominance hierarchy. *Curr. Anthropol.,* 34, 227–254.

Boehm, C. 1999. *Hierarchy in the Forest: The Evolution of Egalitarian Behavior.* Cambridge, MA: Harvard University Press.

Boesch, C. & Boesch-Achermann, H. 2000. *The Chimpanzees of the Taï Forest: Behavioural Ecology and Evolution.* Oxford: Oxford University Press.

Bouquet, M. 1993. *Reclaiming English Kinship: Portuguese Refractions of British Kinship Theory.* Manchester, UK: Manchester University Press.

Brettell, C. B. 2001. Not that lineage stuff: teaching kinship into the twenty-first century. In: *New Directions in Anthropological Kinship* (Ed. by L. Stone), pp. 48–70. Lanham, MD: Rowman and Littlefield.

Brown, D. E. 1991. *Human Universals.* New York: McGraw-Hill.

Carsten, J. 1997. *The Heat of the Hearth: The Process of Kinship in a Malay Fishing Community.* Oxford: Clarendon Press.

Carsten, J. 2000. *Cultures of Relatedness: New Approaches to the Study of Kinship.* Cambridge: Cambridge University Press.

Chagnon, N. A. & Irons, W. 1979. *Evolutionary Biology and Human Social Behavior: An Anthropological Perspective.* North Scituate, MA: Duxbury Press.

Chapman, C. A., White, F. J., & Wrangham, R. W. 1994. Party size in chimpanzees and bonobos: a reevaluation of theory based on two similarly forested sites. In: *Chimpanzee Cultures* (Ed. by R. W. Wrangham, W. C. McGrew, F. B. M. de Waal, & P. G. Heltne), pp. 41–58. Cambridge, MA: Harvard University Press.

Clifford, J. 1988. *The Predicament of Culture: Twentieth-Century Ethnography, Literature, and Art.* Cambridge, MA: Harvard University Press.

Collier, J. F. & Rosaldo, M. Z. 1981. Politics and gender in simple societies. In: *Sexual Meanings: The Cultural Construction of Gender and Sexuality* (Ed. by S. B. Ortner & H. Whitehead), pp. 275–329. Cambridge: Cambridge University Press.

Collins, D. A., Busse, C. D., & Goodall, J. 1984. Infanticide in two populations of savannah baboons. In: *Infanticide: A Comparative and Evolutionary Perspective* (G. Hausfater & S. B. Hrdy), pp. 193–215. New York: Aldine.

de Munck, V. C. 1993. The dialectics and norms of self-interest: reciprocity among cross-siblings in a Sri Lankan Muslim community. In: *Siblings in South Asia: Brothers and Sisters in Cultural Context* (Ed. by C. W. Nuckolls), pp. 143–162. New York: Guilford Press.

de Waal, F. B. M. 2001. Apes from Venus: bonobos and human social evolution. In: *Tree of Origin: What Primate Behavior Can Tell Us About Human Social Evolution* (Ed. by F. B. M. de Waal), pp. 39–68. Cambridge, MA: Harvard University Press.

Draper, P. & Harpending, H. 1988. A sociobiological perspective on the development of human reproductive strategies. In: *Sociobiological Perspectives on Human Development* (Ed. by K. B. MacDonald), pp. 340–372. New York: Springer-Verlag.

Dunbar, R. 1996. *Grooming, Gossip, and the Evolution of Language.* Cambridge, MA: Harvard University Press.

Ember, M. & Ember, C. R. 1971. The conditions favoring matrilocal versus patrilocal residence. *Am. Anthropol.,* 73, 571–594.

Ferguson, R. B. & Whitehead, N. L. 1992. *War in the Tribal Zone: Expanding States and Indigenous Warfare*. Santa Fe, NM: School of American Research Press.

Fortes, M. 1953. The structure of unilineal descent groups. *Am. Anthropol.*, 55, 17–41.

Fox, R. 1967. *Kinship and Marriage*. Harmondsworth, UK: Penguin Books.

Fox, R. 1972. Alliance and constraint: sexual selection in the evolution of human kinship systems. In: *Sexual Selection and the Descent of Man 1871–1971* (Ed. by B. Campbell), pp. 282–331. Chicago: Aldine.

Fox, R. 1975. Primate kin and human kinship. In: *Biosocial Anthropology* (Ed. by R. Fox), pp. 9–35. New York: John Wiley and Sons.

Fox, R. 1979. Kinship categories as natural categories. In: *Evolutionary Biology and Human Social Behavior: An Anthropological Perspective* (Ed. by N. A. Chagnon & W. Irons), pp. 132–144. North Scituate, MA: Duxbury Press.

Fox, R. 1980. *The Red Lamp of Incest: An Inquiry into the Origins of Mind and Society*. Notre Dame, IN: University of Notre Dame Press.

Fox, R. 1993. Sisters' sons and monkeys' uncles: six theories in search of an avunculate. In: *Reproduction and Succession: Studies in Anthropology, Law, and Society* (Ed. by R. Fox), pp. 191–232. New Brunswick, NJ: Transaction Publishers.

Franklin, S. & McKinnon, S. 2001. *Relative Values: Reconfiguring Kinship Studies*. Durham, NC: Duke University Press.

Galvin, K.-L. 2001. Schneider revisited: sharing and radification in the construction of kinship. In: *New Directions in Anthropological Kinship* (Ed. by L. Stone), pp. 109–124. Lanham, MD: Rowman and Littlefield.

Gamble, C. 1998. Paleolithic society and the release from proximity: a network approach to intimate relations. *World Archaeol.*, 29, 426–449.

Gamble, C. 1999. *The Palaeolithic Societies of Europe*. Cambridge: Cambridge University Press.

Geertz, C. 1973. *The Interpretation of Cultures*. New York: Basic Books.

Goodale, J. 1981. Siblings as spouses: the reproduction and replacement of Kaulong society. In: *Siblingship in Oceania* (Ed. by M. Marshall), pp. 225–302. Ann Arbor, MI: University of Michigan Press.

Goodall, J. 1986. *The Chimpanzees of Gombe: Patterns of Behavior*. Cambridge, MA: Harvard University Press.

Hamilton, W. D. 1964. The genetical evolution of social behavior. I and II. *J. Theor. Biol.*, 7, 1–52.

Hartung, J. 1985. Matrilineal inheritance: new theory and analysis. *Behav. Brain Sci.*, 8, 661–688.

Hawkes, K. 1990. Why do men hunt? Some benefits for risky strategies. In: *Risk and Uncertainty in Tribal and Peasant Economies* (Ed. by E. Cashdan), pp. 145–166. Boulder, CO: Westview Press.

Hawkes, K. 1991. Showing off: tests of an hypothesis about men's foraging goals. *Ethol. Sociobiol.*, 12, 29–54.

Hawkes, K. 1993. Why hunter-gatherers work: an ancient version of the problem of public goods. *Curr. Anthropol.*, 34, 341–361.

Hawkes, K. 2000. Big game hunting and the evolution of egalitarian societies. In: *Hierarchies in Action: Cui Bono?* (Ed. by M. Diehl), pp. 59–83. Carbondale, IL: Southern Illinois University. Center for Archaeological Investigations, Occasional Paper No. 27.

Hawkes, K. & Bliege Bird, R. 2002. Showing off, handicap signaling, and the evolution of men's work. *Evol. Anthropol.*, 11, 58–67.

Hawkes, K., O'Connell, J. F., & Blurton Jones, N. G. 2001. Hunting and nuclear families: some lessons from the Hadza about men's work. *Curr. Anthropol.*, 42, 681–709.

Heiser, R. F. & Elsasser, A. B. 1980. *The Natural World of the California Indians*. Berkeley, CA: University of California Press.

Holy, L. 1996. *Anthropological Perspectives on Kinship*. London: Pluto Press.

Hsu, F. L. K. 1965. The effects of dominant kinship relations on kin and non-kin behavior: a hypothesis. *Am. Anthropol.*, 67, 638–661.

Idani, G. 1991. Cases of inter-unit group encounters in pygmy chimpanzees at Wamba, Zaire. In: *Primatology Today: Proceedings of the XIIIth Congress of the International Primatological Society* (Ed. by A. Ehara, T. Kimura, O. Takenaka, & M. Iwamoto), pp. 235–238. Amsterdam: Elsevier.

Jones, D. 2000. Group nepotism and human kinship. *Curr. Anthropol.*, 41, 779–809.

Jones, R. 1984. Hunters and history: a case study from western Tasmania. In: *Past and Present in Hunter-Gatherer Studies* (Ed. by C. Schrire), pp. 27–65. London: Academic Press.

Kano, T. 1992. *The Last Ape: Pygmy Chimpanzee Behavior and Ecology* (Trans. by E. O. Vineberg). Stanford, CA: Stanford University Press.

Kaplan, H., Hill, K., Lancaster, J. B., & Hurtado, A. M. 2000. A theory of human life history evolution: diet, intelligence, and longevity. *Evol. Anthropol.*, 9, 156–185.

Kelly, R. C. 1993. *Constructing Inequality: The Fabrication of a Hierarchy of Virtue Among the Etoro*. Ann Arbor, MI: University of Michigan Press.

Kelly, R. C. 2000. *Warless Societies and the Origin of War*. Ann Arbor, MI: University of Michigan Press.

Kelly, R. L. 1995. *The Foraging Spectrum: Diversity in Hunter-Gatherer Lifeways*. Washington, DC: Smithsonian Institution.

Kensinger, K. 1985. *The Sibling Relationship in Lowland South America*. Bennington, VT: Bennington College. Working papers on South American Indians, No. 7.

Knauft, B. M.1987. Reconsidering violence in simple human societies: homicide among the Gebusi of New Guinea. *Curr. Anthropol.*, 28, 457–500.

Knauft, B. M. 1991. Violence and sociality in human evolution. *Curr. Anthropol.*, 32, 391–428.

Knauft, B. M. 1994. Culture and cooperation in human evolution. In: *The Anthropology of Peace and Nonviolence* (Ed. by L. Sponsel & T. Gregor), pp. 37–68. Boulder, CO: Lynne Rienner.

Kohl, S. & Bennett, J. W. 1965. Kinship, succession, and the migration of young people in a Canadian agricultural community. In: *Kinship and Geographical Mobility* (Ed. by R. Paddington), pp. 95–116. Leiden: E. J. Brill.

Kolenda, P. 1993. Sibling relations and marriage practices: a comparison of north, central, and south India. In: *Siblings in South Asia: Brothers and Sisters in Cultural Context* (Ed. by C. W. Nuckolls), pp. 103–141. New York: Guilford Press.

Kopytoff, I. 1977. Matrilineality, residence and residential zones. *Am. Ethnol.*, 4, 539–558.

Kroeber, A. L. 1952. Basic and secondary patterns of social structure. In: *The Nature of Culture* (Ed. by A. L. Kroeber), pp. 210–218. Chicago: University of Chicago Press. (Article originally published 1938.)

Kuper, A. 1982. Lineage theory: a critical retrospect. *Annu. Rev. Anthropol.*, 11, 71–95.

Kuper, A. 1988. *The Invention of Primitive Society: Transformations of an Illusion*. London: Routledge.

Kuper, A. 1999. *Culture: The Anthropologists' Account*. Cambridge, MA: Harvard University Press.

Lancaster, J. B. 1997. The evolutionary history of human parental investment in relation to population growth and social stratification. In: *Feminism and Evolutionary Biology* (Ed. by P. A. Gowaty), pp. 466–488. London: Chapman and Hall.

Lancaster, J. B. & Lancaster, C. 1987. The watershed: change in parental-investment and family formation strategies in the course of human evolution. In: *Parenting Across the Human Lifespan: Biosocial Dimensions* (Ed. by J. B. Lancaster, J. Altmann, A. S. Rossi, & L. R. Sherrod), pp. 187–205. Hawthorne, NY: Aldine.

Langness, L. L. 1964. Some problems in the conceptualization of Highlands social structure. *Am. Anthropol.*, 66(4), part 2, special publication, pp. 130–158.

Leach, E. R. 1961. *Rethinking Anthropology*. London: Athlone Press. London School of Economics Monographs on Social Anthropology, No. 22.

Lee, R. B. & DeVore, I. 1968. Problems in the study of hunters and gatherers. In: *Man the Hunter* (Ed. by R. B. Lee & I. DeVore), pp. 3–12. New York: Aldine.

Lesser, A. 1961. Social fields and the evolution of society. *Southwestern J. Anthropol.*, 17, 40–48.

Lovejoy, C. O. 1981. The origin of man. *Science*, 211, 341–350.

MacDonald, D. H. & Hewlett, B. S. 1999. Reproductive interests and forager mobility. *Curr. Anthropol.*, 40, 501–523.

Maine, H. 1861. *Ancient Law*. London: Murray.

Malinowski, B. 1913. *The Family Among the Australian Aborigines*. London: Hodder.

Manson, J. H. & Wrangham, R. W. 1991. Intergroup aggression in chimpanzees and humans. *Curr. Anthropol.*, 32, 369–390.

Marriott, M. 1955. Little communities in an indigenous civilization. In: *Village India: Studies in the Little Community* (Ed. by M. Marriott), pp. 171–222. Chicago: University of Chicago Press.

Marshall, M. 1981. *Siblingship in Oceania*. Ann Arbor, MI: University of Michigan Press.

Mesnick, S. L. 1997. Sexual alliances: evidence and evolutionary implications. In: *Feminism and Evolutionary Biology: Boundaries, Intersections, and Frontiers* (Ed. by P. A. Gowaty), pp. 207–260. New York: International Thomson.

Michaelson, E. J. & Goldschmidt, W. 1971. Female roles and male dominance among peasants. *Southwestern J. Anthropol.*, 27, 330–352.

Morgan, L. H. 1871. *Systems of Consanguinity and Affinity of the Human Family*. Washington, DC: Smithsonian Institution.

Murdock, G. P. 1949. *Social Structure*. New York: Free Press.

Murphy, R. F. 1957. Intergroup hostility and social cohesion. *Am. Anthropol.*, 59, 1018–1036.

Nuckolls, C. W. 1993. An introduction to the cross-cultural study of sibling relations. In: *Siblings in South Asia: Brothers and Sisters in Cultural Context* (Ed. by C. W. Nuckolls), pp. 19–41. New York: Guilford Press.

Otterbein, K. F. 1968. Internal war: a cross-cultural study. *Am. Anthropol.*, 70, 277–289.

Otterbein, K. F. & Otterbein, C. S. 1965. An eye for an eye, a tooth for a tooth: a cross-cultural study of feuding. *Am. Anthropol.*, 67, 1470–1482.

Paige, K. E. & Paige, J. M. 1981. *The Politics of Reproductive Ritual*. Berkeley, CA: University of California Press.

Parish, A. R. 1996. Female relationships in bonobos (*Pan paniscus*): evidence for bonding, cooperation, and female dominance in a male-philopatric species. *Hum. Nat.*, 7, 61–96.

Parish, A. R. & de Waal, F. B. M. 2000. The other "closest living relative": how bonobos (*Pan paniscus*) challenge traditional assumptions about females, dominance, intra- and inter-sexual interactions, and hominid evolution. *Ann. N.Y. Acad. Sci.*, 907, 97–113.

Peletz, M. G. 1995. Kinship studies in late twentieth-century anthropology. *Annu. Rev. Anthropol.*, 24, 343–372.

Pollock, D. 1985. Looking for a sister: Culina siblingship and affinity. In: *The Sibling Rela-*

tionship in Lowland South America (Ed. by K. Kensinger), pp. 8–16. Bennington, VT: Bennington College. Working papers on South American Indians, No. 7.

Pusey, A. E. & Packer, C. 1987. Dispersal and philopatry. In: *Primate Societies* (Ed. by B. B. Smuts, D. L. Cheney, R. M. Seyfarth, R. W. Wrangham, & T. T. Struhsaker), pp. 250–266. Chicago: University of Chicago Press.

Radcliffe-Brown, A. R. 1924. The mother's brother in South Africa. *S. Afr. J. Sci.*, 21, 542–555. Reprinted in Radcliffe-Brown, A. R. 1952. *Structure and Function in Primitive Society,* pp. 15–31. New York: Free Press.

Radcliffe-Brown, A. R. 1940. On social structure. *J. Royal Anthropol. Inst.*, 70, 1–12. Reprinted in Radcliffe-Brown, A. R. 1952. *Structure and Function in Primitive Society*, pp. 188–204. New York: Free Press.

Rodseth, L. 1998. Distributive models of culture: a Sapirian alternative to essentialism. *Am. Anthropol.*, 100, 55–69.

Rodseth, L. & Novak, S. A. 2000. The social modes of men: toward an ecological model of human male relationships. *Hum. Nat.,* 11, 335–366.

Rodseth, L. & Novak, S. A. In press. The impact of primatology on the study of human society. In: *Missing the Revolution: Darwinism for Social Scientists* (Ed. by J. H. Barkow). New York: Oxford University Press.

Rodseth, L., Wrangham, R. W., Harrigan, A. M., & Smuts, B. B. 1991. The human community as a primate society. *Curr. Anthropol.*, 32, 221–254.

Sahlins, M. D. 1972. The domestic mode of production: the structure of underproduction. In: *Stone Age Economics* (Ed. by M. D. Sahlins), pp. 41–99. New York: Aldine.

Sahlins, M. D. 1976. *The Use and Abuse of Biology*. Ann Arbor, MI: University of Michigan Press.

Scheffler, H. W. 1991. Sexism and naturalism in the study of kinship. In: *Gender at the Crossroads of Knowledge: Anthropology in the Postmodern Era* (Ed. by M. di Leonardo), pp. 361–382. Berkeley, CA: University of California Press.

Scheffler, H. W. 2001. *Filiation and Affiliation*. Boulder, CO: Westview Press.

Schneider, D. M. 1961. Introduction: the distinctive features of matrilineal descent groups. In: *Matrilineal Kinship* (Ed. by D. M. Schneider & K. Gough), pp. 1–29. Berkeley, CA: University of California Press.

Schneider, D. M. 1972. What is kinship all about? In: *Kinship Studies in the Morgan Centennial Year* (Ed. by P. Reining), pp. 32–63. Washington, DC: Anthropological Society of Washington.

Schneider, D. M. 1977. Kinship, nationality, and religion in American culture: toward a definition of kinship. In: *Symbolic Anthropology: A Reader in the Studies of Symbols and Meanings* (Ed. by J. L. Dolgin, D. S. Kemnitzer, & D. M. Schneider), pp. 63–71. New York: Columbia University Press.

Schneider, D. M. 1980. *American Kinship: A Cultural Account*, 2nd ed. Chicago: University of Chicago Press.

Schneider, D. M. 1984. *A Critique of the Study of Kinship*. Ann Arbor, MI: University of Michigan Press.

Schneider, D. M. 1995. *Schneider on Schneider: The Conversion of the Jews and Other Anthropological Stories* (Ed. by R. Handler). Durham, NC: Duke University Press.

Schweitzer, P. P. 2000. *Dividends of Kinship: Meanings and Uses of Social Relatedness*. London: Routledge.

Smith, A. D. 1984. Ethnic myths and ethnic revivals. *Eur. J. Sociol.*, 25, 283–305.

Smith, A. D. 1991. *National Identity*. Harmondsworth, UK: Penguin.

Smith, A. D. 1999. *Myths and Memories of the Nation*. Oxford: Oxford University Press.

Smuts, B. B. 1985. *Sex and Friendship in Baboons*. Cambridge, MA: Harvard University Press.

Smuts, B. B. 1992. Male aggression against women: an evolutionary perspective. *Hum. Nat.*, 3, 1–44.

Smuts, B. B. 1995. The evolutionary origins of patriarchy. *Hum. Nat.*, 6, 1–32.

Smuts, B. B. & Smuts, R. W. 1993. Male aggression and sexual coercion of females in nonhuman primates and other mammals: evidence and theoretical implications. *Adv. Study Behav.*, 22, 1–63.

Sterck, E. H. M., Watts, D. P., & van Schaik, C. P. 1997. The evolution of female social relationships in nonhuman primates. *Behav. Ecol. Sociobiol.*, 41, 291–309.

Stone, L. 2001. *New Directions in Anthropological Kinship*. Lanham, MD: Rowman and Littlefield.

Strathern, A. 1973. Kinship, descent and locality: some New Guinea examples. In: *The Character of Kinship* (Ed. by J. Goody), pp. 21–33. Cambridge: Cambridge University Press.

Strathern, M.. 1992a. *After Nature: English Kinship in the Late Twentieth Century*. Cambridge: Cambridge University Press.

Strathern, M. 1992b. *Reproducing the Future: Anthropology, Kinship and the New Reproductive Technologies*. New York: Routledge.

Testart, A. 1982. The significance of food storage among hunter-gatherers: patterns, population densities, and social inequalities. *Curr. Anthropol.*, 23, 523–537.

Tindale, N. B. 1974. *Aboriginal Tribes of Australia*. Los Angeles: University of California Press.

Titiev, M. 1943. The influence of common residence on the unilateral classification of kin. *Am. Anthropol.*, 45, 511–530.

Trautmann, T. R. 1987. *Lewis Henry Morgan and the Invention of Kinship*. Berkeley, CA: University of California Press.

Tylor, E. B. 1889. On a method of investigating the development of institutions: applied to the laws of marriage and descent. *J. Anthropol. Inst.*, 18, 245–272.

van Schaik, C. 1989. The ecology of social relationships amongst female primates. In: *Comparative Socioecology: The Behavioral Ecology of Humans and Other Mammals* (Ed. by V. Standen & G. R. A. Foley), pp. 195–218. Oxford: Blackwell Scientific Publications.

van Schaik, C. 1996. Social evolution in primates: the role of ecological factors and male behaviour. In: *Evolution of Social Behaviour Patterns in Primates and Man* (Ed. by W. Runciman, J. Maynard-Smith, & R. Dunbar), pp. 9–31. London: British Academy.

van Velzen, H. U. E. T. & van Wetering, W. 1960. Residence, power groups, and intra societal aggression. *Int. Arch. Ethnol.*, 49, 169–200.

Verdery, K. 1999. *The Political Lives of Dead Bodies: Reburial and Postsocialist Change*. New York: Columbia University Press.

Washburn, S. L. & Lancaster, C. 1968. The evolution of hunting. In: *Man the Hunter* (Ed. by R. B. Lee & I. DeVore), pp. 293–303. Cambridge, MA: Harvard University Press.

Weisner, T. S. 1989. Comparing sibling relationships across cultures. In: *Sibling Interaction Across Cultures: Theoretical and Methodological Issues* (Ed. by P. Zukow), pp. 11–25. New York: Springer-Verlag.

Westermarck, E. 1922. *The History of Human Marriage, Vol. II*. 5th ed. New York: Allerton.

Whiting, J. W. M. & Whiting, B. B. 1975. Aloofness and intimacy of husbands and wives: a cross-cultural study. *Ethos*, 3, 183–207.

Whitten, P. 1987. Infants and adult males. In: *Primate Societies* (Ed. by B. B. Smuts, D. L.

Cheney, R. M. Seyfarth, R. W. Wrangham, & T. T. Struhsaker), pp. 343–357. Chicago: University of Chicago Press.

Wiessner, P. 1982. Risk, reciprocity and social influences on !Kung San economics. In: *Politics and History in Band Societies* (Ed. by E. Leacock & R. Lee), pp. 37–59. Cambridge: Cambridge University Press.

Wiessner, P. 1998. Indoctrinability and the evolution of socially defined kinship. In: *Indoctrinability, Ideaology, and Warfare: Evolutionary Perspectives* (Ed. by I. Eibl-Eibesfeldt & F. K. Slater), pp. 133–250. New York: Berghahn Books.

Wilmsen, E. N. 1989. *Land Filled with Flies: A Political Economy of the Kalahari.* Chicago: University of Chicago Press.

Wolf, A. P. 1995. *Sexual Attraction and Childhood Association: A Chinese Brief for Edward Westermarck.* Stanford, CA: Stanford University Press.

Wolf, E. R. 1982. *Europe and the People Without History.* Berkeley, CA: University of California Press.

Wrangham, R. W. 1979. On the evolution of ape social systems. *Soc. Sci. Info.*, 18, 335–368.

Wrangham, R. W. 1980. An ecological model of the evolution of female-bonded groups of primates. *Behaviour*, 75, 262–300.

Wrangham, R. W. 1982. Mutualism, kinship and social evolution. In: *Current Problems in Sociobiology* (Ed. by King's College Sociobiology Group), pp. 269–289. Cambridge: Cambridge University Press.

Wrangham, R. W. 1987a. The evolution of social structure. In: *Primate Societies* (Ed. by B. B. Smuts, D. L. Cheney, R. M. Seyfarth, R. W. Wrangham, & T. T. Struhsaker), pp. 282–296. Chicago: University of Chicago Press.

Wrangham, R. W. 1987b. The significance of African apes for reconstructing human social evolution. In: *Primate Models of Hominid Evolution* (Ed. by W. G. Kinzey), pp. 51–71. Albany, NY: SUNY Press.

Wrangham, R. W. 1999. Evolution of coalitionary killing. *Ybk. Phys. Anthropol.*, 42, 1–30.

Wrangham, R. W. 2000. Why are male chimpanzees more gregarious than mothers? A scramble competition hypothesis. In: *Male Primates* (Ed. by P. Kappeler), pp. 248–258. Cambridge: Cambridge University Press.

Wrangham, R. W. 2001. Out of the *Pan*, into the fire: how our ancestors' evolution depended on what they ate. In: *Tree of Origin: What Primate Behavior Can Tell Us About Human Social Evolution* (Ed. by F. B. M. de Waal), pp. 119–144. Cambridge, MA: Harvard University Press.

Wrangham, R. W., Jones, J. H., Laden, G., Pilbeam, D., & Conklin-Brittain, N. L. 1999. The raw and the stolen: cooking and the ecology of human origins. *Curr. Anthropol.*, 40, 567–594.

Wrangham, R. W. & Peterson, D. 1996. *Demonic Males: Apes and the Origins of Human Violence.* Boston: Houghton Mifflin.

Wrangham, R. W. & Smuts, B. B. 1980. Sex differences in the behavioural ecology of chimpanzees in Gombe National Park, Tanzania. *J. Repro. Fert. Suppl.*, 28, 13–31.

18

Residence Groups Among Hunter-Gatherers: A View of the Claims and Evidence for Patrilocal Bands

Helen Perich Alvarez

The idea that patrilocal, exogamous bands are the simplest and oldest human social system was elaborated in influential detail in the early 1900s. Although challenges emerged, tallies of the residence rules among modern hunter-gatherers (Ember 1975, 1978) showing most to be patrilocal seemed convincing support. When evidence of female natal dispersal in chimpanzees and gorillas began to accumulate, the similarities between the male philopatry of African apes and the patrilocal bands of human hunter-gatherers stimulated influential hypotheses about phylogenetic features of the clade we share. On the grounds that features shared by living descendants were likely present in their common ancestor, researchers hypothesized that the ancestors of the African ape clade and all hominids were characterized by female natal dispersal, male philopatry, and defense of female ranges by male kin coalitions (Ghiglieri 1987, Wrangham 1987, Foley & Lee 1989, Rodseth et al. 1991). If true, this pattern places strong constraints on hypotheses about what happened in human evolution and the evolution of other hominoids.

Here I examine the origins of the patrilocal band model of human social organization, beginning with a review of the ideas and disputes within the field of social anthropology, followed by a brief review of the evidence from the conferences on hunting and gathering societies where participants described flexible patterns of group formation contingent upon individual age and circumstance, and varying with the seasonal round. These descriptions were challenged by Carol Ember (1978), who showed that 62% of the sample of hunting-gathering bands from the *Ethnographic Atlas* (Murdock 1967) was patrilocal. I review the ethnographies for the coding of the sample analyzed in Ember (1975) and find that, in clear contrast to her widely cited conclusion, bilocal residence patterns are most common. The conclusion that female dispersal and male kinship alliances are a species-specific trait for humans is not justified by the evidence. Instead, strategic flexibility, for both males and females, is clearly indicated among modern human foragers. Recent data also increasingly underline the strategic flexibility of other living primates, especially chimpanzees.

These patterns of flexible association suggest that, especially for species in which social relationships are so important to daily welfare, more emphasis should be given to hypotheses for explaining variability under specific circumstances, and less to characterizing species-typical sex-biased dispersal or species-typical mating systems. Work by Ryne Palombit (1994) recording multiple movements and shifting pair bonds in gibbons demonstrates how labels such as "monogamous pairs" and "nuclear families" obscure the dynamic nature of social relationships in gibbons. Unifying labels and simple assumptions often provide insight into complex processes, but when those labels are burdened by multiple unspoken associations, they direct our attention away from important elements of the behavior we seek to understand. Here I show that the concepts of patrilineage and patrilocality, as applied to the evolution of human sociality, have focused attention away from the actual, strategically varying behavior of individuals, and fostered erroneous ideas about the social organization of hunter-gatherers.

Ideas and Disputes in Social Anthropology

Since the 1930s, ideas about the social organization of human foragers have been influenced by an analysis of Australian Aboriginal kinship systems by A. R. Radcliffe-Brown (1930). Radcliffe-Brown conducted his investigations, termed "salvage ethnology" by both his defenders and his critics (Birdsell 1968), among the Aranda and Kariera through interviews with elderly men settled around missions and cattle stations who described life before colonial disruption. From descriptions and visits to their home territories, Radcliffe-Brown developed a model of social organization based on residential groups composed of the adult males of a patriline, their wives, and dependent children living on a territory ritually associated with the male members. At marriage, women left their natal families and joined the families of their husbands.

The Australian system described by Radcliffe-Brown was composed of closed, patrilineal, patrilocal groups occupying bounded, defended territories, a model that was generalized over hunting and gathering peoples across many different ecosystems. Julian Steward (1936), whose own work was distinguished by close attention to the links between ecology and social organization, adopted the patrilocal model in his band typology. Steward concluded that the patrilineal band, a politically autonomous, communally land owning, exogamous, patrilocal group, was the most common because it "is produced by recurring ecological and social factors" (Steward 1936: 331). Both Radcliffe-Brown and Steward explained the patrilocal organization of human societies with arguments for the importance of male provisioning, big game hunting, and the sexual division of labor.

Elman Service (1962) extended Steward's model over a worldwide sample of hunter-gatherers. But he rejected the proposition that patrilineages (groups defined by claims of common descent) were characteristic of band organization, proposing that they emerged when band societies became more complex tribal societies, usually with the adoption of agriculture. He agreed with Steward that patrilocality was a typical pattern based on the solidarity of male kin. Service thought that exceptions to the patrilocal type were due to disturbance by European contact. Service had it both ways; where patrilocal bands existed, they provided important information about basic human patterns. Where they did not, it was because of depopulation due to introduced diseases, wage labor, trapping contracts, and colonial occupation that pushed aboriginal people onto marginal lands, disturbing typical patterns of organization.

In these arguments, Service anticipated recent objections to the use of modern hunting and gathering groups as archetypes of prehistoric human adaptations (Schrire 1984, Wilmsen & Denbow 1990). All known groups had long histories of association with agricultural societies, early traders, and colonial occupation that had the effect of erasing or at the least changing characteristic subsistence practices and social structures. These arguments and the responses to them by social anthropologists (reviewed by Barnard 1983) have dominated the social anthropology of modern hunter-gatherers in recent decades, largely overshadowing the work of behavioral ecologists who use the opportunity provided by modern humans living on wild resources to explore constraints and opportunities faced by our ancestors (Hawkes & O'Connell 1992, Smith & Winterhalder 1992).

The formulations of Radcliffe-Brown, Steward, and Service were not in full agreement with each other, and none were universally adopted by their contemporaries. As noted above, the distribution and importance of descent groups was one issue in dispute. Cross-cultural variation in the presence and composition of descent groups has long been of interest to ethnologists. Service argued that they represented a level of organizational complexity not found in band societies, but when present they could be important in politics and economics, as described by the social anthropologists studying African farming and herding societies. Descent groups often hold corporate rights in land, leadership positions, and ritual property. Where important rights and duties are ascribed to patrilineal or matrilineal descent groups, residence is sometimes assumed to be unilocal as well, children then growing up in close proximity to others in their lineage.

But descent groups defined by explicit claims of common ancestry are not the same as residence groups. In a 1938 essay, Alfred Kroeber attempted to establish the principle that residence is as important as lineage in the structure of society. "Instead of considering the clan, moiety, totem, or formal unilateral group as primary in social structure and function, the present view conceives them as secondary and often unstable embroideries on the primary patterns of group residence and subsistence associations" (Kroeber 1938: 308).

George Peter Murdock, who became a central figure in the development of evidence from human societies, sided with Kroeber (Murdock 1949). Early in his career, Murdock had determined to establish a science of human behavior by compiling data on a variety of cultural traits from a review of published ethnographies. His goal was to compile information that could be used for cross-cultural, statistical comparisons of attributes. This work was the beginning of a coding effort by Murdock and his students that culminated in the publication of the *Ethnographic Atlas* (hereafter the atlas), an encyclopedia of coded attributes that initially included 862 societies (Murdock 1967).

Murdock compiled and coded evidence over many cultures and many traits in spite of disagreement among ethnographers and ethnologists over both the conceptual issues and the evidence. The focus on abstract descent groups, which were a prominent feature of the atlas subsequently used by many investigators to study the correlation between traits and social organization, had many critics within social anthropology. Edmund Leach, for example, argued that comparisons "must start from a concrete reality—a local group of people—

rather than from an abstract reality—such as the concept of lineage or the notion of kinship system" (Leach 1961: 104). However, it turns out that studying social organization from residence groups is no easier than inferring it from descent.

The Fischer-Goodenough debates are an especially well-published illustration of this problem. Ward Goodenough (1956) and J. L. Fischer (1958) worked on the island of Romomon in Truk three years apart, each making careful census counts of the households on the island, but coming to different conclusions over the results. Both counted a number of households where married couples lived with the husband's parents, the parents living matrilocally. Goodenough classified these as avunculocal on the grounds that the couples were living with the husband's matrilineage, but Fischer classified these as patrilocal because "I did not want to involve matrilineages in the definition of residence form" (Fischer 1958: 509). Fischer's interpretation was reasonable; the son was living with his father and no mother's brothers lived in the household, but he himself admitted "clearly this form of residence is different from the classical case of patrilocal residence, since the husband's father is in a dependent position in the residence group, while the husband's mother is the parent belonging to the family core" (Fischer 1958: 509).

Fischer recommended that residence be defined in terms of individuals instead of couples, that the residence of all individuals in the community be recorded, not just that of married couples, and that a person's residence be defined in terms of the time he entered the household, not in terms of the present. The latter injunction came from the realization that residency changes through the life span. Goodenough and Fischer agreed on this point. In his ethnography of the people of Lakali, Goodenough (1962) affirmed that decisions affecting residence stem from individual self-interest worked out within the rules of the society and the options available to each person in terms of sponsors, a view from a cultural anthropologist that human behavioral ecologists should find very congenial.

Murdock was well aware of the Goodenough-Fischer debate when he started the project of coding the atlas, but he deliberately set aside the issues they raised, arguing that the society is the unit of study: "We are not greatly concerned with distinguishing rules from practice in our definitions, since we assume that in *integrated societies*, for which our system is primarily designed, rules are ordinarily reasonably consistent with prevailing practice" (Murdock 1962: 117, emphasis added). But "prevailing practice" was hardly ever consistent with the rules, a problem that bothered social anthropologists who insisted on getting beyond the rules. John Barnes charged that an entire generation of anthropologists trained in the models of Radcliffe-Brown and the African patterns of kinship described from the Nuer (Evans-Pritchard 1940), the Tiv (Bohannon & Bohannon 1953), and the Tallensi (Fortes 1949) were unable to appreciate the evidence they found in Highland New Guinea where people talked of patriclans, but the villages he studied were full of "matrilateral kin, affines, refugees and casual visitors" (Barnes 1962: 5).

The African patrilineal, patrilocal organizations were exemplified by Evans-Prichard's description of the Nuer. Contrary to the models, these descriptions showed that the African systems were just as variable on the ground as were the patterns in the Highlands of New Guinea. Evans-Pritchard reported the composition of real Nuer villages, revealing the Nuer paradox: they are so patrilocal they need not always practice it. In spite of alleged patrilineal descent and patrilocal rules, the headman in any local community advanced his own power and prestige by attracting a wide variety of supporters related by patrilineal, matrilineal, and affinal ties under his sphere of influence, while his brothers and male cousins split off to further their own ambitions in other villages (Evans-Pritchard 1951, Kelley 1974). As with Goodenough's description cited above, Evans-Pritchard found that prevailing practice revealed individual strategists choosing among available alternatives to serve their own ends.

The International Conferences on Hunter-Gatherers

This pattern of opportunistic alliance formation in humans was evident to the ethnographers who assembled at the international conferences on band organization and current research among hunting and gathering societies initiated in Ottawa in 1965 (Damas 1969) and Chicago in 1966 (Lee & DeVore 1968). In these conferences, ethnographers reported patterns of flexible joining, leaving, and regrouping among the people they studied.

The presentations and discussions at the Chicago conference prompted Julian Steward to reexamine unpublished material from his field notes and those of Isabel Kelly (1932). Attending to his own quantitative evidence forced Steward to radically revise his ideas on band-level societies, a revision published in a festschrift for Sven S. Liljeblad and reprinted in 1977 (Steward 1977). The census data collected by Steward and Kelly confirmed the individual nature of residential choices. In Fish Lake Valley, Steward counted 24 marriages, over three generations, in which 10 of the men and 11 of the women came from sites 5 to 50 miles distant. In spite of an expressed preference for matrilocal residence in the Southern Paiute, Kelly's census data show 12 family clusters around brothers, 2 around sisters, and 11 around siblings of both sexes.

Steward was also influenced by the work and ideas of June Helm, who studied groups of Athabaskan people in the Northwest Territories of Canada. Helm tallied all the census data for the Mackenzie Dene in historical records from 1829 to 1977, showing that the population prior to the 1950s was "almost static" (Helm 2000: 210), thereby falsifying the hypotheses of Service (1962) and Williams (1974) that the Mackenzie Dene people lost patrilocal band organization as a consequence of depopulation during the European colonization of the region.

To characterize the actual social organization of the small groups she studied, Helm devised a method of quantifying links within the group based on Fischer's (1958) idea that a person is "sponsored" into a group through an affinal or kin link to an existing member. Examining kinship links in nine groups, she found that conjugal pairs often had multiple links to the group, and when all these links were tallied there were an equal number of male versus female ends (Helm 1965, 1969). The pattern of group formation through sponsorship was consistent over the time "slices," 1911 to 1956, examined by Helm; quantitative data demonstrated that the studied groups were organized on bilateral principles. In the larger group more than half the marriages were endogamous, but in the smallest groups available partners were limited, and exogamy was the practice. The child-to-parent kin links in these groups were biased toward daughters over sons (Helm 1965: table 1). Helm demonstrated that "for men, not only kin ties to father and brothers, but affinal links through wife and married sisters serve as the mechanisms of potential affiliation with one group or another" (Helm 2000: 12). Helm's evidence is a challenge to the Radcliffe-Brown, Service, and early Steward conception of human social organization structured on male kinship ties. Helm, who was an active participant at the conferences in Ottawa and Chicago, summarized the work of the presenting ethnographers.

Among contiguous groups in their habitual ranges, membership is fluid. Individuals and families move from one group to another. This fluidity is even more pronounced in respect to residential camp groups or task groups. Finally, with the exception of the local bands of the Birhor of northeastern India, no field ethnologist encountered the patrilocal band, of whatever size, either as an actual on-the-ground coresidential camp group or as the exclusive region-occupying or region-exploiting group. Specifically, the Australian field workers demonstrated that in actuality socioterritorial groups of any size are composed of men of *different* patriclans, whether in coassemblage as camp task groups or as scattered residents of a range. (Helm 2000: 12)

The Australian fieldworkers were especially interested in examining features of social life, since Radcliffe-Brown had claimed the patrilocal horde and all that entailed as a nearly universal feature of Australian social organization. Hiatt (1962) reviewed the evidence from many of these studies, reporting that Meggitt found in some Walbiri communities roughly half the marriages were endogamous, and that the composition and size of groups varied with the seasons as groups moved across the landscape over ranges that were identified with other hordes. In his own work, Hiatt encountered Gidjingali community groups that included men from as many as four to six neighboring land-owning units. No community group represented a closed unit comprised of patrilineally related males, their wives, and dependent children ranging over a bounded, owned territory.

Subsequent ethnographic studies among hunter-gatherers are consistent with these conclusions. The variability in social ties is confirmed by studies among the Pintupi, hunter-gatherers of the Western desert of Australia, who were one of the last aboriginal groups to encounter Europeans. They were first studied by Long (1971), who conducted population surveys of the Western desert between 1957 and 1964. When Long questioned the travelers he met, he found that groups were often living great distances from the "clan country" of the male members of the group; four of the six adult males in a group north of Lake Mackay were up to 160 or 250 km from their county and had been there for some period of time. Husbands and wives often traveled separately, and women often traveled with one or both parents. Long found no evidence of marked preference for family residence in the husband's country, nor that the so-called mother-in-law avoidance rule had the effect of separating married daughters from their parents.

Fred Myers (1986) conducted field studies in the same region from 1973 to 1984 after the Labor government had inaugurated a policy of Aboriginal self-determination, and many Aborigines left the government settlements and returned to the bush. Myers, who studied groups in the bush and those who remained in camps around the mission at Hermannsburg, reported that he found "domestic arrangements" little different in the two contexts. Because of the ecology of the Western desert, there had been little white encroachment into the country, but the Pintupi had been driven out of the bush to the mission by prolonged drought in the past. Myers reports that when they dispersed again in the late 1970s, a group of Pintupi who moved west met a group of nine relatives who had never encountered whites. Myers's accounts are consistent with those of Long; Pintupi groups were "aggregations of individuals based on complex bilateral ties" (Myers 1986: 453).

Numerous reports from many time periods and from all over the world describe patterns of flexible joining, leaving, and regrouping among hunter-gatherers; yet the assumption that patrilocal bands are the most basic human organization persisted in hypotheses about the phylogenetically primitive social organization of the hominids (Ghiglieri 1987, Wrangham

1987, Foley & Lee 1989, Rodseth et al. 1991). Investigators, drawing phylogenetic inferences from cross-species comparisons, were powerfully influenced by chimpanzees at Gombe and Mahale, where violent conflicts were observed between groups, with lethal aggression posing extreme danger to single males on the boundaries of their ranges. These initially surprising observations invited comparisons with human patterns of patrilocality and warfare by researchers who ignored the flexibility described by so many ethnographers, and relied instead upon evidence presented by Carol Ember (1975, 1978) that supported the view prevailing in ethnology before the hunter-gatherer conferences.

Ember's Counterarguments and the *Ethnographic Atlas*

In two essays, Ember (1975, 1978) sought to replace anecdotal, personal, ethnographic accounts with a comprehensive, statistical survey from a large number of societies based on selected attributes coded in the *Ethnographic Atlas* (Murdock 1967). She was concerned that the papers presented at the various conferences on hunting and gathering peoples were a biased sample of the larger universe, and particularly that Richard Lee's (1972) conclusion that most hunting and gathering societies were characterized by bilocal residence was biased by his failure to include proportionally as many samples from North America as he included from the rest of the world. Her tabulations from the atlas showed that most hunting, fishing, and gathering societies were patrilocal, exogamous bands. Her essays are widely cited in part because she replaced anecdotal information with statistical analyses, compiling variable data from a large number of sources into a few general categories that could be easily assessed.

Even though Murdock had designed the atlas for just such a purpose, there are many reasons why the material from the atlas should not replace evidence from the conferences. Murdock deliberately chose the earliest sources for coding the atlas in order to capture as much information as possible about precontact life. In many cases, the ethnographies cited by Murdock represent salvage ethnography, the investigator collecting information from a few surviving elders in an effort to record all possible aspects of societies that were in danger of extinction or had already ceased to exist in an aboriginal state. Many of these ethnographies are simple tabulations of traits, absent detail, discussion, or quantitative analyses. In contrast, the reports at the conferences came from ethnographers who lived and traveled with small groups who continued to gain a large portion of their daily subsistence from wild resources.

A Reexamination

I have reviewed nearly all the ethnographies cited by Murdock for the hunting and gathering societies included in Ember (1975). With a few notable exceptions, the ethnographies from which the atlas is coded are totally devoid of the actual number of cases for any described behavior. At their best, the early ethnographies are good descriptions of behavior observed by ethnographers who spent time living with their informants in the bush where the people hunted and gathered (Turney-High 1941, Hart & Pilling 1960, Turnbull 1961, Henry 1964, Marshall 1976, Holmberg 1985). Others contain simple declarative statements with no supporting material. In salvage ethnographies assembled from the recollections of one or a few informants, nothing was measured. There are no numbers of animals caught, roots gathered, berries eaten, fish dried, or seeds collected. There are no genealogies, no counts of individuals in camps, nor their relationships to each other. In 48 of these references, I found only five instances where camp sites were diagrammed, residence types counted, and genealogical links between the individuals in camp recorded (Schebesta 1929, Driver 1936, Hill 1956, Dunning 1959, Turnbull 1965).

In the atlas, residence patterns are coded in column 16. Murdock's 1949 definition of bilocal residence focused on a flexible alternative governed by existing conditions, but for the atlas Murdock substituted the term "ambilocal," designated B, and added the notion of proportion in the society.

> Ambilocal, i.e., residence established optionally with or near the parents of either the husband or the wife, depending upon circumstances or personal choice, where neither alternative exceeds the other in actual frequency by a ratio greater than two to one. If the differential frequency is greater than this, the symbols Uv or Vu are used to denote, respectively, a marked preponderance of uxorilocal or virilocal practice.
>
> Lowercase letters following a capital indicate either culturally patterned alternatives to, or numerically significant deviations from, the prevailing profile. Lowercase letters preceding a capital indicate the existence of a different rule or profile for the first years or so of marriage, e.g., uP for initial uxorilocal residence followed by permanent patrilocal residence. (Murdock 1967: 48)

Murdock substituted U and V for M and P respectively for societies where matrikin and patrikin are "not aggregated" in matrilocal and matrilineal or patrilocal and patrilineal kin groups. He specifically defines uxorilocal: "Equivalent to 'matrilocal' but confined to instances where the wife's matrikin are not aggregated in matrilocal and matrilineal kin groups" (Murdock 1967: 48). The precision of Murdock's definitions, which may be accurate for many of the agricultural societies, belies the fact that numerical censuses are, with few exceptions, totally absent from the early ethnographies of hunting and gathering societies included in the atlas and tallied by Ember (1975). Reading the ethnographies, it is nearly impossible to determine how the individual coding the data was to decide the difference.

Yet, using data weighted by the theoretical problems that worried many investigators and the methodological problems detailed above, Ember concluded: "patrilocal residence was the most typical form of residence among recent hunter-gatherers" (Ember 1978: 443). In this essay, she used the residence codes for all the societies in the atlas coded zero for agriculture and zero for herding. To examine the actual patterns coded in the atlas, I gathered data from the ethnographies for the more limited sample of hunter-gatherers listed in the 1975 essay (Ember 1975). In that essay, Ember selected from the atlas those societies also represented in the Human Relations Area Files, giving a sample size of 36, to which she added 14 cases randomly selected from the atlas. Since she was interested in comparing variables correlated with patrilocality versus matrilocality, she excluded all societies coded as "being prevailingly or alternatively avunculocal, duolocal or neolocal (capital or small A, C, D, N, O)" (Ember 1975: 200).

For her sample, I went to the original sources cited by Murdock to determine the basis for the residential coding. The interested reader can find the original citations for each reference by going to the specific volume and page of *Ethnology* cited for each society in Murdock (1967: 8–45). In table 18.1, I give Murdock's ID for the group, the name, the

Table 18.1. Societies Classified in Ember 1975[a]

ID[b]	Name[c]	Source[d]	Quotations from the Citations[e]	M[f]	E[g]	A[h]
Id2	Murngin	Warner 1937	"All sisters and daughters, since the rule of marriage is patrilocal after the first year or two, go to their husband's clan to live and rear their families," p. 19.	uP	P	P
Nd4	Sanpoil[i]	Ray, V. F. 1933	"Residence was patrilocal," p. 124.	V	P	P
Nd43	White Knife	Harris 1940	"Residence was generally matrilocal, but certain visible factors, such as personal desire, size of family, circumstances of food, or the personality of the mother-in-law disrupted this general form, so that even membership in the camp group was unstable, and no winter community was a lineage group," p. 47.	U	P	Mn
Sg1	Yaghan	Gusinde 1937	"They begin their independent life, free of their parents," p. 428. "The individual family consisting of parents and children is basically independent and a compact union," p. 387.	V	P	N[j]
Na9	Aleut	Jochelson 1933	"Amix, the brother of a woman, was the educator of her sons. They lived in his house and worked for him. The reason was that a woman's children are more clearly related to her brother than to her husband," p. 71.	uV	P	M
Nd14	Coeur d'Alêne	Teit 1930	"A family might winter at the father's or husband's village, or again at the mother's or wife's," p. 150. "Often they wintered with one set of relatives and summered with another," p. 151. "The woman generally followed her husband and lived among his people," p. 161.	V	P	B
Ne4	Crow	Lowie 1912, 1935	"A young couple were not obliged to settle either with the bride's or the groom's parents but at the beginning of wedlock patrilocal residence was usual," p. 57. "Subsequently, independent households were apparently the rule," p. 58.	V	P	N
Aa2	Dorobo	No references in Murdock	"A further development is the extended patrilocal family," Huntingford 1951: 22. Cited from the HRAF.	P	P	P
Nb16	Klallam	Gunther 1927	"Patrilocal residence is the general rule. A couple frequently visits the village of the wife's parents, especially when the visits can be arranged as part of fishing or root digging expeditions. Sometimes a newly married couple stays with the bride's family for a short time," p. 244. "The mother of this bride presents one of the few cases of matrilocal residence which I have recorded," p. 245.	V	P	B

Na41	Micmac	Denys 1908, NA; information from HRAF, Wallis & Wallis	"Each boy takes his wife to his wigwam. He must live in his own settlement; the chief would not consent to his quitting it permanently. Some informants, however asserted that at marriage the boy went to live in the *udan*, or settlement, of the girl. In the rare cases of extra tribal marriage the man joined his wife's tribe," p. 239.	uV	P	Pm
Na32	Montagnais	Burgesse 1944	"This reason, however, would seem to conflict with the custom whereby a son joins the family of his father-in-law. When this was pointed out to Germain he said that there is no fixed rule which demands that sons join the family of their father-in-law and that those who did so, acted from choice and because the young man had no hunting grounds of his own," p. 4	V	P	B
Na28	Sekani	Jenness 1937	"He could then take his wife wherever he wished. Generally he returned to his father and kinsmen, but occasionally he remained with his wife's parents for a year or two longer, though on a more equal basis," p 54.	uV	P	B
Ej3	Semang	All quotes from Evans 1937. See text for details from Schebesta 1937.	"The woman is assisted by the midwife and by her female relatives, there being midwives in every camp," p. 246. "A bachelor takes a wife from another camp of the tribe and brings her home with him, but the couple visit the girl's people after a while and stop with them for some time, and other visits are paid subsequently," p. 250. "Among the Jehai it is customary for the son-in-law, after marriage, to stay for about two years near his father-in-law and work for him. After this he returns to his father's camp to live, but visits his father-in-law at intervals," p. 254.	uV	P	B
Nd2	S. Ute	Opler 1940	"Despite the looseness of these rules governing marital residence, the first household of the married couple was usually matrilocal," p. 128. "Sometimes it was mutually agreed by the families of the marrying couple that residence should be patrilocal, especially if the boy's parents needed his help or his wife's. When the affinal relatives on both sides of the newly formed family made no strong bid for aid, then the young couple was perfectly free to fend for itself or to visit back and forth between the two sets of parents-in-law," p. 129.	Uv	P	B

(continued)

429

Table 18.1. Continued

ID[b]	Name[c]	Source[d]	Quotations from the Citations[e]	M[f]	E[g]	A[h]
Nb13	Squamish	Barnett 1955	"Residence after marriage was patrilocal, but sometimes desirable young men were taken into the households of their fathers-in-law." "He remained in his father-in-law's household for life unless he found it possible to set up a ménage of his own. The residence of such men in the houses of their fathers-in-law was accepted as natural." "It was common for a dissatisfied individual to shift residence from his father's people to his mother's," p. 193.	V	P	Pm
Sg4	Tehuelche	Cooper 1946	"Residence was generally patrilocal," p. 149.	V	P	P
Nc3	Yokuts	Driver 1937, Gayton 1948	"Without census data there is no way of determining how correct generalizations of informants are. Almost all informants admitted many exceptions to these rules. However, the fact that tribal, moietal, and lineage affiliation west of the Sierras were all three patrilineal supports the patrilocal bias of this area." "Kroeber, Hdbk. 493, states generically that the Yokuts were matrilocal," Driver p. 128. "A young man went to a girl's house and remained there if she accepted him. For about a year he hunted and helped his father-in-law. Then the couple removed to his parents' home where, if there was room, they remained permanently," Gayton p. 30.	uP	P	P
Ec7	Ainu	Sugiura 1962	"Although the contemporary Ainu family tends to be extended, in aboriginal times the nuclear family seems to have been the preponderant pattern, with occasional coresidence of a widowed kinsman," p. 287. "At marriage a man left his family of orientation and established a separate household near his patrikin, and his wife came to live with him," p. 288.	Vu	Pm	N
Nc12	Maidu[k]	Riddell 1978	"Before residing permanently in the husband's village, the married couple lived for a time with the bride's family, and the new husband rendered service to them by providing food," p. 380.	bVu	Pm	P
Nb11	Nooka	Drucker 1951	Women traveled to the home of their husbands after marriage but there was another type of marriage. "In certain cases a marriage with matrilocal residence (mawiʼitph) was arranged. The young man's family brought him to the bride's house, gave presents to all her people, and announced they were 'giving him to her father.' Then he stayed there permanently. This was a perfectly proper arrangement which carried no stigma," p. 299.	Vu	Pm	Pm

Id3	Tiwi	Hart & Pilling 1960	"There has been a prolonged argument in the anthropological journals about the Australian territorial unit, especially as to whether it was patrilineal or matrilineal. For the Tiwi, such a problem could not arise. A father bestowed his daughters where he wished and at puberty they joined their husbands. Where his sons found wives was no concern of the father, and hence where they established their households was of no interest to him either. The father would wish, however, that they establish their households as far away from his as possible since then he would not have to worry about them interfering with his young wives," p. 32.	Vu	Pm	?
Nc2	Tubatulabal	Smith 1978	"Each household comprised a single, biological, bilateral family, neither markedly paternal nor maternal," Voegelin 1938: 43, cited in Smith. "The family also included any widowed parents, and depending upon the marriage form, sons-in-law or, less frequently, daughters-in-law," Smith p. 439.	Vu	Pm	B
Nb4	Yurok	Kroeber 1925	"Half marriage was not rare. The bridegroom paid what he could and worked out a reasonable balance in services to his father-in-law. Of course he lived in the old man's house and was dependent on him for some years, whereas the full-married man took his wife home at once—in fact had her brought to him," p. 29.	Vu	Pm	Pm
Na24	Angmagsalik	NA		Vu	Pm	
Nb9	Bellacoola	McIlwraith 1948	"Y usually lives in the house of X's father, though economic circumstances may compel her husband to live in her home," p. 399.	Vu	Pm	Pm
Ne1	Gros Ventre	Flannery 1953	"There was no hard and fast rule in regard to residence. Matrilocal residence in the same lodge was incompatible with the observance of the mother-in-law taboo and it was said that if a man wished his son-in-law to remain with him, he provided the separate lodge and all the equipment. The tendency seems to have been toward patrilocal residence in general, although it was not unusual for a couple to move back and forth," p. 176.	Vu	Pm	B
Nd7	Kutenai	Turney-High 1941	"It has been mentioned that the Kutenai family is bilateral. The bilaterality consisted in an attempt to count the kinship on both sides of the family. It was also weakly patrilinear inasmuch as a person was primarily a member of his father's line." "Despite the bilateral character of Kutenai kin counting and actual functioning, it was the father's side which carried the emotional content. In spite of this, the family was also definitely matrilocal," p. 134.	uVu	Pm	B

(continued)

431

Table 18.1. Continued

ID[b]	Name[c]	Source[d]	Quotations from the Citations[e]	M[f]	E[g]	A[h]
Na34	Pekangekum	Durning 1959	"In all past cases, however, informants claimed that no formal rule of residence after marriage was recognized. It seems to have been a matter of arrangement between the two families or between the prospective bridegroom and his affines," p. 810. See p. 811 for an actual census of winter and summer residence, in winter approximately 2/3 virilocal and 1/5 uxorilocal, but in summer approximately 1/2 virilocal, 1/6 uxorilocal, and 1/5 alternating.	uPu	Pm	Pm
Eh1	Andamanese	Radcliffe-Brown 1948	"If either a man or a woman lives in a local group other than that of his or her parents, he or she pays frequent visits to them," p. 79.	B	B	
Nc18	E. Pomo	Loeb 1926	"The married couple kept moving from one family to the other, but when a child was expected they always went to live with the wife's family." "Where the wife and husband each came from a different village, Gifford claims that statistics show the immediate residence of the couple to have been 75% matrilocal," p. 279.	B	B	B
Nd5	Hukundika	Hoebel 1939, Steward 1943	"Shoshone residence was bilocal in extended family households or groupings within the band," Hoebel p. 446. "Residence was usually matrilocal for a year or so, which was more or less a period of bride service. After that, it was independent, the location being determined by circumstances of subsistence and relationship," Steward p. 279.	B	B	B
Aa1	!Kung	Marshall 1965	"When the daughters of a !Kung couple marry, and the daughters' husbands come to live with their brides' parents to give them bride service, the unit becomes an extended family. In a second extended phase, the married sons may have returned from bride service with their wives and offspring. In this phase the married daughters with their husbands and offspring may stay or they may go to live with their husband's people," p. 258. From the HRAF, code 591.	uB	B	B
Aa5	Mbuti	Turnbull 1956	Murdock 1962: 393 reports that the entry in column 16 follows Schebesta; Czekanowski reports uxorilocal. Turnbull reports that women leave their band and join the bands of their husbands. However, in a footnote he adds, "In my own experience, among both archers and net hunters, patrilineal descent systems and patrilocality are no more than trends, most bands being effectively cognatic territorial units," p. 176.	B	B	?

432

Nd52	Shivwits	Lowie 1924	"It was customary for the young couple to stay with the bride's parents for a long time, then they would live with the husband's family; nowadays this custom is not followed and the couple settle wherever they please," p. 275.	uB	B	Mp
Nc20	Wappo	Driver 1936	"Summary: 10 instances are patrilocal, 9 matrilocal, 4 are neutral. This evidence, combined with that above, indicates that no rule of residence existed," p. 205.	B	B	B
Nd6	Washo	Freed 1976	"Washo residence patterns are difficult to characterize since there is no residence rule which must be strictly followed," p. 360.	uBn	B	B
Nd66	Yavapai	Gifford 1936	"After marriage, couple lived with bride's parents for 6 months or year. Husband hunted and performed man's work. Then he took wife to live with his parents for 6 months or year. Thereafter couple could move around by themselves," p. 296.	uBn	B	N
Sj3	Aweikoma	Henry 1941	"For there are any number of men with whom an individual may join forces, and if he hunts with his father today he may go away with his wife's father tomorrow and with his sister's husband the day after. Nothing compels him to fix his allegiance," p. 37.	B	B	B
Nb33	Chimariko	Driver 1937	In the cultural element list (p. 346) all categories under Postnuptial Residence are coded absent except First Residence variable and Final (permanent) Residence variable.	B	B	B
Ne3	Comanche	Hoebel 1939, Gladwin 1948	"Comanche residence tended to be more patrilocal, but equally subject to easy change," Hoebel p. 446. "Thus the comparatively frequent matrilocal residence is reflected in the fact that it is the maternal grandparents who are considered the closest," Gladwin p. 90.	B	B	B
Nd24	Kidutokado	Kelly 1932	"Initial matrilocal residence was the rule." "Independent residence seems to have been established after the birth of one or two children. Patrilocal residence occasionally obtained if the youth were an only child," p. 166.	uB	B	B
Na12	Nunamiut	Pospisil & Laughlin 1963	"A newly married couple was expected to establish residence, at least for a while, with the bride's family. Sometimes this arrangement endured for life, but many couples adopted neolocal residence after a time or, exceptionally, joined the husband's aging parents," p. 181.	uB	B	Mn
Nh1	Chiricahua	Opler 1941	"This emphasis upon matrilocal residence calls attention to one of the fundamental themes of the culture, for the organization of the economy assumes the presence and the closest cooperation of the sons-in-law," p. 162.	Uv	Mp	M

(continued)

Table 18.1. Continued

ID[b]	Name[c]	Source[d]	Quotations from the Citations[e]	M[f]	E[g]	A[h]
Sh1	Mataco	Karsten 1932	"Closely connected with this idea is the custom prevailing among most matriarchal peoples, that when a man marries, he stays with his wife's family, becoming a member of her clan," p. 49.	uUv	Mp	M
Ec6	Yukaghir	Jochelson 1910, 1926 NA		uUv	Mp	
Na29	Beaver	Goddard 1916	"The Young man seems invariably to have made his home with the bride's parents," p. 221.	U	M	M
Si1	Bororo	Oberg 1953	"After marriage a young man lives with his wife's family until his father-in-law dies. He then builds a house near his own brothers," p. 108.	M	M	Mn
Sh5	Choroti	Karsten 1932	"That when a man marries, he stays with his wife's family, becoming a member of her clan," p. 49.	U	M	M
Se1	Siriono	Holmberg 1985	"Because of matrilocal residence a woman is able to avoid most direct contacts with her parents-in-law, but a man, while in the house, is almost constantly thrown into contact with his parent-in-law by virtue of the fact that his (and his wife's) hammock hangs not three feet from theirs, with nothing more than a few embers of fire to separate them," p. 143.	U	M	M
Sc1	Warrau, synonym Warao	Wilbert 1958	"The extended family, which usually lives under one roof, typically contains all the women in a direct line of descent, as well as their spouses and unmarried children," p. 6.	U	M	M

[a]The order of societies follows Ember (1975).

[b]ID: the identification number from the *Ethnographic Atlas* (Murdock 1967).

[c]Name: the name of the society as listed in the atlas.

[d]Source: the ethnographer cited by Murdock.

[e]Quotations from the citations: with page numbers from the ethnographies cited by Murdock. The full citation and date of publication of the original can be found in *Ethnology*, volume and page for each society cited in the *Ethnographic Atlas* (Murdock 1967: 8–45). NA indicates that the source quoted by Murdock was not available.

[f]M from column 16 of the *Ethnographic Atlas* (Murdock 1967).

[g]E from Ember (1975).

[h] A (Alvarez): my coding from the descriptions.

[i]Sanpoil: Murdock cited a paper by Ray that was not available. I used Ray (1933).

[j]N (Neolocal): i.e., "normal residence apart from the relatives of both spouses or at a place not determined by the kin ties of either" (Murdock 1967: 48). With the exception of N, the codes in the last three columns are defined in the text.

[k]Murdock's cited sources not available; Riddell (1978) quotes from those sources.

quoted source I consulted, and direct quotations from the ethnography. In these descriptions, I have tried to extract the relevant information, including both quotes that corroborate the atlas coding and additional material that illustrates a more complex pattern. Where the description is very brief, it represents all the information I found in the cited source. In the last three columns, I give the codes from the atlas, Column 16: Marital Residence; the classification from Ember, who dispensed with Murdock's patrilocal/virilocal and matrilocal/uxorilocal distinctions; and my own coding from the descriptions (column A). Ember ignored Murdock's attempt to distinguish between lineage and residence; instead she collapsed the atlas coding into five residential types: Patrilocal (P), Patrilocal with a Matrilocal Alternative (Pm), Bilocal (B), Matrilocal with a Patrilocal Alternative (Mp), and Matrilocal (M).

Given the absence of quantitative information for most of the societies, I assumed that any indication of choice of residence with either the husband's parents or the wife's parents represented bilocality. If the ethnographer specifically mentioned that the first years of marriage were matrilocal followed by patrilocality, I coded the society patrilocal.

For the 17 societies that Ember (1975) classified as P, I find unequivocal evidence for patrilocality in only 6: Murngin, Sanpoil, Dorobo, Tehuelche, Yokuts, and Maidu. Ember's designation of the White Knife as patrilocal is a coding error. In her appendix column 3: Postmarital Residence (atlas column 16) she coded the White Knife V (Ember 1975: 224), but in Murdock (1967: 110, column 16) they are coded Uv. I code the White Knife Mn where n represents the option of neolocal residence, the Yaghan and Crow N, the Aleut M, the Micmac and Squamish PM, and the Coeur de Alêne, Klallam, Montaganais, Sekani, Semang, and Southern Ute B.

Ember coded 11 societies Pm, Patrilocal with a Matrilocal Alternative. As noted above, I add the Micmac and Squamish to this category. I agree with her on the Nootka, Yurok, Bellacoola, and Pekangekum, but change the Ainu to N, the Maidu to P, and the Tubatulabal, Gros Ventre, and Kutenai to B. I do not code the Tiwi for reasons quoted from the ethnographers, table 18.1, and the extended discussion below.

In the coding of bilocality, I agree with Ember for 10 of the societies she coded as B, add the 6 that she coded P, and 6 that she coded Pm. In addition, I change the Shivwits from B to Mp, the Yavapai from B to N, and the Nunamiut from B to Mn. I do not code the Mbuti because of the confusion reflected in the quotes from Turnbull.

For Ember's last two categories, Matrilocal with a Patrilocal Alternative and Matrilocal, I find no distinction in the ethnographies. To the five societies Ember coded as M, I add the Aleut, Chiricahua, and Mataco. I change the code for Bororo from M to Mn.

In table 18.2, I summarize the results of the three coding schemes: the *Ethnographic Atlas*, Ember, and my own try from the descriptions in table 18.1. My sample contains only 48 societies, as I did not find the original ethnographies for the Angmagsalik and the Yukaghir.

If the cases Ember coded as more frequently patrilocal (Pm) are added to those coded patrilocal only (P), then her tally shows 56% (28/50) of cases are patrilocal, making that the most frequent pattern. If her Pm cases are counted as bilocal, then with her coding patrilocality drops to 34% (17/50); but her coding is not supported by the original sources. When these cases are recoded from the ethnographic descriptions, only 12.5% (6/48) of them are patrilocal. If Pm and Mp in table 18.2 are collapsed into B, the bilocal tally is 58% (29/50) for the atlas, 56% (28/50) for Ember, and 54% (26/48) for my coding. Bilocality

Table 18.2. Summary of Marital Residence in 50 Societies[a]

	P	Pm	B	Mp	M	N	?	Sum
Murdock[b]	15	8	17	4	6	0		50
Ember	17	11	14	3	5	0		50
Alvarez[c]	6	6	19	1	10	4	2	48

[a]Summary of categories from table 18.1. Symbols are defined in the text.

[b]Murdock (1967) codes from the atlas are combined as follows—P: all P, uP, V, uV; Pm: all Vu; B: all bVu, uVu, uPu, uB, uBn; Mp: all Uv, uUv, M: all U.

[c]Mn was tallied with M.

is the most frequent pattern in the collapsed table as it was in the descriptions from the conferences.

The differences between my coding scheme, the atlas scheme, and Ember's interpretation illustrate the difficulty of coding from qualitative data. Only the Pekangekum, where Dunning (1959) took a census of summer and winter camps, are simple to code from the cited reference.

Yet, qualitative data do provide a wealth of information about human social organization if many sources can be combined. Schebesta (1929) provides a description of the Semang that gives a sense of the multiple options open to individuals. He diagrams a camp of 15 huts, 12 headed by males and 3 by females. Two of the senior males have sons and daughters-in-law in the camp. Each man also has a daughter; one widowed, one living with her husband and children in the hut next to her father, mother, and the young daughter of her deceased older sister. The widow lives next to her father's hut and near the hut of her daughter and son-in-law. The son-in-law has no relatives in the camp except his sister, who is married to the son of a senior male. A second brother-sister pair lives in the camp, the wife married in, but the brother having no other relatives in the camp. Individuals that came from some distance occupy three of the huts in this camp. Schebesta notes that one of the senior males and his sons are Sabubn who are living in the camp by virtue of marriage to Jehai women. The details of this encampment show how difficult it is to classify residence in hunter-gatherers.

The Tiwi are another example of the problem of classifying. The Tiwi have been isolated on Melville and Bathurst islands off the northwestern coast of Australia probably since the first occupation of the islands. They were studied by Hart in 1928 and 1929 and by Pilling in 1953 and 1954. The two, without previous collaboration, met in 1957 to begin work on the book published in 1960. Through their collaboration, they were surprised at how little they differed on their understanding of the Tiwi, in spite of studying them decades apart (Hart & Pilling 1960).

As indicated by the short quote included in table 18.1 and a more extensive reading of the ethnography, Ember's designation of the Tiwi as patrilocal is entirely unwarranted. I quote extensively from the ethnographers to illustrate the complex behavior that is lost when a richly textured description from a good ethnography is condensed in a classification system derived from the classic categories of social anthropology. In this case, the behavior that is lost in the coding bears on the question of male alliances in our lineage. Quite unlike the hypothesized system of alliances based on male kinship (Ghiglieri 1987, Wrangham

1987, Foley & Lee 1989, Rodseth et al. 1991), male alliances among the Tiwi are based in the rights of bestowal of women, sometimes biological daughters but more often unrelated females. The male alliances based on rights and obligations in women may include male kin, but they necessarily cut across clan lines.

Three features distinguish the Tiwi social system: the right of naming and bestowing women in marriage resides in the mother's husband; every woman must be married; and the males of a matriline have no rights in the bestowal of the women of their lineage, although they may have some influence in the remarriage of their elderly mothers. Infant daughters are betrothed at birth, and widows, including the elderly, are remarried as soon as their husbands are buried. Since women are betrothed at birth to powerful men, they are sent at puberty to join the households of men at least thirty years older, and thus are apt to be widowed several times through their lives. Each time a woman is widowed and remarried, a new "father" acquires the right to rename and bestow her widowed daughters. Since girls are "married" at birth, women may be both widowed and "father orphaned" before they reach the age of puberty, becoming available for bestowal again by a replacement father. This likelihood makes an elderly widow a valuable asset for men between the ages of 32 to 37 attempting to begin the political career which is their only avenue to acquisition of a young wife by the age of 40. Prevented from marrying a young woman by obligations and relationships among elderly men, a young man marrying an elderly widow begins his own network of alliances with other men by virtue of his "power" to bestow the widow's daughters.

> To get a start in life as a household head and thus to get his foot on the first rung of the prestige ladder, a Tiwi man in his thirties has first of all to get himself married to an elderly widow, preferably one with married daughters. This is the beginning of his career as a responsible adult. The widow does several things for him. She becomes his food provider and housekeeper. She serves as a link to ally him with her sons. As her husband, he acquires some rights in the future remarriages of her daughters when they became widowed. And she, as the first resident wife in his household, stands ready to be the teacher, trainer, and guardian of his young bestowed wives when they begin to join him after they reach puberty (Hart & Pilling 1960: 25).

I include this lengthy discussion of the Tiwi not as a general model, but as an example of the capacity of both males and females to negotiate complex social relationships that further self-interest in a variety of contexts quite different from our notions of lineage organization, and to illustrate the tension between observance of the rules of the society and the promotion of self-interest.

> Every Tiwi male tried to beat the system, especially as he became older and more influential, by intriguing in the remarriages of his mother and elderly sisters—matters in which, according to the strict letter of the law, he had no right to interfere, since bestowal rights resided with the "father" or "fathers" of these women (Hart & Pilling 1960: 24).

In these manipulations, Tiwi men are very similar to the Yanomamo, where males "create" wives by "moving" females from one kin category to another to increase their number of eligible female cross cousins (Chagnon 1988). According to Chagnon, ethnographers focusing on structural systems notice that individuals often fudge genealogies to their own benefit, but the observers sweep these deviations from the rules under the carpet. From his own extensive studies, he concluded: "what's under the carpet *is* the system" (Chagnon

1988: 25, emphasis original). The idea that our phylogenetic heritage is characterized by individual initiative over adherence to the rules of the society is the important point missed by those who define the phylogenetic origins of human social organization from the evidence in Ember's (1975, 1978) tallies.

Ember has also been criticized for including cases with marked sociopolitical inequalities (Knauft 1991). Knauft argues that this practice obscures characteristics of simple hunter-gatherers by including cases that are more like farming and herding societies, in which inequities based on property can readily develop. He abstracted his own sample from the 563 represented in *Atlas of World Cultures* (Murdock 1981), eliminating those societies with agriculture, "significant animal domestication," primary reliance on fishing, and "significant sociopolitical class distinctions," leaving 39 societies.

> Of the resulting 39 societies, 71.8% (28/39) were rated by Murdock as having an "absence of patrilineal kin groups and also of patrilineal exogamy" (0 in column 20). In contrast, 59.0% had cognatic kin groups recognizing ambilineal or bilateral descent (B or K in column 24). Only 25.6% (10/39) were rated as exhibiting patrilocal residence (P in column 16), the remainder being characterized as practicing various combinations of amibilocal and uxorilocal residence or virilocal residence "confined to instances where the husband's patrikin are not aggregated in patrilocal and patrilineal kin groups." These trends would probably have been still more pronounced if cases based on somewhat questionable older data had been excluded and if a less rigid typology of residential types had been used. These findings are consistent with ethnographic suggestions above that social organization and residence tend to be shifting, open, and flexible among nonintensive foragers. Available evidence thus supports the conclusion that fraternal interest groups and the kind of violence associated with them are seldom of importance in simple human societies (Knauft 1991: 405).

Knauft also notes a problem with including the Australians in the sample. Contrary to the evidence from the conferences and the work of the Australian investigators cited above, in the older ethnographies, influenced by Radcliffe-Brown's interpretations, these societies are designated patrilocal. Knauft finds that if the Australian cases are excluded from his sample, "the percentage *lacking* patrilocal residence increases to 87.1% (27/31)" (Knauft 1991: 406, emphasis added).

Conclusion

The assumptions of patrilocality derived from the chimpanzee models and the Radcliffe-Brown/Service/Ember emphasis on patrilocality places strong constraints on development and evaluation of alternative hypotheses about human and ape evolution. The most influential hypotheses about human evolution have been drawn around a suite of traits that is assumed to accompany patrilocality. Among these, Ghiglieri (1987) nominates as characteristic of human social structure: (1) females disperse, (2) bonds between females are weak and interfemale cooperation limited, (3) females do not mate promiscuously, (4) male kin cooperate in communal reproductive strategies, (5) male groups are stable and semiclosed with mating competition more intense between than within groups, and finally (6) extensive male provisioning from hunting.

But promising alternative hypotheses have also been suggested. One of these, the grandmother hypothesis, which explains unique features of human life histories, assumes that females of childbearing age reside near older female kin who help provision weaned offspring, thereby giving humans a higher birth rate than expected for our age of maturity. Selection can favor longevity past childbearing age because these long-lived females gain reproductive benefits through younger kin (Hawkes et al. 1998, Alvarez 2000). Elements of the grandmother hypothesis can be tested from the fossil evidence (O'Connell et al. 1999, Hawkes et al. 2003), but social organization can only be inferred from modern behavior, comparative studies, and genetic analyses. If the assumption that female natal dispersal and male philopatry characterized all hominids is correct, then hypotheses assigning a central role in human evolution to cooperation between aging mothers and maturing daughters cannot be entertained. We now know that contemporary chimpanzee females may stay with their mothers when the benefits for doing so outweigh the costs. There are no grounds for assuming less capacity for strategic adjustments among ancestral hominids.

The phylogenetic models were consistent with the chimpanzee evidence at the time, but are undercut by newer studies (Mitani et al. 2002). The evidence, developed by de Waal (1982, 1996) from Arnhem and other captive communities and van Schaik (1999) in his reappraisal of orangutan sociality, is testimony to the capacity of apes to build social alliances in novel circumstances. The capacity for flexible sociality casts doubt on the idea that ape communities are structured on male kinship ties. The authors of two studies of kinship in wild chimpanzee communities using mitochondrial DNA sequences come to the same conclusion. "The results suggest that kin selection is weaker than previously thought as a force promoting intra-community affiliation in chimpanzees" (Goldberg & Wrangham 1997: 559). "These findings add to a growing body of empirical evidence that suggest kinship plays an ancillary role in structuring patterns of wild chimpanzee behavior within social groups" (Mitani et al. 2000: 885).

The assumption that female natal dispersal and male philopatry is a conserved trait in our lineage is further challenged by microsatellite genotyping showing that males in the Taï community are no more closely related than females (Vigilant et al. 2001), that males can leave their natal groups and immigrate successfully into other groups (Sugiyama 1999), and that young females often breed in their natal groups (Pusey & Packer 1997, Constable et al. 2001), a suggestion tentatively made by Harcourt et al. (1981) for female gorillas years before DNA analyses became a tool for hypothesis testing. Instead of conforming to one pattern of social affiliation, it appears that individuals, both male and female, develop and maintain social ties that promote self-interest. For a chimpanzee female, the decision to leave or stay may depend more upon the ranking of her mother, the quality of the range, and the number of males in the community than upon the danger of inbreeding (Pusey et al. 1997, Hrdy 2000).

While data showing that chimpanzees are much more flexible than previously thought are only now appearing, evidence against the patrilocal band model from human foragers has been available for decades. The ethnographic record points to important characteristics of human sociality: (1) residence is most often bilocal with individuals linked to groups through both female and male ties, (2) residential choices vary by circumstance and age, (3) men and women negotiate social networks widely, (4) male cooperation cuts across ties of biological kinship, and (5) foraging groups are of variable size depending on the season and resources targeted. Evolutionary hypotheses that are consistent with these characteristics should have our attention.

Acknowledgments I thank Bernard Chapais, Carol Berman, Kristen Hawkes, and an outside reviewer for suggestions that improved this manuscript.

References

Alvarez, H. P. 2000. Grandmother hypothesis and primate life histories. *Am. J. Phys. Anthropol.*, 113, 435–450.

Barnard, A. 1983. Contemporary hunter-gatherers: current theoretical issues in ecology and social organization. *Annu. Rev. Anthropol.*, 12, 193–214.

Barnes, J. A. 1962. African models in the New Guinea Highlands. *Man*, 62, 5–9.

Birdsell, J. B. 1968. The magic numbers "25" and "500": determination of group size in modern and Pleistocene hunters. In: *Man the Hunter* (Ed. by R. B. Lee & I. DeVore), pp. 245–248. Chicago: Aldine.

Bohannon, L. & Bohannon, P. 1953. *The Tiv of Central Nigeria*. London: International African Institute.

Chagnon, N. A. 1988. Male Yanomano manipulations of kinship classifications of female kin for reproductive advantage. In: *Human Reproductive Behaviour: A Darwinian Perspective* (Ed. by L. Betsig, M. Bogerhoff Mulder, & P. Turke), pp. 23–48. Cambridge: Cambridge University Press.

Constable, J., Ashley, M., Goodall, J., & Pusey, A. 2001. Noninvasive paternity assignment in Gombe chimpanzees. *Mol. Ecol.*, 10, 1279–1300.

Damas, D. 1969. *Band Societies: Proceedings of the Conference on Band Organization, Ottawa, 30 August to 2 September 1965*. Ottawa: National Museums of Canada. Anthropological Series No. 84.

de Waal, F. 1982. *Chimpanzee Politics: Power and Sex Among the Apes*. New York: Harper and Row.

de Waal, F. 1996. *Good Natured*. Cambridge, MA: Harvard University Press.

Driver, H. E. 1936. Wappo ethnography. *Univ. Calif. Pub. Am. Archaeol. Ethnol.*, 36, 179–220.

Dunning, R. W. 1959. Rules of residence and ecology among the Northern Ojibwa. *Am. Anthropol.*, 61, 806–816.

Ember, C. 1975. Residential variation among hunter-gatherers. *Behav. Sci. Res.*, 3, 199–227.

Ember, C. 1978. Myths about hunter-gatherers. *Ethnology*, 17, 439–448.

Evans-Pritchard, E. E. 1940. *The Nuer*. Oxford: Oxford University Press.

Evans-Pritchard, E. E. 1951. *Kinship and Marriage Among the Nuer*. Oxford: Oxford University Press.

Fischer, J. L. 1958. The classification of residence in censuses. *Am. Anthropol.*, 60, 508–517.

Foley, R. & Lee, P. C. 1989. Finite social space, evolutionary pathways, and reconstructing hominid behavior. *Science*, 243, 901–906.

Fortes, M. 1949. *The Web of Kinship Among the Tallensi*. London: Oxford University Press.

Ghiglieri, M. 1987. Sociobiology of the great apes and the hominid ancestor. *J. Hum. Evol.*, 16, 319–358.

Goldberg, T. & Wrangham, R. 1997. Genetic correlates of social behaviour in wild chimpanzees: evidence from mitochondrial DNA. *Anim. Behav.*, 54, 559–570.

Goodenough, W. H. 1956. Residence rules. *Southwestern J. Anthropol.*, 12, 22–37.

Goodenough, W. H. 1962. Kindred and hamlet in Lakalai, New Britain. *Ethnology*, 1, 5–12.

Harcourt, A. R., Fossey, D., & Sabater-Pi, J. 1981. Demography of *Gorilla gorilla*. *J. Zool. Soc. Lond.*, 195, 215–233.

Hart, C. W. M. & Pilling, A. R. 1960. *The Tiwi of North Australia*. New York: Holt, Rinehart and Winston.

Hawkes, K. & O'Connell, J. F. 1992. On optimal foraging models and subsistence transitions. *Curr. Anthropol.*, 33, 63–66.

Hawkes, K., O'Connell, J. F., & Blurton Jones, N. G. 2003. The evolution of human life histories: primate tradeoffs, grandmothering socioecology, and the fossil record. In: *The Role of Life Histories in Primate Socioecology* (Ed. by P. Kappeler & M. Pereira), pp. 204–231. Chicago: University of Chicago Press.

Hawkes, K., O'Connell, J. F., Blurton Jones, N. G., Alvarez, H., & Charnov, E. L. 1998. Grandmothering, menopause, and the evolution of human life histories. *Proc. Natl. Acad. Sci. USA*, 95, 1336–1339.

Helm, J. 1965. Bilaterality in the socio-territorial organization of the Artic drainage Dene. *Ethnology*, 4, 361–385.

Helm, J. 1969. Remarks on the methodology of band composition analysis. In: *Contributions to Anthropology: Band Societies* (Ed. by D. Damas), pp. 212–217. Ottawa: National Museums of Canada.

Helm, J. 2000. *The People of Denendeh: Ethnohistory of the Indians of Canada's Northwest Territories*. Iowa City, IA: University of Iowa Press.

Henry, J. 1964. *Jungle People: A Kaingang Tribe of the Highlands of Brazil*. New York: Vintage Books.

Hiatt, L. R. 1962. Local organization among the Australian Aborigines. *Oceania*, 32, 267–286.

Hill, G. W. 1956. *The Warao of the Amacuro Delta*. Caracas: Universidad Central de Venezuela. English translation Human Area Relations File.

Holmberg, A. R. 1985. *Nomads of the Longbow: The Siriono of Eastern Bolivia*. Prospect Heights, IL: Waveland Press.

Hrdy, S. B. 2000. The optimal number of fathers: evolution, demography, and history in the shaping of female mate preferences. In: *Evolutionary Perspectives on Human Reproductive Behavior* (Ed. by D. LeCroy & P. Moller), pp. 75–96. New York: New York Academy of Sciences.

Kelley, R. C. 1974. *Etoro Social Structure*. Ann Arbor, MI: University of Michigan Press.

Kelly, I. T. 1932. Ethnography of the Surprise Valley Paiute. *Univ. Calif. Pub. Am. Archaeol. Ethnol.*, 31, 67–210.

Knauft, B. 1991. Violence and sociality in human evolution. *Curr. Anthropol.*, 35, 391–428.

Kroeber, A. L. 1938. Basic and secondary patterns of social structure. *J. Royal Anthropol. Inst.*, 68, 299–309.

Leach, E. R. 1961. *Rethinking Anthropology*. London: University of London, Athlore Press.

Lee, R. B. 1972. !Kung spatial organization: an ecological and historical perspective. *Hum. Ecol.*, 1, 125–148.

Lee, R. B. & DeVore, I. 1968. *Man the Hunter*. Chicago: Aldine.

Long, J. P. M. 1971. Arid region Aborigines: the Pintubi. In: *Aboriginal Man and Environment in Australia* (Ed. by D. J. Mulvaney & J. Golson), pp. 262–270. Canberra: Australian National University Press.

Marshall, L. 1976. *The !Kung of Nyae Nyae*. Cambridge: Harvard University Press.

Mitani, J. C., Merriwether, D. A., & Zhang, C. 2000. Male affiliation, cooperation and kinship in wild chimpanzees. *Anim. Behav.*, 59, 885–893.

Mitani, J. C., Watts, D. P., & Muller, M. N. 2002. Recent developments in the study of wild chimpanzee behavior. *Evol. Anthropol.*, 11, 9–25.

Murdock, G. P. 1949. *Social Structure*. New York: Macmillan.

Murdock, G. P. 1962. Ethnographic atlas. *Ethnology*, 1, 113–134.

Murdock, G. P. 1967. *Ethnographic Atlas*. Pittsburgh: University of Pittsburgh Press.

Murdock, G. P. 1981. *Atlas of World Cultures*. Pittsburgh: University of Pittsburgh Press.

Myers, F. 1986. *Pintubi Country, Pintupi Self*. Washington, DC: Smithsonian Institution.

O'Connell, J. F., Hawkes, K., & Blurton Jones, N. G. 1999. Grandmothering and the evolution of *Homo erectus*. *J. Hum. Evol.*, 36, 461–485.

Palombit, R. A. 1994. Dynamic pair bonds in hylobatids: implications regarding monogamous social systems. *Behaviour*, 128, 65–101.

Pusey, A. & Packer, C. 1997. The ecology of relationships. In: *Behavioural Ecology: An Evolutionary Approach* (Ed. by J. R. Krebs & N. B. Davies), pp. 254–283. Oxford: Blackwell Science.

Pusey, A., Williams, J., & Goodall, J. 1997. The influence of dominance rank on the reproductive success of female chimpanzees. *Science*, 277, 828–831.

Radcliffe-Brown, A. R. 1930. The social organization of Australian tribes. *Oceania*, 1, 34–63, 322–341, 426–456.

Ray, V. F. 1933. The Sanpoil and Nespelem: Salishan peoples of north eastern Washington. *Univ. Washington Pub. Anthropol.*, 5, 1–237.

Riddell, F. A. 1978. Maidu and Konkow. In: *Handbook of North American Indians, Vol. 8*, (Ed. by W. C. Sturtevant), pp. 370–386. Washington, DC: Smithsonian Institution.

Rodseth, L., Wrangham, R., Harrington, A., & Smuts, B. 1991. The human community as a primate society. *Curr. Anthropol.*, 32, 221–254.

Schebesta, P. 1929. *Among the Forest Dwarfs of Malay*. London: Hutchinson and Co.

Schrire, C. 1984. Wild surmises on savage thoughts. In: *Past and Present in Hunter Gatherer Studies* (Ed. by C. Schrire), pp. 1–21. New York: Academic Press.

Service, E. A. 1962. *Primitive Social Organization: An Evolutionary Perspective*. New York: Random House.

Smith, E. A. & Winterhalder, B. 1992. Natural selection and decision-making: some fundamental principles. In: *Evolutionary Ecology and Human Behavior* (Ed. by E. A. Smith & B. Winterhalder), pp. 25–60. New York: Aldine de Gruyter.

Steward, J. H. 1936. The economic and social basis of primitive bands. In: *Essays in Anthropology* (Ed. by R. H. Lowie), pp. 321–347. Berkeley, CA: University of California Press.

Steward, J. H. 1977. The foundations of Basin-Plateau Shoshonean society. In: *Evolution and Ecology: Essays on Social Transformations* (Ed. by J. C. Steward & R. F. Murphy), pp. 366–406. Urbana, IL: University of Illinois Press.

Sugiyama, Y. 1999. Socioecological factors of male chimpanzee migration at Bossou, Guinea. *Primates*, 40, 61–68.

Turnbull, C. M. 1961. *The Forest People*. New York: Simon and Schuster.

Turnbull, C. M. 1965. The Mbuti pygmies: an ethnographic survey. *Anthropol. Pap. Am. Mus. Nat. Hist.*, 50, 145–281.

Turney-High, H. H. 1941. Ethnography of the Kutenai. *Am. Anthropol. Suppl.*, 43, 1–202.

van Schaik, C. P. 1999. The socioecology of fission-fusion sociality in orangutans. *Primates*, 40, 69–86.

Vigilant, L., Hofreiter, M., Siedel, H., & Boesch, C. 2001. Paternity and relatedness in wild chimpanzee communities. *Proc. Natl. Acad. Sci. USA*, 98, 12890–12895.

Williams, B. J. 1974. A model of band society. *Mem. Soc. Am. Archaeol.*, No. 29.

Wilmsen, E. N. & Denbow, J. R. 1990. Paradigmatic history of San-speaking peoples and current attempts at revision. *Curr. Anthropol.*, 31, 489–524.

Wrangham, R. 1987. The significance of African apes for reconstructing human social evolution. In: *The Evolution of Human Behavior: Primate Models* (Ed. by W. Kinzey), pp. 55–71. Albany, NY: SUNY Press.

19

Mating, Parenting, and the Evolution of Human Pair Bonds

Kristen Hawkes

Human pair bonds are widely assumed to have arisen when ancestral females began to rely on subsistence support from mates (e.g., Lovejoy 1981; Lancaster & Lancaster 1983, 1987). The most influential version of this argument is the "hunting hypothesis." Its elements have a long history in Western thought (Cartmill 1994). August Westermarck in his 1891 book, *The History of Human Marriage*, provided a nineteenth-century example:

> When the human race passed beyond its frugivorous stage and spread over the earth living chiefly on animal food, the assistance of an adult male became still more necessary for the subsistence of the children. Everywhere the chase devolves on the man, it being a rare exception among savage peoples for a woman to engage on it. Under such conditions a family consisting of mother and young only would probably, as a rule, have succumbed. (p. 39)

The hunting hypothesis was elaborated with especially wide influence by Sherwood Washburn in the mid-twentieth century. Using what was then known of nonhuman primate behavior, modern hunter-gatherers, and the paleoanthropological record, Washburn linked tool use, bipedalism, brain expansion, nuclear families, and an array of other features that distinguish modern humans from our closest living primate relatives to the "hunting adaptation" (e.g., Washburn & Avis 1958, Washburn 1960, Washburn & DeVore 1961, Washburn & Lancaster 1968). Building on Raymond Dart's arguments tying bipedality to hunting with weapons (Dart 1953), Washburn proposed that hunting large animals favored bipedalism, tool use, and bigger brains, which in turn created an obstetrical dilemma for mothers. To be born at all, babies had to be born less well developed and so more dependent, longer, on maternal care. Maternal duties kept women from hunting, so they required provisioning by hunting mates. Pair bonds and a sexual division of labor arose as reliance on large-animal hunting made paternal provisioning essential. This need for biparental care was the foundation of the evolution of human pair bonds and nuclear families.

Paradoxically, the volume in which Washburn's most widely cited essay on this topic appeared (Washburn & Lancaster 1968), *Man the Hunter*, also contained ethnographic reports of hunting and gathering societies in which subsistence was much more dependent on women's work and on plant foods than meat. Ethnographers found that, although people claimed that big-game hunting was of primary importance, systematic observation of actual food consumption showed heavy reliance on plants—except at high latitudes where few plant foods are available. Richard Lee's (1968) chapter in that volume was especially important in challenging many assumptions about hunter-gatherers that were widespread at the time. Lee reported his observations of !Kung-speaking Bushmen in southern Africa and also summarized patterns from the ethnographic literature on foragers worldwide to show that aspects of the !Kung case were broadly typical of modern foragers in temperate and tropical habitats.

Since the early 1960s, there have been important additions to the descriptive record of ethnology and primate socioecology. The paleoanthropological evidence now shows bipedalism, brain expansion, and archaeological evidence for big-game hunting to be separated from each other by millions of years (Klein 1999). The theoretical perspective and modeling tools used to investigate evolutionary questions have also changed dramatically. Nevertheless, the hunting hypothesis, much as it was articulated by Washburn before these changes occurred, still continues to be favored (e.g., Tooby & DeVore 1987, Deacon 1997, Kaplan et al. 2000, Horrobin 2001, Calvin 2002).

Here I review concepts, models, and data in three domains that have implications for the hypothesis that human pair bonds evolved due to the dependence of mothers on subsistence support from hunting husbands. I begin with the history of ideas about links between parenting and sexual selection. Starting with Darwin's theory, I consider George Williams' (1966) elaboration, Robert Trivers' (1972) definition of parental investment, and subsequent broadening of that definition (Kleiman & Malcom 1981, Clutton Brock 1991). This broadening may be useful for some questions, but it obscures distinctions between mating and parenting effort that are central to the theory of sexual selection and were explicitly maintained by Trivers (1972). The errors that arise from the broader definition confuse evolutionary explanations for behavior, especially among primates—humans in particular.

The second domain of special relevance includes both theoretical and empirical work since the mid-1970s distinguishing parenting from mating, primarily among nonhuman primates. As Darwin himself noted, the distinction between these things is not always straightforward. But it is fundamental to the theory of sexual selection. The recognition that females may seek copulations to improve their parenting success (Hrdy 1979, 1981, 1999), and that males may care for infants and juveniles to improve their mating success (Smuts 1983a, b, 1985, 1987), underlined the importance of distinguishing the form of behavior from its adaptive function. These and other findings about the ways that reproductive strategies of males and females can lead to relationships between them have implications for social patterns in many species. Their implications for the evolution of human pair bonds are of special interest here.

The third domain I consider is the behavioral record for our own species, with particular emphasis on the ethnology and behavioral ecology of sexual divisions of labor among hunter-gatherers. Washburn and many others assumed that the basis for human pair bonds must be women's dependence for subsistence on a provisioning mate. This implies a "sex contract" (Fisher 1981), in which women trade increased probability of paternity to a man in exchange for his subsistence support (cf. Beckerman & Valentine 2002). However, recent hunter-gatherer ethnography shows that family provisioning goals do not account for the work men usually do. Women and children often get most of the meat they eat from the hunting of men outside their own family, and the amount they get is generally unrelated to the hunting success of their own husbands and fathers. More recent quantitative data on this topic actually echo and elaborate qualitative claims that have long been made about household economics in foraging societies (Leacock 1972, Sahlins 1972, Marshall 1976, Kelly 1995).

This material undermines the basis for scenarios that make human pair bonds the consequence of increased paternal effort. A different suite of hypotheses based explicitly on sexual selection and dangers posed by male mating competition have been advanced to explain special relationships between males and females in nonhuman species. These hypotheses apply to other primates that live in multimale social groups, making them particularly well suited on both phylogenetic and socioecological grounds to identify the evolutionary foundation for pair bonds in humans.

Parenting and Sexual Selection

Darwin's theory of sexual selection depends on a distinction between characteristics that affect parenting success and those that affect success in competition for mates. Beginning with a review of Darwin's argument, I summarize some of the subsequent developments that both clarified implications and highlighted complications.

Darwin's Reasons for Sexual Selection

Many differences between the sexes, Darwin argued, can be explained by "ordinary" natural selection.

> When the two sexes differ in structure in relation to different habits of life . . . they have no doubt been modified through natural selection. . . . So again the primary sexual organs, and those for nourishing or protecting the young, come under this same head: for those individuals which generated and nourished their offspring best, would leave, *ceteris paribus*, the greatest number to inherit their superiority; whilst those which generated or nourished their offspring badly, would leave but few to inherit their weaker powers. ([1871] 1981: 256)

Other differences between the sexes, such as male armaments and ornaments, cannot be explained by "ordinary" natural selection, because

> the males have acquired their present structure, not from being better fitted to survive in the struggle for existence, but from having gained an advantage over other males. . . . That these characters are the result of sexual selection and not of ordinary selection is clear, as unarmed, unornamented, or unattractive males would succeed equally well in the battle for life and in leaving numerous progeny, if better endowed males were not present. (Darwin [1871] 1981: 257–258)

While Darwin noted that it was not always easy to tease apart the operation of sexual selection from other forms of natural selection, the distinction between parenting on one hand and competition for mates on the other was central to his theory. An important basis for this distinction was the evident asymmetry in the operation of sexual selection on males and females. "On the whole," Darwin observed, "there can be no doubt that with almost all animals in which the sexes are separate, there is a constantly recurrent struggle between the males for the possession of the females" ([1871] 1981 ii: 260). Because of this "it is the males that fight together and sedulously display their charms before the females" (p. 272). He recognized two components of sexual selection, male-male combat and female choice, both of which shape the morphology, physiology, and behavior of males, so that "it is the male which, with rare exceptions, has been chiefly modified" (p. 272).

Recognizing this asymmetry, "we are naturally led to enquire why the male in so many and such widely distinct classes has been rendered more eager than the female, so that he searches for her and plays the more active part in courtship . . . why should the male almost always be the seeker?" (Darwin [1871] 1981 ii: 273). Malte Andersson (1994) pointed out that Darwin came close to the anisogamy explanation for stronger sexual selection on males. But it was not until A. J. Bateman's (1948) experiments on sex differences in reproductive success among *Drosophila*, and then George Williams' (1966) use of this example, that the anisogamy explanation was widely appreciated. Bateman's experiments showed that reproductive success varied more among males, each additional mating increasing the reproductive success of males but not females. The anisogamy explanation for this result is based on differences in the two-gamete types. Since each zygote requires one gamete from each sex, and female gametes are fewer, produced more slowly, it is female gametes that limit offspring production. With more male gametes, produced at a faster rate, a male can increase his reproductive success by fertilizing more females than other males do. This makes paternity competition a zero-sum game, as any male's gain subtracts from the paternity available for others. Williams (1966) started with anisogamy, and then moved beyond it to include additional kinds of parental contributions, extracting increased explanatory power from sexual selection by linking it to asymmetries in parental expenditure.

Jeffrey Baylis (1981) drew attention to one reason why Darwin himself did not make that link: patterns of ornamentation and parental care in teleost fish. Darwin's survey of reproductive patterns was characteristically wide ranging. Summarizing observations of many species of fish, he concluded that it is "manifest that the fact of the eggs being protected or unprotected has had little or no influence on the differences in color between the sexes. It is further manifest, [that in many of the very cases where] the males take exclusive charge of the nests and young . . . the males are more conspicuously colored than the females" (Darwin [1871] 1981 ii: 21). Thus Darwin saw the greater adornment of males even though they appeared to be more actively engaged in parenting than females as evidence against any general dependence of sexual selection on relative parental expenditure.

Links Between Parenting and Sexual Selection

Williams (1966), building on Darwin and of course Fisher (1930), was also characteristically wide ranging. He did not take up this riddle of the fish, but, reviewing many other examples, he showed that the evolutionary battle of the sexes arises from differences in the allocation of reproductive effort to parenting and mating. This conflict of interest was further clarified by Trivers (1972), who linked parenting to sexual selection by defining parental investment so as to highlight a necessary interdependence among the trade-offs individuals must face given finite reproductive effort. Defining parental investment as the cost that a contribution to one offspring's fitness imposes on "a parent's ability to invest in other offspring" (Trivers 1972: 139), Trivers implied that the sum of all the parental investments an individual makes would equal its total parental effort (Low 1978). If both sexes expend about the same average overall total reproductive effort (Williams 1966), and that total is composed of parental and mating effort, then differences in the parental investment of the sexes must imply inverse differences in their mating effort. Members of the sex investing less in parenting compete with each other for the greater parental investment of the opposite sex.

The currency Trivers identified for measuring parental investment clarified some basic trade-offs. But actually measuring the cost imposed on a parent's ability to invest in other offspring is not so easy. A measure of this cost for each sex is required for calculating the differences between them in species where both sexes contribute some parental care. An alternative index, more readily measured, is potential reproductive rate (Clutton Brock 1991, Clutton Brock & Parker 1992). As recognized by R. A. Fisher, and maybe Darwin himself (Edwards 1998), the fact that (among most sexual reproducers) everybody has a mother and a father makes sex ratios likely to be even most of the time. Since overall reproductive rates depend on the rate of the slower sex, members of the faster sex, ready and waiting sooner, must compete for each reproductive opportunity.

Another issue here, also partly one of measurement, can involve the fundamental question of distinguishing parenting from mating. Trivers focused on the costs to a parent's ability to invest in other offspring. Some contributions that improve offspring welfare may benefit multiple recipients jointly. For example, a nest may cost the nest builder the same amount of effort and benefit each offspring the same whether it is used by one or several. This collective aspect of certain goods has been recognized with various labels, depreciable and nondepreciable care among them (Clutton Brock 1991). All goods and services are not the private property of individual parents and offspring.

In the case of the egg-guarding fish that may have kept Darwin from linking relative parental effort to sexual selection, the collective nature of the benefits to eggs has fundamental consequences for the fitness-related benefits to the guarding males. If a unit of guarding provided to one egg or nest does not reduce the guarding benefits available to another egg, a male can guard many clutches of eggs with the same effort with which he could guard a few. If females gain parental advantages by leaving eggs where they are safer, females may prefer to leave their eggs in the presence of many other eggs. Moreover, conspecific males may release sperm to compete with the sperm of the bright defending male, so the guarding male's probability of paternity for the eggs laid at his site can be substantially less than 100% (Gross & Sargent 1985). Guarding may thus have little effect on relative parenting success of the guarding male, that is, the relative welfare of his eggs versus those of other males. Under these circumstances, selection could maintain brightness and guarding because of its relative effects on male mating success. Whether an aspect of reproductive effort affects fitness through parenting or mating is of great importance in explaining both its origin and maintenance.

Why Indirect Care May Not Be Parental Effort

Considering specifically the problem of identifying male parental effort in mammals, Devra Kleiman and James Malcolm (1981: 348) developed a very broad alternative to Trivers' definition. They "conceive male parental investment as *any increase in a prereproductive mammal's fitness attributable to the presence or action of a male*" (original italics). This includes not only nondepreciable benefits, as defined above, but also "indirect care."

> Indirect male parental investment includes those acts a male may perform in the absence of the young which increase the latter's survivorship. These acts may have delayed effects on survivorship of young and include such behaviors as the acquisition, maintenance, and defense of critical resources within a home range or territory by the elimination of competitors, the construction of shelters, and actions which improve the condition of pregnant or lactating females. Many forms of male parental investment that are indirect are also incidental to the species' breeding system, ecology, or social organization. These are activities which males would perform regardless of the presence of young . . . behaviors such as scent marking and long distance vocalizations which aid in the spacing of individuals or groups, and thus may maintain critical resources for eventual use by young, should also be considered as indirect forms of male parental investment. (Kleiman & Malcom 1981: 348–349)

This very inclusive definition seems to imply, incorrectly, that parenting is the adaptive function of any male behavior that could affect the welfare of young. While parenting *could* be the adaptive function of many of these behaviors, that depends not on whether the young benefit, but whether the net fitness benefits that maintain the behavior in the males come through differential welfare of the performer's offspring compared to the young of males who act otherwise. Many of the activities in this list supply collective goods. Once supplied, they could benefit not only the male's offspring, but also others besides. Jeffrey Kurland and Steven Gaulin (1984: 285) pointed out one of the problems this presents for the specific case of troop defense in baboons (*Papio* spp.) or macaques (*Macaca* spp.).

> Not only does indiscriminate troop defense not change the relative fitness of dominant, resident males, it also allows transient low ranking males who father one or a few offspring in each of several troops to effectively parasitize the parental efforts of the resident males. The young of such "floating," "cheater" males would be defended and the males themselves would incur none of the attendant risks. Thus, transient, nonparental males might reproductively do as well as, if not better than the resident males . . . [rendering] the resident male's parental behavior "evolutionarily unstable"

On those grounds, parental benefits to the defending male (differential welfare for his own offspring) are poor candidates to explain troop defense by resident males. Kurland and Gaulin surmise that the persistence of this behavior is more likely due to effects on the defending males' mating success. The speculations about egg-guarding fish above parallel their suggestion: "Much of what often passes for male 'parental care' may, in primates and mammals in general, in fact be more parsimoniously interpreted as male mating effort rather than male parental investment" (Kurland & Gaulin 1984: 285).

Distinguishing Mating from Parenting, Especially in Primates

Bobbi Low (1978: 200) pointed out that "Different apportionment of [reproductive effort] between mating and parental effort in the two sexes may occur even when the same structure or behavior is involved. For example, in any mammalian species in which males use horns solely in dominance displays and fights to secure matings while females use theirs in defense of young, horns represent mating effort for males and parental effort for females." The lesson applies generally. Care for eggs or infants may be parenting, and copulation may be mating effort, but not necessarily. Whether the adaptive function of a behavior is mating or parenting depends on the character of the net fitness effects on the performer compared to a nonperformer. This includes strategies that result in pair bonds.

Darwin's cross-species review of reproductive behavior supported his conclusion that males are generally more eager to mate than females. He linked this to sexual selection and so to the benefits of mating effort for males. Pointing to the usual difference in parental expenditure of the sexes, Williams (1966: 183) said "the traditional coyness of the female is thus easily attributed to adaptive mechanisms by which she can discriminate the ideal moment and circumstances for assuming the burdens of motherhood." Trivers (1972) elaborated this argument with illustrative examples of ardent males and coy females. Paralleling Williams (1966), Trivers (1972) noted that "sex role reversed" species provide a test of whether the sex investing less in parenting spends more effort competing for the greater parental investment of the opposite sex. His examples were species where males contribute more parental effort than females, and where, as predicted, males are coy and females compete ardently for matings.

But What About Primate Females Who Are Not Coy?

Our own order, however, presents clear challenges to the generalization that the more parental sex is coy. Like other mammals, primate females do the bulk of parenting. Yet there are numerous examples in which females are anything but shy (Rowell 1988, Berkovitch 1995). Smuts (1987: 392) went so far as to say, "Females have been observed soliciting copulations in virtually every primate species that has been studied, and in the majority of species, females initiate the majority of copulations."

Sarah Hrdy faced the puzzle of libidinous females initially with hanuman langurs (*Presbytis entellus*). Having shown that sexual selection could favor the infanticidal behavior of langur males, she turned to the question of female counterstrategies. Exploring the role that reproductive competition among females plays in the evolutionary battle of the sexes (Hrdy 1979, 1981), she documented the eagerness of primate females to copulate with strange males.

> At issue here are behaviors exhibited by the majority of species in the order primates, the best studied order of animals in the world, and the order specifically included by Bateman in his extrapolation from coyness in arthropods to coyness in anthropoids. Furthermore, females engaged in such "promiscuous matings" entail obvious risks, ranging from retaliatory attacks by males, venereal disease, the energetic costs of multiple solicitations, predation risks from leaving the troop, all the way to the risk of lost investment by a male consort who has been selected to avoid investing in other males' offspring (Trivers 1972). In retrospect one really does wonder why it was nearly 1980 before promiscuity among females attracted more than cursory theoretical interest. (Hrdy 1986: 126–127)

Once researchers began to investigate the patterns, an array of hypotheses were suggested to explain them. Most nominate benefits in offspring welfare that females may earn by copulating with many males. Hrdy developed the hypothesis that primate females solicited stranger males in response to the danger that males pose to the infants fathered by other males. If males are less likely to kill infants they may have fathered, a female might reduce the dangers to her own infants by spreading the possibility of paternity widely. This was known initially as the "manipulation" hypothesis, but self-defense might have been emphasized instead. As Hrdy observed more recently (1999: 87–88), "pejorative-sounding words like 'promiscuous' only make sense from the perspective of the males. . . . From the perspective of the female however her behavior is better understood as 'assiduously maternal.'"

Hrdy's hypothesis was about relationships. She surmised that females might establish relationships with strange males and so reduce the danger those males otherwise posed. Increasing interest in exploring the character and consequences of relationships in primate social groups expanded from the late 1970s (Hinde 1979, 1983; Harcourt 1989, 1992; Chapais 1992, 1995). One result was to draw further attention to the fact that more was going on between the sexes than copulation. As Barbara Smuts (1983a: 112) noted at about that time, "most studies of male-female interactions in nonhuman primates have focused on sexual behavior. Several recent studies, however, have shown that in savannah baboons and macaques, adult males and females may form long term friendly bonds that persist in the absence of any immediate sexual relationship."

Special Relationships

The importance of these friendships to both sexes is indicated by the effort they invest in them. "Both males and females compete for these relationships (e.g., both males and females sometimes threaten potential rivals away from their friends)" (Smuts 1987: 398). Females may interrupt others' copulations and solicit the male themselves. The fitness-related benefit of this "female mating competition" was proposed to be protection that male friends provide a female and her infant (Smuts 1985). In terms of the reproductive effort typology, the females gain parental benefits, so the effort they expend on these relationships is maternal effort.

What kind of reproductive effort are the males expending? Hrdy's infanticide protection hypothesis proposed that males behaved differently toward the offspring of previous copulation partners because of the possibility they had fathered the infants. Some probability of paternity could have favored and maintained this strategic adjustment in males because of its net parenting benefits to them.

Other hypotheses are also worth entertaining. The problem of paternity confidence was highlighted by Trivers (1972) in his exploration of conflicts of interest between the sexes. He argued that a history of selection should design males to be sensitive to probable paternities and to avoid "misdirecting" paternal effort to the offspring of other males. This "danger of cuckoldry" argument was a basis for labeling solicitation of copulations by females as "deceptive" or "manipulative." The proposition that males are generally in danger of being "cuckolded by deceptive females" resonates with long-held Western views (Beckerman & Valentine 2002), including those about men's ownership rights over women (Wilson & Daly 1992) and about the basis of male sexual jealousy (Daly et al. 1982, Daly & Wilson 1987, Pinker 1997). But uncertain paternity arises from male mating competition. Male behavior in mammals generally, including other primates where little paternal effort is ever dispensed, shows that the dangers males pose for each other provide sufficient reason for male jealousy.

Both modelers (e.g., Maynard Smith 1977) and fieldworkers (e.g., Smuts 1985, Paul et al. 1992; reviews in Whitten 1987, Wright 1990) have shown that paternity certainty is not necessarily a good predictor of male behavior toward infants. Smuts (1987: 393–394) summarized some findings of her own and others this way:

> In a number of species living in multimale groups, males appear to contribute to offspring survival through babysitting, protection, occasional carrying, and other affiliative behaviors. Male parental investment does not fully explain these behaviors because the infants are in many cases unlikely to be the male's own offspring. . . . [Observers have] suggested that, by developing an affiliative relationship with an infant, male savannah baboons might be able to improve their chances of mating with the mother in the future. If this hypothesis is correct, then male care of infants sometimes represents mating effort rather than parental investment.

For olive baboons (*Papio anubis*), Smuts (1985) showed that friendship with the mother was a better predictor of male care for the infants than likely fatherhood. Consistent with her hypothesis that female choice favored males who formed these relationships, she found that females preferred their friends as mating partners. Smuts and Gubernick (1992) assembled data from a number of primate species that were consistent with this hypothesis. Van Schaik and Paul (1996: 153), summarizing additional work on the variation in male care among many species of nonhuman primates, affirmed Smuts' assessment that this variation "is not adequately explained by parentage. Indeed, much of what has traditionally been considered paternal behavior may be better explained as mating effort in situations where females can exert control over their choice of mates."

Other Males Are a Problem for Both Males and Females

The picture is not, however, one of female choice favoring the evolution of ever greater gentleness and affability in males (Hrdy 1981). The advantage of these friendships for females is "an ally who because of his larger size and superior fighting ability, may make a significant contribution to the fitness of the female and her offspring" (Smuts 1983b: 263). Large size and fighting ability are especially valuable because an important source of danger to the welfare of infants and females themselves is posed by other males. Richard Wrangham (1979) hypothesized that females may establish and maintain relationships with aggressive males because of the protection this provides from other dangerous males (Wrangham & Rubenstein 1986).

The importance of this danger was chronicled by Smuts and Smuts (1993: 2) as male coercion: "the use by a male of force, or threat of force, that functions to increase the chances that a female will mate with him at a time when she is likely to be fertile, and to decrease the chances that she will mate with other males, at some cost to the female." Clutton Brock and Parker (1995) provided formal modeling to show how male coercion could play a large role in sexual selection in many species, including primates. Smuts (1987: 396) characterized the effects of female strategizing in the context of male coercion this way: "by cooperating with some potential mates and resisting others, females alter the costs and benefits of competition among males and thus influence the form and frequency of male-male competition."

Reviewing primatology up to the early 1980s, Hrdy and Williams (1983: 5) pointed out "a strong tendency among field workers to focus on male behavior." However, only certain aspects of male behavior had received much attention. Smuts noted in her 1987 review that "the relationship between male dominance and male mating activity has probably received more attention than any other aspect of primate social behavior" (p. 388). But other aspects of male social behavior were not studied much. More recently, Peter Kappeler (1999: 26) could justifiably conclude that "social relationships among males remain the most poorly studied aspect of primate socioecology." The same errors that Hrdy and Williams (1983: 8) blamed for insufficient attention to female social behavior biased the attention paid to males: "The fallacy of measuring reproductive success in both sexes by zygote production. . . . Even worse is the assumption that it can be measured by mere copulatory performance, and that this will be the focus of selection on reproductive fitness."

Appreciation of the important effects of other behaviors on both parenting and mating success has steadily increased since then. Carel van Schaik (1996) saw that the dangers posed by infanticidal males might do more than spur females to seek protectors: this danger might be fundamental to the shape of primate social systems. Van Schaik and Peter Kappeler (1997) linked aspects of primate life history to the benefits for infanticidal males. They proposed that the extreme sociality of our order, evident in the year-round associations of males and females, results from the strategies used by both males and females to counter this danger. This provides a framework for understanding both adult social behavior and reproductive physiology (van Schaik et al. 1999, 2000), as well as the unusually high frequency and wide diversity in male-infant interactions that distinguish primates from most other mammals (Whitten 1987, Paul et al. 2000).

Ryne Palombit (1999) summarized observational and experimental data in savannah baboons and mountain gorillas (*Gorilla gorilla beringei*) consistent with the hypothesis that females maintain special relationships with particular males, that is, pair bonds, to reduce the danger of infanticide by other males. As he noted, following Hrdy and Smuts, appreciation of these patterns involves an important shift in thinking about the evolution of pair bonds. From the perspective of Trivers (1972), pair bonds are expected only with monogamy, which in turn is assumed to entail both sexes devoting substantial reproductive effort to parenting. Devra Kleiman (1977) had pointed out that in contrast to this prediction, monogamy did not necessarily imply either pair bonds or biparental care. Work over the past two decades has shown that the converse is also true. Strong male-female attachments that persist outside estrus do not depend on monogamy.

Long appreciation of these relationships in baboons, combined with the variation in social organization both within and among species, makes genus *Papio* especially interesting for questions about pair bonds. While females benefit from the infanticide protection supplied by friends (Palombit et al. 2000), the benefits males earn from these friendships may be more variable. Smuts (1985) found mating benefits among olive baboons, but data on other savannah populations challenges the mating effort hypothesis (Palombit 2000). Observations and playback experiments among chacma baboons (*Papio cynocephalus ursinus*) (Palombit et al. 1997) showed that males responded differentially to the screams of their friends only when infants were threatened by infanticidal males. The males, who were possible fathers in 68% of the friendships in this study, stopped responding to their (former?) friend's screams after the death of an infant. Just when the females would be returning to estrus and so presenting a mating opportunity, the friends responded even less to the playbacks than control males did.

It could be that infant loss brings an end to friendships. During this study, infant mortality was 76%. At least half of this loss was due to infanticides committed by a recently immigrated alpha male with whom the mothers of the victims subsequently mated. As the researchers note (Palombit et al 1997: 611), females may have ended friendships under these circumstances because their friends were unable to protect them.

The variation among and within baboon populations over time is an invitation to test socioecological hypotheses about the interplay of male and female strategies. Robert Barton (2000, Barton et al. 1996) proposed a model for some of the variation that includes characteristics of the food resources which shape feeding competition among females, the dangers posed by infanticidal males, and the mixture of coercion and affiliation that males use to keep females from associating with other males.

Changing Views of Monogamy

Coincident with these developments in understanding the importance of pair relationships in nonmonogamous primates, the 1980s and 1990s saw radical changes in the understanding of monogamy as a "mating system." David Gubernick (1994) listed 17 hypotheses for the evolution of monogamy and noted that they all involved various combinations of pair bonds, mating exclusivity, and male care, each of which might, in principle, evolve independently. The most dramatic impetus to a revised perspective on monogamy came first in ornithology, where the traditional view was "turned on its head" (Black 1996) with the discovery that "social monogamy" persisted in the absence of "genetic monogamy."

In his definitive review of avian breeding patterns, David Lack (1968) explained the extremely high frequency of monogamy in birds as the result of the importance of paternal provisioning. Pairs usually persisted through a breeding season, sometimes longer, he explained, because male parental effort increased the reproductive success of the family. Lack was not the only influential figure to favor this hypothesis about pair bonds. As Patricia Gowaty (1996a: 23) noted, "almost everyone's (Darwin 1871, Williams 1966, Lack 1968, Orians 1969, Trivers 1972) ideas about selective pressures accounting for the evolution of mating systems pivot around the necessity (or not) of male parental care."

Lack's hypothesis was challenged initially by a combination of theoretical developments that highlighted conflicts of interest between the sexes (Williams 1966, Trivers 1972, Maynard Smith 1977), and later by technological developments that allowed investigators to discover that extra-pair paternities were much more frequent than observers had previously guessed (Birkhead & Moller 1992). New questions arose about the fitness-related benefits to each partner for maintaining a pair bond. Investigators found that while male care sometimes varied with probable paternity, more often it did not (Wright 1990, Houston 1995). With experimental manipulation of pairs, investigators measured the effects of male care on female parenting success. Reviewing the data from removal experiments (Bart & Tornes 1989), Gowaty (1996b: 489) concluded that while "for some females there exist important advantages for male care . . . For many females male parental care has small or negligible effects on female reproductive success, suggesting that as a general explanation for social monogamy, the Male Care is Essential Hypothesis is inadequate."

The discoveries that avian pair bonds were not necessarily based on mating exclusivity, that a male's care for infants often did not go to his own offspring, and that females sometimes fledged no more offspring with a male partner than without one meant that long-favored hypotheses did not explain partnerships in birds (Black 1996). Male-female relationships that persisted over a breeding season had been assumed to imply mutual parenting and so require sexual exclusivity. But, in fact, they did not. Combined with the work on male-female relationships in primates, this has provided the basis for revisions in assumptions about the character and evolution of pair bonds generally, and about the evolution of human pair bonds in particular.

Pair Bonds, Marriage, and Human Evolution

It is widely assumed that, among modern hunter-gatherers, men's hunting is paternal effort. This is the major source of support for the claim that substantial paternal investment is the key to human pairing relationships and major transitions in human evolution. The assumption persists even though other widely accepted claims about men's reproductive strategies contradict it. I review some of these claims, noting that they have generally incorporated the contradictions uncritically. They often provide better justification for the alternative hypothesis that hunting, like many other male occupations, is largely shaped by mating competition. The discussion clears the ground to return to hypotheses about the evolution of human pair bonds that arise from work on other primates.

Before tackling these issues, some concerns about placing human pair bonds in a comparative perspective require attention. We all have firsthand knowledge about emotional dimensions and behavioral patterns associated with human pair bonds. As noted in the preceding section, our ideas about them shape hypotheses about other species. They may, however, be wrong for other species. They may not even apply generally to our own. Reviewing the human variation recorded in both history and ethnography, Sarah Hrdy (1999: 232) concluded that "earlier commentators failed to consider how unusual are the particular environmental and demographic conditions that make long term monogamy advantageous for both sexes." Duran Bell (1997: 241) recently noted that "contemporary Western 'marriage' is a poor vantage point from which to consider the ethnographic universe of marriage." Donald Symons (1979: 141), more than 20 years ago, also warned behavioral biologists against relying too much on their own personal experience: "Intuitions about marriage based on the extremely artificial circumstances of modern industrial societies may be somewhat misleading, since in industrial societies, unlike face-to-face, kin based societies in which the overwhelming majority of human evolution occurred, one's mate is often one's only hope for establishing an intimate, durable relationship with another adult."

Social anthropologists, from the beginning of the discipline (Morgan 1870), have struggled with both describing and explaining the wide variation observed in human marriage practices. Disputes over the definition of pair bonds in behavioral biology are but a whisper compared to the loud volume of debate devoted to the definition of marriage in sociocultural anthropology (e.g., summary in Goodenough 1970). Often, those studying the ethnographic variation have been skeptical that studies of other species could have any relevance to understanding the wide variability in our own.

Edmund Leach (1988: 91) memorably claimed that "mating and marriage are totally different concepts, as different as chalk from cheese." His definition of marriage focused on inheritance of relatively imperishable property. "Marriage provides a set of legal rules under which such items of property are handed down from generation to generation" (p. 93). This made marriage "exclusively a feature of human societies and not all human societies at that." Questions of the universality of marriage in human societies continue to be debated in social anthropology (e.g., Bell 1997). In many cultural settings, the legitimate inheritance of property is of great importance, and sometimes it explicitly depends on a special kind of legally recognized union in combination with other ties among potential claimants. But sometimes, especially among mobile foragers, people do not hold much material property and little, if any, of that is imperishable enough to pass to descendants. Property concepts, as argued below, have many important uses, including the help they provide for distinguishing parenting from mating effort. They can also be useful for analyzing strategies of mate defense. But these were not the property issues Leach had in mind. He focused on marriage to the explicit disregard of mating, assuming that mating involved little variation of interest since "the 'prevailing mating system' is a free-for-all cuddle in the dark" (Leach 1988: 107).

Others, prepared to see a relationship between mating and marriage, might still second Leach's view that the differences between us and other animals are greater than the similarities. Symons (1979: 108) made this observation:

> The lexicon of English is woefully inadequate . . . [for] describing the thoughts, feelings, and behaviors associated with marriage and with other relations among men and women. . . . No doubt complexity and subtlety of thought, feeling, and action inevitably must be sacrificed if the written record is to be made at all . . . but to shrink the present vocabulary to one phrase—pair bond—and to imagine that in doing so one is being scientific—subsuming humans under principles that account for data on nonhuman animals—is simply to delude oneself.

The danger could be even greater: We may underestimate the experience of other animals—especially our primate cousins—as well as our own.

But along with costs come benefits from simplification. While language reveals aspects of experience that can only be appreciated in humans, other dimensions can be studied in comparable ways in both human and nonhuman animals. Some of the human behavioral variation may be due to the same processes that govern the variation in other species and the wider cross-species differences. The many physiological and specifically neurological and endochronological processes we share with other primates are a foundation for the emotional architecture that shapes behavior in us all (Darwin [1872] 1965, Hrdy 1999). Similarities between us and other animals can correct erroneous conclusions about the uniqueness of human patterns. The similarities are especially important for any investigation of human evolution.

Sex Differences and the "Sexual Division of Labor"

Symons (1979) assembled a wide range of ethnographic and sociological data showing marked differences in the sexual preferences and behavior of men and women. He persuasively linked the differences to the much greater importance of mating competition among men, concluding, "Humans then are typical mammals in that selection has favored greater male-male reproductive competition" (p. 144) and "The evidence suggests that in hunting, as in fighting, human males are effectively in competition for females, and that there are substantial differences among males in competitive abilities" (p. 162).

Yet this conclusion presented him with a serious problem. Recognizing that the importance of mating competition among men is consistent with Darwin's predictions from sexual selection, Symons also recognized the links that had been made between parenting and sexual selection. He cited Trivers (1972) for the expectation that structures and behaviors associated with mating competition are favored in members of the sex that expends less in parenting. The problem was that Symons, following Washburn, also "knew" that high paternal effort was characteristic of humans under ethnographic circumstances most like those of deepest antiquity:

> The basic social unit of human hunter-gatherers is the nuclear family in which men hunt, women gather vegetable foods, and the results are shared and given to their offspring. (p. 130)

> Obligations and rights entailed by marriage vary among societies, but marriage is fundamentally a political, economic, and child raising institution, based on a division of labor by sex and on economic cooperation between the spouses. (p. 121)

Symons assumed the hunting hypothesis to be generally correct. However, according to theory (Williams 1966, Trivers 1972), one sex has more to spend on mating competition only if it spends less on parenting. The problem, then, is that substantial parenting effort from men, supposed to characterize our species and to be key to our evolution, should mean reduced mating effort compared to our close primate relatives. Symons dealt with this problem in the following way:

> It is not then a simple question of high female parental investment and male competition for females: males and females invested in different ways. Not only did males hunt while females gathered, but, if warfare was often over land and other scarce resources from which the winning males' offspring benefited, male fighting was in part paternal investment; that is, like hunting and gathering, fighting and nurturing were part of the human division of labor by sex. (Symons 1979: 163)

The trade-off assumptions of reproductive effort models are thus suspended for men. Symons had assembled evidence to support the claim that "throughout most of human evolutionary history, hunting, fighting, and that elusive activity, 'politics,' were highly competitive, largely male domains" (p. 163). Then, following widespread usage, he classified these activities as paternal effort. Instead of more parental effort resulting in less mating effort, the activities of human males are more of both.

It does sometimes happen that a single activity maximizes two things at once, but that is rare in a finite world. Optimality models have proven to be powerful tools for explaining the diversity of life because most of the time trade-offs are inescapable (Maynard Smith 1978, 1982; Seger & Stubblefield 1996). In the case at hand, the adaptive function of men's hunting, a trick of language draws attention away from the trade-offs faced by individuals. Characterizing many activity differences between men and women as a "sexual division of labor" is the legacy of a long history of talking about human behavior from a societal point of view. As Symons (1979: 147) himself noted, there is a tendency to use the passive voice when talking of the sexual division of labor, as for example, in the claim, "In all known societies the defensive role is assigned to adult males. . . . The passive construction conceals the subject, the agent who did the assigning. When the subject does materialize in such statements, almost invariably it turns out to be 'society'." The implication is that work is divided to serve the production goals of some larger entity, the family, domestic group, or household, which assigns tasks by sex and age. If families are assumed to be units of common interest, the conflicts of interest between and within the sexes (in theory the heart of the story) become, at best, secondary complications.

In spite of noting this problem, Symons still talked of a "sexual division of labor," but he also considered some likely reproductive benefits to the individuals involved. He followed others, however, in claiming that "if warfare was often over land and other scarce resources from which the winning males' offspring benefited, male fighting was in part paternal investment because the offspring of the winning males benefit." The objection raised by Kurland and Gaulin to a similar explanation for troop defense in baboons applies here. Resource benefits that go to the winner also go to all on the winning side, including the offspring of males who did not pay the cost of the fight.

Public Goods and Collective Action Problems

Parental benefits are insufficient to explain community defense because it is a public good, consumed by all group members whether or not they pay to supply it. Public goods are distinguished from private goods by two features usefully labeled excludability and subtractability (Ostrom & Ostrom 1977). Consumers cannot be excluded from using a public good; and consumption by one does not subtract from the benefits available to others. With private goods, on the other hand, owners incur no cost in excluding other users, and any benefits consumed subtract from those remaining.[1] Few goods or services are perfectly public or perfectly private, but some are more like public goods than others. The cost of exclusion is higher for some goods, and consumption of a unit subtracts less from the remaining benefits of some things than others. In the language cited earlier, goods and services can be more or less depreciable.

The more public a good, the more likely that the value it has for consumers will not motivate commensurate supply. Eliciting financial support from public radio listeners exemplifies the undersupply problem long recognized by economists (Samuelson 1954): Why pay for what you get free? An array of collective action problems (Olson 1965) that arise around these issues have engaged political philosophers for centuries (Hardin 1982). Some of the same problems have been recognized by evolutionary biologists under the heading of individual versus group selection (Williams 1966). Free-rider problems, game theoretic payoff structures like the prisoner's dilemma (e.g., Schelling 1978), and Garrett Hardin's (1968) well known "tragedy of the commons" have come to be frequent illustrations of collective action problems (Hawkes 1992a). These arise whenever the pursuit of individual interests does not promote the welfare of the collective. As Russell Hardin (1982) notes, Adam Smith extolled the beneficial effects of the famous invisible hand, but the back of that hand is just as ubiquitous—and paradoxically both coincidence and conflicts of interest can operate to elevate as well as to depress economic productivity (Hardin 1982, Hirshleifer 2001).

The collective action problem with troop defense that Kurland and Gaulin flagged has two implications. On one hand, since all the troop infants and juveniles get protection, there is no differential benefit for the defenders' own offspring. Parental benefits are thus unlikely to explain the continuing expenditure any male puts into defense. On the other hand, males are observed to "defend the troop," a costly behavior that begs to be explained. Other fitness-related benefits that, unlike offspring protection, are private gains for the defenders may be important in the explanation. Kurland and Gaulin hypothesized that instead of parenting, the defending males earn mating benefits. A parallel argument applies to the human case. As noted above, Symons (1979: 162) makes it: "the evidence suggests that in hunting, as in fighting, human males are effectively in competition for females, and that there are substantial differences among males in competitive abilities."

Collective action problems around community defense can also arise with hunting large prey. Although food is the classic illustration of a private good, some food resources are much more like public than private goods. While a morsel of food goes into only one stomach, and each bite subtracts from the remainder available, large prey come only in big packages, tens or even hundreds of kilos of meat at a time. Acquired by anyone, a large carcass can then be consumed by many. If the prey are taken unpredictably, one hunter may succeed when all others fail to make a capture. Then the cost to the hunter (or anyone else) for excluding other hungry users may be substantial. If the hungry claimants are armed with lethal weapons, the cost of defense can be especially high and the wisdom of hospitality especially clear. The bigger the prey, the higher the cost of not sharing, and the less any additional consumer subtracts from available consumption benefits. As costs go up and benefits go down, marginal gains for trying to exclude other claimants disappear altogether (Blurton Jones 1984, 1987). Hunting creates a collective action problem whenever hunted resources are like public goods (Hawkes et al. 1991, Hawkes 1992a, 1993).

Symons (1979: 158) concluded that "among all hunter-gatherers as well as among many other peoples, the primary economic activity of adult males is hunting, and nowhere do men hunt only for themselves, the fruits of the hunt are always shared with women and children." A long and rich ethnographic record confirms that food, especially meat, is often widely shared (e.g., Sahlins 1972, Kelly 1995, Wiessner 1996). The sharing is usually labeled exchange and/or reciprocity by social anthropologists, a usage that leads biologically trained readers to assume that something like Trivers' (1971) "reciprocal altruism" might explain the sharing—hunters doling out shares of meat in return for shares repaid in future when hunting fortunes are reversed. But the sharing that is common in hunter-gatherer societies (Sahlins 1972) is not what Karl Polanyi (1957) called "market exchange." As two disciplines studying social behavior, sociobiology and social anthropology are "divided by a common language." Trivers' model is essentially a market model of private goods and services exchanged by owners with negligible externalities, that is, no effects external to the exchanging parties. Individuals can benefit from this kind of reciprocity as long as they "keep score" and terminate transfers to any who fail to repay. They benefit as long as sharing is contingent on repayment.

The quid pro quo accounting required for such exchange strategies to be evolutionarily stable (Axlerod & Hamilton 1981) is explicitly denied in a multitude of ethnographies that detail food sharing in kinship societies (e.g., compilations in Dowling 1968, Lévi-Strauss 1969, Sahlins 1972, Kelly 1995). Insistence that the meat of large prey is not the hunter's private property is repeatedly confirmed (e.g., Marshall 1976, Barnard & Woodburn 1988, Wiessner 1996, Woodburn 1998). It is possible, of course, that ethnographers have been mistaken, or that their subjects' descriptions of their own norms do not reflect actual behavior. Sometimes, however, ethnographers see and report that the successful hunter plays no part in the distribution, so is in no position to direct shares according to his personal accounts (e.g., Hill & Kaplan 1988).

Ethnographers have also used observations of actual distributions to investigate whether food sharing is contingent on food repayments. In a few cases, for some foods, researchers report contingency (Gurven et al. 2000, Hames 2000). But more frequently, and especially for large game animals, there is little indication that the distribution of shares depends on meat repayments to the hunter (e.g., Marshall 1976, Lee 1979, Kaplan & Hill 1985a, Bliege Bird & Bird 1997, Woodburn 1998, Hawkes et al. 2001a). The quantitative investigations add to the evidence against the view that meat is the hunter's private property (Hawkes 2001). The more often large prey are like public goods, the more likely it is that parenting benefits do not provide a general explanation for why men hunt them (Hawkes 1990, 1991, 1993; Hawkes et al. 2001b). The same collective action problem that undercuts parenting explanations for community defense applies to hunting big animals: the food benefits for the hunter's effort go not only to his wife and offspring but to the wives and offspring of less successful and less hardworking hunters as well. This is of course an outcome the hunter can anticipate. Knowing that most of the meat will be claimed by others, he still sets out to hunt.

These observations recall an old idea in cultural anthropology that emphasized women's lack of economic dependence on husbands among hunter-gatherers. Westermarck's near contemporary Lewis Henry Morgan (1870) speculated that monogamy was actually a late development in the evolution of human society. Frederick Engels ([1884] 1972) relied on Morgan in his scenario of *The Origin of the Family, Private Property and the State*. In the mid-twentieth century, Eleanor Leacock, an ethnographer and ethnohistorian of the North American Montagnais, agreed with Engels. Other cultural anthropologists have developed and defended similar views (e.g., Collier & Rosaldo 1981). Leacock (1972: 29) surmised that civilization transformed "the nuclear family into the basic economic unit of society, within which a woman and her children became dependent on an individual man." Leacock hypothesized that, in "primitive communal society . . . the economy did not involve the dependence of the wife and children on the husband. All major food supplies, large game and produce from the fields, were shared among a group of families" (p. 33). Like other hunter-gatherer ethnographers, Leacock was especially impressed with the constant sharing and egalitarian character of hunter-gatherer social life.

Morton Fried (1967: 33) classically defined an egalitarian society as one in which "there are as many positions of prestige in any given age-sex grade as there are persons capable of filling them." Arguments long favored in cultural anthropology to explain such patterns pointed to the group benefits of wide sharing which leveled differences and prevented wealth accumulation (e.g., Fried 1967, Sahlins 1972, Wiessner 1996). By contrast, most behavioral ecologists ever since Williams (1966), and some social scientists ever since Schelling (1960) and Olson (1965), have seen fatal flaws in explanations for social behavior that ignore payoffs to individuals and rely only on group-level functions. David S. Wilson and colleagues (e.g., Wilson 1983, 1998; Sober & Wilson 1998) argue that these critiques of group selection led to premature rejection by sociobiologists of explanations for human social behavior in terms of group benefits. Emphasizing that selection is a multilevel process, the "new group selectionists" argue that well-known patterns of human cooperation remain inexplicable when the focus is improperly restricted to within-group effects on individuals. Christopher Boehm (1993, 1999a, b) agrees with this criticism. He has revived the group benefit explanations for egalitarian societies previously advanced by an earlier generation of cultural anthropologists, concluding that patterns of distinctly cooperative human behavior defy the "standard evolutionary paradigm" (Boehm 1999b: 209).

The counterargument is central to the issues of this chapter: It turns on a full tally of the fitness-related costs and benefits to individuals for both meat sharing and hunting. My disagreement with Wilson and Boehm is not about the logic of multilevel selection models, but about the empirical assessment of the costs and benefits to the individual actors. Boehm (1999b: 209) notes that "among many mobile hunter-gatherers the most able hunters willingly acquire game for the entire group, and this meat is widely distributed with a minimum of bickering even though unrelated families are sharing it." I agree with this ethnological generalization, but Boehm (1999b: 210) surmises that "if band members are disposed to assist nonkin in the band, this is likely to require a group selection argument". His explicit inference is that sharing imposes a net cost on sharers compared to nonsharers, and that the "able hunters" do not get differential fitness benefits for their effort. There is evidence to the contrary.

Ethnography shows a considerable cost to *not* sharing in these communities (Blurton Jones 1984, 1987; Petersen 1993). If the prey shared are like public goods, then those claiming shares are appropriating from the public domain. Under these circumstances, anyone trying to exclude claimants is interfering with their "rights" and is likely to pay a cost for that interference. As to the work invested in procuring the prey in the first place, the point I underline here is that the hunter gets benefits other than the meat. The magnitude and character of both his costs and his benefits should be assessed empirically. Getting the correct estimate of the costs and benefits for both sharing and hunting is just as important under the banner of multilevel selection as it is within a "standard evolutionary paradigm."

Why Do Men Hunt?

A collective action framework directs attention to benefits that go only to the individual hunters. What Olson (1965) called "selective incentives," private gains that go only to the suppliers of public goods, are the thing to look for. If men in hunter-gatherer communities are often choosing to specialize in resources that go mostly to others, this should not obscure the benefits for this effort that go only to hunters themselves.

Like egg guarding or troop defense, the meat of large prey is consumed by many, not just the supplier's own family. But the hunter is the one who gets credit for supplying it. Distinctions between the credit and the meat are important. Examining hunter-gatherer ethnology, Fried (1967: 34) concluded that conventions for assigning "ownership" of the prey are "all techniques by which credit for bringing game to camp is randomized." He, and others, saw these as "leveling mechanisms," with leveling further reflected in the famously self-effacing style of hunters (Lee 1969, Sahlins 1972, Harris 1977, Hawkes 1992b). But neither practice interferes with widespread interest in the actual events of a hunt.

Among foragers, the behavior of hunters and their prey and the circumstances of the death of particular animals are usually topics of endless interest (Blurton Jones & Konner 1976, Marshall 1976, Lee 1979, Hawkes 2000). In contrast to Fried's claim about "randomizing credit," the evidence indicates that repeated storytelling assures that details of each hunt are widely known and well remembered. Ethnographers can collect lifetime tallies of the large prey killed by each hunter (Lee 1979), and rankings of hunters' success rates that closely correlate with records of kills that the ethnographers observed (Kaplan & Hill 1985a, Blurton Jones et al. 1997). Conventions that assign "ownership" of prey animals delegate duties associated with distributing the meat (Marshall 1976), but the carcass is no more this "owner's" private property than it is the hunter's (Marshall 1976, Barnard & Woodburn 1988, Woodburn 1998, Hawkes 2001). And such conventions do not interfere with eager interest in who killed the animal.

Since everyone pays attention to hunters' successes, credit to successful hunters develops into reputations. Like other costly signals of quality (Zahavi 1975, 1977, 1995; Zahavi & Zahavi 1997; Grafen 1990), hunting reputations benefit hunters as this information is used by others in the many decisions of social life. Because people are already interested in the meat, hunting large animals can be a particularly effective way for men to display how desirable as allies and dangerous as competitors they are. Everyone has two reasons to be pay attention to hunters' successes: they get meat and they get information about the hunter's qualities (Hawkes & Bleige Bird 2002).

The nonfood benefits for the hunters themselves can be substantial. By hunting, a man maintains or improves his social standing among other men. His social position affects whether and how much other men defer to him, and that affects his value as an ally to both men and women—with consequences for his mating success (Kaplan & Hill 1985b; Hawkes 1993; Hill & Hurtado 1996; Blurton Jones et al. 1997, 2000; Bliege Bird et al. 2001; Hawkes et al. 2001b). Costly signaling models can help highlight mating benefits that explain why men hunt while women gather, as well as other aspects of foraging differences between the sexes (Bliege Bird 1999, Bliege Bird et al. 2001). The same framework applies, as Thorstein Veblen ([1899] 1922) proposed more than a century ago, to many other puzzles of human behavior in which social benefits can explain what is mysterious on more "utilitarian" grounds (e.g., Boone 1998, Neiman 1998, Roberts 1998, Frank 1999, Miller 2000, Smith & Bliege Bird 2000, Gintis et al. 2001).

Arguments about the nonfood benefits men earn from hunting make the emerging evidence about hunting among chimpanzees (*Pan troglodytes*) of special interest. When Washburn elaborated his version of the hunting hypothesis, chimpanzees were not known to hunt. Now evidence from a wide array of study sites shows chimpanzees to be active and effective hunters (summaries in Stanford 1996, 1999; Mitani et al. 2002). As among humans, chimpanzee hunting is a male specialty and prey are more widely shared than other foods, although little is eaten by females and juveniles. Several lines of evidence now indicate that hunting and meat sharing are costly, with little nutritional gain for the time spent. Instead of a feeding strategy, chimpanzee hunting appears to be motivated by male status competition (Mitani & Watts 2001).

In chimpanzees and in other primates, male activities have large effects on females. As Smuts (1992: 5) noted, "when we look closely, we find that in many primates, hardly an aspect of female existence is not constrained in some way by the presence of aggressive males." Yet there is no temptation with nonhuman species to describe these aspects of male behavior as part of a "sexual division of labor." In humans, lethal weapons change the cost of aggressive encounters, making the potential effects of male behavior even greater. Armed opponents, whatever their relative physical strength, can be much more dangerous. Models that include contest costs predict a tendency to adopt conventional solutions to contests more quickly as those costs increase (Maynard Smith 1982, Blurton Jones 1987, Clutton Brock & Parker 1995). This line of argument points toward the greater use of conventional solutions to contests among men than among chimpanzees (Hawkes 2000). But it does not make the male competition less important.

In addition to the use of lethal weapons, men also differ from chimpanzees by often hunting prey larger than their own body size. Large prey also mean more meat with each kill. More meat means more consumers and more general interest in the success of hunters. The more important hunting reputations are to a man's social standing, the more men are likely to hunt. Resulting levels of meat procurement can be high enough that meat becomes a substantial component of the average diets of women and children.

Frank Marlowe (2001) has linked cross-cultural variation in men's average subsistence contribution to variation in women's average reproductive success. Using a worldwide sample of foraging societies, Marlowe found that increases in men's relative contribution to the diet are associated with increases in the average number of surviving children per woman. While the reasons for variation in male economic production between societies are not clear, Marlowe's result may imply that increased production from men means more food available to women, allowing them to increase fertility without commensurate losses in offspring survival. If so, this is an especially interesting instance of a general primate pattern in which more food—from whatever sources—means more surviving offspring.

Large effects on the number of children that women can rear are temptation to classify the economic production from men as parental effort (or indirect paternal investment, e.g., Kaplan et al. 2000, Marlowe 2000). Of course men can and sometimes do expend parental effort. But even when hunting has large effects on average food consumption, it is not paternal effort if the usual hunter-gatherer patterns hold. As long as the wives and children of other men get about the same amount of meat from a hunter's kills as his own children do, his fitness payoffs cannot be differential nutrition for his offspring (Kaplan & Hill 1985a; Hawkes et al. 1991, 2001a, b). If the differential benefit to the hunter himself, the private benefit that only he gets for his work, is the credit for his kills, then his payoff depends on the effects of his hunting reputation. High status could bring differential treatment of his children by others (Kaplan & Hill 1985b, Hawkes 1990). But the evidence available shows that hunting reputations affect men's fitness largely through mating advantages. Both female choice (Kaplan and Hill 1985b) and male competition are implicated (Hill & Hurtado 1996, Hawkes et al. 2001b).

What About Pair Bonds?

The tenacity of the assumption that men's contribution to subsistence is paternal effort turns partly on the ubiquity of human pair bonds. Disagreements about definitions of marriage continue, but in all ethnographic reports of human communities, men and women form special relationships with mating partners that involve more than copulation. People may not mate for life and partnerships may not be sexually exclusive (Beckerman & Valentine 2002), but men and women do form persistent emotional attachments. The hunting hypothesis has long been the favored explanation for this human tendency. The contrary hypothesis, that hunting is driven by male status competition, cancels the provisioning reason for women to pair with hunters. If women can consume food procured by men whether or not they are married to them, why marry?

Special relationships in other primates where males supply no provisioning indicate that females can gain other things from partners, especially protection from other males. Benefits for males are initially less clear. But modeling results show that male mating competition alone can make pair bonds advantageous to males. One set of models (Hawkes et al. 1995) focused only on male strategies, with mating effort and parental effort assumed to be mutually exclusive to clarify the relative strength of their fitness effects. In these simulations, pairing was the usual outcome, but not because males specialized in parenting. Under a wide array of parameter conditions, including potentially large effects on offspring survival, the model males earned higher fitness payoffs for mate guarding than for parental effort. No female choice was included, yet pair bonds—each male putting all his reproductive effort into guarding a female—were the usual result of the simulations.

Pairing patterns among hunter-gatherers show some parallels with these results. Nicholas Blurton Jones and colleagues (2000) examined variation in pair bond stability among four hunter-gatherer societies for which both operational sex ratio and the effects of father's presence on child survival could be estimated. In this sample, the stability of pair bonds varied directly with the intensity of male mating competition and was unrelated to variation in "father effects." Divorce rates were lower where there were fewer paternity opportunities per male, highlighting the mate-guarding advantages that men may earn from marriage. Among the Hadza, foragers in East Africa and one of the cases in this sample, differences in father's hunting success had no direct effect on children's nutritional welfare, as expected given the wide sharing of meat. Better hunters were, however, found to be married to harder working wives whose children's nutrition reflected the differential work of their mothers and grandmothers. These results highlighted the advantages that marriage may provide to husbands more than any advantages to their wives (Hawkes et al. 2001b).

That analysis may underestimate the value of a protector (Wrangham et al. 1999). Because human children are more dependent, longer on provisioning by mothers and grandmothers (Hawkes et al. 1998, 2003), harassment by males may impose higher costs on women than it does on other female primates (Blurton Jones et al. 2000). Hypotheses about partners as protectors have been more fully developed by those investigating pair bonds in other primates. These hypotheses, already relevant on phylogenetic grounds, become more likely on grounds of socioecology as well. Smuts (1992: 9–10) said:

> Most reconstructions of human evolution have assumed that pair bonds evolved to facilitate the exchange of resources between the sexes . . . , often with a particular emphasis on the need for increased male parental investment in the form of meat. . . . These scenarios assume that females benefited from pair bonds because they gained meat from males. Given the importance of male sexual coercion among non-human primates, and especially among our closest living relatives (chimpanzees, gorillas, and orangutans) however, we should carefully consider the alternative hypothesis that pair bonds benefited females initially because of the protection mates provided against other males (including protection from infanticide).

Smuts developed a hypothetical scenario that reflects the importance of coalitions in male mating competition in many other primate species.

> I suggest that, among hominids, the kind of tolerance we see among male allies in nonhuman primates became formalized as each male began to develop a long term mating association with a particular female or females (a trend foreshadowed in savanna baboons).
> . . . Viewed in this light, human pair bonds, and therefore human marriage, can be considered a means by which cooperating males agree about mating rights, respect (at least in principle) one another's possession of particular females, protect their mates and their mates' children from aggression by other men, and gain rights to coerce their own females with reduced interference by other men. (Smuts 1992: 10–11)

Sarah Mesnick (1997) assembled data on a wide array of taxa, including her own on elephant seals, to further document the high costs that male aggression can impose on females. She formulated the "bodyguard hypothesis," in which protection is a primary criterion of female choice. Agreeing with Smuts, she showed that "alliances with protective males can be an effective female behavior that reduces vulnerability to aggression from other, conspecific males. It is also a factor to consider in explaining . . . human pairbonding" (Mesnick 1997: 207). Wilson and Mesnick (1997) tested predictions of the bodyguard hypothesis for humans on Canadian records of sexual assault homicides and reported nonlethal sexual aggression. They found, as predicted, that married women were less at risk from other men.

Helen Fisher (1992) suggested a scenario for the evolution of human pair bonds that deserves special attention for its use of some of these ideas. She argued that while pair bonds are "the hallmark of the human animal" (p. 66), these relationships are often neither lifelong nor fully exclusive. While assuming that the hunting hypothesis in which ancestral males were paternal provisioners was generally correct, Fisher proposed that "our ancestors only needed to form pair bonds long enough to rear their young through infancy" (p. 153). Assembling data from a wide array of sources, she showed that marriages most frequently break up after about four years, and surmised an ancient tendency to pair just long enough "to raise a single dependent child through infancy" (p. 154). This "four-year itch," however, would make fathers an unreliable source of help at just the time that human mothers need it most. One of the salient ways that human offspring differ from other primates is that our children are unable to feed themselves at weaning. It is when a mother shifts her effort to the next baby that help in caring for the still-dependent toddler is so crucial.

Others have attributed the evolution of our extended juvenile dependence to help from provisioning fathers. But the general primate patterns, and the data and arguments about hunting and sharing among modern hunter-gatherers are evidence against that scenario. An alternative hypothesis links our overlapping dependents to life history shifts resulting from changes in female foraging strategies that include increased longevity and delayed maturity (Hawkes et al. 1998, 2003; O'Connell et al. 1999, 2002; Alvarez 2000; Hawkes 2003). The general health and vigor of peri- and postmenopausal women and the late maturity of human adolescents distinguishes us from other apes. Both grandmothers and older siblings provide a source of help to weanlings when mothers bear newborns. This means our pattern of cooperative breeding (Hrdy 1999, 2001) does not depend on fathers. Though they may sometimes be enlisted, other help is available when fathers trade-off parenting for mating.

Human mothers are well equipped to deal with the daily care and feeding of infants, so much so that, as with other primates, an infant's death advances the possible time of a next conception. This means we share a vulnerability to infanticidal males, since, under some circumstances, that could increase the paternity chances for a man unlikely to be the infant's father. Helen Fisher recognized the dangers of infanticide to primates but made less of this than have others cited here. Nevertheless she recognized that special relationships between males and females in other primates are the likely evolutionary foundation for human pair bonds. "Olive baboons provide a . . . model . . . for how pairbonding, the nuclear family, and divorce could have evolved" (Fisher 1992: 154).

We know that the hominid radiation included genera unlike any now living, that modern humans are a very recent species, and that other members of our own genus were different from us. Since all extinct hominids not only differed from modern humans but also from each other in ways largely unknown, their mating arrangements may have been quite diverse. But that does not make all possibilities equally likely. Like us, they were all large-bodied primates. We know that primates have bigger brains and slower life histories for body size than other mammals; and that marked sociality, with continuous year-round mixed-sex groups is a (related) hallmark of our order. Individuals interact repeatedly over long time periods with the same others, who can be their most important competitors as well as potential allies. Consequently, capacities and strategies for managing social relationships are especially well developed in primates (Harcourt 1992). Conflicts of interest between (and within) the sexes have especially complex repercussions in the context of these life histories and this sociality (van Schaik & Janson 2000).

In light of what we now know about other primates, and our own species—especially, but not only, in hunting and gathering communities—the hypothesis that human pair bonds developed as a consequence of paternal provisioning by ancestral males should be viewed with serious skepticism. Numerous lines of theory and evidence stand against it. At the same time, the special relationships between males and females in other primates and their links to dangers posed by male mating competition are increasingly well described. On phylogenetic grounds alone, those patterns should be the first place to look for hypotheses about the evolution of our emotional attachments. Increasing understanding of the socioecology of those relationships makes them an even more promising foundation for hypotheses about the evolution of human pair bonds.

Acknowledgments   I thank Helen Alvarez, Carol Berman, Doug Bird, Rebecca Bird, Monique Borgerhoff-Mulder, Bernard Chapais, Nick Blurton Jones, Eric Charnov, Robert Hinde, Sarah Hrdy, Mark Jeffries, Jim O'Connell, Barbara Smuts, Craig Stanford, and Carel van Schaik for their useful comments and excellent advice.

Note

1. Excludability and subtractability are seen as independent dimensions, so their intersection defines four classes of goods (Ostrom & Ostrom 1977). The two remaining are common pool resources, which are subtractable but not excludable, and toll (or club) goods, which are excludable but not subtractable. The consequences and the interplay of these dimensions are topics of a large and diverse literature in public choice and property rights economics.

References

Alvarez, H. P. 2000. Grandmother hypothesis and primate life histories. *Am. J. Phys. Anthropol.*, 113, 435–450.

Andersson, M. 1994. *Sexual Selection*. Princeton, NJ: Princeton University Press.

Axelrod, R. & Hamilton, W. D. 1981. The evolution of cooperation. *Science*, 211, 1390–1396.

Barnard, A. & Woodburn, J. 1988. Property power and ideology in hunting and gathering societies: an introduction. In: *Hunters and Gatherers 2: Property, Power and Ideology* (Ed. by T. Ingold, D. Riches, & J. Woodburn), pp. 4–31. New York: Berg.

Bart, J. & Tornes, A. 1989. Importance of monogamous male birds in determining reproductive success: evidence for house wrens and review of male-removal experiments. *Behav. Ecol. Sociobiol.*, 24, 109–116.

Barton, R. A. 2000. Socioecology of baboons: the interaction of male and female strategies. In: *Primate Males: Causes and Consequences of Variation in Group Composition* (Ed. by P. M. Kappeler), pp. 97–107. Cambridge: Cambridge University Press.

Barton, R. A., Byrne, R. W., & Whitten, A. 1996. Ecology, feeding competition and social structure in baboons. *Behav. Ecol. Sociobiol.*, 38, 321–329.

Bateman, A. J. 1948. Intrasexual selection in *Drosophila*. *Heredity*, 2, 349–368.

Baylis, J. R. 1981. The evolution of paternal care in fishes, with reference to Darwin's rule of male sexual selection. *Environ. Biol. Fishes*, 6, 223–251.

Beckerman, S. & Valentine, P. 2002. Introduction: the concept of partible paternity among native South Americans. In: *Cultures of Multiple Fathers: The Theory and Practice of Partible Paternity in Lowland South America* (Ed. by S. Beckerman & P. Valentine), pp. 1–13. Gainesville, FL: University Press of Florida.

Bell, D. 1997. Defining marriage and legitimacy. *Curr. Anthropol.*, 38, 237–253.

Berkovitch, F. 1995. Female cooperation, consortship maintenance, and male mating success in savanna baboons. *Anim. Behav.*, 50, 137–149.

Birkhead, T. & Moller, A.1992. *Sperm Competition in Birds: Its Evolutionary Causes and Consequences*. London: Academic Press.

Black, J. M. 1996. *Partnerships in Birds*. Oxford: Oxford University Press.

Bliege Bird, R. B. 1999. Cooperation and conflict: the behavioral ecology of the sexual division of labor. *Evol. Anthropol.*, 8, 65–75.

Bliege Bird, R. & Bird, D. 1997. Delayed reciprocity and tolerated theft: the behavioral ecology of food sharing strategies. *Curr. Anthropol.*, 38, 49–78.

Bliege Bird, R., Smith, E. A., & Bird, D. 2001. The hunting handicap: costly signaling in male foraging strategies. *Behav. Ecol. Sociobiol.*, 50, 9–19.

Blurton Jones, N. G. 1984. A selfish origin for food sharing: tolerated theft. *Ethol. Sociobiol.*, 5, 1–3.

Blurton Jones, N. G. 1987. Tolerated theft, suggestions about the ecology and evolution of sharing, hoarding, and scrounging. *Soc. Sci. Info.*, 26(1), 31–54.

Blurton Jones, N. G., Hawkes, K., & O'Connell, J. F. 1997. Why do Hadza children forage? In: *Uniting Psychology and Biology: Integrative Perspectives on Human Development* (Ed. by N. Segal, G. E. Weisfeld, & C. C. Weisfeld), pp. 279–313. Washington, DC: American Psychological Association.

Blurton Jones, N. G. & Konner, M. J. 1976. !Kung knowledge of animal behavior (or: the proper study of mankind is animals). In: *Kalahari Hunters: Studies of the !Kung San and Their Neighbors* (Ed. by R. B. Lee & I. DeVore), pp. 325–348. Cambridge, MA: Harvard University Press.

Blurton Jones, N. G., Marlowe, F., Hawkes, K., & O'Connell, J. F. 2000. Paternal investment and hunter-gatherer divorce. In: *Adaptation and Human Behavior: An Anthropological Perspective* (Ed. by L. Cronk, N. Chagnon, & W. Irons), pp. 61–90. New York: Aldine de Gruyter.

Boehm, C. 1993. Egalitarian society and reverse dominance hierarchy. *Curr. Anthropol.*, 35, 178–180.

Boehm, C. 1999a. *Hierarchy in the Forest: The Evolution of Egalitarian Behavior*. Cambridge, MA: Harvard University Press.

Boehm, C. 1999b. The natural selection of altruistic traits. *Hum. Nat.*, 10, 205–252.

Boone, J. L. 1998. The evolution of magnanimity: when is it better to give than to receive? *Hum. Nat.*, 9, 1–21.

Calvin, W. H. 2002. *A Brain for All Seasons: Human Evolution and Abrupt Climate Change*. Chicago: University of Chicago Press.

Cartmill, M. 1994. *A View to Death in the Morning: Hunting and Nature Through History.* Cambridge, MA: Harvard University Press.

Chapais, B. 1992. The role of alliances in the social inheritance of rank among female primates. In: *Coalitions and Alliances in Humans and Other Animals* (Ed. by A. H. Harcourt & F. B. M. de Waal), pp. 29–60. Oxford: Oxford University Press.

Chapais, B. 1995. Alliances as a means of competition in primates: evolutionary, developmental, and cognitive aspects. *Ybk. Phys. Anthropol.*, 38, 115–136.

Clutton Brock, T. H. 1991. *The Evolution of Parental Care.* Princeton, NJ: Princeton University Press.

Clutton Brock, T. & Parker, G. A. 1992. Potential reproductive rates and the operation of sexual selection. *Q. Rev. Biol.*, 67, 437–456.

Clutton Brock, T. & Parker, G. 1995. Sexual coercion in animal societies. *Anim. Behav.*, 49, 1345–1365.

Collier, J. F. & Rosaldo, M. Z. 1981. Politics and gender in simple societies. In: *Sexual Meanings* (Ed. by S. Ortner & H. Whitehead), pp. 275–329. New York: Cambridge University Press.

Daly, M. & Wilson, M. 1987. The Darwinian psychology of discriminative parental solicitude. *Nebr. Symp. Motivation*, 35, 91–144.

Daly, M., Wilson, M., & Weghorst, S. 1982. Male sexual jealousy. *Ethol. Sociobiol.*, 3, 11–27.

Darwin, C. [1871] 1981. *The Descent of Man and Selection in Relation to Sex.* Reprint, Princeton, NJ: Princeton University Press.

Darwin, C. [1872] 1965. *The Expression of the Emotions in Man and Animals.* Reprint, Chicago: University of Chicago Press.

Dart, R. A. 1953. The predatory transition from ape to man. *Int. Anthropol. Ling. Rev.*, 1, 201–217.

Deacon, T. W. 1997. *The Symbolic Species: The Co-evolution of Language and the Brain.* New York: Norton.

Dowling, J. H. 1968. Individual ownership and the sharing of game in hunting societies. *Am. Anthropol.*, 70, 502–507.

Edwards, A. W. F. 1998. Natural selection and the sex ratio: Fisher's sources. *Am. Nat.*, 151, 564–569.

Engels, F. [1884] 1972. *Origin of the Family, Private Property and the State.* Reprint, with an introduction by Eleanor Burke Leacock, New York: International Publishers.

Fisher, H. E. 1981. *The Sex Contract: The Evolution of Human Behavior.* New York: William Morrow.

Fisher, H. E. 1992. *Anatomy of Love: The Natural History of Monogamy, Adultery, and Divorce.* New York: Norton.

Fisher, R. A. 1930. *The Genetical Theory of Natural Selection.* Oxford: Oxford University Press.

Frank, R. 1999. *Luxury Fever: Why Money Fails to Satisfy in an Era of Excess.* New York: Free Press.

Fried, M. H. 1967. *The Evolution of Political Society: An Essay in Political Anthropology.* New York: Random House.

Gintis, H., Smith, E. A., & Bowles, S. 2001. Costly signaling and cooperation. *J. Theor. Biol.*, 213, 103–119.

Goodenough, W. H. 1970. *Description and Comparison in Cultural Anthropology.* Chicago: Aldine.

Gowaty, P. A. 1966a. Battle of the sexes and the origins of monogamy. In: *Partnerships in Birds* (Ed. by J. M. Black), pp. 21–52. Oxford: Oxford University Press.

Gowaty, P. A. 1996b. Field studies of parental care in birds: new data focus questions on variation among females. *Adv. Study Behav.*, 25, 477–531.

Grafen, A. 1990. Biological signals as handicaps. *J. Theor. Biol.*, 144, 517–46.

Gross, M. & Sargent, R. 1985. The evolution of male and female parental care in fishes. *Am. Zool.*, 25, 807–822.

Gubernick, D. J. 1994. Biparental care and male-female relations in mammals. In: *Infanticide and Parental Care* (Ed. by S. Parmigiani & F. S. vom Saal), pp. 427–463. Chur, Switzerland: Harwood Academic.

Gurven, M., Hill, K., Hurtado, A., & Lyles, R. 2000. Food transfers among Hiwi foragers of Venezuela: tests of reciprocity. *Hum. Ecol.*, 28, 171–214.

Hames, R. 2000. Reciprocal altruism in Yanomamo food exchange. In: *Adaptation and Human Behavior: An Anthropological Perspective* (Ed. by L. Cronk, N. Chagnon, & W. Irons), pp. 397–416. New York: Aldine de Gruyter.

Harcourt, A. H. 1989. Social influences on competitive ability: alliances and their consequences. In: *Comparative Socioecology of Mammals and Man* (Ed. by V. Standon & R. Foley), pp. 223–242. London: Blackwell.

Harcourt, A. H. 1992. Coalitions and alliances: are primates more complex than non-primates? In: *Coalitions and Alliances in Humans and Other Animals* (Ed. by A. H. Harcourt & F. B. M. de Waal), pp. 445–472. Oxford: Oxford University Press.

Hardin, G. 1968. The tragedy of the commons. *Science*, 162, 1243–1248.

Hardin, R. 1982. *Collective Action*. Baltimore, MD: Johns Hopkins University Press.

Harris, M. 1977. *Cannibals and Kings: The Origins of Cultures*. New York: Random House.

Hawkes, K. 1990. Why do men hunt? Some benefits for risky strategies. In: *Risk and Uncertainty in Tribal and Peasant Economies* (Ed. by E. Cashdan), pp. 145–166. Boulder, CO: Westview Press.

Hawkes, K. 1991. Showing off: tests of an hypothesis about men's foraging goals. *Ethol. Sociobiol.*, 12, 29–54.

Hawkes, K. 1992a. Sharing and collective action. In: *Evolutionary Ecology and Human Behavior* (Ed. by E. Smith & B. Winterhalder), pp. 269–300. New York: Aldine de Gruyter.

Hawkes, K. 1992b. On sharing and work (a comment on Bird-David). *Curr. Anthropol.*, 33(4), 404–407.

Hawkes, K. 1993. Why hunter-gatherers work: an ancient version of the problem of public goods. *Curr. Anthropol.*, 34(4), 341–361.

Hawkes, K. 2000. Big game hunting and the evolution of egalitarian societies. In: *Hierarchies in Action: Cui Bono?* (Ed. by M. Deihl), pp. 59–83. Carbondale, IL: Southern Illinois University. Center for Archaeological Investigations, Occasional Paper No. 27.

Hawkes, K. 2001. Is meat the hunter's property? Ownership and explanations of hunting and sharing. In: *Meat-Eating and Human Evolution* (Ed. by C. Stanford & H. Bunn), pp. 219–236. Oxford: Oxford University Press.

Hawkes, K. 2003. Grandmothers and the evolution of human longevity. *Am. J. Hum. Biol.*, 15, 380–400.

Hawkes, K. & Bliege Bird, R. 2002. Showing-off, handicap signaling, and the evolution of men's work. *Evol. Anthropol.*, 11, 58–67.

Hawkes, K., O'Connell, J. F., & Blurton Jones, N. G. 1991. Hunting income patterns among the Hadza: big game, common goods, foraging goals, and the evolution of the human diet. *Phil. Trans. Royal Soc. Lond., B*, 334, 243–251.

Hawkes, K., O'Connell, J. F., & Blurton Jones, N. G. 2001a. Hadza meat sharing. *Evol. Hum. Behav.*, 22, 1–30.

Hawkes, K., O'Connell, J. F., & Blurton Jones, N. G. 2001b. Hunting and nuclear families: some lessons from the Hadza about men's work. *Curr. Anthropol.*, 42, 681–709.

Hawkes, K., O'Connell, J. F., & Blurton Jones, N. G. 2003. Human life histories: primate tradeoffs, grandmothering socioecology, and the fossil record. In: *The Role of Life Histories in Primate Socioecology* (Ed. by P. Kappeler & M. Pereira), pp. 204–227. Chicago: University of Chicago Press.

Hawkes, K., O'Connell, J. F., Blurton Jones, N. G., Alvarez, H., & Charnov, E. L. 1998. Grandmothering, menopause, and the evolution of human life histories. *Proc. Natl. Acad. Sci. USA*, 95(3), 1336–1339.

Hawkes, K., Rogers, A. R., & Charnov, E. L. 1995. The male's dilemma: increased offspring production is more paternity to steal. *Evol. Ecol.*, 9, 662–677.

Hill, K. & Hurtado, A. M. 1996. *Ache Life History: The Ecology and Demography of a Foraging People*. New York: Aldine de Gruyter.

Hill, K. & Kaplan, H. 1988. Tradeoffs in male and female reproductive strategies among Ache foragers. In: *Human Reproductive Effort* (Ed. by L. Betzig, M. Borgerhoff-Mulder, & P. Turke), pp. 277–306. Cambridge: Cambridge University Press.

Hinde, R. A. 1979. *Towards Understanding Relationships*. London: Academic Press.

Hinde, R. A. ed. 1983. *Primate Social Relationships: An Integrated Approach*. Sunderland, MA: Sinauer Associates.

Hirshleifer, J. 2001. *The Dark Side of the Force: Economic Foundations of Conflict Theory*. Cambridge: Cambridge University Press.

Horrobin, D. 2001. *The Madness of Adam and Eve: How Schizophrenia Shaped Humanity*. London: Bantam Press.

Houston, A. I. 1995. Parental effort and paternity. *Anim. Behav.*, 50, 1635–1644.

Hrdy, S. B. 1979. Infanticide among animals: a review, classification, and examination of the implications for the reproductive strategies of females. *Ethol. Sociobiol.*, 1, 13–40.

Hrdy, S. B. 1981. *The Woman That Never Evolved*. Cambridge, MA: Harvard University Press.

Hrdy, S. B. 1986. Empathy, polyandry, and the myth of the coy female. In: *Feminist Approaches to Science* (Ed. by R. Bleier), pp. 119–146. New York: Pergemon Press.

Hrdy, S. B. 1999. *Mother Nature: A History of Mother's Infants and Natural Selection*. New York: Pantheon Books.

Hrdy, S. B. 2001. Mothers and others. *Nat. Hist.*, 110(4), 50–64.

Hrdy, S. B. & Williams, G. C. 1983. Behavioral biology and the double standard. In: *Social Behavior of Female Vertebrates* (Ed. by S. K. Wasser), pp. 1–17. New York: Academic Press.

Kaplan, H. & Hill, K. 1985a. Hunting ability and reproductive success among male Ache foragers: preliminary results. *Curr. Anthropol.*, 26, 131–133

Kaplan, H. & Hill, K. 1985b. Food sharing among Ache foragers: tests of explanatory hypotheses. *Curr. Anthropol.*, 26, 223–246.

Kaplan, H., Hill, K., Lancaster, J., & Hurtado, A. M. 2000. A theory of human life history evolution: diet, intelligence, and longevity. *Evol. Anthropol.*, 9, 156–185.

Kappeler, P. M. 1999. Primate socioecology: new insights from males. *Naturwissenschaften*, 85, 18–29.

Kelly, R. L. 1995. *The Foraging Spectrum: Diversity in Hunter-Gatherer Lifeways*. Washington, DC: Smithsonian Institution Press.

Kleiman, D. G. 1977. Monogamy in mammals. *Q. Rev. Biol.*, 5, 39–69.

Kleiman, D. G. & Malcolm, J. R. 1981. The evolution of male parental investment in mammals. In: *Parental Care in Mammals* (Ed. by D. J. Gubernick & P. H. Klopfer), pp. 347–387. New York: Plenum Press.

Klein, R. G. 1999. *The Human Career: Human Biological and Cultural Origins,* 2nd ed. Chicago: University of Chicago Press.

Kurland, J. & Gaulin, S. 1984. The evolution of male parental investment: effects of genetic relatedness and feeding ecology on the allocation of reproductive effort. In: *Primate Paternalism* (Ed. by D. M. Taub), pp. 259–308. New York: Van Nostrand Reinhold.

Lack, D. 1968. *Ecological Adaptations for Breeding in Birds.* London: Methuen.

Lancaster, J. B. & Lancaster, C. 1983. Parental investment: the hominid adaptation. In: *How Humans Adapt: Biocultural Odyssey* (Ed. by D. J. Ortner), pp. 33–56. Washington, DC: Smithsonian Institution Press.

Lancaster, J. B. & Lancaster, C. 1987. The watershed: change in parental-investment and family formation strategies in the course of human evolution. In: *Parenting Across the Life Span: Biosocial Dimensions* (Ed. by J. B. Lancaster, J. Altmann, A. S. Rossi, & L. R. Sherrod), pp. 187–205. Hawthorne, NY: Aldine de Gruyter.

Leach, E. 1988. The social anthropology of marriage and mating. In: *Mating and Marriage* (Ed. by V. Reynolds & J. Kellett), pp. 91–110. Oxford: Oxford University Press.

Leacock, E. B. 1972. Introduction and notes. In: *The Origins of the Family, Private Property and the State* (By Frederick Engels), pp. 7–67. New York: International Publishers.

Lee, R. B. 1968. What hunters do for a living, or how to make out on scarce resources. In: *Man the Hunter* (Ed. by R. B. Lee & I. DeVore), pp. 30–48. Chicago: Aldine.

Lee, R. B. 1969. Eating Christmas in the Kalahari. *Nat. Hist.,* 14–22, 60–63.

Lee, R. B. 1979. *The !Kung San: Men, Women and Work in a Foraging Society.* Cambridge: Cambridge University Press.

Lévi-Strauss, C. 1969. *The Elementary Structures of Kinship.* Revised and translated from the French edition (1949). New York: Beacon Press.

Lovejoy, C. O. 1981. The origin of man. *Science,* 211, 341–350.

Low, B. S. 1978. Environmental uncertainty and the parental strategies of marsupials and placentals. *Am. Nat.,* 112(983), 197–213.

Marlowe, F. 2000. Paternal investment and the human mating system. *Behav. Proc.,* 51, 45–61.

Marlowe, F. 2001. Male contribution to diet and female reproductive success. *Curr. Anthropol.,* 42, 755–760.

Marshall, L. 1976. *The !Kung of Nyae Nyae.* Cambridge, MA: Harvard University Press.

Maynard Smith, J. 1977. Parental investment: a prospective analysis. *Anim. Behav.,* 25, 1–9.

Maynard Smith, J. 1978. Optimization theory in evolution. *Annu. Rev. Ecol. Sys.,* 9, 31–56.

Maynard Smith, J. 1982. *Evolution and the Theory of Games.* Cambridge: Cambridge University Press.

Mesnick, S. L. 1997. Sexual alliances: evidence and evolutionary implications. In: *Feminism and Evolutionary Biology: Boundaries, Intersections, and Frontiers* (Ed. by P. A. Gowaty), pp. 207–257. New York: Chapman Hall.

Miller, G. F. 2000. *The Mating Mind: How Sexual Choice Shaped the Evolution of Human Nature.* New York: Doubleday.

Mitani, J. D. & Watts, D. P. 2001. Why do chimpanzees hunt and share meat? *Anim. Behav.,* 61, 915–924.

Mitani, J. C., Watts, D. P., & Muller, M. N. 2002. Recent developments in the study of wild chimpanzee behavior. *Evol. Anthropol.,* 11, 9–25.

Morgan, L. H. 1870. *Systems of Consanguinity and Affinity of the Human Family.* Washington, DC: Smithsonian Institution Press.

Neiman, F. D. 1998. Conspicuous consumption as wasteful advertising: a Darwinian perspective on spatial patterns in the Classic Maya terminal monument dates. In: *Rediscovering Darwin: Evolutionary Theory and Archaeological Explanation* (Ed. by C. M.

Barton & G. A. Clark), pp. 267–290. Washington, DC: Archaeological Papers of the American Anthropological Association.

O'Connell, J. F., Hawkes, K., & Blurton Jones, N. G. 1999. Grandmothering and the evolution of *Homo erectus*. *J. Hum. Evol.*, 36, 461–485.

O'Connell, J. F., Hawkes, K., Lupo, K. D., & Blurton Jones, N. G. 2002. Male strategies and Plio-Pleistocene archaeology. *J. Hum. Evol.*, 43, 831–872.

Olson, M. 1965. *The Logic of Collective Action: Public Goods and the Theory of Groups*. Cambridge, MA: Harvard University Press.

Orians, G. H. 1969. On the evolution of mating systems in birds and mammals. *Am. Nat.*, 103, 589–603.

Ostrom, V. & Ostrom, E. 1977. Public goods and public choices. In: *Alternatives for Delivering Public Services: Toward Improved Performance* (Ed. by E. S. Savas), pp. 7–49. Boulder, CO: Westview Press.

Palombit, R. A. 1999. Infanticide and the evolution of pair bonds in nonhuman primates. *Evol. Anthropol.*, 7, 117–129.

Palombit, R. A. 2000. Infanticide and the evolution of male-female bonds in animals. In: *Infanticide by Males and Its Implications* (Ed. by C. P. van Schaik & C. H. Janson), pp. 239–268. Cambridge: Cambridge University Press.

Palombit, R. A, Cheney, D. L., Fischer, J., Johnson, S., Rendall, D., Seyfarth, R. M., & Silk, J. B. 2000. Male infanticide and defense of infants in chacma baboons. In: *Infanticide by Males and Its Implications* (Ed. by C. P. van Schaik & C. H. Janson), pp. 123–152. Cambridge: Cambridge University Press.

Palombit, R. A, Seyfarth, R. M., & Cheney, D. L. 1997. The adaptive value of "friendships" to female baboons: experimental and observational evidence. *Anim. Behav.*, 54, 599–614.

Paul, A., Kuester, J., & Arnemann, J. 1992. DNA fingerprinting reveals that infant care by male Barbary macaques (*Macaca sylvanus*) is not parental investment. *Folia Primatol.*, 58, 93–98.

Paul, A., Preuschoft, S., & van Schaik, C. P. 2000. The other side of the coin: infanticide and the evolution of affiliative male-infant interactions in Old World primates. In: *Infanticide by Males and Its Implications* (Ed. by C. P. van Schaik & C. Janson), pp. 269–292. Cambridge: Cambridge University Press.

Peterson, N. 1993. Demand sharing: reciprocity and the pressure for generosity among foragers. *Am. Anthropol.*, 95, 860–874.

Pinker, S. 1997. *How the Mind Works*. New York: Norton.

Polanyi, K. 1957. The economy as instituted process. In: *Trade and Market in the Early Empires: Economies in History and Theory* (Ed. by K. Polanyi, C. Arensberg, & H. Pearson), pp. 243–270. Chicago: Henry Regnery.

Roberts, G. 1998. Competitive altruism: from reciprocity to the handicap principle. *Proc. Royal Soc. Lond., B*, 265, 427–431.

Rowell, T. E. 1988. The social system of guenons, compared with baboons, macaques, and mangabeys. In: *A Primate Radiation: Evolutionary Biology of the African Guenons* (Ed. by F. Gautier-Hion, H. Bourlier, J. P. Gautier, & J. Kingdon), pp. 347–351. Cambridge: Cambridge University Press.

Sahlins, M. D. 1972. *Stone Age Economics*. Chicago: Aldine.

Samuelson, P. A. 1954. The pure theory of public expenditure. *Rev. Econ. Stat.*, 36, 387–389.

Schelling, T. C. 1960. *The Strategy of Conflict*. Cambridge, MA: Harvard University Press.

Schelling, T. C. 1978. *Micromotives and Macrobehavior*. New York: Norton.

Seger, J. & Stubblefield, J. W. 1996. Optimization and adaptation. In: *Adaptation* (Ed. by G. Lauder & M. R. Rose), pp. 93–123. New York: Academic Press.

Smith, E. A. & Bliege Bird, R. 2000. Turtle hunting and tombstone opening: public generosity as costly signaling. *Evol. Hum. Behav.*, 21, 245–261.

Smuts, B. B. 1983a. Dynamics of special relationships between adult male and female olive baboons. In: *Primate Social Relationships: An Integrated Approach* (Ed. by R. A. Hinde), pp. 112–120. Sunderland, MA: Sinauer Associates.

Smuts, B. B. 1983b. Special relationships between adult male and female olive baboons: selective advantages. In: *Primate Social Relationships: An Integrated Approach* (Ed. by R. A. Hinde), pp. 262–266. Sunderland, MA: Sinauer Associates.

Smuts, B. B. 1985. *Sex and Friendship in Baboons*. New York: Aldine.

Smuts, B. B. 1987. Sexual competition and mate choice. In: *Primate Societies* (Ed. by B. B. Smuts, D. L. Cheney, R. Seyfarth, R. W. Wrangham, & T. T. Struhsaker), pp. 385–399. Chicago: University of Chicago Press.

Smuts, B. B. 1992. Male aggression against women: an evolutionary perspective. *Hum. Nat.*, 3, 1–44.

Smuts, B. B. & Gubernick, D. 1992. Male-infant relationships in nonhuman primates: paternal investment or mating effort? In: *Father-Child Relations: Cultural and Biosocial Contexts* (Ed. by B. Hewlett), pp. 1–30. New York: Aldine de Gruyter.

Smuts, B. B. & Smuts, R. T. 1993. Male aggression and sexual coercion of females in nonhuman primates and other mammals: evidence and theoretical implications. *Adv. Study Behav.*, 22, 1–63.

Sober, E. & Wilson, D. S. 1998. *Unto Others: The Evolution and Psychology of Unselfish Behavior*. Cambridge, MA: Harvard University Press.

Stanford, C. B. 1996. Hunting ecology of chimpanzees. *Am. Anthropol.*, 98, 96–113.

Stanford, C. B. 1999. *The Hunting Apes*. Princeton, NJ: Princeton University Press.

Symons, D. 1979. *The Evolution of Human Sexuality*. Oxford: Oxford University Press.

Tooby, J. & DeVore, I. 1987. The reconstruction of human behavioral evolution through strategic modeling. In: *Primate Models of Human Behavior* (Ed. by W. Kinzey), pp. 183–237. Albany, NY: SUNY Press.

Trivers, R. L. 1971. The evolution of reciprocal altruism. *Q. Rev. Biol.*, 46, 35–57.

Trivers, R. L. 1972. Parental investment and sexual selection. In: *Sexual Selection and the Descent of Man* (Ed. by B. Campbell), pp. 139–179. Chicago: Aldine.

van Schaik, C. P. 1996. Social evolution in primates: the role of ecological factors and male behavior. *Proc. Br. Acad.*, 88, 9–31.

van Schaik, C. P., Hodges, J. K., & Nunn, C. L. 2000. Paternity confusion and the ovarian cycles of female primates. In: *Infanticide by Males and Its Implications* (Ed. by C. P. van Schaik & C. Janson), pp. 361–387. Cambridge: Cambridge University Press.

van Schaik, C. P. & Janson, C. 2000 *Infanticide by Males and Its Implications*. Cambridge: Cambridge University Press.

van Schaik, C. P. & Kappeler, P. M. 1997. Infanticide risk and the evolution of male-female associations in primates. *Proc. Royal Soc. Lond., B*, 64, 1687–1694.

van Schaik, C. P. & Paul, A. 1996. Male care in primates: does it ever reflect paternity? *Evol. Anthropol.*, 5, 152–156.

van Schaik, C. P., van Noordwijk, M. A., & Nunn, C. L. 1999. Sex and social evolution in primates. In: *Comparative Primate Socioecology* (Ed. by P. C. Lee), pp. 204–240. Cambridge: Cambridge University Press.

Veblen, T. [1899] 1992. *The Theory of the Leisure Class* (Ed. by C. W. Mills). Reprint, New Brunswick, NJ: Transaction Publishers.

Washburn, S. 1960. Tools and human evolution. *Sci. Am.*, 203, 63–75.

Washburn, S. & Avis, V. 1958. Evolution and human behavior. In: *Behavior and Evolution*

(Ed. by A. Roe & G. G. Simpson), pp. 421–436. New Haven, CT: Yale University Press.

Washburn, S. & DeVore, I. 1961. Social behavior of baboons and early man. In: *Social Life of Early Man* (Ed. by S. Washburn), pp. 91–105. Chicago: Aldine.

Washburn, S. L. & Lancaster, C. S. 1968. The evolution of hunting. In: *Man the Hunter* (Ed. by R. B. Lee & I. DeVore), pp. 293–303. Chicago: Aldine.

Westermarck, E. 1891. *The History of Human Marriage*. London: Macmillan.

Whitten, P. 1987. Males and infants. In: *Primate Societies* (Ed. by B. B. Smuts, D. L. Cheney, R. Seyfarth, R. W. Wrangham, & T. T. Struhsaker), pp. 343–357. Chicago: University of Chicago Press.

Wiessner, P. 1996. Leveling the hunter: constraints on the status quest in foraging societies. In: *Food and the Status Quest: An Interdisciplinary Perspective* (Ed. by P. Wiessner & W. Schiefenhovel), pp. 171–191. Providence: Berghahn Books.

Williams, G. C. 1966. *Adaptation and Natural Selection*. Princeton, NJ: Princeton University Press.

Wilson, D. S. 1983. The group selection controversy: history and current status. *Annu. Rev. Ecol. Syst.*, 14, 159–87.

Wilson, D. S. 1998. Hunting, sharing, and multilevel selection: the tolerated theft model revisited. *Curr. Anthropol.*, 39, 73–97.

Wilson, M. & Daly, M. 1992. The man who mistook his wife for a chattel. In: *The Adapted Mind: Evolutionary Psychology and the Generation of Culture* (Ed. by J. H. Barkow, L. Cosmides, & J. Tooby), pp. 289–322. New York: Oxford University Press.

Wilson, M. & Mesnick, S. 1997. An empirical test of the bodyguard hypothesis. In: *Feminism and Evolutionary Biology: Boundaries, Intersections, and Frontiers* (Ed. by P. A. Gowaty), pp. 505–511. New York: Chapman Hall.

Woodburn, J. 1998. Sharing is not a form of exchange: an analysis of property sharing in immediate return hunter-gatherer societies. In: *Property Relations: Renewing the Anthropological Tradition* (Ed. by C. M. Hann), pp. 48–63. Cambridge: Cambridge University Press.

Wrangham, R. W. 1979. On the evolution of ape social systems. *Soc. Sci. Info.*, 18, 334–368.

Wrangham, R. W., Jones, J. H., Laden, G., Pilbeam, D., & Conklin-Brittain, N. L. 1999. The raw and the stolen: cooking and the ecology of human origins. *Curr. Anthropol.*, 40, 567–594.

Wrangham, R. W. & Rubenstein, D. I. 1986. Social evolution in birds and mammals. In: *Ecological Aspects of Social Evolution: Birds and Mammals* (Ed. by R. W. Wrangham & D. I. Rubenstein), pp. 452–470. Princeton, NJ: Princeton University Press.

Wright, P. 1990. Patterns of paternal care in primates. *Int. J. Primatol.*, 11, 89–102.

Zahavi, A. 1975. Mate selection: selection for a handicap. *J. Theor. Biol.*, 53, 205–214.

Zahavi, A. 1977. The cost of honesty: further remarks on the handicap principle. *J. Theor. Biol.*, 67, 603–605.

Zahavi, A. 1995. Altruism as a handicap: the limitations of kin selection and reciprocity. *Avian Biol.*, 26, 1–3.

Zahavi, A. & Zahavi, A. 1997. *The Handicap Principle*. Oxford: Oxford University Press.

Conclusion

20

Variation in Nepotistic Regimes and Kin Recognition: A Major Area for Future Research

Bernard Chapais
Carol M. Berman

In our introductory chapter, we described the topic of kinship and behavior in primates as a sort of black box. While researchers have come to recognize the critical importance of kinship for understanding primate social relationships and social systems from both empirical and theoretical perspectives, there are still fundamental issues about which we remain in the dark. In the intervening chapters of this book, we have attempted, through our chapter authors, to provide comprehensive reviews of many areas of inquiry concerning kinship and behavior including (1) current knowledge, (2) useful and promising methodologies, and (3) remaining important areas of the kinship black box that beg for illumination. Since each contribution ends with a summary and conclusion of its particular subject matter, we will not attempt to provide a watered-down summary here. Instead, we would like to draw together some of the findings in the book to briefly discuss one aspect of primate kinship study that has not been covered explicitly so far in this volume, but that we consider to be a most intriguing and important corner of the kinship black box—the question of interspecies variation in nepotistic regimes and kin recognition.

Although primatologists have long been concerned with explaining how kinship affects behavior in particular societies, they have just begun to look into explanations for variation between species. However, we expect investigation into the processes producing variation in nepotism and kin recognition to become a major research area over the next decade, thanks in part to the accelerating application of molecular genetic techniques to describing kinship structures among a wide range of species. In this concluding chapter, we briefly discuss some of the factors likely to be critical in understanding the nature and origins of the variation and attempt to highlight some of the most relevant questions for future research.

An Outline of the Issues

Primate societies create opportunities for particular categories of kin to live together, but not all kin act nepotistically. Of all the kin that meet each other on a regular basis, some act as if they are oblivious to their genetic relatedness, while others engage in preferential relationships. There are two issues here. The first concerns the factors that determine the numbers and categories of kin that live together—in other words, the kinship structure of primate groups. The most basic factors involved are those that determine a group's social organization, defined here as its size, sexual composition, and spatiotemporal cohesion (Kappeler & van Schaik 2002). Indeed, whether the basic social unit is a single male-female pair, a one-male group, or a multimale-multifemale group, it determines the number of individuals of each sex that meet each other on a regular basis and have frequent opportunities for interaction. Thus, social organizations define the boundaries of kinship structures. But kinship structures are further affected by a number of additional factors, the most important being the population's dispersal pattern (chapter 4), its actual mating system (chapter 8), and its life history characteristics.

The second issue concerns the determinants of the structure of nepotistic relationships (or nepotistic regimes). When kin (matrilineal and/or patrilineal) co-reside, what factors determine which subsets actually engage in favoritism attributable to kinship, that is, which subsets display kin biases in behavior and why? Several factors are expected to affect the distribution of kin biases, including the extent to which individuals are dependent on social partners (kin or nonkin) to satisfy their various needs, the functional nature of the behaviors involved in the attainment of these needs (i.e., whether the interactions are cooperative or altruistic), the cost-benefit ratio of the behaviors, and the categories and numbers of kin present in the group. Because these factors are expected to vary considerably between species across the primate order, we expect profound variation between groups and species in the relative importance of nepotism in organizing social relationships, in the distribution of behavioral biases among kin categories, and, since kin recognition is inferred from kin biases, in the extent of kin recognition as well.

To illustrate the issues at stake, we focus our discussion on multimale-multifemale societies, because this type of social organization exhibits the most complex kinship structures in terms of the sheer number of co-resident kin categories. And although some multimale-multifemale species exhibit mixed sex-biased patterns of dispersal, such as red howler monkeys (*Alouatta seniculus*), for simplicity's sake we limit our discussion to the two most extreme but not unusual dispersal regimes—female philopatry/male dispersal and male philopatry/female dispersal.

Kinship Structures in Female Philopatric/ Male Dispersal Groups

In groups in which females are typically philopatric and in which males routinely disperse as juveniles or adolescents, kin dyads are composed primarily of females. Females are related to each other both matrilineally (chapter 7) and patrilineally (chapter 8). By having daughters who subsequently have their own daughters within the group, any matriarch may generate a matriline that contains several categories of female kin relationships (e.g., maternal sisters, matrilineal aunt-niece dyads, matrilineal cousins, etc.). Since females remain in their natal groups for life, at any one time, they are likely to co-reside with a variety of matrilineal female kin representing several of these relationships. In addition, females may also be related through the paternal line. However, patrilineal kinship should be much less extensive than matrilineal kinship, being most often limited to father-daughter dyads and paternal sisters. Other classes of patrilineal kin are unlikely because a female's father being born in another group, the female does not co-reside with her father's kin. Since females rarely if ever transfer between groups or fuse with groups of nonkin, all females in female philopatric groups are likely to be related to one another by varying degrees, through the maternal line, the paternal line, or both.

In contrast, kinship relationships between females and males in these groups are much more limited in nature, because males in most female philopatric/male dispersal species emigrate and join other social groups before reproducing. When this occurs, categories of co-resident male-female kin are typically limited to mother-son and father-daughter dyads, grandmother-grandson and grandfather-granddaughter dyads, and brother-sister dyads (paternal or maternal). For the minority of species in which young natal males typically delay dispersing and are sexually active in their natal groups (e.g., Barbary macaques [*Macaca sylvanus*] and Tibetan macaques [*M. thibetana*]) (Paul et al. 1992, Paul & Kuester 1996, Zhao 1996), other categories are also expected. However, in most cases, sisters do not co-reside with the offspring of their (paternal or maternal) brothers, because these offspring are born in other groups. In contrast, brothers may co-reside with the offspring of their (maternal or paternal) sisters, especially if a brother is younger than his sister, and hence still present in the group when she reproduces (or if a nephew transfers into the same group as his uncle). Thus, female philopatry produces an asymmetry in the co-residence of avunculates: nieces are less likely to reside with their father's sisters than they are to reside with their mother's brothers before the latter emigrate.

At present, our knowledge of the kinship structure of specific female philopatric/male dispersal groups is also asymmetrical. We have been able to analyze the structures of matrilineal relatedness in such groups primarily because those structures are behaviorally manifest through mother-infant bonds and often through enduring bonds between mothers and older offspring, but so far we have comparatively little information about the patrilineal component. Furthermore, what we know about the matrilineal component comes mostly from a few long-term studies of provisioned groups belonging to a handful of cercopithecine species. Nevertheless, we expect that the development of molecular genetic methods and of noninvasive genotyping methods will help fill this gap in the next decades (chapters 2, 3).

In addition to dispersal pattern, two other factors are expected to produce marked variation in the kinship structure of groups, namely male reproductive skew (chapters 8, 10) and female reproductive rate. Although we know extremely little about the impact of these two factors, both are expected to affect the number of kin categories residing in the same group at any one time and the number of individuals per category. Consider female reproductive rate first. Female reproductive rates are determined by several other life history factors, including interbirth intervals, age at first reproduction, and female longevity. Female reproductive rates in turn determine the number of generations found within a given social group, and hence the categories of direct descendants (in addition to mother-daughter dyads) likely to be alive simultaneously in the group (e.g., grandmother-granddaughter pairs). The same life history traits also determine the size of maternal sibships, and hence the number of matrilineal aunt-niece dyads, the size of matrilineal "cousinships," and so on. The impact of life history characteristics on the kinship structure of primate groups is an almost untouched research area, but a most promising one. For example, Dunbar (1988) compared hypothetical populations with various reproductive rates, including one with low reproductive rates (first birth at six years and interbirth intervals of three years) and another with high reproductive rates (first birth at four years and interbirth intervals of one year). Dunbar found that mean matriline size, counting only mature females, varied between 1.5 and 5.4 kin between the two populations. Thus, the genealogical environment of individuals is likely to vary considerably depending on their life history characteristics. Clearly, this variation has important consequences for the distribution of kin biases in behavior (chapters 6, 14), kin recognition (chapter 13), and kin selection (chapter 16) that need further exploration (see below).

Male reproductive skew refers to the extent to which reproduction is monopolized by a few males within a group. In multimale groups, dominance is known to produce rank-related reproductive skew of varying degrees (Cowlishaw & Dunbar 1991). When skew is marked in a given year, the resulting paternal sibships among infants are expected to be fewer in number but larger than when skew is mild (Altmann & Altmann 1979). Thus in groups with high degrees of skew, a large proportion of the same-age females will be related. Also, when a reproductively successful male has a relatively long tenure, the sibships he produces will span a larger age bracket than when his tenure is relatively short. Thus paternal sibships may range from single-cohort sibships to multicohort sibships. Following this, factors influencing the length of males' tenures will also determine whether males will co-reside not only with their daughters but also with their granddaughters. Paternity studies have just begun to shed light on patterns of male reproductive skew and their consequences on the composition of paternal sibships in female philopatric groups (Altmann et al. 1996, Widdig et al. 2001, Buchan et al. 2003, Smith et al. 2003). In addition to these sorts of studies, further genetic analyses are needed to assess the whole structure of patrilineal kinship in these species.

Kinship Structure in Male Philopatric/ Female Dispersal Groups

To a large extent, the kinship structures of male philopatric groups can be seen as the converse of those of female philopatric groups described above. In male philopatric groups, kin dyads are mostly composed of males, because females leave their natal groups. Any sexually active male can potentially produce a patriline within the group, consisting of not only father-son dyads but also paternal brothers, grandfather-grandson dyads, patrilineal uncle-nephew and cousin dyads, and so on. Similarly, males may be related matrilineally. However, in most cases maternal kin should be limited to mother-son dyads and maternal brothers because a male's mother being born in another group, the male does not co-reside with his mother's kin. Thus, all males in a male philopatric group are related patrilineally and/or matrilineally. In contrast, males are much less likely to be related to females, because females emigrate and reproduce in other social groups. In species such as muriquis (*Brachyteles arachnoides*) (Strier & Ziegler 2000), in which females virtually never reproduce before dispersal, co-residing male-female kin should be limited to mother-son and father-daughter dyads, grandmother-grandson and grandfather-granddaughter dyads, and brother-sister dyads (paternal and maternal). Other categories may be found, however, among species such as chimpanzees (*Pan troglodytes*), in which some females delay dispersing until after they become sexually active. But for the most part, brothers in male philopatric/female dispersal species do not associate with the offspring of their (paternal or maternal) sisters, because these offspring are born in other groups. In contrast, sisters may associate with the offspring of their (maternal or paternal) brothers, particularly when a sister is younger than her brother, and hence still present in the group when he reproduces. Thus, male philopatry produces an asymmetry in the co-residence of avunculates: nephews generally do not reside with their mother's brothers, but they may reside with their father's sisters before the latter emigrate.

Empirically, we know even less about the kinship structures of specific male philopatric groups than about those of female philopatric groups. In both types of societies, the structure of patrilineal kinship is almost entirely unknown. Moreover, researchers are limited in their abilities to infer matrilineal relationships in male philopatric groups from mother-infant bonds beyond the level of mother-son and maternal sibling relationships. In the absence of genetic data, father-son relationships are not obvious to researchers; hence, kin relationships involving the offspring of co-residing maternal brothers (e.g., maternal uncle-nephew, maternal cousins, etc.) remain hidden. For this reason, our progress in this area will depend critically on genetic analysis, including mtDNA haplotype identity, to infer matrilineality (Goldberg & Wrangham 1997, Mitani et al. 2000). Needless to say, we also know very little about how female reproductive rate and male reproductive skew produce variation in the kinship structures of male philopatric groups (chapter 8). These questions await both quantitative modeling and the accumulation of comparative data on groups differing in female reproductive rates and male reproductive skew.

Kin Biases, Kin Discrimination, and Kin Recognition in Female Philopatric/Male Dispersal Groups

The potential co-residence in female philopatric groups of several categories of matrilineal and patrilineal kin raises the question of what kin among all those present actually behave with one another as such. The ability to recognize kin is generally inferred from the display of kin discrimination. Kin discrimination is inferred in turn from the presence of positive or negative biases in behavior attributable to the effect of kinship. Aunts and nieces, for instance, are inferred to behave as kin if their levels of positive interactions are significantly higher than those of nonkin and if these levels cannot be explained by other possibly confounding variables (such as proximity in rank). However, theoretically, kin recognition can occur even in the absence of kin discrimination (chapters 13, 14) if kin who recognize each other nevertheless refrain from displaying kin bias. Thus kin biases may underestimate the actual limits of kin recognition. Nevertheless, an understanding of kin bias is important for both setting a minimal limit for kin recognition and for understanding the impact of kin discrimination on inter-individual relationships.

Several factors are expected to affect the distribution of kin biases among kin, and hence one's inferences about the extent of kin recognition or discrimination in any particular group. First, not all categories of behavior should be equally kin biased, and therefore not all behaviors are equally good indicators of animals' abilities to recognize or discriminate kin. Second, we expect significant variation between groups and species in the relative importance of kinship as a factor affecting social relationships, hence in the importance of kin biases. For these reasons, our inferences about kin recognition and discrimination are likely to be species specific, if not group specific. Below we examine these issues, focusing on matrilineal kinship. Similar principles should apply to patrilineal kin. However, a comparable exercise would be much more speculative because much less is known about the extent of patrilineal kin discrimination—although recent findings are promising (Widdig et al. 2001, Buchan et al. 2003, Smith et al. 2003, chapter 13).

Matrilineal Kin Biases Should Not Affect All Categories of Behavior Equally

Kin biases in behavior have been reported so often in the primate literature that one can easily assume that all categories of beneficent behavior exhibit kin bias. This should not be expected, however. Consider cooperation between kin. Cooperation (defined here as mutualism or reciprocity) with kin provides both direct (personal) fitness benefits and indirect (inclusive) fitness benefits to each participant (Wrangham 1982). Chapais and Bélisle (chapter 16) have argued that gains in inclusive fitness are maximized by cooperating with one's closest kin (Altmann 1979), but gains in personal fitness are maximized by cooperating with the most competent partner regardless of degree of kinship. For this reason, one should expect cooperative activities requiring specific individual characteristics (such as experience, high rank, or specific skills) to be less strongly kin biased than those not involving such attributes or skills. Examples of attribute-dependent cooperation include coalition formation and group hunting. This reasoning may explain the absence of correlations between matrilineal kinship and cooperation in several behavioral areas among chimpanzees (Mitani et al. 2000, 2002).

In contrast, other cooperative activities such as huddling or mutual grooming in some contexts do not require specific skills or individual characteristics (except body size in the case of huddling). By selecting close kin as partners to perform such attribute-independent cooperative activities, individuals should maximize their inclusive fitness at no cost to their personal fitness. Thus, these forms of cooperation should be more strongly kin biased compared to attribute-dependent cooperation.

Finally, a third category of beneficent behavior, namely altruism, should be even more highly kin biased than attribute-dependent cooperation. Contrary to cooperation, which is profitable regardless of the kinship relationship between the partners, altruism, and more specifically unilateral altruism, is profitable only when it is directed to certain kinds of kin (Hamilton 1964). Examples of unilateral altruistic interactions include protection of juveniles by adult females and tolerance of juveniles by adults at defensible food sources. Such interactions should be strongly kin biased.

Thus, one can think of a gradient in the degree to which behaviors should be kin biased, with unilateral altruism at one extreme, attribute-dependent cooperation at the other, and attribute-independent cooperation between the two (chapter 16). Does that mean that the distribution of unilateral altruism among kin in any particular group is a better indicator of the extent of kin discrimination or recognition, compared to the distribution of cooperation? Not necessarily, because we expect the deployment of such altruism among kin to be severely limited by the cost-benefit ratio (b/c) of the behavior, as predicted by kin selection theory. For any given form of altruism (and corresponding b/c ratio), Hamilton's (1964) equation specifies the degree of relatedness beyond which altruism ceases to be profitable. Behaviors with low b/c ratios should be restricted to close kin, whereas behaviors with larger b/c ratios could extend to more distant kin. Thus the observed limit of a specific form of altruism in any particular group could reflect either the limit of the profitability of that form of altruism or the limit of kin recognition (Chapais 2001, Chapais et al. 2001). The limit of nepotism would indicate the limit of kin recognition only if it could be shown that this limit was the same for several behaviors with different b/c ratios. A few studies point in this direction (Kapsalis & Berman 1996, Chapais et al. 1997, Chapais et al. 2001), but much more work needs to be done. At present it appears safer to assume that primates have the potential to recognize more categories of kin than those that are revealed by the distribution of kin biases.

Another factor that complicates using the distribution of kin biases to make inferences about kin recognition in primates is the effect of matriline size. In groups with low growth rates (i.e., low female reproductive rates and high mortality), the average size of matrilines is likely to be smaller than in groups with higher growth rates (i.e., high female reproductive rates and low mortality). Individuals living in groups with small matrilines have smaller average numbers of close kin living with them, and on this basis alone they should have more time available for interacting with more distant kin. Thus, one expects the deployment of nepotism according to degree of kinship to be more extensive in groups with low growth rates (Chapais 2001). Furthermore, in these groups individuals are likely to have smaller numbers of close kin to satisfy their own needs, compared to individuals living with larger numbers of close kin; hence they are likely to be more dependent on distant kin to satisfy these needs. The two factors—time constraints and need satisfaction—should act in the same direction to either limit or extend the deployment of nepotism depending on group composition (chapter 16). Returning to kin recognition, because groups with low growth rates and small matrilines maximize the opportunities for interacting with more distant kin, they constitute better social contexts for assessing the individual's ability to distinguish between distant kin and nonkin. Thus, other things being equal, the extent of kin recognition would be more aptly studied in groups with low growth rates.

The Relative Importance of Matrilineal Kinship Should Vary Markedly Between Groups

The functional importance of having co-residing kin lies in their potential to be valuable social partners. Among primates, social partners are potentially useful in various contexts: they may warn against predators; they provide grooming, allocare, and protection against males; they may play a prominent role in rank acquisition; they may ally with one another in agonistic and competitive interactions, etc. Preferential relationships, including those based on kinship, exist when animals have needs that are best satisfied through interaction with others. Thus, high levels of nepotism are thought to indicate high levels of social dependence on kin. However, not all societies impose the same levels of inter-individual dependence. For example, one major source of need for social partners is contest competition for food. The predominant model linking food competition to female social relationships was developed by van Schaik (1989) and refined in Sterck et al. (1997) (chapter 4).

According to the model, contest competition for defensible food patches within groups leads to the formation of stable dominance relationships between females, and nepotistic alliances that ultimately give rise to strict matrilineal dominance hierarchies and female philopatry. Sterck et al. (1997) called this category of societies resident-nepotistic. On the other hand, when within-group contest competition is weak and does not lead to the formation of dominance relationships, females might nevertheless face strong between-group competition, in which case they are still expected to stay with their kin. This category of societies is called resident-egalitarian (Sterck et al. 1997).

The comparison of the distribution of kin biases in behavior between these two types of societies should be revealing. In both types, females live with their kin; thus, in theory, the same categories of kin are co-resident. However, in one case (resident-nepotistic), females need allies within their group to compete with other group members. In the other (resident-egalitarian), they need only to form group-level alliances against other groups. In other words, a female's need for allies, including kin allies, in the context of within-group interactions is strong in the case of resident-nepotistic groups, and relatively weak in the case of resident-egalitarian ones. On this basis alone, one may predict profound differences in the relative role of kinship in structuring social relationships, in the behavioral areas that exhibit kin biases, in the precise categories of kin that act nepotistically, and hence in the extent of kin discrimination. Other things being equal (e.g., group composition), kin discrimination is expected to be both more salient and more extensive in resident-nepotistic societies.

A third category of society proposed by Sterck et al. (resident-nepotistic-tolerant) consists of groups that experience high levels of within-group competition as well as other needs requiring group-level cooperation, such as high levels of between-group competition or high risks of predation. In these groups, linear dominance hierarchies are expected, but high-ranking individuals are expected to show more tolerance around defensible resources than those in resident-nepotistic groups, as a way to ensure the continued cooperation of lower ranking group members. In these societies, one may also expect lower levels of kin discrimination than in resident-nepotistic groups. So far, we lack sufficient data on kinship structures, competitive regimes, and the structure of nepotistic relationships for each type of society to test these predictions, but such data are accumulating gradually and should become available in the near future.

Another comparison, for which some data are available, is already revealing. Species classified as resident-nepotistic and resident-nepotistic-tolerant by Sterck et al. (1997) vary considerably in their "dominance style" (sensu de Waal 1989), ranging from "despotic" species with unidirectional and intense aggression between dominants and subordinates to "egalitarian" or "relaxed" species with bidirectional and less severe aggression, and higher levels of tolerance and conciliatory interactions (Thierry 1985, de Waal & Luttrell 1989, Chaffin et al. 1995, Petit et al. 1997, Matsumura 1998). Concomitant with the variation in dominance style is variation in the relative importance of nepotism. Kin biases in behavior are more pronounced in species with despotic dominance styles than in species with egalitarian dominance styles (Butovskaya 1993; Aureli et al. 1997; Thierry 1990, 2000). Two explanations have been proposed to account for differences in the impact of kinship on behavior in these societies. Both explanations have in common the idea that variation in the propensity of kin to form alliances is the key factor causing interspecific differences in dominance style and other aspects of social relationships. However, the two explanations differ in that one model attributes the variation in levels of nepotistic alliances to phylogenetic constraints

(Thierry 1990, 2000; Thierry et al. 2000), whereas the other model links that variation to the strength of contest competition for food and its consequences on the profitability of kin support (Chapais in press). Nevertheless, the two models are not incompatible.

Regardless of which model proves correct, they suggest a developmental pathway through which differences in alliance patterns might produce different patterns of kin discrimination and recognition. In societies where mothers and daughters maintain preferential bonds over their lifetimes, sisters born several years apart nevertheless associate with one another disproportionately, thanks to their proximity to a common mother. Each female also witnesses the interactions between her sister and their mother, and hence may come to recognize her sister by the way her mother treats her. These processes provide sisters with a basis for treating each other preferentially (i.e., for recognizing their sisterhood). Presumably, the same processes allow females to recognize other matrilineal kin, including their aunts (their mother's associates), nieces, and cousins (chapter 14). But suppose mothers and daughters do not maintain long-term preferential relationships. In such a situation, a newborn female would have no means to recognize her older sisters, and hence no means to recognize her nieces and cousins. Thus, in a familiarity-based system of kin recognition, the persistence of preferential relationships between mothers and daughters is a prerequisite for the recognition of other categories of matrilineal kin.

The duration and intensity of mother-daughter bonds may be expected to be highly variable between species. For example, in resident-egalitarian species where females do not form dominance relationships, and in resident-nepotistic-tolerant species with a relatively relaxed dominance style, females would be substantially less dependent on allies to acquire and maintain their rank, hence to acquire resources than females in resident-nepotistic societies. If such alliances are a major factor leading to the persistence of mother-daughter bonds, resident-egalitarian species and resident-nepotistic-tolerant species should exhibit less persistent and less intense mother-daughter bonds than the more despotic resident-nepotistic species. As a result, females should recognize a smaller number of matrilineal kin, and kin biases in behavior should be much less extensive in these societies. The available data on the impact of nepotism in despotic and egalitarian species of macaques (Thierry 2000) are compatible with this explanation, but more detailed analyses are needed. In particular, one needs to know whether the duration, intensity, or other aspects of mother-daughter relationships vary between these species, and whether the patterns of proximity and interactions of sisters in relation with their mother also vary.

Kin Bias, Kin Discrimination, and Kin Recognition in Male Philopatric/Female Dispersal Groups

Although paternity studies have been carried out in male philopatric groups (chapters 2, 8), no studies have yet focused on the differential treatment of patrilineal kin in male philopatric species. Hence, we simply do not know with certainty whether patrilineal kinship translates into behavioral biases.

The impact of matrilineal kinship in male philopatric species has been the object of a few studies (chapter 7). Although preferential relationships have been observed between mothers and adult sons and between maternal brothers in chimpanzees and bonobos (*Pan paniscus*) (Goodall 1986, Furuichi 1997), they have not been observed between maternal brothers in muriquis (Strier et al. 2002). Moreover, among chimpanzees, male matrilineal kinship in general, as assessed by mtDNA haplotype identity, has not been found to translate into preferential relationships (Goldberg & Wrangham 1997; Mitani et al. 2000, 2002). In a male philopatric society, a male would not be expected to recognize categories of matrilineal kin other than his mother and his maternal brothers. First, his mother being born in another group, the male would not witness preferential relationships between his mother and her maternal brothers, nor her father, whom she probably would not recognize herself. Second, because his maternal sisters most often emigrate and reproduce in other groups, the male would not be familiar with his matrilineal nephews. Thus, of all the categories of male matrilineal kin that share the same mtDNA, only maternal brothers would be in position to recognize each other. However, apparently maternal brothers typically do not cooperate preferentially with each other, but choose their social partners using criteria other than matrilineal kinship, for example age and rank similarity (Mitani et al. 2002).

If males in philopatric groups are indeed unable to distinguish most of their patrilineal kin and only a fraction of their matrilineal kin, one would not expect social relationships to be strongly patterned according to degree of kinship. Nevertheless, because all males within a male philopatric group are likely to be related to some extent, males might raise their inclusive fitness by behaving cooperatively with co-residing males, irrespective of their degree of kinship. In this case, nepotism (matrilineal and/or patrilineal) would be apparent at the group level and would likely be manifested in two ways. First, nepotism would be expected to translate into higher levels of cooperation among males than among females, assuming that the average degree of relatedness of males is higher than that of females (Morin et al. 1994; but see Vigilant et al. 2001). Second, one would expect positive correlations between male cooperation and average degrees of relatedness across groups and across species (Moore 1992). However, this reasoning holds for cooperative interactions, and not necessarily for unilateral altruistic ones. Hamilton's (1964) equation sets the limit of profitability for altruism in terms of degree of relatedness; for a given behavior and cost-benefit ratio, altruism ceases to be profitable past a certain r, or relatedness threshold for altruism (Chapais 2001, Chapais et al. 2001). If individuals cannot recognize degrees of kinship, even roughly, they cannot allocate altruism conditionally upon r, except perhaps when the costs of particular forms of altruism are very low. This problem would not apply to cooperation, however, because all cooperating kin benefit, irrespective of their degrees of kinship. Thus between-group comparisons might reveal higher levels of cooperative interaction between males, not necessarily higher levels of altruistic interaction.

Conclusion

We have argued that the influence of kinship on behavior should vary profoundly between primate species, both because kinship structures vary considerably by type of social organization, dispersal regime, mating system, and life history regime, and because the opportunities and needs to interact with kin should also vary markedly in relation to competitive regimes and levels of competition, among other factors. Primatologists have barely begun to explore that variation. Over the coming years, kinship should prove to be no less unitary an independent variable than, for instance, social dominance. Our present state of knowledge about the variable effects of kinship on behavior may be similar to our state of knowledge about the variable effects of dominance on behavior in the 1960s and early 1970s. Over the last 30 years we have come to realize that ecological factors play a major role in determining whether animals establish dominance relationships or not, that the functional consequences of dominance differ fundamentally between the sexes, and that several aspects of dominance vary markedly between species, including the behavioral manifestations of submission, the degree of power asymmetry between ranks, the proximate and developmental processes involved in the acquisition of rank, and the relative impact of dominance on behavior, that is, its explanatory value for understanding social relationships. The interspecific variability of dominance has been uncovered at a faster pace than that of kinship partly because the necessary data have been easier to collect. Indeed, one does not necessarily need long-term background data to assess whether animals form dominance relationships or not, how they rank relative to each other, and the extent to which rank correlates with various behavioral measures, as is the case with kinship studies. As a result, data on dominance have accumulated rapidly for a large number of primate species, whereas in contrast, the equivalent data on kinship relationships are available for only a handful of primate groups. There is no doubt, however, that the accelerating application of molecular genetic techniques is rapidly changing the situation and will allow researchers to rapidly make up the time that has been spent painstakingly establishing kinship relationships. In the coming years, we hope to have a much better idea of the extent to which kin recognition, kin discrimination, and kin biases vary across the primate order.

References

Altmann, J. 1979. Altruistic behaviour: the fallacy of kin deployment. *Anim. Behav.*, 27, 958–962.

Altmann, J., Alberts, S. C., Haines, S. A., Dubach, J., Muruthi, P., Coote, T., Geffen, E., Cheesman, D. J., Mututua, R. S., Saiyalele, S. N., Wayne, R. K., Lacy, R. C., & Bruford, M. W. 1996. Behavior predicts genetic structure in a wild primate group. *Proc. Natl. Acad. Sci. USA*, 93, 5797–5801.

Altmann, S. A. & Altmann, J. 1979. Demographic constraints on behavior and social organization. In: *Primate Ecology and Human Origins* (Ed. by I. S. Bernstein & E. O. Smith), pp. 47–64. New York: Garland Press.

Aureli, F., Das, M., & Veenema, H. C. 1997. Differential kinship effect on reconciliation in three species of macaques (*Macaca fascicularis, M. fuscata*, and *M. sylvanus*). *J. Comp. Psych.*, 111, 91–99.

Buchan, J. C., Alberts, S. C., Silk, J. B., & Altmann, J. 2003. True paternal care in multimate primate society. *Nature*, 425, 179–181.

Butovskaya, M. 1993. Kinship and different dominance styles in groups of three species of the genus macaca (*M. arctoides, M. mulatta, M. fascicularis*). *Folia Primatol.*, 60, 210–224.

Chaffin, C. L., Friedlen, K., & de Waal, F. B. M. 1995. Dominance style of Japanese macaques compared with rhesus and stumptail macaques. *Am. J. Primatol.*, 35, 103–116.

Chapais, B. 2001. Primate nepotism: what is the explanatory value of kin selection? *Int. J. Primatol.*, 22, 203–229.

Chapais, B. In press. How kinship generates dominance structures: a comparative perspective. In: *How Societies Arise, the Macaque Model* (Ed. by B. Thierry, M. Singh, & W. Kaumanns). Cambridge: Cambridge University Press.

Chapais, B., Gauthier, C., Prud'homme, J., & Vasey, P. 1997. Relatedness threshold for nepotism in Japanese macaques. *Anim. Behav.*, 53, 1089–1101.

Chapais, B., Savard, L., & Gauthier, C. 2001. Kin selection and the distribution of altruism in relation to degree of kinship in Japanese macaques. *Behav. Ecol. Sociobiol.*, 49, 493–502.

Cowlishaw, G. & Dunbar, R. I. M. 1991. Dominance rank and mating success in male primates. *Anim. Behav.*, 41, 1045–1056.

de Waal, F. B. M. 1989. Dominance "style" and primate social organization. In: *Comparative Socio-ecology: The Behavioral Ecology of Humans and Other Animals* (Ed. by V. Standen & R. A. Foley), pp. 243–263. Oxford: Blackwell.

de Waal, F. B. M. & Luttrell, L. M. 1989. Toward a comparative socioecology of the genus *Macaca*: different dominance styles in rhesus and stumptail monkeys. *Am. J. Primatol.*, 19, 83–109.

Dunbar, R. I. M. 1988. *Primate Social Systems.* Ithaca, NY: Cornell University Press.

Furuichi, T. 1997. Agonistic interactions and matrifocal dominance rank of wild bonobos (*Pan paniscus*) at Wamba. *Int. J. Primatol.*, 18, 855–875.

Goldberg, T. L. & Wrangham, R. W. 1997. Genetic correlates of social behaviour in wild chimpanzees: evidence from mitochondrial DNA. *Anim. Behav.*, 54, 559–579.

Goodall, J. 1986. *The Chimpanzees of Gombe.* Cambridge, MA: Harvard University Press.

Hamilton, W. D. 1964. The genetical theory of social behavior. I and II. *J. Theor. Biol.*, 7, 1–52.

Kappeler, P. M. & van Schaik, C. P. 2002. Evolution of primate social systems. *Int. J. Primatol.*, 23, 707–740.

Kapsalis, E. & Berman, C. M. 1996. Models of affiliative relationships among free-ranging rhesus monkeys, *Macaca mulatta*. I. Criteria for kinship. *Behaviour*, 133, 1209–1234.

Matsumara, S. 1998. Relaxed dominance relations among female moor macaques (*Macaca maurus*) in their natural habitat, South Sulawesi, Indonesia. *Folia Primatol.*, 69, 345–356.

Mitani, J. C., Merriwether, D. A., & Zhang, C. 2000. Male affiliation, cooperation, and kinship in wild chimpanzees. *Anim. Behav.*, 59, 885–893.

Mitani, J. C., Watts, D. P., Pepper, J. W., & Merriwether, A. D. 2002. Demographic and social constraints on male chimpanzee behaviour. *Anim. Behav.*, 64, 727–737.

Moore, J. 1992. Dispersal, nepotism, and primate social behavior. *Int. J. Primatol.*, 13, 61–378.

Morin, P. A., Moore, J. J., Chakraborty, R., Jin, L., Goodall, J., & Woodruff, D. S. 1994. Kin selection, social structure, gene flow, and the evolution of chimpanzees. *Science*, 265, 1193–1201.

Paul, A., Kuester, J., & Arnemann, J. 1992. Maternal rank affects reproductive success of male Barbary macaques (*Macaca sylvanus*): evidence from DNA fingerprinting. *Behav. Ecol. Sociobiol.*, 30, 337–341.

Paul, A., Kuester, J., & Arnemann, J. 1996. The sociobiology of male-infant interactions in Barbary macaques, *Macaca sylvanus. Anim. Behav.*, 51, 155–170.

Petit, O., Abegg, C., & Thierry, B. 1997. A comparative study of aggression and conciliation in three cercopithecine monkeys (*Macaca fuscata, Macaca nigra, Papio papio*). *Behaviour*, 434, 415–432.

Smith, K., Alberts, S. C., & Altmann, J. 2003. Wild baboons bias their social behaviour towards paternal half-sisters. *Proc. Royal Soc. Lond., B*, 270, 503–510.

Sterck, E. H. M., Watts, D. P., & van Schaik, C. P. 1997. The evolution of female social relationships in nonhuman primates. *Behav. Ecol. Sociobiol.*, 4, 291–309.

Strier, K. B., Dib, L. T., & Figueira, J. E. C. 2002. Social dynamics of male muriquis (*Brachyteles arachnoides hypoxanthus*). *Behaviour*, 139, 315–342.

Strier, K. B. & Ziegler, T. E. 2000. Lack of pubertal influences on female dispersal in muriqui monkeys, *Brachyteles arachnoides*. *Anim. Behav.*, 59, 849–860.

Thierry, B. 1985. Patterns of agonistic interactions in three species of macaque (*Macaca mulatta, M. fascicularis, M. tonkeana*). *Agg. Behav.*, 2, 223–33.

Thierry, B. 1990. Feedback loop between kinship and dominance: the macaque model. *J. Theor. Biol.*, 145, 511–521.

Thierry, B. 2000. Covariation of conflict management patterns across macaque species. In: *Natural Conflict Resolution* (Ed. by F. Aureli & F. B. M. de Waal), pp. 106–128. Berkeley, CA: University of California Press.

Thierry, B., Iwaniuk, A. N., & Pellis, S. M. 2000. The influence of phylogeny on the social behavior of macaques (Primates: Cercopithecidae, genus *Macaca*). *Ethology*, 106, 713–728.

van Schaik, C. P. 1989. The ecology of social relationships amongst female primates. In: *Comparative Socio-ecology: The Behavioral Ecology of Humans and Other Animals* (Ed. by V. Standen & R. A. Foley), pp.195–218. Oxford: Blackwell.

Vigilant, L., Hofreiter, M., Siedel, H., & Boesch, C. 2001. Paternity and relatedness in wild chimpanzee communities. *Proc. Natl. Acad. Sci. USA*, 98, 12890–12895.

Widdig, A., Nürnberg, P., Krawczak, M., Jürgen Streich, W., & Bercovitch, F. B. 2001. Paternal relatedness and age proximity regulate social relationships among adult female rhesus macaques. *Proc. Natl. Acad. Sci. USA*, 98, 13769–13773.

Wrangham, R. W. 1982. Mutualism, kinship and social evolution. In: *Current Problems in Sociobiology* (Ed. by King's College Sociobiology Group), pp. 269–289. Cambridge: Cambridge University Press.

Zhao, Q.-K. 1996. Etho-ecology of Tibetan macaques at Mount Emei, China. In: *Evolution and Ecology of Macaque Societies* (Ed. by J. Fa & D. G. Lindburg), pp. 263–289. Cambridge: Cambridge University Press.

Species Index

Subject Index

Affiliative behaviors and kinship, 154–56, 158–60, 162, 192, 300, 319–337. *See also* Alloparental care; Co-feeding and kinship; Co-sleeping; Grooming; Kin biases; Play and kinship; Proximity and kinship; Reconciliation and kinship

Affinal relationships, 394, 408–09

AFLP (amplified fragment length polymorphism), 17, 21, 22, 32, 50. *See also* Genotyping

Aggression
between females, 76–78, 81–83, 144, 156–57, 163, 350
and kinship, 320–21 (*see also* Dominance)
See also Competition, food; Dispersal; Group living

Aiding. *See* Coalitions and alliances

Allelic dropout, 17, 27, 35, 53

Alliances. *See* Coalitions and alliances

Alliances, sexual. *See* Pair bonds

Alloparental care
in callitrichids, 226, 232–33
and kinship, 320
in nongregarious prosimians, 208, 210–11

Altruism, 178, 180
fallacy of proportional deployment among kin, 369
relatedness threshold for, 158, 181, 189, 280–82, 308, 369, 373–76
unilateral, 373–74, 486
See also Kin selection; Reciprocal altruism

Altruism, distribution of, 368–76
and inferences about kin recognition, 482–83

effect of availability of kin on, 372–73
effect of cost-benefit ratio on, 369–70, 373–76
effect of matriline size on, 370–71
effect of recipients' needs on, 371–72
effect of time constraints on, 369–71

Amplified fragment length polymorphism. *See* AFLP

Assignment tests, 32, 116. *See also* Dispersal and population genetics

Attachment, 324

Avunculate, 406–7

Bands
in hamadryas baboons, 248
in hunters-gatherers, 396–98
reexamination of the patrilocal band model, 420–42

Biomarket theory, 244

Bodyguard/male protection hypothesis. *See* Pair bonds

CATS (comparative anchor tagged sequences), 29. *See also* Genotyping

Clan in hamadryas baboons, 248

Coalitions and alliances
among men, 395–98, 436–37, 438
comparison between apes and humans, 398–400, 404
expected distribution among kin, 370
kin biases in primates, 72–76, 143–44, 158–59, 162, 163, 189–92, 193, 256, 258–59, 320–21, 367, 373–74, 378–79
and kin recognition patterns, 485

Coercion, male
and the evolution of human pair bonds, 462–63